PROBLEMS AND SOLUTIONS IN QUANTUM CHEMISTRY AND PHYSICS

Charles S. Johnson, Jr.
Lee G. Pedersen

Department of Chemistry
University of North Carolina

Dover Publications, Inc., New York

To
Ellen and Barbara

Published in Canada by General Publishing Company, Ltd., 30 Lesmill Road, Don Mills, Toronto, Ontario.
Published in the United Kingdom by Constable and Company, Ltd.

This Dover edition, first published in 1986, is an unabridged and further corrected verson of the corrected third printing of the work originally published in 1974 by Addison-Wesley Publishing Company, Inc., Reading, Massachusetts.

Manufactured in the United States of America
Dover Publications, Inc., 31 East 2nd Street, Mineola, N.Y. 11501

Library of Congress Cataloging-in-Publication Data

Johnson, Charles S. (Charles Sidney), 1936–
Problems and solutions in quantum chemistry and physics.

Reprint. Originally published: Reading, Mass. : Addison-Wesley Pub. Co., 1974.
Bibliography: p.
Includes index.
1. Quantum chemistry—Problems, exercises, etc. 2. Quantum theory—Problems, exercises, etc. I. Pedersen, Lee G. II. Title.
QD462.J63 1986 540.2'8'076 86-16820
ISBN 0-486-65236-X

FOREWORD

We have prepared this book of problems to aid students and teachers of quantum chemistry and physics. The level of the material is similar to that found in advanced physical chemistry and introductory quantum mechanics courses. In developing this book we have been guided by our belief that an exposition of the principles of quantum theory is often not sufficient for the beginning student, and that numerous examples, illustrations, and exercises can be an aid to understanding and to the development of the proper kind of intuition. Education in this subtle field must be to a large extent self education.

We believe that this collection is unique in the variety of problems covered and in the detail with which the solutions are presented. There is a wider range of applications in each chapter than can be found in most standard quantum chemistry texts. Each chapter includes a brief introduction followed by about twenty problems of increasing difficulty for which detailed solutions are provided. These are followed in each chapter by a section of "supplementary" problems, which have answers but not solutions. Ten appendices on mathematical topics have been included for the convenience of the reader. We have attempted to maintain a continuity in each chapter so that the logical development of concepts and techniques can be followed.

For completeness a number of standard topics, *e.g.* the quantum treatment of the harmonic oscillator, have been presented in the problem-and-answer format. Also included are many of the more or less standard problems, which have appeared in textbooks and in problem collections. We have encountered or developed many of the problems in this collection in the course of teaching advanced physical chemistry, spectroscopy, and quantum chemistry. The *American Journal of Physics* and the *Journal of Chemical Education* have been important sources. We have not attempted to itemize all of the problem sources, but have made an effort to refer to useful discussions in the literature. The reference list is, of course, biased somewhat toward those texts which we happen to own.

It is certain that some errors have slipped in as we have modified and recopied old problems, constructed new problems, and converted many solutions to SI units. Conceptual errors may, of course, also be present. We plan to maintain a list of errors, and would appreciate receiving communications concerning suspected errors from readers. We apologize for the errors and for the omission of some important topics. Perhaps the deficiencies can be remedied in a later edition.

We have benefited considerably from discussions with our colleagues, and from courses taught in chemistry and physics at the University of North Carolina. Professors A. D. Buckingham (Cambridge University), S. H. Lin (Arizona State University), and J. L. Whitten (SUNY at Stony Brook) kindly read parts of the manuscript and made useful suggestions. All of the remaining errors are, of course, ours. Part of the manuscript was written at Cambridge where one of us (C.S.J.) was a Guggenheim Fellow (1972-1973).

We wish especially to thank Mrs. Becky Smith for her patient typing of several versions of the manuscript. Without her diligence and accuracy the book would have been greatly delayed.

Finally we thank our wives, Ellen and Barbara, for their encouragement and patience during the last three years.

Chapel Hill 1973

Charles S. Johnson, Jr.
Lee G. Pedersen

In this second printing we have taken the opportunity to correct a number of errors. We would like to thank our colleagues at North Carolina and elsewhere for pointing out these errors and inconsistencies. Special thanks are in order for Professor E. Merzbacher (University of North Carolina) who made numerous suggestions. Again we encourage readers to inform us of suspected errors.

Chapel Hill 1976

Charles S. Johnson, Jr.
Lee G. Pedersen

CONTENTS

Solution: Ref. GO,23

(a) The Lagrangian function for the projectile is:

$$L = T - V = \frac{m}{2}(\dot{x}^2 + \dot{y}^2) - mgy$$

where g is the acceleration of gravity. Lagrange's equation is

$$\frac{d}{dt}\left(\frac{\partial L}{\partial \dot{q}_i}\right) - \frac{\partial L}{\partial q_i} = 0 \text{ with } q_1 = x \text{ and } q_2 = y.$$

For the x equation we find

$$\frac{\partial L}{\partial \dot{x}} = m\dot{x} \quad \text{and} \quad \frac{\partial L}{\partial x} = 0$$

so that

$$\frac{d}{dt}(m\dot{x}) = 0$$

For the y equation we find

$$\frac{\partial L}{\partial \dot{y}} = m\dot{y} \quad \text{and} \quad \frac{\partial L}{\partial y} = -mg$$

The resulting differential equation is

$$\frac{d}{dt}(m\dot{y}) = -mg$$

(b) The Hamiltonian function is

$$H = T + V = (p_x^2 + p_y^2)/2m + mgy$$

Hamilton's canonical equations give

$$\dot{x} = \frac{\partial H}{\partial p_x} = \frac{p_x}{m}, \quad \dot{p}_x = -\frac{\partial H}{\partial x} = 0$$

and

$$\dot{y} = \frac{\partial H}{\partial p_y} = \frac{p_y}{m}, \quad \dot{p}_y = -\frac{\partial H}{\partial y} = -mg$$

In this simple problem the direct application of Newton's law f = ma is clearly the preferred procedure.

1.2 Show that the total kinetic energy of two particles of masses m_1 and m_2 can be written as

$$T = \frac{1}{2} M\underset{\sim}{V}_{CM}^2 + \frac{1}{2}\mu \underset{\sim}{v}_R^2$$

where

$$M = m_1 + m_2 = \text{total mass}, \quad \mu = \frac{m_1 m_2}{m_1 + m_2} = \text{reduced mass}$$

$\underset{\sim}{V}_{CM}$ = the velocity of the center of mass

CHAPTER 1

ATOMIC PHYSICS AND THE OLD QUANTUM THEORY

In this first chapter we emphasize the early experiments and theory which motivated the development of quantum mechanics. By the year 1900 Classical or Newtonian mechanics was available in the powerful formulations of Lagrange and Hamilton, and the classical electromagnetic theory was embodied in the differential equations of Maxwell. Defects were, however, made evident by the failure of the classical theories to explain the frequency dependence of the intensity of radiation emitted by a blackbody. A turning point for the theory came with Planck's explanation of blackbody radiation (1900) and Einstein's description of the photoelectric effect (1905). Both of these "quantum theories" postulated a discreteness of energy. For a blackbody the radiating oscillators were allowed to have only certain discrete energies, and in the photoelectric effect the radiation was assumed to consist of energy quanta or photons. A major success of the quantum idea was Bohr's theory of one-electron atoms (1911).

From 1900 to 1925 quantum phenomena were treated by imposing quantum restrictions on the classical solutions for mechanical systems. The most general formulation of the old quantum theory is the Wilson-Sommerfeld quantization rule (1915), but it is clear that before 1925 there was no satisfactory, general theory. A major step in the direction of quantum mechanics was de Broglie's association of wave properties with matter (1923).

The problems in this chapter concentrate on classical mechanics and the application of simple quantization rules. The properties of waves will be treated in detail in Chapter 2. In all numerical problems SI units are used. For a discussion of units see Appendix 1.

Suggested general reading: VW; JA; HI; PW, Chapters 1 and 2.

PROBLEMS

1.1 Consider a projectile that has been fired from the earth. Take the vertical direction as y and the horizontal direction as x.

(a) Write out the Lagrangian function L for the system and use it with Lagrange's equation to derive the two differential equations which describe the motion of the projectile. (neglect air friction)

(b) Write out the Hamiltonian function H for the system and use it to obtain the canonical equations of motion. ($q_1 = x$, $q_2 = y$)

APPENDICES

Chapter 12

SCATTERING THEORY 358

and

$$\underset{\sim}{v}_R = \underset{\sim}{v}_1 - \underset{\sim}{v}_2 = \text{the velocity of } m_1 \text{ relative to } m_2$$

Solution: Ref. GO,58

The center of mass is defined by: $\underset{\sim}{r}_{CM} = \dfrac{m_1\underset{\sim}{r}_1 + m_2\underset{\sim}{r}_2}{m_1 + m_2}$

The velocity of the center of mass is then given by:

$$\dot{\underset{\sim}{r}}_{CM} = \frac{m_1\dot{\underset{\sim}{r}}_1 + m_2\dot{\underset{\sim}{r}}_2}{m_1 + m_2} = \frac{m_1\underset{\sim}{v}_1 + m_2\underset{\sim}{v}_2}{M} = \underset{\sim}{V}_{CM}$$

Solving for $\underset{\sim}{v}_1$ and $\underset{\sim}{v}_2$ in terms of $\underset{\sim}{V}_{CM}$ and $\underset{\sim}{v}_R$ we find

$$\underset{\sim}{v}_1 = \underset{\sim}{V}_{CM} + \frac{m_2}{M}\underset{\sim}{v}_R, \quad \underset{\sim}{v}_2 = \underset{\sim}{V}_{CM} - \frac{m_1}{M}\underset{\sim}{v}_R$$

Thus

$$T = \frac{1}{2} m_1\underset{\sim}{v}_1{}^2 + \frac{1}{2} m_2\underset{\sim}{v}_2{}^2 = \frac{1}{2} m_1 \left(\underset{\sim}{V}_{CM} + \frac{m_2}{M}\underset{\sim}{v}_R\right)^2 + \frac{1}{2} m_2 \left(\underset{\sim}{V}_{CM} - \frac{m_1}{M}\underset{\sim}{v}_R\right)^2$$

$$= \frac{1}{2}(m_1 + m_2)\underset{\sim}{V}_{CM}{}^2 + \frac{1}{2}\frac{m_1m_2}{(m_1 + m_2)}\underset{\sim}{v}_R{}^2 = \frac{1}{2} M\underset{\sim}{V}_{CM}{}^2 + \frac{1}{2}\mu\underset{\sim}{v}_R{}^2$$

The total kinetic energy is, therefore, the sum of the energy of motion of the center of mass and the energy of motion about the center of mass.

1.3 Show that the rotational kinetic energy of a freely rotating dumbbell consisting of the masses m_1 and m_2 separated by the distance $\underset{\sim}{R} = \underset{\sim}{r}_1 - \underset{\sim}{r}_2$ is

$$E = \frac{1}{2} I\omega^2$$

where $I = \mu\underset{\sim}{R}^2$ is the moment of inertia and ω is the angular velocity of the system. (Hint: Separate the kinetic energy of the motion of the center of mass from the total kinetic energy to obtain the rotational kinetic energy.)

Solution: Ref. GO,59

From Problem 1.2 we have

$$T = \frac{1}{2} M\underset{\sim}{V}_{CM}{}^2 + \frac{1}{2}\mu\underset{\sim}{v}_R{}^2$$

$$= \frac{1}{2} M\underset{\sim}{V}_{CM}{}^2 + \frac{1}{2}\mu(\dot{\underset{\sim}{r}}_1 - \dot{\underset{\sim}{r}}_2)^2$$

$$= \frac{1}{2} M\underset{\sim}{V}_{CM}{}^2 + \frac{1}{2}\mu\dot{\underset{\sim}{R}}^2$$

The first term is the translational kinetic energy of the center of mass while the second term must be the rotational kinetic energy.

Expressing $\underset{\sim}{R}$ in spherical polar coordinates we find

$$R_x = R\sin\theta\cos\phi, \quad R_y = R\sin\theta\sin\phi, \quad R_z = R\cos\theta$$

Taking time derivatives with R constant we find

$$\dot{R}_x = R\cos\theta\cos\phi\dot{\theta} - R\sin\theta\sin\phi\dot{\phi}$$

$$\dot{R}_y = R\cos\theta\sin\phi\dot{\theta} + R\sin\theta\cos\phi\dot{\phi}$$

$$\dot{R}_z = -R\sin\theta\dot{\theta}$$

So that

$$\dot{\tilde{R}}^2 = R^2(\dot{\theta}^2 + \sin^2\theta\dot{\phi}^2)$$

With no loss of generality we can choose $\theta = \pi/2$ and $\dot{\theta} = 0$. Then

$$E = \frac{1}{2}\,\mu\dot{\tilde{R}}^2 = \frac{1}{2}\,\mu R^2\dot{\phi}^2 = \frac{1}{2}\,\mu\omega^2 R^2$$

where ω is the angular velocity. The rotational energy may also be written as

$$E = J^2/2I$$

where the angular momentum J is defined as $I\omega$.

1.4 In a simple harmonic oscillator two point masses m_1 and m_2 are connected by a spring so that Hooke's Law is obeyed.

$$\text{restoring force} = \tilde{f} = -k(\tilde{R} - \tilde{R}_e)$$

where

k = the "force" constant

R = the instantaneous separation of the masses

R_e = the equilibrium separation

(a) Show that the total vibrational energy can be written as

$$E_{vib} = \frac{1}{2}\,\mu\dot{\tilde{R}}^2 + \frac{1}{2}\,k(\tilde{R} - \tilde{R}_e)^2$$

where μ is the reduced mass.

(b) Solve Lagrange's equation of motion for this system and show that the natural frequency of vibration is

$$\nu = \frac{1}{2\pi}\sqrt{\frac{k}{\mu}}$$

Solution: Ref. BL,64

(a) The total energy of the system is

$$E_T = \frac{1}{2}\,m_1\dot{r}_1{}^2 + \frac{1}{2}\,m_2\dot{r}_2{}^2 + V(R) \qquad (1)$$

Since Hooke's law is obeyed we have

$$f = -\frac{\partial V}{\partial x} = -\,kx \qquad (2)$$

where

$$x = |\tilde{R} - \tilde{R}_e|$$

Integration of (2) yields

$$V = \frac{1}{2} kx^2 = \frac{1}{2} k(\underset{\sim}{R} - \underset{\sim}{R}_e)^2 \tag{3}$$

where the constant of integration is chosen so that

$$V(R_e) = 0$$

This fixes the zero of energy. If T_{CM} is taken as the translational kinetic energy of the center of mass, we have

$$E_T = T_{CM} + E_{rot} + E_{vib} \tag{4}$$

Since

$$E_{rot} = \frac{1}{2} \mu \dot{\underset{\sim}{R}}^2 \quad \text{(Problem 1.3)}$$

we may further write

$$E_T - T_{CM} = \frac{1}{2} \mu \dot{\underset{\sim}{R}}^2 + V(R)$$

In contrast to the dumbbell problem $\dot{\underset{\sim}{R}}^2$ not only has contributions from $\dot{\theta}$ and $\dot{\phi}$ but also from $\dot{R}$. Thus, using the same procedure as in Problem 1.3 we find

$$\dot{\underset{\sim}{R}}^2 = R^2(\dot{\theta}^2 + \sin^2\theta\dot{\phi}^2) + \dot{R}^2$$

Since E_{rot} results from $R^2(\dot{\theta}^2 + \sin^2\dot{\phi}^2)$, we then have for the vibrational energy

$$E_{vib} = \frac{1}{2} \mu\dot{R}^2 + \frac{1}{2} k(\underset{\sim}{R} - \underset{\sim}{R}_e)^2 = \frac{1}{2} \mu\dot{x}^2 + \frac{1}{2} kx^2 = H_{vib}$$

(b) The Lagrangian function for this system is

$$L = T - V = \frac{1}{2} \mu\dot{x}^2 - \frac{1}{2} kx^2$$

and Lagrange's equation of motion is

$$\frac{d}{dt} \frac{\partial L}{\partial \dot{x}} - \frac{\partial L}{\partial x} = \frac{d}{dt} \frac{\partial T}{\partial \dot{x}} + \frac{\partial V}{\partial x} = 0 \quad \text{or} \quad \mu\ddot{x} + kx = 0$$

The general solution of the differential equation

$$\ddot{x} = -\frac{k}{\mu} x$$

is

$$x = A\sin \sqrt{\frac{k}{\mu}}\, t + B\cos \sqrt{\frac{k}{\mu}}\, t$$

or

$$x = A'\cos\left(\sqrt{\frac{k}{\mu}}\, t + \alpha\right)$$

The natural frequency of the oscillation is clearly given by

$$\nu = \frac{1}{2\pi} \sqrt{\frac{k}{\mu}} \quad \text{or} \quad \omega = \sqrt{\frac{k}{\mu}}$$

1.5 A sphere of radius a with a total charge of +e has a uniform distribution of positive charge throughout. When an electron is released inside this sphere, what frequency of oscillation is predicted by classical mechanics? Compute the numerical value of the frequency assuming that a = 1 Å. This problem is based on the model of the atom that was proposed by Thomson in 1903. (Hint: First show that the force on the electron is $-kr$ where k is a constant and r is the distance of the electron from the center of the sphere.)

Solution: Ref. JA,70; S3,27

The potential energy of interaction of the point charges q_1 and q_2 is

$$V(r) = \frac{q_1 q_2}{4\pi\varepsilon_o r}$$

where r is their separation. Here we are using the rationalized MKS system of units. A discussion of units is given in Appendix 1. Now consider the interaction of the charge q_1 with the shell of thickness dr_s and charge density ρ shown in Fig. 1.5.

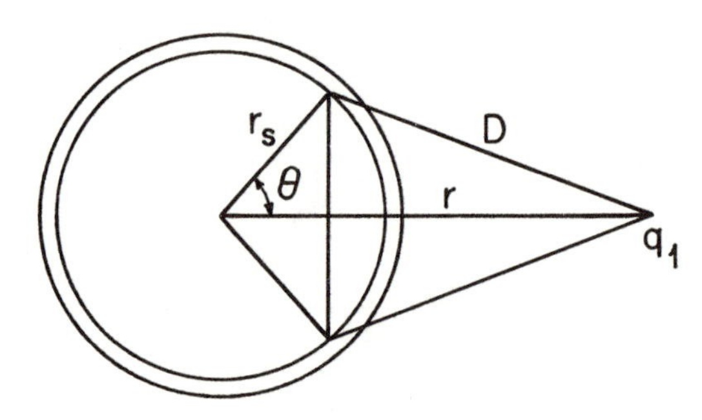

Fig. 1.5

The charge element dq_2 in the ring r_s to $r_s + dr_s$ and θ to $\theta + d\theta$ is given by

$$dq_2 = \rho(2\pi r_s \sin\theta) r_s d\theta dr_s$$

and the potential energy of interaction between q_1 and dq_2 is

$$\frac{q_1}{(4\pi\varepsilon_o)} \cdot \frac{\rho 2\pi r_s^2 dr_s \sin\theta d\theta}{D}$$

The contribution from the entire shell of radius r_s is obtained by integrating over θ:

$$dV = \frac{q_1 \rho 2\pi r_s^2 dr_s}{4\pi\varepsilon_o} \int_o^\pi \frac{\sin\theta d\theta}{D}$$

But we know that:

$$D^2 = r_s^2 + r^2 - 2r_s r\cos\theta, \quad DdD = r_s r\sin\theta d\theta$$

So

$$\int_0^{\pi} \frac{\sin\theta d\theta}{D} = \int_{r-r_s}^{r+r_s} \frac{dD}{r_s r} = \frac{2}{r} \; ; \quad r > r_s$$

$$= \int_{r_s-r}^{r_s+r} \frac{dD}{r_s r} = \frac{2}{r_s} \; ; \; r < r_s$$

Therefore each shell having $r_s < r$ contributes

$$dV = \frac{q_1(\rho 4\pi r_s^2 dr_s)}{(4\pi\varepsilon_o)r} \tag{1}$$

to the total potential while each shell having $r_s > r$ contributes

$$dV = \frac{q_1(\rho 4\pi r_s^2 dr_s)}{(4\pi\varepsilon_o)r_s} \tag{2}$$

If the charge q_1 is situated a distance r from the center of a sphere of uniform charge density ρ, the total potential energy becomes

$$V = \frac{q_1(4\pi\rho)}{(4\pi\varepsilon_o)} \left\{ \int_o^r \frac{r_s^2 dr_s}{r} + \int_r^a \frac{r_s^2 dr_s}{r_s} \right\}$$

where a is the radius of the sphere and $r_s < a$. Integration leads to the result

$$V = \frac{q_1(\rho 4\pi)}{4\pi\varepsilon_o} \left[\frac{-r^2}{6} + \frac{a^2}{2} \right]$$

And the force on q_1 is

$$F = -\frac{\partial V}{\partial r} = + \frac{q_1(\rho 4\pi)r}{(4\pi\varepsilon_o)3}$$

For the Thomson atom $q_1 = -e$ and $\rho = +e/(4\pi a^3/3)$. The force equation becomes

$$F = -\frac{e^2}{(4\pi\varepsilon_o)a^3} r = -kr$$

and we are dealing with a simple harmonic oscillator. From Problem 1.4 we have

$$\nu = \frac{1}{2\pi}\sqrt{\frac{k}{m_e}} = \frac{1}{2\pi}\left[\frac{e^2}{(4\pi\varepsilon_o)a^3 m_e}\right]^{1/2}$$

$$= \frac{1}{2\pi}\left[\frac{(1.602 \times 10^{-19})^2}{4\pi(8.854 \times 10^{-12})(10^{-10})^3(9.109 \times 10^{-31})}\right]^{1/2}$$

$$\cong 2.53 \times 10^{15}\text{Hz}$$

Thus the radiation from an electron in this model of the atom (Thomson's atom) is expected to lie in the ultraviolet part of the spectrum. This model does not agree with the results of atomic spectroscopy where many different frequencies are found even for the hydrogen atom.

It is interesting to note that the analogous problem can be carried through for the gravitational attraction of masses m_1 and m_2. A solid sphere with a hole drilled through its center and a small test mass released in the hole could in principle be used to measure the gravitational constant G. Of course, the experiment would have to be carried out under free fall conditions.

1.6 A mass is suspended by a massless elastic rod having a rectangular cross-section as shown in Figure 1.6.

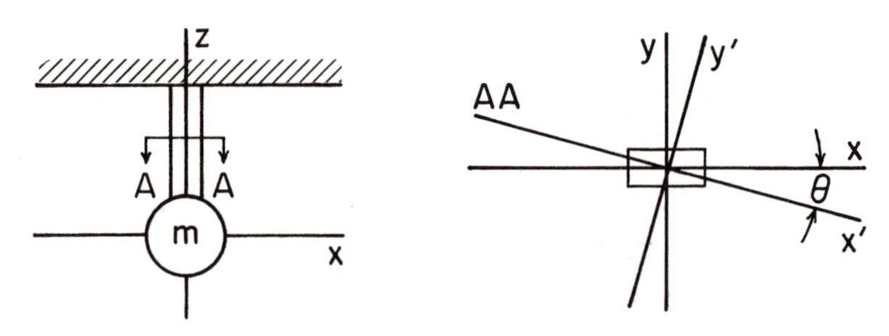

Fig. 1.6

A small displacement

$$\underset{\sim}{r} = x\hat{i} + y\hat{j}$$

of the mass perpendicular to the axis of the rod leads to the restoring force

$$\underset{\sim}{F} = -k_x x\hat{i} - k_y y\hat{j}$$

(a) Set up the Lagrangian function for this system in the x'y' axis system.

(b) Derive the differential equations which describe the motion of the mass in the x'y' axis system after a small displacement from equilibrium and solve for the natural frequencies of the motion.

<u>Solution</u>: Ref. HZ3,62

(a) From Problem 1.4 we see that the potential energy is given by

$$V(x, y) = \frac{k_x}{2} x^2 + \frac{k_y}{2} y^2$$

The displacements in the xy frame can be written in terms of the displacements in the x'y' frame using the relations

$$x = x'\cos\theta + y'\sin\theta, \quad y = -x'\sin\theta + y'\cos\theta$$

Therefore

$$V(x', y') = \frac{k_x}{2}(x'\cos\theta + y'\sin\theta)^2 + \frac{k_y}{2}(-x'\sin\theta + y'\cos\theta)^2$$

The kinetic energy T has the form

$$T = \frac{m}{2}[(\dot{x}')^2 + (\dot{y}')^2]$$

and

$$L(\dot{x}',\dot{y}',x',y') = T(\dot{x}',\dot{y}') - V(x',y')$$

(b) Lagrange's equation has the form

$$\frac{d}{dt}\left(\frac{\partial L}{\partial \dot{q}_i}\right) - \frac{\partial L}{\partial q_i} = 0$$

where in this problem $q_1 = x'$ and $q_2 = y'$. Therefore, we obtain the equations

$$m\ddot{x}' + \cos\theta(x'\cos\theta + y'\sin\theta)k_x - \sin\theta(-x'\sin\theta + y'\cos\theta)k_y = 0$$

$$m\ddot{y}' + \sin\theta(x'\cos\theta + y'\sin\theta)k_x + \cos\theta(-x'\sin\theta + y'\cos\theta)k_y = 0$$

The natural frequencies can be obtained by assuming that the displacements carry out simple harmonic motions with the angular frequency ω so that

$$x' = x_o'\cos(\omega t + \phi), \quad y' = y_o'\cos(\omega t + \phi)$$

where ϕ is a constant phase factor. We then have

$$[-\omega^2 m + (k_x\cos^2\theta + k_y\sin^2\theta)]x' + [(k_x - k_y)\sin\theta\cos\theta]y' = 0$$

$$[(k_x - k_y)\sin\theta\cos\theta]x' + [-\omega^2 m + (k_x\sin^2\theta + k_y\cos^2\theta)]y' = 0$$

This set of equations has non-trivial solutions for x' and y' only when the determinant of coefficients vanishes. This becomes evident when a solution is attempted by Cramer's rule. The only unknown is the ratio x'/y' and there are two equations. Hence, the numerators vanish and one is left with the trivial solution x' = y' = 0 except when the denominator vanishes.

Setting the determinant of coefficients equal to zero we have

$$\begin{vmatrix} [-\omega^2 m + (k_x\cos^2\theta + k_y\sin^2\theta)] & (k_x - k_y)\sin\theta\cos\theta \\ (k_x - k_y)\sin\theta\cos\theta & [-\omega^2 m + (k_x\sin^2\theta + k_y\cos^2\theta) \end{vmatrix} = 0$$

And expanding the determinant we find

$$\omega^4 m^2 - \omega^2 m(k_x + k_y) + k_x k_y = 0$$

or

$$(\omega^2 m - k_x)(\omega^2 m - k_y) = 0$$

and the two roots are

$$\omega_x = \sqrt{\frac{k_x}{m}}\ , \quad \omega_y = \sqrt{\frac{k_y}{m}}$$

The natural frequencies ω_x and ω_y are of course independent of the orientation of the coordinate system. In the special case that $\theta = 0$, the equations for $\ddot{x}'$ and $\ddot{y}'$ are uncoupled and can be solved separately to obtain ω_x and ω_y.

1.7 In the classical planetary model of a hydrogen-like atom an electron with mass m_e and charge $-e$ is assumed to move in a circular orbit around a nucleus which has a mass M and a charge $+Ze$. The only forces acting on the particles arise from the coulombic attraction of the charges and it is assumed that $M >> m_e$.

(a) Obtain an equation for the total energy E of the atom in terms of the radius r of a stable orbit. Neglect the radiation of energy by the accelerating electron.

(b) Derive an equation for the radius r of a stable orbit in terms of m_e, e, and the orbital angular momentum $\mathcal{L}$ of the electron. Show that the radius r for a stable orbit can be written as

$$r = \frac{(4\pi\varepsilon_o)\mathcal{L}^2}{mZe^2}$$

Solution: Ref. KP,20; HZ1,15; PW,36; S1,395

(a) The coulombic attraction provides the centripetal force to maintain a circular orbit. Therefore

$$\frac{Ze^2}{(4\pi\varepsilon_o)r^2} = \frac{mv^2}{r} \tag{1}$$

The total energy is given by

$$E = T + V = \frac{mv^2}{2} - \frac{Ze^2}{(4\pi\varepsilon_o)r} \tag{2}$$

and using Eq. (1) we have

$$E = -\frac{Ze^2}{(4\pi\varepsilon_o)2r} \tag{3}$$

where r is the radius of a stable orbit.

(b) From Eqs. (1) and (2) we have

$$E = -\frac{Ze^2}{(4\pi\varepsilon_o)2r} = -\frac{mv^2}{2} \tag{4}$$

Since the orbital angular momentum is defined as $\mathcal{L} = mvr$, v^2 can be eliminated from Eq. (4) to give

$$r = \frac{(4\pi\varepsilon_o)\mathcal{L}^2}{mZe^2}$$

The solutions given here are based on the assumption that no radiation is emitted even though the electron is accelerating toward the nucleus. It should be realized that an electron moving in a circular orbit with the fixed radius r and the angular frequency ω has the acceleration $\omega^2 r$ toward the center. Since an accelerating charge does radiate, the classical planetary model of the atom fails.

1.8 Two particles having the same mass are connected by springs with equal force constants as shown in Figure 1.8. If we choose the displacements of the masses from their equilibrium positions to be x_1 and x_2 in the same direction, then we may write the potential energy as

$$V = \frac{1}{2}kx_1^2 + \frac{1}{2}k(x_1 - x_2)^2 + \frac{1}{2}kx_2^2$$

Assuming that the particles move on a frictionless plane, show that the natural frequencies for motion of the particles are

$$\nu_1 = \frac{1}{2\pi}\sqrt{\frac{k}{m}} \quad \text{and} \quad \nu_2 = \frac{1}{2\pi}\sqrt{\frac{3k}{m}}$$

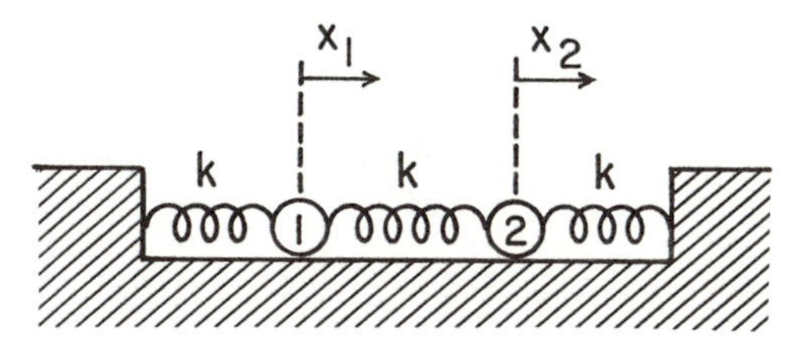

Fig. 1.8

Solution: Ref. WDC,28

The Lagrangian T-V is

$$L = T-V = \frac{1}{2}m(\dot{x}_1^2 + \dot{x}_2^2) - \frac{1}{2}k[x_1^2 + (x_1 - x_2)^2 + x_2^2]$$

The two equations of motion are then

$$\frac{d}{dt}\frac{\partial L}{\partial \dot{x}_1} - \frac{\partial L}{\partial x_1} = m\ddot{x}_1 + kx_1 + k(x_1 - x_2) = m\ddot{x}_1 + k(2x_1 - x_2) = 0 \qquad (1)$$

$$\frac{d}{dt}\frac{\partial L}{\partial \dot{x}_2} - \frac{\partial L}{\partial x_2} = m\ddot{x}_2 + kx_2 + k(x_2 - x_1) = m\ddot{x}_2 + k(2x_2 - x_1) = 0 \qquad (2)$$

Choosing

$$x_1 = A_1\cos(\omega t + \alpha) \quad \text{and} \quad x_2 = A_2\cos(\omega t + \alpha)$$

we find by substitution into (1) and (2)

$$\begin{aligned} (2k - m\omega^2)A_1 - kA_2 &= 0 \\ -kA_1 + (2k - m\omega^2)A_2 &= 0 \end{aligned} \tag{3}$$

This equation has nontrivial solutions only if the determinant of the 2x2 matrix vanishes. Thus

$$\begin{vmatrix} 2k - m\omega^2 & -k \\ -k & 2k - m\omega^2 \end{vmatrix} = m^2\omega^4 - 4km\omega^2 + 3k^2 = 0$$

The roots are

$$\omega^2 = \frac{k}{m}, \quad \frac{3k}{m}$$

Since

$$\omega_i = 2\pi\nu_i$$

we have

$$\nu_1 = \frac{1}{2\pi}\sqrt{\frac{k}{m}}, \quad \nu_2 = \frac{1}{2\pi}\sqrt{\frac{3k}{m}}$$

Substitution of ν_1 and ν_2 back into (3) allows determination of the relative amplitudes:

for ν_1: $A_1 = A_2$, thus both masses are moving to the right (see Fig. 1.8)

for ν_2: $A_1 = -A_2$, the two balls are moving in opposite directions (see Fig. 1.8)

1.9 The term "black-body" radiation refers in practice to the radiation emitted from a small orifice of a heated hollow cavity. The experimental dependence of the energy density of such radiation on temperature and wavelength is shown in Fig. 1.9. The energy density $\rho(\nu)$ in the range ν to $\nu + d\nu$ was shown by Planck to be

$$\boxed{\rho(\nu)d\nu = \frac{\langle\varepsilon\rangle N(\nu)d\nu}{V} = \frac{8\pi\nu^3}{c^3}\left(\frac{h}{e^{h\nu/k_BT} - 1}\right)d\nu}$$

where

$\langle\varepsilon\rangle$ is the average energy of the oscillators which are assumed to make up the walls of the cavity and which are in equilibrium with the radiation field inside the cavity,

$N(\nu)d\nu$ is the number of modes of the radiation field in a cavity of volume V in the frequency range ν to $\nu + d\nu$

This problem is concerned with the evaluation of $\langle\varepsilon\rangle$.

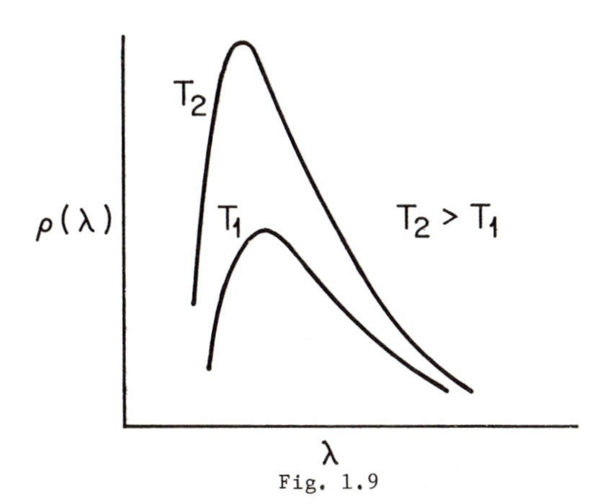

Fig. 1.9

(a) According to classical statistical mechanics, the average energy $\langle\varepsilon\rangle$ for systems in thermal equilibrium is

$$\langle\varepsilon\rangle = \frac{\iint \varepsilon e^{-\varepsilon/k_B T} dpdx}{\iint e^{-\varepsilon/k_B T} dpdx}$$

where k_B is Boltzmann's constant. For a simple harmonic oscillator

$$\varepsilon = \frac{p^2}{2m} + \frac{1}{2} kx^2$$

Evaluate the integral to obtain an equation for $\langle\varepsilon\rangle$.

(b) If the energy of an oscillator can only take on the values

$$\varepsilon_i = ih\nu$$

where i is an integer, then the integral in part (a) must be replaced by

$$\langle\varepsilon\rangle = \frac{\sum_{i=0} \varepsilon_i e^{-\varepsilon_i/k_B T}}{\sum_{i=0} e^{-\varepsilon_i/k_B T}}$$

Obtain a closed form equation for $\langle\varepsilon\rangle$. (Hint: It is convenient to define $y = \exp(-h\nu/k_B T)$ and to use the identity $(1 - y)^{-1} = \sum_{i=0}^{\infty} y^i$).

Solution: Ref. PI1,3; EI,57; S3,253; WS,450

(a) For a continuous distribution of energy

$$\langle\varepsilon\rangle = \frac{\iint \varepsilon e^{-\varepsilon/k_BT}\,dpdx}{\iint e^{-\varepsilon/k_BT}\,dpdx}$$

$$= \frac{\iint \left(\frac{p^2}{2m} + \frac{1}{2}kx^2\right) e^{-\frac{p^2}{2mk_BT}} e^{-\frac{kx^2}{2k_BT}}\,dpdx}{\iint e^{-\frac{p^2}{2mk_BT}} e^{-\frac{kx^2}{2k_BT}}\,dpdx}$$

$$= \frac{\frac{1}{2m}\int_{-\infty}^{+\infty} p^2 e^{-\frac{p^2}{2mk_BT}}\,dp}{\int_{-\infty}^{+\infty} e^{-\frac{p^2}{2mk_BT}}\,dp} + \frac{\frac{1}{2}k\int_{-\infty}^{+\infty} x^2 e^{-\frac{kx^2}{2k_BT}}\,dx}{\int_{-\infty}^{+\infty} e^{-\frac{kx^2}{2k_BT}}\,dx}$$

$$= \frac{(2mk_BT)^{3/2}(\sqrt{\pi}/4)}{2m(2mk_BT)^{1/2}(\sqrt{\pi}/4)} + \frac{k(2k_BT/k)^{3/2}(\sqrt{\pi}/4)}{2(2k_BT/k)^{1/2}(\sqrt{\pi}/4)}$$

$$= \frac{k_BT}{2} + \frac{k_BT}{2} = k_BT$$

Thus each quadratic term in ε contributes an amount $k_BT/2$ to $\langle\varepsilon\rangle$. The generalization of this result is called the "Principle of Equiparition of Energy."

(b) For a discrete distribution of energy

$$\langle\varepsilon\rangle = \frac{E_{TOTAL}}{N_{TOTAL}} = \frac{\sum_{i=0} \varepsilon_i n_i}{\sum_{i=0} n_i}$$

$$= \frac{\sum_{i=0} ih\nu N_o e^{-\frac{ih\nu}{k_BT}}}{\sum_{i=0} N_o e^{-\frac{ih\nu}{k_BT}}}$$

$$= h\nu \frac{\sum_i iy^i}{\sum_i y^i}$$

where

$$y = e^{-\frac{h\nu}{k_B T}}$$

Since

$$\sum_{i=0} y^i = 1 + y + y^2 + \cdots = (1 - y)^{-1}$$

and

$$\sum_{i=0} iy^i = y + 2y^2 + 3y^3 + \cdots = y(1 + 2y + 3y^2 + \cdots) = y(1 - y)^{-2}$$

we have

$$\langle\varepsilon\rangle = h\nu \frac{y(1 - y)^{-2}}{(1 - y)^{-1}} = h\nu \frac{y}{(1 - y)}$$

$$= h\nu \frac{e^{-h\nu/k_B T}}{1 - e^{-h\nu/k_B T}} = \frac{h\nu}{e^{h\nu/k_B T} - 1}$$

In the case that $k_B T >> h\nu$

$$e^{h\nu/k_B T} = 1 + \left(\frac{h\nu}{k_B T}\right) + \cdots \approx 1 + \frac{h\nu}{k_B T}$$

so that

$$\langle\varepsilon\rangle \to \frac{h\nu}{1 + \frac{h\nu}{k_B T} - 1} = k_B T$$

Thus as the spacing between the levels goes to zero ($k_B T >> h\nu$), the classical result of part (a) is reclaimed.

1.10 A string is attached between two supports that are separated by a distance L as shown in Fig. 1.10. Standing waves are then excited in the string so that $\lambda/L << 1$. The wave velocity in the string is $v = \lambda\nu$.

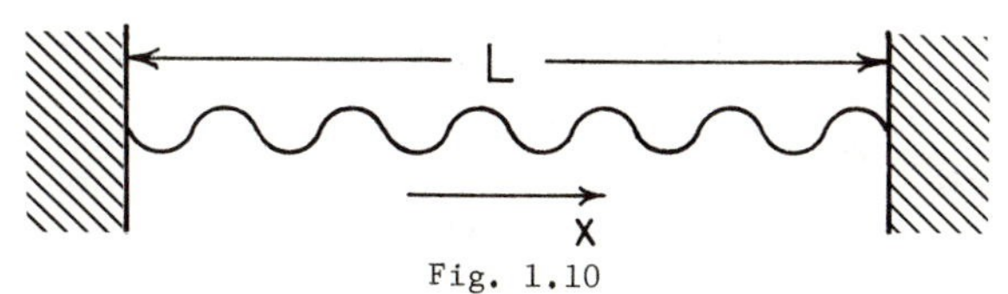

Fig. 1.10

(a) Obtain an expression for the number of vibrational modes dn that have wavelengths between λ and $\lambda + d\lambda$.

(b) Obtain an expression for the number of vibrational modes that have frequencies between ν and $\nu + d\nu$.

Solution:

(a) Waves in the string have the form

$$\psi(x) = A\sin\left(\frac{2\pi x}{\lambda} + \phi\right)$$

with the boundary conditions,

$$\psi(0) = \psi(L) = 0$$

These conditions are met by setting $\phi = 0$ and requiring that

$$(2\pi L/\lambda) = n\pi$$

where n is an integer. Therefore, the number of half wavelengths in the distance L is $n = 2L/\lambda$ and each value of n specifies a "mode" of vibration. The number of modes which occur in the range λ to $\lambda + d\lambda$ is given by

$$dn = 2L|d(1/\lambda)| = (2L/\lambda^2)d\lambda$$

where n has been treated as a continuous variable since dn/n is assumed to be vanishingly small.

(b) From part (a) we have

$$dn = (2L/v)|d(v/\lambda)| = (2L/v)d\nu$$

for the number of modes in the range ν to $\nu + d\nu$.

1.11 Consider a cavity of volume V which contains black-body radiation. Show that the number of modes of the radiation field with frequencies between ν and $\nu + d\nu$ is given by:

$$dN = \frac{8\pi\nu^2}{c^3} d\nu \cdot V$$

In counting the number of modes it is convenient to use a three dimensional "k" space with axes: $k_x = (1/\lambda_x)$, $k_y = (1/\lambda_y)$, and $k_z = (1/\lambda_z)$. The number of modes is proportional to a volume in this space and the calculation proceeds easily in spherical polar coordinates.

Solution: Ref. S4,225; S3,254

From Problem 1.10 we have that

$$dn_x = 2L_x d(1/\lambda_x) = 2L_x dk_x$$

Now consider a cavity with dimensions L_x, L_y, and L_z so that $L_xL_yL_z = V$. For three dimensions we write

$$dn_x dn_y dn_z = 8(L_xL_yL_z)dk_x dk_y dk_z$$

Since $\nu = c/\lambda = ck$ and $d\nu = cdk$, we need to determine the number of modes in the range k to k + dk.

It is convenient to work in k-space with the quantity

$$k = \sqrt{k_x^2 + k_y^2 + k_z^2}$$

which is invariant to the orientation of the coordinate system. In spherical polar coordinates we can immediately write

$$dk_x dk_y dk_z = k^2 \sin\theta d\theta d\phi dk$$

The total number of modes in the range k to k + dk is then given by

$$dN' = \frac{1}{8}\int_0^{\pi}\int_0^{2\pi} 8Vk^2 \sin\theta d\theta d\phi dk = V4\pi k^2 dk$$

The factor of 1/8 arises because we only want positive values of the components k_x, k_y, and k_z (see Fig. 1.11).

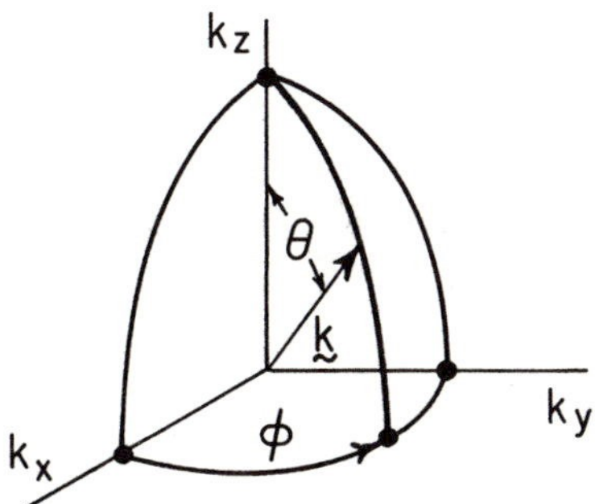

Fig. 1.11

Converting to frequency we can write

$$dN = 2dN' = \left(\frac{8\pi\nu^2}{c^3}\right) d\nu \cdot V$$

An extra factor of 2 has appeared in this step to take into account the two directions of polarization for photons. In a classical treatment this factor arises because there are two transverse waves associated with electromagnetic radiation.

1.12 The Stefan-Boltzmann law states that the total energy density in blackbody radiation at all frequencies is proportional to the fourth power of the temperature.

$$W = \int_0^{\infty} \rho(\nu) d\nu = \sigma T^4$$

Use the Planck radiation equation to obtain this result and determine the constant σ.

Solution: Ref. PI1,3; PA1,576; LI,137

Inserting Planck's equation for $\rho(\nu)$ from Problem 1.9 we obtain

$$W = \int_0^{\infty} \frac{8\pi\nu^3}{c^3}\left(\frac{h}{e^{h\nu/k_BT} - 1}\right) d\nu$$

For convenience let $x = h\nu/k_BT$. Then

$$W = \int_0^\infty \frac{8\pi}{c^3}\left(\frac{xk_BT}{h}\right)^3 \frac{h}{(e^x - 1)}\left(\frac{k_BT}{h}\right) dx$$

$$= \frac{8\pi k_B{}^4T^4}{c^3h^3}\int_0^\infty \frac{x^3dx}{(e^x - 1)} = \sigma T^4 \qquad (1)$$

The integral can be written as

$$\int_0^\infty \frac{x^3e^{-x}}{1 - e^{-x}}\,dx = \int_0^\infty x^3e^{-x}(1 + e^{-x} + e^{-2x} + \cdots)dx$$

$$= \int_0^\infty x^3(e^{-x} + e^{-2x} + e^{-3x} + \cdots)dx$$

Then integration by parts or use of a table of integrals shows that

$$\int_0^\infty x^3e^{-nx}dx = \frac{6}{n^4}$$

The integral becomes

$$\int_0^\infty \frac{x^3dx}{e^{-x} - 1} = 6\sum_{n=1}^\infty n^{-4} = 6\left(\frac{\pi^4}{90}\right) = \frac{\pi^4}{15}$$

In this step we have used the definition of the Bernoulli numbers:

$$B_2 = \frac{1}{30} = \frac{4!}{\pi^4 2^3}\sum_{n=1}^\infty n^{-4}$$

Finally from Eq. (1) we obtain

$$W = \left(\frac{8\pi^5k_B{}^4}{15c^3h^3}\right)T^4 = \sigma T^4$$

and

$$\sigma = 7.565\cdot10^{-16}J\cdot m^{-3}\cdot K^{-4}$$

The rate of emission from a surface at the temperature T is given by

$$e(T) = \frac{c}{4}W = \frac{c\sigma T^4}{4} = \sigma' T^4$$

Here $\sigma' = 5.669\text{x}10^{-8}Wm^{-2}K^{-4}$. This is the Stefan-Boltzmann constant that is often quoted in the literature.

1.13 A plot of the energy density $\rho(\lambda)$ of the radiation in equilibrium with a blackbody versus λ shows a maximum at $\lambda = \lambda_{max}$. The Wien displacement law states that $\lambda_{max} = A/T$ where A is a constant and T is the absolute temperature. Show that Planck's radiation equation is consistent with this law and determine the constant A.

Solution: Ref. MM1,492

Planck's radiation law (Problem 1.9) can be written as

$$\rho(\lambda) = \frac{8\pi ch}{\lambda^5}\left(\frac{1}{e^{hc/\lambda k_B T} - 1}\right)$$

The maximum in $\rho(\lambda)$ is then located by setting $d\rho(\lambda)/d\lambda$ equal to zero.

$$\frac{d\rho(\lambda)}{d\lambda} = 8\pi ch\left[\frac{-5}{\lambda^6(e^{hc/\lambda k_B T} - 1)} + \frac{hc}{k_B T\lambda^7}\,\frac{e^{h\nu/\lambda k_B T}}{(e^{hc/\lambda k_B T} - 1)^2}\right] = 0$$

$$0 = -5 + \frac{hc}{k_B T\lambda}\,\frac{e^{h\nu/k_B T}}{(e^{hc/\lambda k_B T} - 1)}$$

Letting $x = hc/\lambda k_B T$ this equation becomes

$$5 = xe^x/(e^x - 1), \quad e^x = 5/(5 - x)$$

This transcendental equation can be solved by numerical methods to obtain $x = 4.965$. Therefore,

$$T\lambda_{max} = \frac{hc}{4.965\ k_B} = 2.90\text{x}10^{-3}\text{m}\cdot\text{K}$$

1.14 An Einstein crystal is defined to be a collection of N three dimensional oscillators in thermal equilibrium at temperature T. Use the equation for the average energy of an oscillator to derive an equation for the heat capacity C_v of such a crystal. Determine the high and low limits of C_v and plot C_v versus $k_B T/h\nu$.

Solution: Ref. DA3,354; RU,30

The average energy of an oscillator from Problem 1.9 is

$$<\varepsilon> = \frac{h\nu}{e^{h\nu/k_B T} - 1}$$

For 3N oscillators we have $E = 3N<\varepsilon>$. By definition the heat capacity at constant volume is

$$C_v = \left(\frac{\partial E}{\partial T}\right)_v = \left(\frac{\partial <E>}{\partial T}\right)_v$$

$$= 3N\frac{h\nu}{(e^{h\nu/k_B T} - 1)^2}\left(e^{h\nu/k_B T}\right)\left(\frac{h\nu}{k_B}\right)\left(\frac{1}{T^2}\right)$$

If N_A = Avogadro's number = N so that $k_B N_A = R$, the ideal gas constant, then

$$C_v = 3R\left(\frac{h\nu}{k_B T}\right)^2 \frac{e^{h\nu/k_B T}}{(e^{h\nu/k_B T} - 1)^2} = 3R\frac{u^2 e^u}{(e^u - 1)^2}$$

where

$$u = h\nu/k_BT$$

(see Figure 1.14).

<u>High Temperature Limit</u>

$$\lim_{T\to\infty} C_v = 3R \lim_{u\to 0} \frac{u^2e^u}{(e^u - 1)^2}$$

$$= 3R \lim_{u\to 0} \frac{u^2(1 + \cdots)}{(1 + u + \cdots - 1)^2} = 3R$$

<u>Low Temperature Limit</u>

$$\lim_{T\to 0} C_v = 3R \lim_{u\to\infty} \frac{u^2e^u}{(e^u - 1)^2} = 3R \lim_{u\to\infty} u^2e^{-u} = 0$$

The high and low temperature limits agree with the experimental findings. However, an extension of the theory to allow for a distribution of frequencies is needed to improve the fit to experimental data over the entire temperature range.

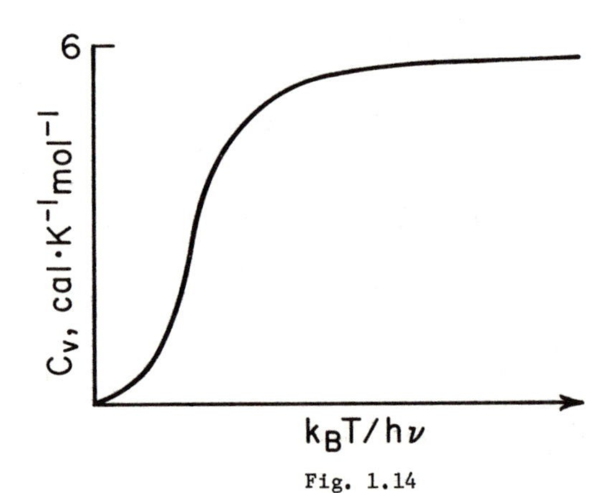

Fig. 1.14

1.15 Consider a molecule having the two low lying energy levels ε_1 and ε_2 with the degeneracies g_1 and g_2, respectively, i.e. there are g_1 states having the energy ε_1 and g_2 states having the energy ε_2. Show that the specific heat for a collection of N such molecules at thermal equilibrium is given by:

$$C_v = Nk_B\left(\frac{\varepsilon}{k_BT}\right)^2 \frac{ge^{-\varepsilon/k_BT}}{(1 + ge^{-\varepsilon/k_BT})^2}$$

where $\varepsilon = \varepsilon_2 - \varepsilon_1$ and $g = g_2/g_1$. Plot C_v versus k_BT/ε in the range 0 to 2 for g = 1.

Solution: Ref. RU,97

From Problem 1.9 we have the total energy

$$E = N\langle\varepsilon\rangle = N \sum_{i=1}^{2} g_i \varepsilon_i e^{-\varepsilon_i/k_BT} \Big/ \sum_{i=1}^{2} g_i e^{-\varepsilon_i/k_BT}$$

We choose to measure the energy from the level ε_1 (i.e. let $\varepsilon_1 = 0$). Therefore

$$E = \frac{Ng_2\varepsilon e^{-\varepsilon/k_BT}}{(g_1 + g_2 e^{-\varepsilon/k_BT})} = \frac{Ng\varepsilon e^{-\varepsilon/k_BT}}{1 + ge^{-\varepsilon/k_BT}}$$

By definition the heat capacity at constant volume is $C_v = dE/dT$ or

$$C_v = \frac{d}{dT}\left[\frac{Ng\varepsilon}{e^{\varepsilon/k_BT} + g}\right] = Ng \frac{\varepsilon^2}{k_BT^2} \frac{e^{\varepsilon/k_BT}}{(e^{\varepsilon/k_BT} + g)^2}$$

Multiplying numerator and denominator by $e^{-2\varepsilon/kT}$ we obtain

$$C_v = Nk_B\left(\frac{\varepsilon}{k_BT}\right)^2 \frac{ge^{-\varepsilon/k_BT}}{(1 + ge^{-\varepsilon/k_BT})^2}$$

A plot of C_v versus k_BT/ε is shown in Fig. 1.15. The significant feature of this plot is that the maximum heat capacity occurs when the energy separation between the levels is about equal to k_BT. This behavior should be compared with that of the harmonic oscillator where an infinite number of levels are available (see Problems 1.9 and 1.14).

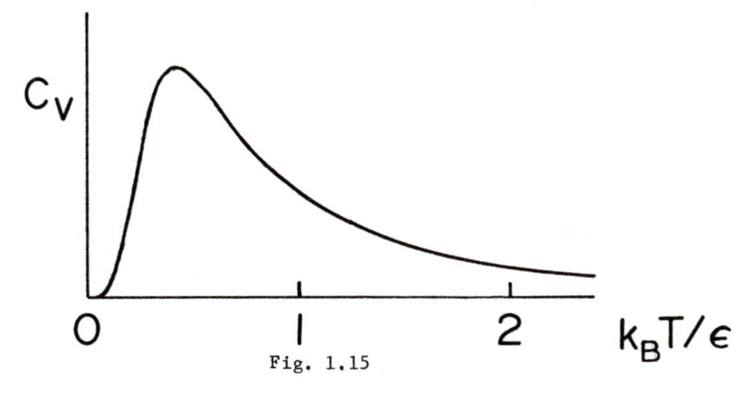

Fig. 1.15

1.16 Figure 1.16 illustrates the essential features of the photoelectric effect experiment. A photon of energy $h\nu$ strikes the surface of a metal. If the energy of the photon is greater than the work function $e\phi$, the electron may be ejected with a maximum kinetic energy of $\frac{1}{2}mv^2$. A "repeller" potential can be used to measure $\frac{1}{2}mv^2$ by finding the minimum voltage V_o necessary to stop the electrons. Show how Planck's constant may be measured from a plot of stopping voltage versus ν.

Solution: Ref. SE,110

From Fig. 1.16 we see that energy balance gives

$$h\nu \quad = \quad \frac{1}{2}mv^2 \quad + \quad e\phi$$

(impingent photon) (kinetic energy) (work function)

Thus

$$\frac{1}{2}mv^2 = h\nu - e\phi$$

The repeller potential energy necessary to just stop all of the electrons ("stopping voltage") must be equal in magnitude to the kinetic energy of the ejected electron. Thus

$$eV_o = \frac{1}{2}mv^2 = h\nu - e\phi,$$ if $e\phi$ is the same for the collector and the emitter.

This equation is then in slope-intercept form. A plot of eV_o versus ν gives a straight line, the slope of which is h and the intercept on the ν axis gives the work function $e\phi = h\nu_o$ (ν_o is the threshold frequency). See: J. Rudnick and D. S. Tannhauser, Am. J. Phys. 44, 796(1976).

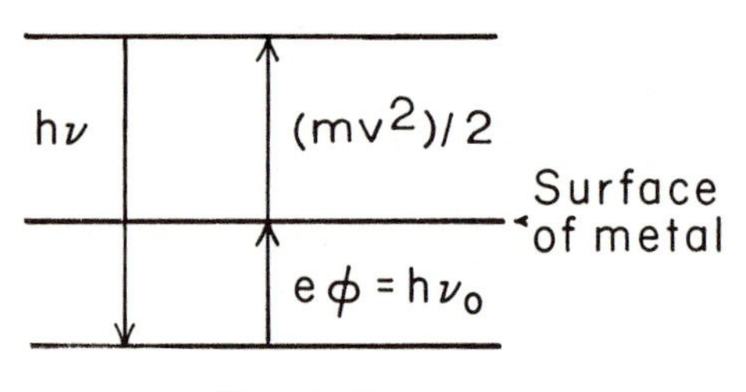

Fig. 1.16

1.17 Bohr's theory for a hydrogen-like atom is based on the classical planetary model with the following additional assumptions:

1. The atom can exist only in "stationary" states having one of the discrete energies E_1, E_2, $\cdots$
2. Only radiation having a frequency $\nu = (E_n - E_m)/h$ can be emitted or absorbed by the atom where E_m and E_n are the energies associated with the stationary states n and m.

The theory correctly predicts that the lines in the absorption and emission spectra of the hydrogen atom have frequencies that are given by the Rydberg equation:

$$\nu = cR(n_i^{-2} - n_f^{-2}) \tag{1}$$

where c is the speed of light, R is the Rydberg constant, and n_i, n_f are integers.

(a) The most direct way to derive the Rydberg equation in the context of the Bohr theory is to quantize the orbital angular momentum of the electron so that $\mathcal{L} = n\hbar$ where n is an integer.

Use this assumption with the classical planetary model (Problem 1.7) to obtain equations for the radius of an allowed orbit, the allowed energies, and the frequencies of emission and absorption.

(b) Show that the quantity $(E_{n+1} - E_n)/h$ approaches the classical rotation frequency ν_{rot} as n becomes large. This is an illustration of the Correspondence Principle.

Solution: Ref. KP,29; MO,588; JA,78

(a) From the classical planetary model of the atom we have

$$r = \frac{(4\pi\varepsilon_o)\mathcal{L}^2}{m_e Ze^2}$$

and

$$E = -\frac{Ze^2}{(4\pi\varepsilon_o)2r}$$

from Problem 1.7. The quantization of angular momentum gives $\mathcal{L} = n\hbar$ and therefore predicts that

$$\boxed{r_n = \frac{(4\pi\varepsilon_o)n^2h^2}{4\pi^2 m_e Ze^2}}$$

and

$$\boxed{E_n = -\frac{2\pi^2 Z^2 e^4 m_e}{(4\pi\varepsilon_o)^2 n^2 h^2}}$$

where n = 1, 2, 3, ··· for allowed states. The frequencies of emission and absorption as given by:

$$\nu_{fi} = (E_f - E_i)h^{-1} = cZ^2R_\infty(n_i^{-2} - n_f^{-2})$$

where

$$\boxed{R_\infty = \frac{2\pi^2 e^4 m_e}{(4\pi\varepsilon_o)^2 h^3 c} = 1.09737\text{x}10^{+7}\text{m}^{-1}}$$

(b) We wish to show that for large values of N

$$\lim_{n\to N} \nu_{n+1,n} = \nu_{rot} = v/2\pi r$$

But from part (a)

$$\lim_{n\to N} \nu_{n+1,n} = cR \lim_{n\to N} [n^{-2} - (n+1)^{-2}]Z^2$$

$$= cR \lim_{n\to N} n^{-2}[1 - (1 + 1/n)^{-2}]Z^2$$

$$= (cR2/N^3)Z^2$$

$$= \frac{4\pi^2 Z^2 e^4 m_e}{(4\pi\varepsilon_o)^2 h^3 n^3} = \frac{v}{2\pi r} = \nu_{rot}$$

where in the last step we have used the equations

$$(Ze^2/4\pi\varepsilon_o) = r m_e v^2$$

and

$$2\pi m_e v r = nh$$

This result is an illustration of the Correspondence Principle which states that the quantum mechanical solutions must approach classical solutions as $n \to \infty$. Bohr's original treatment of the hydrogen atom was based on this principle instead of on the quantization of angular momentum. The angular momentum treatment is in fact inconsistent since the ground state of the hydrogen atom is known to have zero orbital angular momentum. For a discussion of the treatment by means of the Correspondence Principle see: J. I. Shonle, Am. J. Phys. 39, 779 (1971).

1.18 The emission spectrum of the H atom is analyzed between 1000 Å and 4000 Å. What lines are found in this region?

Solution: Ref. Problem 1.17

Since

$$\Delta E(\text{cm}^{-1}) = R_\infty \left(\frac{1}{n_1^2} - \frac{1}{n_2^2}\right)$$

$$\lambda(\text{Å}) = \frac{1}{\Delta E(\text{cm}^{-1})} \frac{1\text{Å}}{10^{-8}\text{cm}} = \frac{1}{R_\infty 10^{-8}} \frac{n_1^2 n_2^2}{(n_2^2 - n_1^2)} = 911.3 \frac{n_1^2 n_2^2}{n_2^2 - n_1^2}$$

For the Lyman series

$n_1 = 1$, $n_2 = 2, 3, \cdots$:

$n_1 = 1$, $n_2 = 2$: $\lambda = 1215$ Å

$n_1 = 1$, $n_2 = 3$: $\lambda = 1025$ Å; these lines will be seen

All other transitions will have a wavelength $< 10^3$ Å and will not be seen

For the Balmer series

$n_1 = 2$, $n_2 = 3, 4, \cdots$:

$n_1 = 2$, $n_2 = 3$: $\lambda = 6561$ Å
$n_1 = 2$, $n_2 = 4$: $\lambda = 4860$ Å
$n_1 = 2$, $n_2 = 5$: $\lambda = 4339$ Å
$n_1 = 2$, $n_2 = 6$: $\lambda = 4101$ Å
} not seen

$n_1 = 2$, $n_2 = 7$: $\lambda = 3969$ Å, will be seen as well as remaining Balmer lines since $n_2 = \infty$ gives $\lambda = 3645$ Å

For the Ritz-Paschen series

$n_1 = 3,\ n_2 = 4, 5, \cdots$

$n_1 = 3,\ n_2 = 4:\ \lambda = 18746\ \text{Å}$

$n_1 = 3,\ n_2 = \infty:\ \lambda = 8201\ \text{Å}$

Thus no intervening lines in this series or any later series will be seen.

1.19 When the nucleus in a hydrogen-like atom is assumed to be fixed in space the allowed energies are given by

$$E_n = -hcR_\infty n^{-2}$$

Derive the correction factor that must be used to account for the finite mass of the nucleus.

Solution:

When the electronic mass is m_e and the nuclear mass is M the rotational kinetic energy of the Bohr atom is given by

$$T = \frac{1}{2}\mu v^2 \quad \text{where} \quad \mu = m_e M/(m_e + M)$$

(see Problem 1.3). Therefore, m should be replaced in the final energy equation by μ to give

$$E_n = -hcR_\infty \left(\frac{M}{m_e + M}\right) n^{-2} = -hcR_\infty \left(1 + \frac{m_e}{M}\right)^{-1} n^{-2}$$

1.20 The electron with charge -e and the positron with charge +e form a bound state analogous to the hydrogen atom. This system is called positronium. Calculate the ionization potential and the wavelength of the Lyman α line (n = 1 to n = 2) for positronium.

Solution: Ref. Problem 1.19; A. E. Ruark, Phys. Rev. 278 (1945)

From the Bohr theory we have

$$\Delta E = R_\infty hc(1 + m_e/m_{positron})^{-1}(n_i^{-2} - n_f^{-2})$$

For ionization $n_f = \infty$ and

$$I_p = R_\infty hc(1/2) = \frac{(1.0973\text{x}10^7\text{m}^{-1})(6.626\text{x}10^{-34}\text{J}\cdot\text{s})(2.998\text{x}10^{+8}\text{m/s})}{2}$$

$$= 10.899\text{x}10^{-18}\text{J}$$

$$= (10.899\text{x}10^{-18}\text{J})(1.6021\text{x}10^{-19}\text{J/eV})^{-1} = 6.80\text{ eV}$$

For the Lyman α line we have:

$$\bar{\nu} = (R_\infty/2)(1^{-2} - 2^{-2}) = R_\infty(3/8)$$

$$= 4.115\text{x}10^6\text{m}^{-1}$$

The wavelength is therefore:

$$\lambda = (\bar{\nu})^{-1} = 2.430\text{x}10^{-7}\text{m}$$

$$= 2,430\ \text{Å}$$

1.21 Estimate the separation in wave numbers for

(a) the 2 → 1 transition in hydrogen and deuterium (^2H).

(b) the 2 → 1 transition in hydrogen and tritium (^3H).

Solution: Ref. Problem 1.19

(a) For hydrogen:

$$\bar{\nu}_{2\to1} = R_\infty \frac{\mu}{m_e}\frac{3}{4}, \quad \text{where } \mu = \frac{m_e m_p}{m_p + m_e}$$

For deuterium:

$$\bar{\nu}_{2\to1} = R_\infty \frac{\mu}{m_e}\frac{3}{4}, \quad \text{where } \mu = \frac{m_e m_D}{m_D + m_e}$$

Let us assume

$$m_p \approx m_n \approx 1838\ m_e \quad \text{and} \quad m_D \approx 2\ m_p$$

thus

$$\lambda_H - \lambda_D = \left(\frac{1}{\bar{\nu}_H} - \frac{1}{\bar{\nu}_D}\right) = \frac{4}{3}\frac{1}{R_\infty} m_e \left[\frac{m_p + m_e}{m_e m_p} - \frac{2m_p + m_e}{2m_e m_p}\right]$$

$$= 1.215\text{x}10^{-7}\left[\frac{m_p + m_e}{m_p} - \frac{2m_p + m_e}{2m_p}\right]$$

$$= 1.215\text{x}10^{-7}\left[\frac{m_e}{2m_p}\right] = 3.30\text{x}10^{-11}\text{m}$$

$$= 0.330\ \text{Å}$$

(b) Similarly for the 2 → 1 transition of the hydrogen-tritium pair we have

$$\lambda_H - \lambda_T \approx 1.215\text{x}10^{-7}\left[\frac{m_p + m_e}{m_p} - \frac{3m_p + m_e}{3m_p}\right]$$

$$= 1.215\text{x}10^{-7}\left[\frac{2m_e}{3m_p}\right] = 4.37\text{x}10^{-11}\text{m}$$

$$= 0.437\ \text{Å}$$

1.22 Data for the earth-sun system:

mass of the earth $M_\oplus$ = $5.983\text{x}10^{24}$kg

mass of the sun $M_\odot$ = $1.971\text{x}10^{30}$kg

Gravitational constant $G = 6.673\text{x}10^{-11}\text{N}\cdot\text{m}^2/\text{kg}^2$

earth-sun separation = $9.3\text{x}10^7$miles = $1.497\text{x}10^{11}$m

(a) Determine the allowed energy levels for the earth-sun system by quantizing the angular momentum.

(b) What is the approximate principal quantum number of the current earth-sun system?

(c) What is the energy of the transition $\Delta n = 1$ for the current configuration?

Solution: Ref. Problems 1.7 and 1.17

(a) As in Problem 1.7 we equate the centrifugal force to the gravitational force and then make use of the angular momentum quantization condition to eliminate the velocity:

$$\frac{M_{\oplus} v^2}{r} = \frac{GM_{\oplus}M_{\odot}}{r^2}$$

and

$$M_{\oplus} vr = n\hbar \text{ so that } r = \frac{n^2\hbar^2}{M_{\oplus}^2 M_{\odot} G}$$

The energy is then

$$E = T + V = \frac{1}{2} M_{\oplus} v^2 - \frac{GM_{\oplus}M_{\odot}}{r}$$

$$= -\frac{GM_{\oplus}M_{\odot}}{2r} = -\frac{G^2 M_{\oplus}^3 M_{\odot}^2}{2n^2\hbar^2}$$

(b) Since the current radius is approximately $1.497\text{x}10^{11}$m

$$n = \left(r \cdot \frac{M_{\oplus}^2 M_{\odot} G}{\hbar^2}\right)^{1/2} = 2.52\text{x}10^{74}$$

(c) $$\Delta E = E_{n_2} - E_{n_1} = \frac{G^2 M_{\oplus}^3 M_{\odot}^2}{2\hbar^2}\left(\frac{1}{n_1^2} - \frac{1}{n_2^2}\right)$$

$$\approx \frac{G^2 M_{\oplus}^3 M_{\odot}^2}{2\hbar^2}\left(\frac{2}{n_1^3}\right) \approx 2.08\text{x}10^{-41}\text{J}$$

1.23 Assume that the hydrogen atom is held together only by gravitational forces. What would be the radius of this hydrogen atom in the ground state?

Solution:

From Problem 1.22 we have

$$r = \frac{n^2\hbar^2}{m_e^2 m_p G}$$

If n = 1 for the lowest state, then

$$r = \frac{(6.626\text{x}10^{-34}\text{J}\cdot\text{s})^2}{4\pi^2(9.108\text{x}10^{-31}\text{kg})^2(1.672\text{x}10^{-27}\text{kg})(6.673\text{x}10^{-11}\text{N}\cdot\text{m}^2/\text{kg}^2)}$$

$$= 1.201\text{x}10^{29}\text{m} \qquad \text{(one light year} = 9.464\text{x}10^{15}\text{m)}$$

1.24 A certain dumbbell having a moment of inertia I is rotating about its center of gravity in a plane as shown in Fig. 1.24. Use the Wilson-Sommerfeld quantization rule to determine the allowed energies of this rotator.

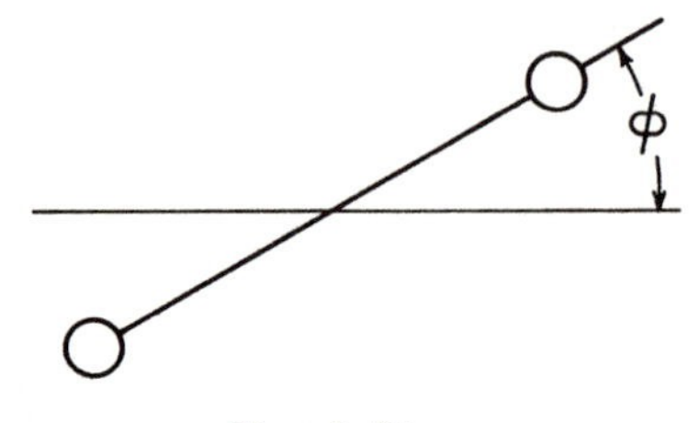

Fig. 1.24

Solution: Ref. PW,31

The action integral to be evaluated is

$$\oint p_\phi d\phi = \int_0^{2\pi} p_\phi d\phi = nh$$

Therefore

$$p_\phi = n\hbar$$

and the allowed values of the energy are

$$E = \frac{p_\phi{}^2}{2I} = \frac{n^2\hbar^2}{2I}, \quad n = 0,\ 1,\ 2,\ \cdots$$

Only values greater than zero are assigned to n in this problem since p_ϕ represents the total angular momentum. The projection of p_ϕ on some axis would have the allowed values $M\hbar$ where $M = -n,\ -n+1,\ \cdots,\ n-1,\ n$.

The rotating dumbbell is treated by quantum mechanics in Problem 4.15.

1.25 Apply the Wilson-Sommerfeld quantization rule to a particle of mass m that is held to its equilibrium position by a linear restoring force, i.e. $f = -kx$ where k is the force constant and x is the displacement from equilibrium. Obtain an equation for the allowed energies of this simple harmonic oscillator.

Solution: Ref. PW,30

The potential energy for this system is given by

$$V(x) = -\int_0^x f dx = +\frac{kx^2}{2}$$

Therefore, the total energy is

$$E = T + V = \frac{p^2}{2m} + \frac{kx^2}{2} \tag{1}$$

The maximum amplitude occurs when $p = 0$ so

$$x_{max} = \sqrt{\frac{2E}{k}}$$

Finally, the momentum p required in the action integral is obtained from Eq. (1)

$$p = \sqrt{\left(\frac{2E}{k} - x^2\right) km}$$

Then we must evaluate

$$\oint pdx = 4\int_0^{\sqrt{2E/k}} \sqrt{km} \sqrt{\left(\frac{2E}{k} - x^2\right)}\, dx$$

where the factor of 4 is required to account for a complete cycle. The result is

$$\oint pdx = 2\pi E \sqrt{\frac{m}{k}} = nh \tag{2}$$

where we have used integral

$$\int \sqrt{a^2 - x^2}\, dx = \frac{1}{2}\left[x\sqrt{a^2 - x^2} + a^2 \sin^{-1}\left(\frac{x}{a}\right)\right]$$

obtained by the substitution $x = a\sin\theta$ or from tables.

From Problem 1.4 we have

$$\nu_o = (1/2\pi)\sqrt{k/m}$$

which with Eq. (2) gives

$$E = nh\nu_o \;;\quad n = 0, 1, 2, \cdots \tag{3}$$

This equation correctly predicts a separation of $h\nu_o$ between the allowed energy levels; however as we shall see in Problem 4.10 the correct equation is

$$E = (n + 1/2)h\nu_o \;;\quad n = 0, 1, 2, \cdots \tag{4}$$

The important difference between Eqs. (3) and (4) is the "zero point" energy $h\nu_o/2$ which exists with $n = 0$.

1.26 Apply the Wilson-Sommerfeld quantization rule to a particle of mass m moving under the influence of the potential shown in Fig. 1.26 to determine the allowed energies E for $E < V_o$.

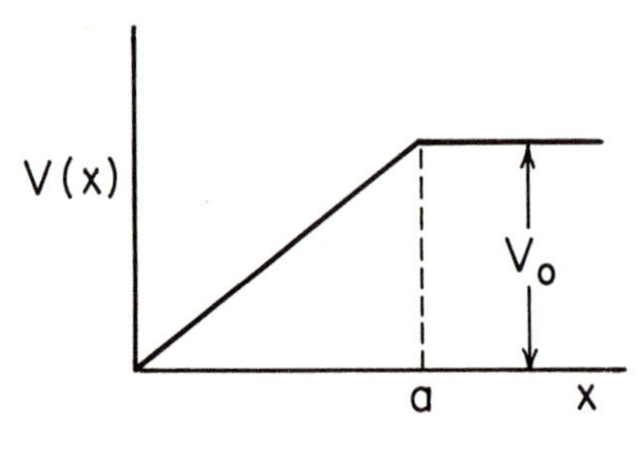

Fig. 1.26

Solution: Ref. TH,86

From Fig. 1.26 we have

$$V(x) = (V_o/a)x; \; 0 \lessgtr x \lessgtr a$$

$$V(x) = V_o; \quad a < x$$

The total energy for $0 \lessgtr x \lessgtr a$ is given by

$$E = \frac{p^2}{2m} + \frac{V_o}{a} x$$

and the magnitude of the linear momentum is

$$p = \sqrt{2m[E - (V_o/a)x]}$$

The classical turning point for the particle occurs when $p = 0$ or

$$x_{max} = aE/V_o$$

Therefore, the action integral over a cycle of motion is

$$2\int_0^{aE/V_o} \sqrt{2m} \; \sqrt{E - (V_o/a)x} \; dx = nh$$

$$= \frac{4}{3} \frac{aE^{3/2}}{V_o} \sqrt{2m}$$

where we have used the integral formula

$$\int \sqrt{a + bx} \; dx = \frac{2}{3b} \sqrt{(a + bx)^3}$$

The energy is then given by

$$E = \left(\frac{3}{4} \frac{nhV_o}{a\sqrt{2m}} \right)^{2/3}$$

It should be noticed that for $E < V_o$ the magnitude of V_o only serves to specify the slope of the potential. The form of the solution will not change as $V_o \to \infty$ if V_o/a remains constant. This problem is equivalent to that for a ball bouncing on a flat surface in a gravitational field. For the bouncing ball the potential is $V = +mgx$ where g is the acceleration of gravity and x is the height. The solution then becomes

$$E = \left[\frac{9n^2}{16}\left(\frac{h^2mg^2}{2}\right)\right]^{1/3} \simeq 2.81\left(\frac{n^2m\hbar^2g^2}{2}\right)^{1/3}$$

The correct quantum mechanical solution is discussed in Problem 4.14.

1.27 De Broglie suggested that wave like properties should be associated with particles as well as with radiation. The "de Broglie wavelength" is given by

$$\lambda = \frac{h}{mv}$$

where h is Planck's constant, m the mass of a particle and v the velocity of the particle ($v \ll c$). Evaluate the wavelength of

(a) an electron with kinetic energy 1 eV, 100 eV

(b) a proton with kinetic energy 1 eV

(c) a UF_6 molecule with kinetic energy 1 eV

(d) a baseball with velocity 100 mph (mass = 0.14 kg)

(Note: A more detailed treatment of the wave properties of matter is given in Chapter 2.)

Solution:

In general (if $v \ll c$)

$$E = \frac{p^2}{2m} = \frac{h^2}{2m\lambda^2}$$

Thus

$$\lambda = h(2mE)^{-1/2}$$
$$= 1.170\text{x}10^{-24}\ [m(\text{kg})\text{x}E(\text{eV})]^{-1/2}$$

(a) 1 eV electron: $\lambda = 1.226\text{x}10^{-9}$ m

10 eV electron: $\lambda = 3.878\text{x}10^{-10}$ m

(b) 1 eV proton: $\lambda = 2.862\text{x}10^{-11}$ m

(c) 1 eV UF_6 (mass = $5.855\text{x}10^{-25}$ kg/molecule): $\lambda = 1.529\text{x}10^{-12}$ m

(d) 100 mph = 44.7 m/s, $\lambda = \frac{h}{mv} = 1.059\text{x}10^{-34}$ m

1.28 Consider the diffraction of a beam of thermalized He atoms by a simple cubic lattice ($d \sim 2\text{Å}$). At what temperature for the He atoms would diffraction be appreciable.

Solution: Ref. SE,171

$$\lambda = \frac{h}{mv} = \frac{h}{p}$$

If the He atoms are thermalized, then the kinetic theory of gases predicts that the average kinetic energy of the atoms will be

$$E = \frac{3}{2} k_B T$$

Thus

$$E = \frac{3}{2} k_B T = \frac{1}{2} mv^2 = \frac{p^2}{2m} = \frac{h^2}{2m\lambda^2}$$

or

$$T = \frac{h^2}{3mk_B\lambda^2}$$

For the diffracted beam to have a reasonable intensity, the lattice spacing d and the wavelength of the radiation must be roughly of the same magnitude. Thus

$$T \approx \frac{(6.626\text{x}10^{-34}\ \text{J}\cdot\text{s})^2(6.023\text{x}10^{23}\ \text{mol}^{-1})}{3\ \text{x}\ (4.0026\text{x}10^{-3}\ \text{kg/mol})(1.38\text{x}10^{-23}\ \text{J/K})(2\text{x}10^{-10}\text{m})^2}$$

$$\simeq 39\ \text{K}$$

SUPPLEMENTARY PROBLEMS

1.29 A simplified model for a symmetrical linear triatomic molecule, e.g. CO_2, is shown in Fig. 1.29. Apply Lagrangian mechanics to show that the normal frequencies for motion along the bond axis are given by

$$\nu_1 = \frac{1}{2\pi}\sqrt{\frac{k}{m_1}}\ ; \quad \nu_2 = \frac{1}{2\pi}\sqrt{\frac{(2m_1 + m_2)k}{m_1 m_2}}$$

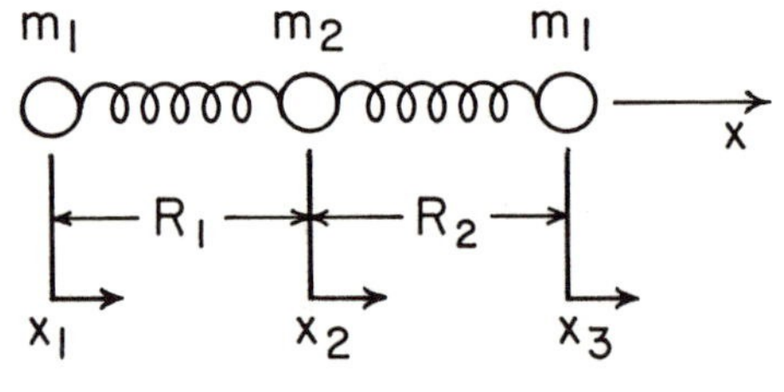

Fig. 1.29

Ref. GO,333

1.30 When the energy density $\rho(\nu)$ of blackbody radiation is plotted versus ν a maximum is found at a frequency ν_{max} which is proportional to the temperature T. One form of Wien's displacement law states that ν_{max} = constant x T. Show that Planck's equation for $\rho(\nu)$ is consistent with this law and determine the proportionality constant. Notice that ν_{max} is not equal to c/λ_{max} when λ_{max} is determined by the method of Problem 1.13. Why?

Ans:

By the method of successive approximations we find

$$\nu_{max} \cong \left(2.82 \frac{k_B}{h}\right) T = (5.88\text{x}10^{10}\text{K}\cdot\text{sec}^{-1})T$$

Notice that $d\lambda \neq d\nu$.

1.31 The maximum intensity of radiation from a certain star occurs at a wavelength of 4000 Å. Ignoring the Doppler Shift of the radiation, estimate the temperature of the star.

Ans:

$T \simeq 7{,}250$ K

1.32 The distribution of energy for a collection of oscillators is that given by Boltzmann

$$n_i = z^{-1} e^{-\varepsilon_i/k_B T}$$

where n_i is the number of oscillators having energy ε_i and

$$z = \sum_i e^{-\varepsilon_i/k_B T}$$

At temperature T what fraction of the oscillators will have the ground state energy?

Ans: Ref. DA3,91

$1 - e^{-h\nu/k_B T}$

1.33 What is the minimum voltage necessary to stop all electrons from reaching the repeller when the incident radiation on a copper surface ($e\phi$ = 0.24 electron-volts) has a wavelength of 2500 Å?

Ans: Ref. SE,110

4.72 volts

1.34 Express the energy of the n = 1 to n = 2 transition for the H atom in

(a) ergs

(b) electron-volts

(c) kcal/mol

(d) cm^{-1}

(e) Angstroms

(f) kJ/mole

Ans:

(a) $1.635\text{x}10^{-11}$ergs

(b) 10.21 eV

(c) 235.4 kcal/mol

(d) 82,302 cm^{-1}

(e) 1,215 Å

(f) 985 kJ/mol

1.35 From the simple Bohr theory, predict the second ionization potential of He, the third ionization potential for Li, and the fourth ionization potential of Be. [The experimental numbers are: He, 54.40 eV; Li, 122.42 eV; and Be, 217.657 eV.]

Ans:

He^+, 54.43 eV; Li^{++}, 122.47 eV; Be^{+++}, 217.7 eV (Using R_∞)

1.36 For periodic motion involving the coordinate x and its canonically conjugate momentum p_x the Wilson-Sommerfeld quantization rule requires that

$$\oint p_x \, dx = nh$$

where the definite integral is taken over a complete orbit of the motion and n is an integer. Apply this rule to a system consisting of a particle in a potential box having infinite walls as shown in Fig. 1.36 in order to obtain an equation for the allowed values of the energy.

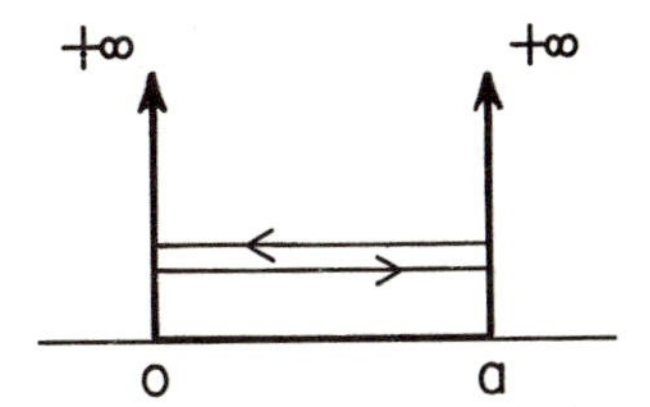

Fig. 1.36

Also, show that de Broglie's relation $p = h/\lambda$ leads to the same result.

Ans:

$$E_n = \frac{n^2h^2}{8ma^2}, \quad n = 1, 2, 3, \cdots$$

The same result is obtained for the solution of the Schrödinger equation.

CHAPTER 2

WAVES AND SUPERPOSITION

In this chapter we explore the consequences of associating waves with particles. First we present some necessary definitions and a brief discussion of orthogonal expansions. The problems which follow provide illustrations and elaborations.

1. Waves

The de Broglie hypothesis associates a wavelength λ with each particle by the equation

$$\lambda = h/p \tag{1}$$

where p is the momentum. The de Broglie waves, however, do not have the same velocity as the particle. The wave or phase velocity v_p is defined by

$$v_p = \nu\lambda = c^2/v \tag{2}$$

Conceptually we imagine the particle as existing inside a group of waves constituting a wave packet. The group velocity v_g of the de Broglie waves is equal to the velocity v of the particle, whereas the individual component waves move with the velocity v_p. The general relationships between group, particle and phase velocities are

$$\boxed{\begin{aligned} v_g &= v_p - \lambda dv_p/d\lambda \\ v &= v_g \end{aligned}} \tag{3}$$

2. Interference

In quantum theory, as in classical optics, we associate the probability P of a given event with a probability amplitude ϕ through the equation

$$P = |\phi|^2 \tag{4}$$

If the event can occur in two ways, the probability amplitude is given by

$$\phi = \phi_1 + \phi_2 \tag{5}$$

and the probability by

$$P = |\phi_1 + \phi_2|^2 = |\phi_1|^2 + |\phi_2|^2 + 2|\phi_1||\phi_2|\cos\delta \tag{6}$$

where δ is the phase difference between ϕ_1 and ϕ_2 and defines the "interference" term. The superposition expressed in (5) then leads to interference effects.

3. Orthogonal Expansions

If the integral $\int |f_i|^2 dx < \infty$ we say that f_i is square integrable. A set g of n functions f_i is called linearly dependent if there exist n constants c_1, c_2, ..., c_n, not all zero, for which

$$c_1 f_1 + c_2 f_2 + \ldots + c_n f_n = 0$$

If such constants do not exist, the n functions are called linearly independent.

We say a set of functions is orthonormal in the set g if

$$\int_g f_i^* f_j \, dx = \delta_{ij} \quad \text{where} \quad \delta_{ij} \begin{array}{l} = 1 \text{ if } i = j \\ = 0 \text{ if } i \neq j \end{array} \tag{7}$$

Given, then, n linearly independent functions f_1, f_2, ..., f_n, there exist in the set g n linear combinations which are orthonormal.

In physics many times we have the problem of finding a linear combination

$$c_1 f_1 + c_2 f_2 + \ldots + c_n f_n$$

of n square integrable functions which are a minimum "distance" from another function F which is also square integrable in g. The constants of the linear combination which best approximates in the mean to F are called the Fourier coefficients of F, and they may be determined by

$$c_n = \int_g F f_n^* \, dx \tag{8}$$

A periodic function f(x)[‡] of period 2L can then be represented by the Fourier series

$$f(x) = \frac{1}{2} a_o + \sum_{n=1}^{\infty} \left[a_n \cos\left(\frac{n\pi x}{L}\right) + b_n \sin\left(\frac{n\pi x}{L}\right) \right] \tag{9}$$

where the coefficients are given by

$$a_n = \frac{1}{L} \int_{-L}^{+L} f(x) \cos\left(\frac{n\pi x}{L}\right) dx, \quad b_n = \frac{1}{L} \int_{-L}^{+L} f(x) \sin\left(\frac{n\pi x}{L}\right) dx$$

It is often convenient to rewrite (9) in terms of complex functions. The complex Fourier series becomes

$$f(x) = \sum_{n=-\infty}^{+\infty} c_n e^{+ik_n x} \tag{10}$$

where

$$c_n = \frac{1}{2L} \int_{-L}^{+L} f(x) e^{-ik_n x} \, dx$$

[‡]The Dirichlet conditions on f(x) must be satisfied. See, for example, WY,248.

and $k_n = n\pi/L$; $n = 0, \pm 1, \pm 2, \ldots$. The c_n's are called the spectrum of f(x).

If the period becomes infinite, we may then define the Fourier Transform pair

$$f(x) = \frac{1}{2\pi}\int_{-\infty}^{+\infty} F(k)e^{+ikx}dk \tag{11a}$$

$$F(k) = \int_{-\infty}^{+\infty} f(x)e^{-ikx}dx \tag{11b}$$

f(x) and its spectrum F(k) are thus related by a Fourier Transformation.

The interference (or superposition) properties of waves can conveniently be handled in terms of orthogonal expansions. An important example is the localized wave packet which is expressed in terms of periodic functions. We frequently find the solutions to many differential equations, which represent physical phenomena, expressed as orthogonal expansions.

PROBLEMS

2.1 Young's two slit experiment is shown below,

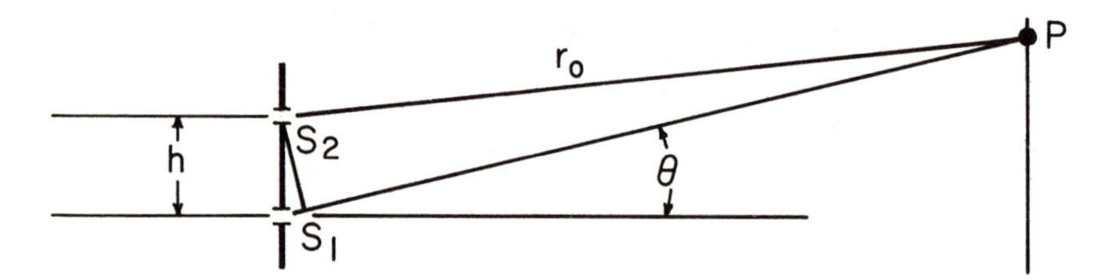

where b is the height of each slit and h is their separation. The screen is assumed to be sufficiently far away that both rays can be described by the angle θ, i.e. $h/r_o << 1$.

(a) Obtain an equation for the intensity distribution on the screen as a function of θ assuming that $\theta << \lambda/h$.

(b) Explain how the intensity distribution will change if either slit is closed.

Solution: Ref. FE3,3; FO,63; BW,257

(a) If a plane wave with the wave number $k = 2\pi/\lambda$ is incident on the slits, the amplitudes at S_1 and S_2 will be given by

$$\psi_1 = |\psi_1|e^{i(\phi - \omega t)}, \quad \psi_2 = |\psi_2|e^{i(\phi - \omega t)}$$

where t is the time of observation and ϕ is the phase angle. At the point P the amplitudes of these waves will differ because the path lengths are different. We find

at P:

$$\psi_1 = |\psi_1| e^{i(\phi + kr_o - \omega t)}$$

$$\psi_2 = |\psi_2| e^{i[\phi + k(r_o + h\sin\theta) - \omega t]}$$

Then

$$(\psi_1 + \psi_2) = e^{i(\phi + kr_o - \omega t)} (|\psi_1| + |\psi_2| e^{ikh\sin\theta})$$

and

$$I(\theta) = |\psi_1 + \psi_2|^2 = |\psi_1|^2 + |\psi_2|^2 + 2|\psi_1||\psi_2|\cos(kh\sin\theta)$$

(b) If slit S_1 is closed, the intensity becomes $|\psi_2|^2$ which for $\theta < \lambda/h$ has no dark spots. When S_2 is closed the intensity becomes $|\psi_1|^2$ with the same general appearance as $|\psi_2|^2$. The interference term $2|\psi_1||\psi_2|\cos(kh\sin\theta)$ is missing in either of these cases.

2.2 Consider the diffraction of an incident plane wave by the single slit shown below where $b/r_o \ll 1$.

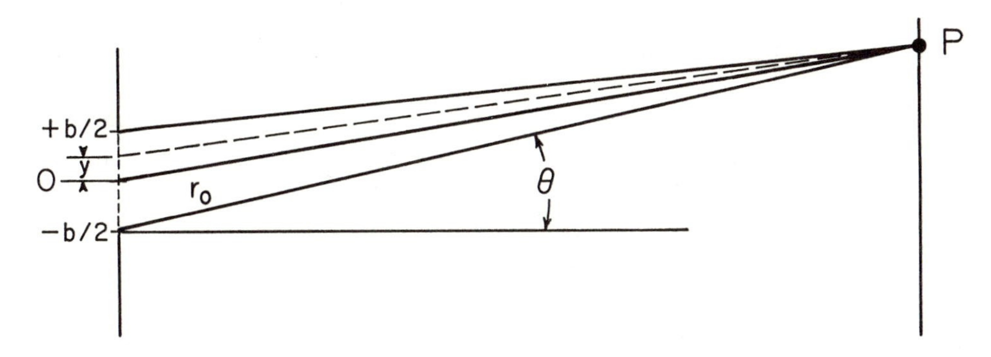

(a) As θ increases from zero the intensity observed on a screen at the extreme right decreases to a minimum and then increases. Determine the magnitude of θ that corresponds to the first dark spot.

(b) Suppose that the incident wave is a de Broglie wave associated with a particle having momentum $p_x = mv_x = h/\lambda$. If this experiment is used to determine the position y and the corresponding momentum p_y, show that $\Delta y \Delta p_y \gtrsim h$ where Δy is the uncertainty in the position, Δp_y is the uncertainty in the momentum, and h is Planck's constant. (Hint: If the particle reaches the screen with $\theta > 0$, the velocity v_y must have been modified by the diffraction process.)

<u>Solution</u>: Ref. BW,393; FO,106; HE,23

(a) An elementary solution to the diffraction problem can be obtained with <u>Huygens' Principle</u>

which states that every point on an advancing wave front can be considered as a source of secondary waves. In the present problem we imagine a series of secondary waves originating on the dotted line across the aperture in the figure. Destructive interference will occur at the point P on the screen at the right when θ is adjusted so that the path length for waves from the upper half of the slit differs by λ/2 from the path length of waves from the lower portion. This assumes that P is sufficiently far away that all of the waves arriving from the slit can be described by the same angle θ. Therefore, for the first dark spot:

$$\frac{b}{2}\sin\theta = \frac{\lambda}{2} \quad \text{or} \quad \theta = \sin^{-1}(\lambda/b)$$

Huygens' Principle predicts the points of maximum destructive interference but does not give the proper intensity distribution. A rigorous treatment of diffraction leads to the Fresnel-Kirchhoff formula. In the case of Fraunhoffer diffraction where the incident and diffracted waves are effectively plane waves this formula reduces to the simple equation

$$U_p = C\iint_A e^{ikr}dA$$

Here U_p is the amplitude of the wave, C is a constant, and A is the area of the aperature. For the single slit shown above having width W and area Wb, dA = Wdy. Then $r = r_o + y\sin\theta$ where r_o is the distance to point P when y = 0 and

$$U_p = Ce^{ikr_o}\int_{-b/2}^{+b/2} e^{iky\sin\theta}W\,dy$$

$$U_p = WCe^{ikr_o}\left(\frac{e^{iky\sin\theta}}{ik\sin\theta}\right)\Bigg|_{-b/2}^{+b/2} = Be^{ikr_o}\,\frac{\sin\beta}{\beta}$$

where

$$\beta = \frac{kb\sin\theta}{2}$$

The intensity is given by $|U_p|^2$ or

$$I(\theta) = I_o\left(\frac{\sin\beta}{\beta}\right)^2 \quad \text{with} \quad I_o = B^2$$

This function is shown in Fig. 2.2.

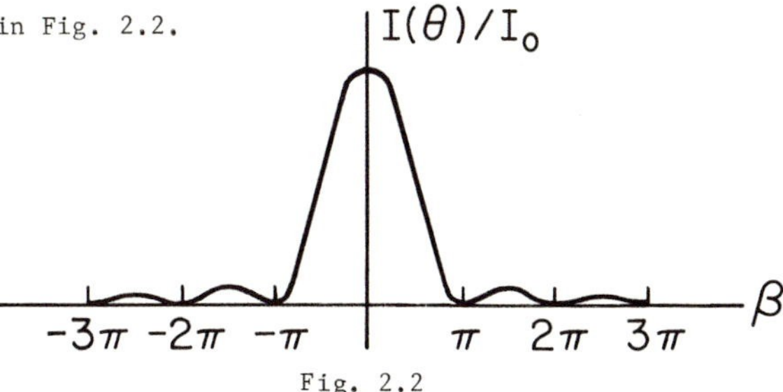

Fig. 2.2

The first minimum in the intensity occurs at $\beta = \pi$ or $\sin\theta = \lambda/b$ as found by Huygens' Principle.

(b) In determining y and p_y for the particle first we see that the slit only defines the position of the particle between $-b/2$ and $+b/2$ so that the uncertainty in y is $\Delta y = b$. The diffraction pattern indicates that the uncertainty in θ is at least λ/b for small values of θ where $\sin\theta \cong \theta$. Therefore, the uncertainty in the y component of the velocity must be $\Delta v_y \gtrsim v_x\lambda/b$ so that $\Delta p_y \gtrsim p_x\lambda/b = h/b$ since $p_x = h/\lambda$. Finally we obtain

$$\Delta y \Delta p_y \gtrsim h$$

which is a statement of Heisenberg's Uncertainty Principle. The exact number appearing on the right side of this inequality depends somewhat on the definition of the uncertainty, but the minimum value is always of order h. The essence of this experiment is that a measurement of the coordinate disturbs the system so that the momentum becomes uncertain. Numerous other "thought experiments" lead to the same conclusion.

2.3 Consider the construction of a one-dimensional wave packet having a non-zero amplitude only in a region of length Δx by superimposing sinusoidal waves with wavelengths near λ_o. The number of wavelengths that can be fit in Δx is then $n = \Delta x/\lambda_o$. If there is to be destructive interference of the component waves outside the region Δx, we must at least allow a wavelength spread $\Delta\lambda$ between the various waves so that

$$\frac{\Delta x}{\lambda_o - \Delta\lambda} \gtrsim n + 1$$

Show that this inequality is sufficient to conclude that

$$\Delta x \Delta p \gtrsim h$$

Solution: Ref. HE,13

The inequality gives

$$\frac{\Delta x}{\lambda_o - \Delta\lambda} - \frac{\Delta x}{\lambda_o} \gtrsim 1$$

or

$$\frac{\Delta x}{\lambda_o}\left(\frac{1}{1 - \Delta\lambda/\lambda_o} - 1\right) \gtrsim 1$$

Now assuming that $\Delta\lambda/\lambda_o << 1$ we obtain

$$\frac{\Delta x}{\lambda_o}\left(1 + \frac{\Delta\lambda}{\lambda_o}\right) - \frac{\Delta x}{\lambda_o} = \frac{\Delta x \Delta\lambda}{\lambda_o^2} \gtrsim 1$$

But

$$|d(1/\lambda)| = d\lambda/\lambda^2 \quad \text{so} \quad \Delta x \Delta(1/\lambda) \gtrsim 1$$

Using $\lambda = h/p$ then

$$\Delta x \Delta p \gtrsim h$$

Heisenberg's uncertainty relation turns out to be a natural consequence of the wave nature of particles. Localization in space can only be achieved at the expense of uncertainty in wavelength and momentum. A more precise treatment of the wave packet will be given in Problem 2.19.

2.4 A wave packet representing a particle moves with a group velocity $v_g = d\nu/d(1/\lambda)$ which corresponds to the classical velocity of the particle. The phase velocity of the contributing waves has no direct interpretation in classical mechanics and may, in fact, exceed the speed of light. Derive the above equation for the group velocity by associating this velocity with the velocity of movement of the "beats" obtained by superposing two waves moving in the x-direction. This is conveniently done with the waves $\exp\{i[(k_o \pm dk)x - (\omega_o \pm d\omega)t]\}$.

Solution: Ref. S1,405; FO,13; BW,18

Superposition of the two waves immediately gives

$$\psi_1 + \psi_2 = \left\{e^{i[(k_o + dk)x - (\omega_o + d\omega)t} + e^{i[(k_o - dk)x - (\omega_o - d\omega)t]}\right\}$$

$$= e^{i(k_o x - \omega_o t)}\left[e^{i(xdk - td\omega)} + e^{-i(xdk - td\omega)}\right]$$

$$= 2e^{i(k_o x - \omega_o t)}\cos(xdk - td\omega)$$

The slowly varying cosine function provides a modulation of the rapidly varying exponential function. (See Fig. 2.4) A crest of the envelope appears when $xdk - td\omega = 0$, and the velocity of this crest or beat is

$$v_g = \frac{x}{t} = \frac{d\omega}{dk}$$

Since $k = 2\pi/\lambda$ and $\nu = \omega/2\pi$, this equation becomes

$$v_g = \frac{x}{t} = \frac{d\nu}{d(1/\lambda)}$$

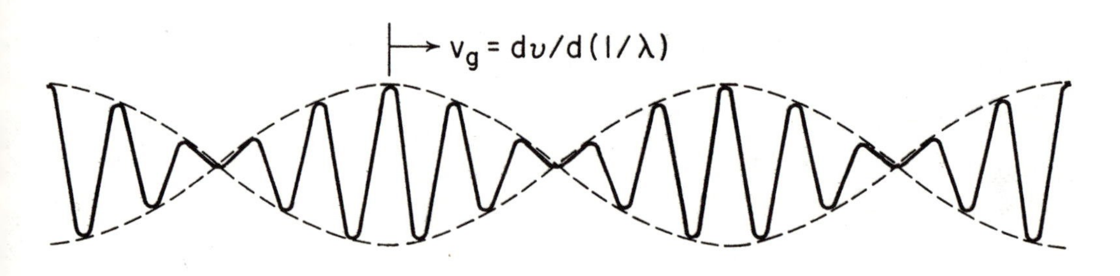

Fig. 2.4

2.5 The function $\psi(x)$ can be expanded as follows:

$$\psi(x) = \sum_{n=1}^{\infty} c_n\phi_n(x)$$

where the ϕ_n's form a complete, orthogonal set. Derive an equation for the coefficient c_m.

Solution: Ref. AN2,15

In deriving coefficients for orthogonal expansions the trick is to multiply through by the complex conjugate of a member of the basis set for the expansion and then to integrate. Orthogonality will eliminate all but one of the terms in the expansion. Thus:

$$\int \phi_m^*(x)\psi(x)\,dx = \sum_{n=0}^{\infty} c_n \int \phi_m^*(x)\phi_n(x)\,dx$$

then

$$\boxed{c_m = \int \phi_m^*(x)\psi(x)\,dx \Big/ \int \phi_m^*(x)\phi_m(x)\,dx}$$

In the special case that the set $\{\phi_n(x)\}$ is orthonormal so that

$$\int \phi_n^*(x)\phi_m(x)\,dx = \delta_{nm}$$

we obtain

$$\boxed{c_m = \int \phi_m^*(x)\,\psi(x)\,dx}$$

2.6 In dealing with the orthogonal expansions it is convenient to define the Projection Operator P_i so that

$$P_i\psi(x) = \phi_i(x)\int \phi_i^*(x)\psi(x)\,dx = \phi_i(x)(\phi_i, \psi)$$

The quantity (ϕ_i, ψ) is called the Scalar Product of ϕ_i and ψ. In Dirac notation it becomes $\langle i|\psi\rangle$.

(a) Show that $P_m\psi(x) = c_m\phi_m(x)$ where c_m is an expansion coefficient.

(b) Show that $P_m^2 = P_m$.

(c) Show that $\sum_{n=0}^{\infty} P_n = 1$.

Solution: Ref. AN2,90; ME,300

(a) $\psi(x)$ can be expanded as:

$$\psi(x) = \sum_{n=0}^{\infty} c_n\phi_n(x)$$

Then operating on both sides of this equation with P_m gives

$$P_m\psi(x) = \phi_m(x)\sum_{n=0}^{\infty} c_n\int\phi_m^*(x)\phi_n(x)dx = c_m\phi_m(x)$$

P_m projects out the mth component of $\psi(x)$.

In Dirac notation the expansion is

$$|\psi\rangle = \sum_{n=0}^{\infty} c_n|n\rangle$$

and the projection operator is written as $P_m = |m\rangle\langle m|$. Repeating the steps above we find

$$P_m|\psi\rangle = |m\rangle\langle m|\psi\rangle = |m\rangle c_m$$

(b) From part (a)

$$P_m\psi(x) = c_m\phi_m(x)$$

Applying P_m again gives

$$P_m{}^2\psi(x) = c_mP_m\phi_m(x) = c_m\phi_m(x)$$

or

$$P_m{}^2 = P_m$$

Dirac notation gives the simple result:

$$P_m{}^2 = P_m|m\rangle\langle m| = |m\rangle\langle m|m\rangle\langle m| = |m\rangle\langle m| = P_m$$

where we have used the normalization condition that

$$\langle m|m\rangle = \int\phi_m^*\phi_m dx = 1$$

(c) Consider the summation

$$\sum_{n=0}^{\infty} P_n\psi(x) = \sum_{n=0}^{\infty} c_n\phi_n(x) = \psi(x)$$

From this we conclude that

$$\sum_{n=0}^{\infty} P_n = 1$$

This is an important, general result. The summation must run over all elements of the complete set. It should be realized, however, that some complete sets have only a finite number of elements.

Again in Dirac notation we write

$$|\psi\rangle = \sum_{n=0}^{\infty} c_n|n\rangle = \sum_{n=0}^{\infty} |n\rangle\langle n|\psi\rangle$$

but

$$\sum_{n=0}^{\infty} P_n|\psi\rangle = \sum_{n=0}^{\infty} |n\rangle\langle n|\psi\rangle = |\psi\rangle$$

Therefore

$$\sum_{n=0}^{\infty} |n\rangle\langle n| = 1$$

$\sum_{n=0}^{\infty} P_n$ is equivalent to the identity operator.

2.7 The orthonormal set of functions in Problem 2.25 can be used to represent the arbitrary periodic function f(x) in the interval $(-\pi, +\pi)$ as follows:

$$f(x) = \frac{A_o}{\sqrt{2\pi}} + \sum_{n=1}^{\infty}\left(A_n \frac{\cos nx}{\sqrt{\pi}} + B_n \frac{\sin nx}{\sqrt{\pi}}\right)$$

Derive equations for the coefficients A_o, A_n, and B_n.

Solution:

According to the result of Problem 2.5 we can write

$$A_o = \frac{1}{\sqrt{2\pi}} \int_{-\pi}^{+\pi} f(x)\,dx$$

$$A_n = \frac{1}{\sqrt{\pi}} \int_{-\pi}^{+\pi} \cos nx \cdot f(x)\,dx$$

$$B_n = \frac{1}{\sqrt{\pi}} \int_{-\pi}^{+\pi} \sin nx \cdot f(x)\,dx$$

When compared with Eq. (9) in the Introduction we see that

$$a_o = \sqrt{2/\pi}\, A_o, \quad a_n = A_n/\sqrt{\pi}, \quad b_n = B_n/\sqrt{\pi}$$

2.8 Given the function f(x) with period 2π, suppose that we wish to approximate f(x) with

$$g(x) = \frac{a_o}{2} + \sum_{n=1}^{K} (a_n \cos nx + b_n \sin nx)$$

where K is finite. If the "best approximation" is defined as the one which occurs when

$$E = \int_{-\pi}^{+\pi} [f(x) - g(x)]^2 dx$$

is a minimum, show that coefficients a_o, a_n, and b_n are identical to the Fourier coefficients.

Solution: Ref. HS,14

E, which is a function of the coefficients a_o, a_n, and b_n, must be minimized with respect to each of these parameters. Thus

$$0 = \frac{\partial E}{\partial a_n} = \int_{-\pi}^{+\pi} 2[f(x) - g(x)] \frac{\partial g(x)}{\partial a_n} dx$$

$$= 2\int_{-\pi}^{+\pi} [f(x) - g(x)]\cos nx dx$$

or

$$\int_{-\pi}^{+\pi} f(x)\cos nx dx = \int_{-\pi}^{+\pi} g(x)\cos nx dx$$

$$= \int_{-\pi}^{+\pi} \frac{a_o}{2} \cos nx dx + \sum_{\ell=1}^{K} \int_{-\pi}^{+\pi} (a_\ell \cos \ell x + b_\ell \sin \ell x)\cos nx dx$$

Using the orthogonality relations of Problem 2.25 we find

$$a_o = \frac{1}{\pi} \int_{-\pi}^{+\pi} f(x) dx$$

and

$$a_n = \frac{1}{\pi} \int_{-\pi}^{+\pi} f(x)\cos nx dx \quad (n = 1, 2, \ldots)$$

Proceeding in the same manner with b_n we have

$$0 = \frac{\partial E}{\partial b_n} = \int_{-\pi}^{+\pi} 2[f(x) - g(x)]^2 \frac{\partial g(x)}{\partial b_n} dx$$

and

$$\int_{-\pi}^{+\pi} f(x)\sin nx dx = \int_{-\pi}^{+\pi} g(x)\sin nx dx$$

$$= \int_{-\pi}^{+\pi} \frac{a_o}{2} \sin nx dx + \sum_{\ell=1}^{\infty} \int_{-\pi}^{+\pi} (a_\ell \cos \ell x + b_\ell \sin \ell x)\sin nx dx$$

The orthogonality relations reduce this equation to

$$b_n = \frac{1}{\pi} \int_{-\pi}^{+\pi} f(x)\sin nx dx \quad (n = 1, 2, \ldots)$$

The Fourier coefficients, therefore, give the "best approximation" to f(x) even for a finite series.

2.9 Obtain equations for the Fourier series expansion of the function f(x) which has a period of 2L.

Solution:

The expansion for f(x) can be obtained from the results of Problem 2.8 by the following device. Let f(x) = g(z) where g(z) has a period of 2π and $x = Lz/\pi$. Then

$$g(z) = \frac{a_o}{2} + \sum_{n=1}^{\infty} (a_n \cos nz + b_n \sin nz)$$

where

$$a_n = \frac{1}{\pi}\int_{-\pi}^{+\pi} g(z)\cos nz dx, \quad b_n = \frac{1}{\pi}\int_{-\pi}^{+\pi} g(z)\sin nz dz$$

We now return to the variable x to obtain

$$f(x) = \frac{a_o}{2} + \sum_{n=1}^{\infty} \left(a_n \cos\frac{n\pi x}{L} + b_n \sin\frac{n\pi x}{L}\right)$$

with

$$a_n = \frac{1}{\pi}\int_{-L}^{L} f(x)\cos\frac{n\pi x}{L}\left(\frac{\pi}{L}dx\right) = \frac{1}{L}\int_{-L}^{L} f(x)\cos\frac{n\pi x}{L}dx$$

and

$$b_n = \frac{1}{\pi}\int_{-L}^{L} f(x)\sin\frac{n\pi x}{L}\left(\frac{\pi}{L}dx\right) = \frac{1}{L}\int_{-L}^{L} f(x)\sin\frac{n\pi x}{L}dx$$

2.10 If f(x) has the period 2L show that the Fourier coefficients can be written as

$$a_n = \frac{1}{L}\int_{\alpha}^{\alpha+2L} f(x)\cos\frac{n\pi x}{L}dx, \quad b_n = \frac{1}{L}\int_{\alpha}^{\alpha+2L} f(x)\sin\frac{n\pi x}{L}dx$$

Solution:

The first integral above can be expanded as

$$\frac{1}{L}\int_{\alpha}^{\alpha+2L} f(x)\cos\frac{n\pi x}{L}dx = \frac{1}{L}\int_{\alpha}^{-L} f(x)\cos\frac{n\pi x}{L}dx +$$

$$\frac{1}{L}\int_{-L}^{+L} f(x)\cos\frac{n\pi x}{L}dx + \frac{1}{L}\int_{L}^{\alpha+2L} f(x)\cos\frac{n\pi x}{L}dx \qquad (1)$$

In the last integral on the right make the substitution $x = z + 2L$ so

$$\frac{1}{L}\int_{-L}^{\alpha} f(z + 2L)\cos\frac{n\pi}{L}(z + 2L)dz = -\frac{1}{L}\int_{\alpha}^{-L} f(z)\cos\frac{n\pi z}{L}dz$$

Therefore, the first and third integrals on the right side of Eq. (1) cancel and the second integral is by definition a_n. The argument for b_n is identical. The most commonly used values of α are 0 and $-L$. For $\alpha = 0$

$$a_n = \frac{1}{L}\int_{o}^{2L} f(x)\cos\frac{n\pi x}{L}dx, \quad b_n = \frac{1}{L}\int_{o}^{2L} f(x)\sin\frac{n\pi x}{L}dx$$

2.11 The function $\psi(x)$, which is shown below, is defined by

$\psi(x) = -h \qquad -L < x < 0$

$\psi(x) = +h \qquad 0 < x < L$

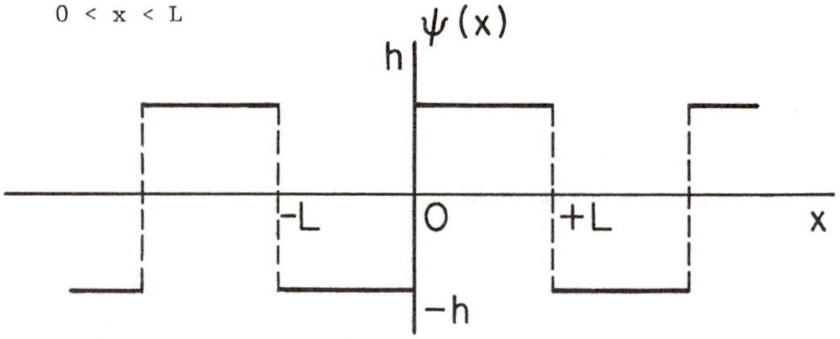

Find the Fourier series expansion for $\psi(x)$.

Solution:

By the definition of the Fourier series we have

$$\psi(x) = \frac{a_o}{2} + \sum_{n=1}^{\infty}\left(a_n\cos\frac{n\pi x}{L} + b_n\sin\frac{n\pi x}{L}\right)$$

and

$$a_n = \frac{1}{L}\int_{-L}^{L}\psi(x)\cos\frac{n\pi x}{L}dx$$

$$= \frac{1}{L}\left(\int_{-L}^{o} -h\cos\frac{n\pi x}{L}dx + \int_{o}^{L} h\cos\frac{n\pi x}{L}dx\right)$$

$$= \frac{1}{L}\left(-\frac{hL}{n\pi}\sin\frac{n\pi x}{L}\Big|_{-L}^{o} + \frac{hL}{n\pi}\sin\frac{n\pi x}{L}\Big|_{o}^{L}\right)$$

$= 0$ for $n \neq 0$

For n = 0

$$a_o = \frac{1}{L}\int_{-L}^{+L} \psi(x)dx = 0$$

Also

$$b_n = \frac{1}{L}\int_{-L}^{L} \psi(x)\sin\frac{n\pi x}{L}\,dx$$

$$= \frac{1}{L}\left(\int_{-L}^{o} -h\sin\frac{n\pi x}{L}\,dx + \int_{o}^{L} h\sin\frac{n\pi x}{L}\,dx\right)$$

$$= \frac{1}{L}\left(\frac{hL}{n\pi}\cos\frac{n\pi x}{L}\Bigg|_{-L}^{o} - \frac{hL}{n\pi}\cos\frac{n\pi x}{L}\Bigg|_{o}^{L}\right)$$

$$= \frac{h}{n\pi}\,[(1 - \cos n\pi) - (\cos n\pi - 1)]$$

$$= \frac{2h}{n\pi}\,(1 - \cos n\pi) = \frac{2h}{n\pi}\,[1-(-1)^n]$$

$\psi(x)$ is an odd function of x since $\psi(x) = -\psi(x)$. The Fourier expansion of odd functions always gives $a_n = 0$ as found above. For even functions $b_n = 0$.

2.12 Find the Fourier series expansion of the function $\psi(x)$ shown below which has a period of 2. $\psi(x)$ is defined by $\psi(x) = (1 - x^2)$ on the interval (-1, +1).

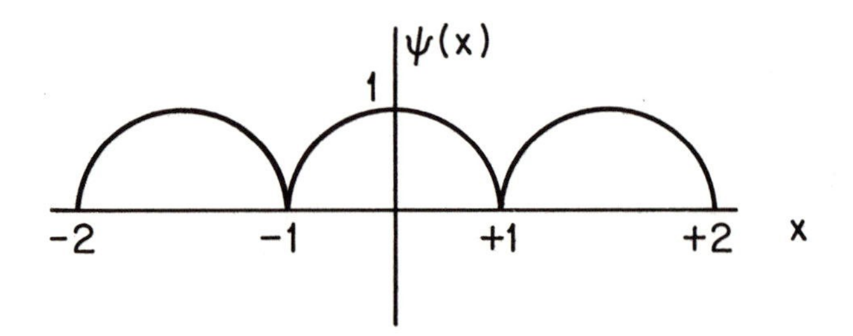

Solution:

Since $\psi(x)$ is an even function of x, the odd terms vanish and we find

$$\psi(x) = \frac{a_o}{2} + \sum_{n=1}^{\infty} a_n \cos n\pi x$$

where

$$a_n = 2\int_o^1 (1 - x^2)\cos n\pi x dx$$

$$= 2\int_o^1 \cos n\pi x dx - 2\int_o^1 x^2 \cos n\pi x dx$$

$$= \frac{2}{n\pi}\sin n\pi x \Big|_o^1 - \frac{2}{(n\pi)^3}\int_o^{n\pi} y^2 \cos y dy, \quad y = n\pi x$$

$$= \left[\frac{-2}{(n\pi)^3}\left(2y\cos y + (y^2 - 2)\sin y\right)\Big|_o^{n\pi}\right]$$

$$= \frac{-4}{(n\pi)^2}\cos n\pi = \frac{-4(-1)^n}{(n\pi)^2}$$

For n = 0

$$a_o = 2\int_o^1 (1 - x^2)dx = \frac{4}{3}$$

Therefore

$$\psi(x) = \frac{2}{3} - \sum_{n=1}^{\infty} \frac{4}{(n\pi)^2}(-1)^n \cos n\pi x$$

2.13 Obtain the Complex Fourier series expansion for the "train of pulses" shown below, and plot the spectrum of coefficients for $d/L = 0.2$ versus k_n.

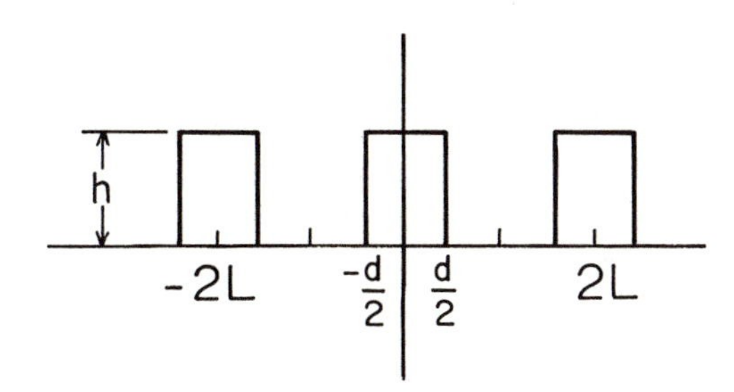

Solution: Ref. HS,58

From Problem 2.26 we have

$$\psi(x) = \sum_{n=-\infty}^{+\infty} c_n e^{ik_n x}$$

where

$$c_n = \frac{1}{2L}\int_{-L}^{L} \psi(x) e^{-ik_n x} dx; \quad k_n = n\pi/L$$

Therefore

$$c_n = \frac{1}{2L}\int_{-d/2}^{d/2} he^{-ik_n x} dx = -\frac{h}{2ik_n L} e^{-ik_n x}\Big|_{-d/2}^{d/2}$$

$$= \frac{h}{2ik_n L}\left(e^{ik_n d/2} - e^{-ik_n d/2}\right)$$

$$= \frac{h}{L}\frac{\sin(k_n d/2)}{k_n} = \frac{hd}{2L}\frac{\sin\beta}{\beta}; \quad \beta = \frac{n\pi d}{2L}$$

The spectrum of coefficients is shown in Fig. 2.13.

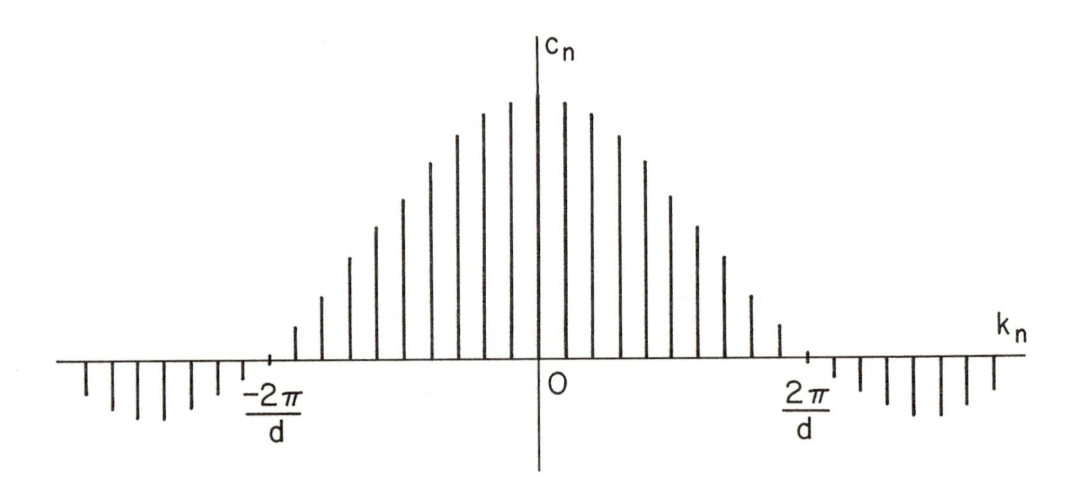

Fig. 2.13

2.14 The complex Fourier series expansion for f(x) is defined by:

$$f(x) = \sum_{n=-\infty}^{n=+\infty} c_n e^{ik_n x}$$

where

$$c_n = \frac{1}{2L}\int_{-L}^{+L} f(x)e^{-ik_n x}dx$$

Show that as $L \to \infty$ ($\Delta k = \Delta n\pi/L \to 0$) the following equations, called the Fourier Transform pair, hold

$$f(x) = \frac{1}{2\pi}\int_{-\infty}^{+\infty} F(k)e^{+ikx}dk$$

$$F(k) = \int_{-\infty}^{+\infty} f(x)e^{-ikx}dx$$

Solution: Ref. WY,275; BR,6; Problem 2.26

Choosing

$$F(k_n) = c_n 2L \quad \text{and} \quad k_n = n\pi/L$$

we may write

$$f(x) = \frac{1}{2L}\sum_{n=-\infty}^{+\infty} F(k_n)e^{ik_n x}$$

and

$$F(k_n) = \int_{-L}^{+L} f(x)e^{-ik_n x}dx$$

Now

$$f(x) = \frac{1}{2\pi}\sum_{n=-\infty}^{+\infty} F(k_n)e^{+ik_n x}\left(\frac{\pi}{L}\right)$$

since

$$\Delta k_n = \Delta n \cdot \frac{\pi}{L} = \frac{\pi}{L} \quad \text{(where } \Delta n = 1\text{)}$$

Thus

$$f(x) = \frac{1}{2\pi}\sum_{n=-\infty}^{+\infty} F(k_n)e^{+ik_n x}\Delta k_n$$

If we now take the limit of f(x) as $|L| \to \infty$ so that $\Delta k_n \to dk$ and $k_n \to k$, we may then write

$$f(x) = \frac{1}{2\pi}\int_{-\infty}^{+\infty} F(k)e^{+ikx}dk, \quad F(k) = \int_{-\infty}^{+\infty} f(x)e^{-ikx}dx$$

The equations are frequently written in the time-frequency domains as

$$f(t) = \int_{-\infty}^{+\infty} g(\omega)e^{i\omega t}d\omega, \quad g(\omega) = \frac{1}{2\pi}\int_{-\infty}^{+\infty} f(t)e^{-i\omega t}dt$$

It is also common to find f(x) and F(k) redefined in a symmetrical form

$$f(x) = \frac{1}{\sqrt{2\pi}}\int_{-\infty}^{+\infty} F(k)e^{+ikx}dk, \quad F(k) = \frac{1}{\sqrt{2\pi}}\int_{-\infty}^{+\infty} f(x)e^{-ikx}dx$$

2.15 Obtain the Fourier Transform F(k) for the square pulse function $\psi(x)$ shown below, and plot F(k) versus k.

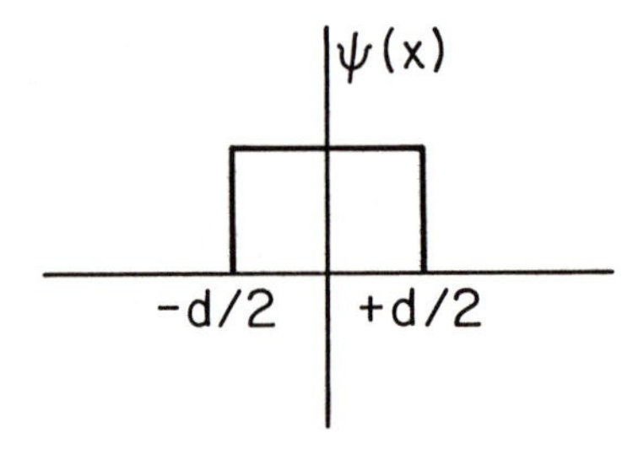

Solution: Ref. Problem 2.14

The Fourier Transform of $\psi(x)$ is defined by

$$F(k) = \int_{-\infty}^{+\infty} f(x)e^{-ikx}dx$$

Therefore

$$F(k) = \int_{-d/2}^{+d/2} he^{-ikx}dx = \left.\frac{he^{-ikx}}{-ik}\right|_{-d/2}^{+d/2}$$

$$F(k) = \frac{2h}{k}\left(\frac{e^{ikd/2} - e^{-ikd/2}}{2i}\right) = hd\,\frac{\sin(kd/2)}{(kd/2)}$$

This function is essentially identical to the envelope of the spectrum obtained in Problem 2.13. Here, however, k is continuous since the period is infinite. F(k) is shown in Fig. 2.15.

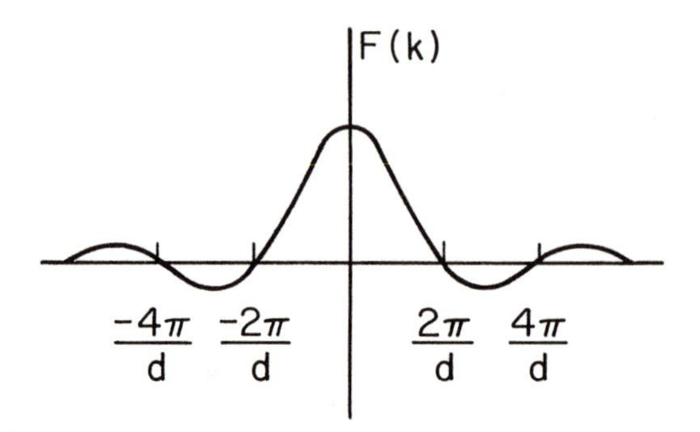

Fig. 2.15

2.16 Obtain the Fourier Transform F(k) for the "wave packet" function ψ(x) defined by:

$$\psi(x) = he^{ik_o x}, \quad -d/2 \lesseqgtr x \lesseqgtr +d/2$$
$$= 0 \quad \text{otherwise}$$

Solution:

By definition

$$F(k) = \int_{-\infty}^{+\infty} \psi(x) e^{-ikx} dx$$

$$= \int_{-d/2}^{d/2} he^{i(k_o - k)x} dx = \frac{he^{i(k_o - k)x}}{i(k_o - k)}\Bigg|_{-d/2}^{d/2}$$

$$F(k) = hd \frac{\sin[(k_o - k)d/2]}{(k_o - k)d/2}$$

The spectrum is similar to that in Problem 2.15 except that it is shifted from the origin by the amount k_o. This function is shown in Fig. 2.16.

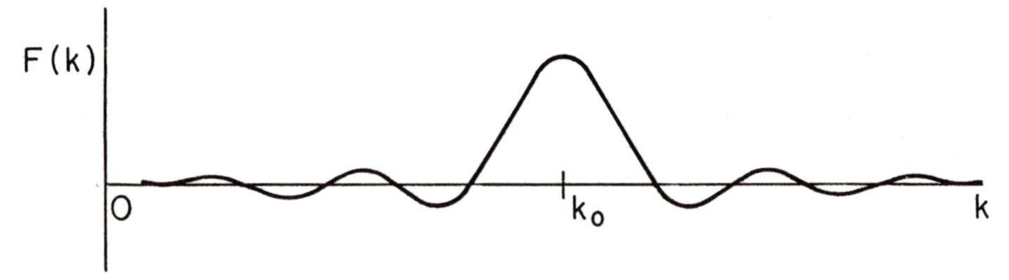

Fig. 2.16

2.17 Obtain the Fourier Transform for the function

$$f(x) = \frac{\delta}{\pi}\frac{1}{\delta^2 + x^2}$$

where δ is a positive non-zero constant. Plot the result.

Solution:

$$F(k) = \int_{-\infty}^{+\infty} f(x)e^{-ikx}dx = \frac{\delta}{\pi}\int_{-\infty}^{+\infty} (\delta^2 + x^2)^{-1}e^{-ikx}dx$$

$$= \frac{2\delta}{\pi}\int_{0}^{\infty} (\delta^2 + x^2)^{-1}\cos kx dx = \frac{2}{\pi}\int_{0}^{\infty} (1 + y^2)^{-1}\cos(k\delta y)dy, \quad y = x/\delta$$

Thus (using integral tables)

$$F(k) = \frac{2}{\pi}\cdot\frac{\pi}{2}\cdot e^{-\delta|k|} = e^{-\delta|k|} \quad \text{if} \quad \delta > 0$$

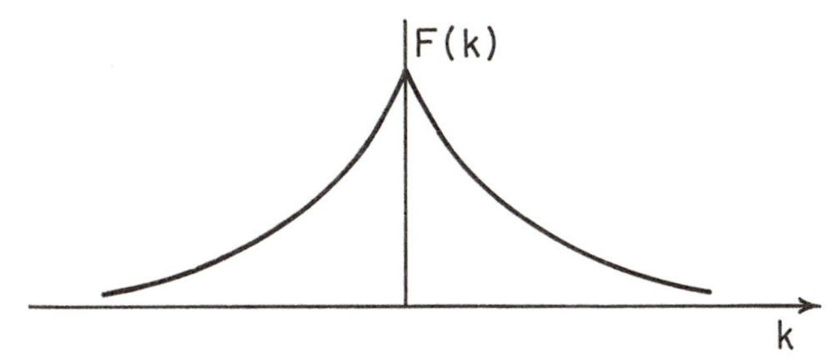

The function $f(x)$ is said to be a "Lorentzian" and is an important spectral line shape formula. The Fourier Transform of a Lorentzian is thus a symmetrical exponential. The quantity δ is one half of the width of $f(x)$ at one half its maximum amplitude.

2.18 A measure of the spread of x values about the average $\langle x\rangle$ is given by $\langle\Delta x^2\rangle$, which is defined for the distribution function $f^2(x)$ by

$$\langle\Delta x^2\rangle = \frac{\int_{-\infty}^{+\infty} (x - \langle x\rangle)^2 f^2(x)dx}{\int_{-\infty}^{+\infty} f^2(x)dx} \tag{1}$$

The Fourier transform of $f(x)$, given by

$$F(k) = \int_{-\infty}^{+\infty} f(x)e^{-ikx}dx$$

implies a similar definition for the spread or uncertainty of k; i.e.

$$<\Delta k^2> = \frac{\int_{-\infty}^{+\infty} (k - <k>)^2 F^2(k)\,dk}{\int_{-\infty}^{+\infty} F^2(k)\,dk} \tag{2}$$

Determine the magnitude of the uncertainty product

$$\delta x \delta k = [<\Delta x^2><\Delta k^2>]^{1/2}$$

for

$$f(x) = \frac{1}{\sqrt{2\pi\sigma^2}} e^{-x^2/2\sigma^2}$$

Solution: Ref. HS,228

The Fourier Transform of the given f(x) was developed in Problem 2.30:

$$F(k) = e^{-k^2\sigma^2/2}$$

we may rewrite Eq. (1) in a more convenient form

$$<\Delta x^2> = <(x - <x>)^2> = <x^2 - 2x<x> + <x>^2>$$

$$= <x^2> - 2<x>^2 + <x>^2 = <x>^2 - <x>^2$$

Similarly,

$$<\Delta k^2> = <k^2> - <k>^2$$

Noting that $<x>^2$ and $<k>^2$ are zero (since f(x) and F(k) are even functions), we then have

$$<\Delta x^2> = <x^2> = \frac{\int_{-\infty}^{+\infty} x^2 f^2(x)\,dx}{\int_{-\infty}^{+\infty} f^2(x)\,dx} = \frac{\sigma^2}{2}$$

and

$$<\Delta k^2> = <k^2> = \frac{\int_{-\infty}^{+\infty} k^2 F^2(k)\,dk}{\int_{-\infty}^{+\infty} F^2(k)\,dk} = \frac{1}{2\sigma^2}$$

Thus

$$\delta x \delta k = [<\Delta x^2><\Delta k^2>]^{1/2} = 1/2$$

This result is consistent with the quantum mechanical Uncertainty Principle as may be seen by noting that

$$\delta k = \hbar^{-1}\delta p_x$$

so that

$$\delta x \delta p_x = \hbar/2$$

In an analogous manner we could determine

$$\delta t \delta\omega = \delta t(2\pi\delta\nu) = 1/2$$

from the Fourier Transform pair defined in the time-frequency notation. Using $E = h\nu$ so that $\delta E = h\delta\nu$ we would then find that

$$\delta t \delta E = \hbar/2$$

2.19 In Problem 2.16, it was shown that the Fourier Transform of the "wave packet" function

$$\psi(x) = he^{ik_o x}, \quad -d/2 \gtrless x \lessgtr +d/2$$
$$= 0 \quad \text{otherwise}$$

is

$$F(k) = \frac{hd\sin[(k_o - k)d/2]}{(k_o - k)d/2} \tag{1}$$

Use these functions and the de Broglie relationship $k = p/\hbar$ to estimate the magnitude of the uncertainty product $\Delta x \Delta p_x$.

Solution: Ref. AN3,136

By examination of Eq. (1) we see that $F(k)$ has its maximum value (see Fig. 2.16) at $k = k_o$ and its first zeroes at

$$|(k_o - k)d/2| = \pi$$

Therefore,

$$\Delta k \cdot d/2 = \Delta p_x/\hbar \cdot d/2 = 2\pi$$

If we choose $d/2$ as the characteristic uncertainty in x, then

$$\Delta p_x \Delta x = h$$

2.20 Consider a length of elastic string under tension T which is stretched between two points on the x axis.

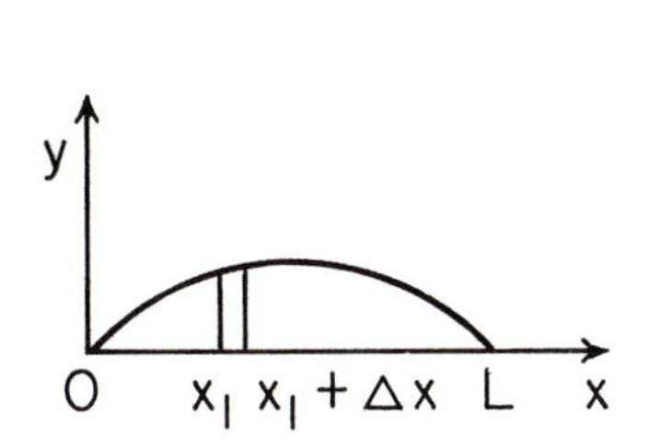

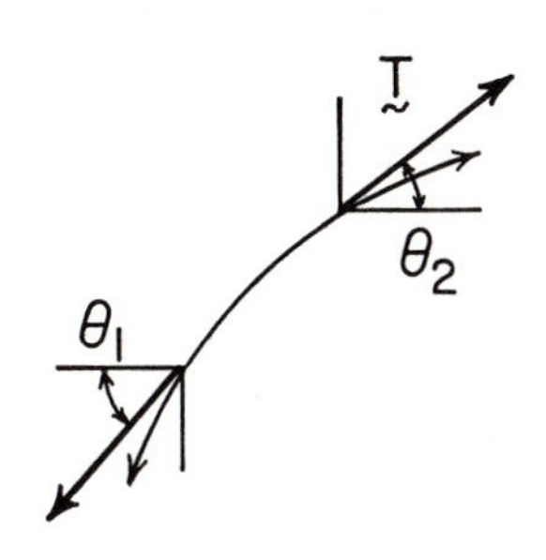

Let the weight per unit length be w. Assume a) that there are no external forces other than tension and gravity, b) the motion is planar, c) the deflection is so small that T is constant, d) force is transmitted in the direction of the tangent to the string, e) that the deflection is so small that $\sin\theta$ can everywhere be replaced with $\tan\theta$.

Give a derivation of the one dimensional wave equation (by use of the above assumptions)

$$\frac{\partial^2 y}{\partial t^2} = a^2 \frac{\partial^2 y}{\partial x^2}$$

where a has the units of velocity. (Hint: Determine the net force on the mass element Δm and apply Newton's Law, $F = (\Delta m)\ddot{y}$.)

Solution: Ref. WY,345

The mass of a small element of length Δx would be $\Delta m = \rho\Delta x$ where ρ is the mass per unit length. The average acceleration which acts over Δx is approximately $\frac{\partial^2 y}{\partial t^2}$ so that a force balance yields

$$\rho\Delta x \frac{\partial^2 y}{\partial t^2} = F_y(x + \Delta x) - F_y(x)$$

$$= T(\sin\theta_2 - \sin\theta_1) \approx T(\tan\theta_2 - \tan\theta_1)$$

$$\approx T(\tan\theta_{x + \Delta x} - \tan\theta_x)$$

Thus

$$\frac{\partial^2 y}{\partial x^2} = \frac{T}{\rho} \frac{(\tan\theta_{x + \Delta x} - \tan\theta_x)}{\Delta x}$$

Noting that $\tan\theta = \frac{\partial y}{\partial x}$, then taking the limit as $\Delta x \to 0$, we have

$$\frac{\partial^2 y}{\partial t^2} = \frac{T}{\rho} \frac{\partial^2 y}{\partial x^2} = a^2 \frac{\partial^2 y}{\partial x^2}$$

The units of $\frac{T}{\rho} = a^2$ are (in cgs units)

$$\frac{\text{(Force)}}{\text{mass/length}} = \frac{(\text{g cm/sec}^2)}{\text{g/cm}} = \frac{\text{cm}^2}{\text{sec}^2} = (\text{velocity})^2$$

2.21 Separate the one dimensional equation developed in Problem 2.20 by assuming $y = \psi(x)T(t)$. Find $T(t)$ and show that

$$\psi'' + \frac{4\pi^2}{\lambda^2}\psi = 0$$

Solution: Ref. BL,79

$$\frac{\partial^2 y}{\partial x^2} = \frac{1}{a^2}\frac{\partial^2 y}{\partial t^2} \qquad (1)$$

Letting $y = \psi(x)T(t)$ we have upon substitution into (1)

$$T\frac{\partial^2\psi}{\partial x^2} = \frac{\psi}{a^2}\frac{\partial^2 T}{\partial t^2}$$

and dividing by $y = \psi \cdot T$

$$\frac{\psi''}{\psi} = \frac{1}{a^2}\frac{\ddot{T}}{T}$$

For this equality to hold for all values of the independent variables x and t, both sides of the equation must equal the same constant. Let us choose this to be $-k^2$; so

$$\frac{\psi''}{\psi} = -k^2 = \frac{1}{a^2}\frac{\ddot{T}}{T} \qquad (2)$$

The solution of the right hand side gives

$$T(t) = A\sin(akt) + B\cos(akt)$$
$$= A\sin(2\pi\nu t) + B\cos(2\pi\nu t)$$

where we have expressed ak as $2\pi\nu$. Since $a = \nu\lambda$ for wave motion, we then may solve for k:

$$k = \frac{2\pi\nu}{a} = \frac{2\pi\nu}{\nu\lambda} = \frac{2\pi}{\lambda}$$

This result, substituted into the left hand side of Eq. (2) then gives us

$$\psi'' + k^2\psi = 0$$

This equation is sometimes called the Helmholtz equation when λ is constant.

2.22 Show that the time-independent Schrödinger equation

$$-\frac{\hbar^2}{2m}\psi'' + V(x)\psi = E\psi$$

can be obtained from the Helmholtz equation by replacing λ with p/h.

Solution: Ref. BL,279

The Helmholtz equation states

$$\psi'' + \frac{4\pi^2}{\lambda^2}\psi = 0 \tag{1}$$

for stationary waves. Particles, via de Broglie, have an associated wavelength of

$$\lambda = \frac{h}{p}$$

Classically, the total energy is given by

$$E = \frac{p_x^{\,2}}{2m} + V(x)$$

so that

$$p_x = \{2m[E - V(x)]\}^{1/2}$$

Thus

$$\frac{1}{\lambda^2} = \frac{p_x^{\,2}}{h^2} = \frac{2m}{h^2}[E - V(x)]$$

Substituting into (1) we have

$$\psi'' + \frac{2m}{\hbar^2}[E - V(x)]\psi = 0 \quad (\hbar = h/2\pi)$$

or

$$-\frac{\hbar^2}{2m}\psi'' + V(x)\psi = E\psi$$

The operator

$$-\frac{\hbar^2}{2m}\frac{\partial^2}{\partial x^2} + V(x)$$

is called the Hamiltonian operator. The same arguments carry over with three dimensions:

$$\psi \to \psi(x, y, z) = \psi(\underset{\sim}{r}), \quad V \to V(x, y, z) = V(\underset{\sim}{r})$$

to give

$$-\frac{\hbar^2}{2m}\nabla^2\psi(\underset{\sim}{r}) + V(\underset{\sim}{r})\psi(\underset{\sim}{r}) = E\psi(\underset{\sim}{r})$$

This is the time independent Schrödinger equation for a single particle moving under the influence of the potential $V(\underset{\sim}{r})$.

2.23 Show that the wave velocity v_p (sometimes called the phase velocity) is given by

$$v_p = \frac{c^2}{v}$$

where c is the speed of light and v is the velocity of a free particle. (Hint: Use the relativistic energy equation $E = mc^2$.)

Solution: Ref. SE,185; PA,87

The wave velocity of the waves associated with a de Broglie particle is given by

$$v_p = \nu\lambda$$

and the total energy is given by

$$E = h\nu = mc^2$$

Since

$$p = mv = \frac{h}{\lambda} \quad \text{(where m = total mass, including rest mass)}$$

for the particle we then have

$$v_p = \nu\lambda = \frac{E}{h}\lambda = \frac{E}{p} = \frac{mc^2}{mv} = \frac{c^2}{v}$$

i.e.

$$\boxed{v_p = \frac{c^2}{v}}$$

The fact that the phase velocity is greater than the particle velocity does not mean that the waves "out run" the particle. The wave packet concept holds that the particle exists inside the group of waves called the wave packet; the packet moves with particle velocity v whereas the waves of which the packet consists individually travel with velocities v_p.

2.24 Show that the wave velocity v_p for a particle with rest mass greater than zero is always greater than c and that v_p is a function of wavelength. (Hint: First show that $E^2 = m_o{}^2c^4 + p^2c^2$, then eliminate E.)

Solution: Ref. SE,186

The total energy of the particle is given by

$$E = mc^2$$

and its momentum is given by

$$p = mv = \frac{Ev}{c^2}$$

In terms of the rest mass, the particle's momentum may also be expressed as

$$p = m_o v\left(1 - \frac{v^2}{c^2}\right)^{-1/2} \quad \text{where } m_o = \text{rest mass}$$

Squaring and eliminating v between the last two equations we have

$$p^2 = m_o{}^2\left(\frac{p^2c^4}{E^2}\right)\left(1 - \frac{p^2c^2}{E^2}\right)^{-1}$$

or

$$E^2 = m_o{}^2c^4 + p^2c^2$$

thus

$$E = h\nu = \frac{hv_p}{\lambda} = (m_o{}^2c^4 + p^2c^2)^{1/2}$$

Since $p = \frac{h}{\lambda}$ we have finally

$$\boxed{v_p = c\left(\frac{m_o{}^2c^2\lambda^2}{h^2} + 1\right)^{1/2}}$$

Thus if $m_o{}^2 > 0$, $v_p > c$.

SUPPLEMENTARY PROBLEMS

2.25 An important complete set of basis functions $\{\phi_n\}$ is: $1/\sqrt{2\pi}$, $(1/\sqrt{\pi})\cos nx$, $(1/\sqrt{\pi})\sin nx$. Show that this set is orthonormal.

Ans:

Evaluation for all values of n and m gives

$$\int_{-\pi}^{+\pi} \phi_n^*\phi_m dx = \delta_{nm}$$

2.26 The complex Fourier series expansion for f(x) is given by

$$f(x) = \sum_{n=-\infty}^{+\infty} c_n e^{ik_n x}$$

where

$$c_n = \frac{1}{2L}\int_{-L}^{+L} f(x)e^{-ik_n x}dx, \quad k_n = n\pi/L$$

Show that the complex series can be obtained from the real series of Problem 2.9 by the use of Euler's formulas:

$$\cos\theta = \frac{e^{i\theta} + e^{-i\theta}}{2}, \quad \sin\theta = \frac{e^{i\theta} - e^{-i\theta}}{2i}$$

2.27 Find the Fourier series expansion of the function $\psi(x)$ shown below

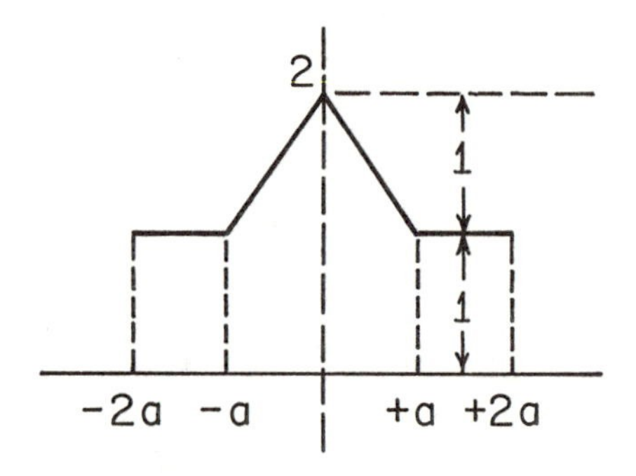

where

$$\psi(x) = 2 - \left|\frac{x}{a}\right| \quad ; \quad -a \leqslant x \leqslant a$$

$$\psi(x) = 1; \quad -2a \leqslant x < -a, \quad a < x \leqslant 2a$$

Ans: Ref. WO,88

$$\psi(x) = \frac{5}{4} + \sum_{n=1}^{\infty} \frac{4}{(n\pi)^2} \left[1 - \cos\left(\frac{n\pi}{2}\right)\right] \cos\left(\frac{n\pi x}{2a}\right)$$

2.28 Show that if the expansion function f(x) is an even function (i.e., f(x) = +f(-x)), the Fourier Transform pair reduces to

$$f(x) = \frac{1}{2\pi} \int_{-\infty}^{\infty} F(k)\cos kx\,dk, \quad F(k) = 2\int_{0}^{\infty} f(x)\cos kx\,dx$$

and if f(x) is an odd function (i.e., f(x) = -f(-x)),

$$f(x) = \frac{1}{2\pi} \int_{-\infty}^{+\infty} F(k)\sin kx\,dk, \quad F(k) = 2\int_{0}^{\infty} f(x)\sin kx\,dx$$

Ref. WY,279

2.29 Obtain the function for which the Fourier Transform is

$$F(k) = e^{-\delta|k|}$$

Ans:

$$f(x) = \frac{\delta}{\pi}\left[\frac{1}{\delta^2 + x^2}\right]$$

2.30 Obtain the Fourier Transform of the Gaussian Function

$$f(x) = \frac{1}{\sqrt{2\pi\Delta^2}} e^{-x^2/2\Delta^2}$$

and plot the result.

Ans: Ref. L. Petrakis, J. Chem. Ed. 44, 432 (1967).

$$F(k) = \exp[-(k\Delta)^2/2]$$

2.31 Given a pulse of the following form

$$\begin{aligned} f(t) &= e^{i\omega_o t} & |t| \leqslant T \\ &= 0 & |t| > T \end{aligned}$$

obtain the frequency spectrum by a Fourier transformation. Using this result and Planck's relation $E = h\nu$, estimate the uncertainty product $\Delta E \Delta t$.

Ans: Ref. AN3,136; Problem 2.15

$$\Delta E \Delta t = h$$

2.32 For one dimensional functions we were able to show that the Fourier series could be expressed as a complex expansion (Problem 2.26):

$$f(x) = \sum_{n=-\infty}^{+\infty} c_n e^{ik_n x}; \quad c_n = \frac{1}{2L}\int_{-L}^{+L} f(x) e^{-ik_n x} dx, \quad k_n = \frac{n\pi}{L}$$

Extend this result to three dimensions.

Ans: Ref. WO,100

$$f(x, y, z) = \sum_{n_x=1}^{\infty} \sum_{n_y=1}^{\infty} \sum_{n_z=1}^{\infty} c_{n_x n_y n_z} e^{i(k_x x + k_y y + k_z z)}$$

and

$$c_{n_x n_y n_z} = \left(\frac{1}{2L}\right)^3 \int_{-L}^{+L} \int_{-L}^{+L} \int_{-L}^{+L} f(x, y, z) e^{-i(k_x x + k_y y + k_z z)} dxdydz$$

2.33 In Problem 2.4, it was shown that by superimposing two waves with slightly different wavelengths and phases the group velocity could be expressed as

$$v_g = \frac{d\nu}{d\left(\frac{1}{\lambda}\right)}$$

Extend this result by showing that

$$v_g = v_p - \lambda \frac{dv_p}{d\lambda}$$

and then prove that the particle velocity v and group velocity v_g are identical.
(Hint: Make use of the result of Problems 2.23 and 2.24.)
Ref. SE,189; PA,90

2.34 If $\hat{i}$ and $\hat{j}$ are chosen as the basis vectors in two-dimensional space, an arbitrary unit vector can be written as

$$\hat{e} = \cos\theta\hat{i} + \sin\theta\hat{j}$$

The projection operator $\hat{P}_e$ is then defined so that $\hat{P}_e \underset{\sim}{A} = \hat{e}(\hat{e}\cdot\underset{\sim}{A})$.
(a) Determine the products $(P_e)_{11} = \hat{i}\cdot\hat{P}_e\hat{i}$, $(P_e)_{21} = \hat{j}\cdot\hat{P}_e\hat{i}$, $(P_e)_{12} = \hat{i}\cdot\hat{P}_e\hat{j}$, and $(P_e)_{22} = \hat{j}\cdot\hat{P}_e\hat{j}$ and thus derive the projection operator matrix.
(b) Show that the matrix product $(\underset{\sim}{P}_e)^2 = \underset{\sim}{P}_e$ using the matrix elements from part (a).
Ans:

(a) $$\underset{\sim}{P}_e = \begin{pmatrix} \cos^2\theta & \cos\theta\sin\theta \\ \sin\theta\cos\theta & \sin^2\theta \end{pmatrix}$$

2.35 Consider the complete set of functions $\{x^n\}$ on the interval x = -1 to x = +1. The first three members are the unnormalized functions $f_o = 1$, $f_1 = x$, $f_2 = x^2$. The functions g_o, g_1, g_3 are constructed from f_o, f_1, f_2 so that the g_n functions are normalized and mutually orthogonal. Starting with $g_o = 1/\sqrt{2}$ derive expressions for g_1 and g_2 using Schmidt orthogonalization or an equivalent scheme. (The appropriate definition of orthonormality here is $\int_{-1}^{+1} g_n g_m dx = \delta_{nm}$.)

Ans: Ref. AN2,27

$$g_1 = \sqrt{\frac{3}{2}}\, x \;; \qquad g_2 = \sqrt{\frac{5}{2}}\left(\frac{1}{2} - \frac{3}{2}x^2\right)$$

CHAPTER 3

POSTULATES AND FORMALISM OF QUANTUM MECHANICS

In this chapter we present problems to illustrate the foundations of quantum mechanics. First, we list a set of Postulates which will provide a framework for developing the theory. This set is not unique and is to some extent redundant; however it is a useful starting point.

I. A state function Ψ is associated with each physical system. Ψ, which must be well-behaved, specifies all that can be known about the system.

II. A linear, Hermitian operator $\hat{Q}$ is associated with each physical quantity q (dynamical variable), and the result of a measurement of q will always be one of the eigenvalues of $\hat{Q}$.

III. The expectation value of the quantity q for a system that is described by the state function Ψ is

$$\langle q\rangle = \frac{\int \Psi^*\hat{Q}\Psi d\tau}{\int \Psi^*\Psi d\tau} = \frac{\langle\Psi|\hat{Q}|\Psi\rangle}{\langle\Psi|\Psi\rangle}$$

IV. The operators $\hat{R}$ and $\hat{S}$ which are associated with the physical quantities (dynamical variables) r and s, respectively, must obey the equation: (see Problems 3.11 and 3.26)

$$[\hat{R}, \hat{S}] = i\hbar\{r, s\}_{PB}$$

V. The time development of the state function Ψ is given by the equation:

$$\hat{\mathcal{H}}\Psi = i\hbar\frac{\partial\Psi}{\partial t}$$

and the time development of the operator $\hat{F}$ is described by Heisenberg's equation of motion

$$\frac{d\hat{F}}{dt} = \frac{\partial\hat{F}}{\partial t} + \frac{i}{\hbar}[\hat{\mathcal{H}}, \hat{F}]$$

The operator $\hat{\mathcal{H}}$ is normally associated with the Hamiltonian function of classical mechanics. This function is essentially the sum of the kinetic and potential energies for a system. For example a conservative system consisting of a particle of mass m moving under the influence of a potential V(x) has the Hamiltonian (total energy)

$$E = \frac{p^2}{2m} + V(x)$$

In the formulation known as wave mechanics the operator substitutions

$$p \rightarrow -i\hbar \frac{\partial}{\partial x}$$

and

$$x \rightarrow x$$

are made to obtain the Hamiltonian operator

$$\hat{\mathcal{H}} = -\frac{\hbar^2}{2m} \frac{\partial^2}{\partial x^2} + V(x)$$

The allowed energies (eigenvalues) are obtained by solving the <u>time independent Schrödinger Equation</u> ($\hat{\mathcal{H}} u = Eu$):

$$\boxed{-\frac{\hbar^2}{2m} \frac{d^2}{dx^2} u(x) + V(x)u(x) = Eu(x)}$$

where u(x) is the time independent wave function.

We rely on the following problems and solutions to define and illustrate the terms introduced above.

PROBLEMS

3.1 If $\hat{\mathcal{H}}$ is an operator, u is a function of x, and E is a constant, the equation $\hat{\mathcal{H}} u = Eu$ indicates that u is an eigenfunction of $\hat{\mathcal{H}}$, and E is the corresponding eigenvalue. <u>In addition to satisfying this eigenvalue equation the function u must also be well-behaved in order to be an acceptable eigenfunction. By well-behaved we mean that u is single valued, finite, and continuous and that the first derivative of u is continuous.</u> Which of the following functions are well-behaved? For those functions that are not well-behaved, indicate the reason.

(a) $u = x$, $x \geq 0$, and $u = 0$ otherwise

(b) $u = x^2$

(c) $u = e^{-|x|}$

(d) $u = e^{-x}$

(e) $u = \cos x$

(f) $u = \sin|x|$

(g) $u = e^{-x^2}$

(h) $u = 1 - x^2$, $-1 \lessgtr x \lessgtr +1$, $u = 0$ otherwise.

Solution:

(a) $u = x$ is not well-behaved. u does not remain finite as $x \to \infty$ and the first derivative at $x = 0$ is not continuous.

(b) $u = x^2$ is not well-behaved since u does not remain finite as $|x| \to \infty$.

(c) $u = e^{-|x|}$ is not well-behaved since the first derivative is not continuous at $x = 0$.

(d) $u = e^{-x}$ is not well-behaved since u does not remain finite as $x \to -\infty$.

(e) $u = \cos x$ is a well-behaved function.

(f) $u = \sin|x|$ is not well-behaved since the first derivative is not continuous at $x = 0$.

(g) $u = e^{-x^2}$ is a well-behaved function.

(h) u is not well-behaved since the first derivative is discontinuous at $x = \pm 1$. However, u may be a good approximation to a well-behaved function which is continuous $|x| = 1$ and very small for $|x| > 1$.

3.2 Consider the operator $\hat{R} = -d^2/dx^2$ and the eigenvalue equation $\hat{R}u = \lambda u$. Write out the possible eigenfunctions and discuss the conditions under which they are well-behaved.

Solution:

We are given a linear differential equation with constant coefficients:

$$\frac{-d^2u}{dx^2} = \lambda u \tag{1}$$

and we assume that λ is a real constant. The general solution of (1) has the form

$$u = Ae^{+\sqrt{-\lambda}\,x} + Be^{-\sqrt{-\lambda}\,x} \quad \text{for } \lambda \neq 0$$

and

$$u = Cx + D \quad \text{for} \quad \lambda = 0$$

We consider three special cases:

$\lambda < 0$: u is not well-behaved since it does not remain finite as $|x| \to \infty$.

$\lambda = 0$: u is well-behaved only if $C = 0$. Otherwise the function will not remain finite.

$\lambda > 0$: $u = (A + B)\cos\sqrt{\lambda}\,x + i(A - B)\sin\sqrt{\lambda}\,x$ which is well-behaved. (Note: $e^{i\theta} = \cos\theta + i\sin\theta$)

3.3 A linear operator $\hat{R}$ has the following properties:

$$\hat{R}(u + v) = \hat{R}u + \hat{R}v$$

$$\hat{R}(cu) = c\hat{R}u$$

where c is a complex number. Which of the following operators are linear?

(a) $\hat{A}u = \lambda u$, λ = constant

(b) $\hat{B}u = u^*$

(c) $\hat{C}u = u^2$

(d) $\hat{D}u = \frac{du}{dx}$

(e) $\hat{E}u = 1/u$

(f) $\hat{L}\hat{P} = [\hat{H}, \hat{P}] = \hat{H}\hat{P} - \hat{P}\hat{H}$

Solution:

(a) $\hat{A}(u + v) = \lambda(u + v) = \hat{A}u + \hat{A}v$

$\hat{A}(cu) = \lambda cu = c\hat{A}u$

Therefore $\hat{A}$ is a linear operator.

(b) $\hat{B}(cu) = c^*\hat{B}u \neq c\hat{B}u$

$\hat{B}$ is not a linear operator.

(c) $\hat{C}(u + v) = (u + v)^2 = u^2 + 2uv + v^2$

but

$\hat{C}u + \hat{C}v = u^2 + v^2 \neq \hat{C}(u + v)$

Therefore, $\hat{C}$ is not a linear operator.

(d) $\hat{D}(u + v) = \frac{du}{dx} + \frac{dv}{dx} = \hat{D}u + \hat{D}v$

$\hat{D}(cu) = c\,\frac{du}{dx} = c\hat{D}u$

$\hat{D}$ is a linear operator.

(e) $\hat{E}(cu) = 1/(cu) = (1/c)\hat{E}u$

$\hat{E}$ is not a linear operator.

(f)
$$\begin{aligned}
\hat{L}(\hat{P} + \hat{Q}) &= \hat{H}(\hat{P} + \hat{Q}) - (\hat{P} + \hat{Q})\hat{H} \\
&= (\hat{H}\hat{P} - \hat{P}\hat{H}) + (\hat{H}\hat{Q} - \hat{Q}\hat{H}) \\
&= [\hat{H}, \hat{P}] + [\hat{H}, \hat{Q}] \\
&= \hat{L}\hat{P} + \hat{L}\hat{Q}
\end{aligned}$$

$$\begin{aligned}
\hat{L}(c\hat{P}) &= \hat{H}c\hat{P} - c\hat{P}\hat{H} = c(\hat{H}\hat{P} - \hat{P}\hat{H}) \\
&= c\hat{L}
\end{aligned}$$

Here $\hat{L}$ is an operator which acts on operators. It is a linear operator.

3.4 An operator is defined to be Hermitian if it satisfies the equation

$$\int \psi_n^* \hat{R}\psi_m d\tau = \int \psi_m (\hat{R}\psi_n)^* d\tau$$

Show that $\hat{p}_x = -i\hbar\, \partial/\partial x$ is Hermitian.

Solution: Ref. EWK,37; AN3,207

If $-i\hbar \frac{\partial}{\partial x}$ is Hermitian then

$$\int_{-\infty}^{+\infty} \psi_n^*(-i\hbar \frac{\partial}{\partial x})\psi_m dx = \int_{-\infty}^{+\infty} \psi_m(-i\hbar \frac{\partial}{\partial x} \psi_n)^* dx$$

To show this, let us integrate the above equation by parts

$$\int_{-\infty}^{+\infty} \psi_n^*(-i\hbar \frac{\partial}{\partial x})\psi_m dx = -i\hbar\ (\psi_n^*\psi_m)\Big|_{-\infty}^{+\infty} + i\hbar \int_{-\infty}^{+\infty} \psi_m \frac{\partial}{\partial x} \psi_n^* dx$$

$$= 0 + \int_{-\infty}^{+\infty} \psi_m(+i\hbar \frac{\partial}{\partial x} \psi_n^*) dx$$

$$= \int_{-\infty}^{+\infty} \psi_m(-i\hbar \frac{\partial}{\partial x} \psi_n)^* dx$$

This proof of course only follows for functions ψ which vanish at the limits.

3.5 Show that eigenvalues of Hermitian operators are real.

Solution: Ref. EWK,27; AN3,209; DI,31

Let $\hat{R}$ be a Hermitian operator with eigenvalue r. Then

$$\hat{R}\psi = r\psi \tag{1}$$

Taking the complex conjugate of both sides we have

$$\hat{R}^*\psi^* = r^*\psi^* \tag{2}$$

Multiplying (1) by ψ^* and integrating, and multiplying (2) by ψ and integrating yields

$$\int \psi^*\hat{R}\psi d\tau = \int \psi^* r\psi d\tau = r\int \psi^*\psi d\tau \tag{3}$$

and

$$\int \psi\hat{R}^*\psi^* d\tau = \int \psi r^*\psi^* d\tau = r^*\int \psi^*\psi d\tau \tag{4}$$

Since $\hat{R}$ is Hermitian

$$\int \psi^*\hat{R}\psi d\tau = \int \psi\hat{R}^*\psi^* d\tau$$

thus

$$r\int \psi^*\psi d\tau = r^*\int \psi^*\psi d\tau$$

Therefore $r = r^*$, i.e. eigenvalues of Hermitian operators are real.

In Dirac notation this proof takes a particularly simple form:

$$\langle r_1|\hat{R}|r_2\rangle = \langle r_2|\hat{R}^+|r_1\rangle^*$$

defines $\hat{R}^+$ as the adjoint of $\hat{R}$. If $\hat{R}$ is Hermitian (self-adjoint) and $|r\rangle$ is the eigenket of $\hat{R}$, then

$$\hat{R}|r\rangle = r|r\rangle, \quad \langle r|\hat{R}|r\rangle = \langle r|\hat{R}|r\rangle^*, \quad r = r^*$$

since

$$\langle r|r\rangle = 1$$

3.6 Show that the eigenfunctions of any Hermitian operator are orthogonal.

Solution: Ref. EWK,32; AN3,209; DI,32

If $\hat{R}$ is a Hermitian operator then the corresponding eigenvalues for ψ_1 and ψ_2 may be determined as

$$\hat{R}\psi_1 = r_1\psi_1 \tag{1}$$

and

$$\hat{R}\psi_2 = r_2\psi_2 \tag{2}$$

Multiplying (1) by ψ_2^* and integrating, we have

$$\int \psi_2^*\hat{R}\psi_1 d\tau = r_1\int \psi_2^*\psi_1 d\tau \tag{3}$$

Then applying the definition of a Hermitian operator and using the fact that the eigenvalues must be real we have

$$\int \psi_2^*\hat{R}\psi_1 d\tau = \int \psi_1\hat{R}^*\psi_2^* d\tau = r_2^*\int \psi_1\psi_2^* d\tau = r_2\int \psi_1\psi_2^* d\tau$$

That is

$$r_1\int \psi_2^*\psi_1 d\tau = r_2\int \psi_1\psi_2^* d\tau$$

or

$$(r_1 - r_2)\int \psi_2^*\psi_1 d\tau = 0$$

If $r_1 \neq r_2$ then

$$\int \psi_2^*\psi_1 d\tau = 0$$

This proof does not apply for degenerate eigenvalues (i.e. if $r_1 = r_2$) (see Problem 3.7 for consideration of the degenerate case).

In Dirac notation the proof goes as follows:

$$\hat{R}|r_1\rangle = r_1|r_1\rangle \quad (1)$$

$$\hat{R}|r_2\rangle = r_2|r_2\rangle \quad (2)$$

From the first equation we find:

$$\langle r_2|\hat{R}|r_1\rangle = r_1\langle r_2|r_1\rangle \quad (3)$$

By taking the Hermitian conjugate we convert the second equation into the form

$$\langle r_2|\hat{R} = \langle r_2|r_2$$

using

$$\hat{R}^+ = \hat{R}, \quad r_2^* = r_2, \quad \text{and} \quad (|r_2\rangle)^+ = \langle r_2|$$

Then

$$\langle r_2|\hat{R}|r_1\rangle = r_2\langle r_2|r_1\rangle \quad (4)$$

Subtracting (4) from (3) we have

$$(r_1 - r_2)\langle r_2|r_1\rangle = 0$$

As before $\langle r_2|r_1\rangle = 0$ unless $r_1 = r_2$ which demonstrates orthogonality for nondegenerate eigenkets.

3.7 It can be shown that the eigenfunctions of an Hermitian operator which belong to different eigenvalues are orthogonal (see Problem 3.6). When k eigenfunctions correspond to the same eigenvalue they are said to belong to a k-fold degenerate set. Suppose that the functions ϕ_1 and ϕ_2 are degenerate eigenfunctions of $\hat{\mathcal{H}}$ which are linearly independent and normalized but are not orthogonal.

(a) Show that the function $u = c_1\phi_1 + c_2\phi_2$ is an eigenfunction of $\hat{\mathcal{H}}$ which has the same eigenvalue as ϕ_1 and ϕ_2.

(b) Construct two linear combinations of ϕ_1 and ϕ_2 which are orthogonal to each other, and hence demonstrate that there is always sufficient freedom to choose degenerate functions which are orthogonal.

Solution: Ref. AN1,282

(a) We are given:

$$\hat{\mathcal{H}}\phi_1 = \lambda\phi_1$$

$$\hat{\mathcal{H}}\phi_2 = \lambda\phi_2$$

Therefore:

$$\hat{\mathcal{H}}(c_1\phi_1 + c_2\phi_2) = c_1\lambda\phi_1 + c_2\lambda\phi_2 = \lambda(c_1\phi_1 + c_2\phi_2)$$

or

$$\hat{\mathcal{H}}u = \lambda u$$

Since linear combinations of degenerate functions give additional degenerate eigenfunctions, we see that the initial degenerate set is not unique.

(b) We are given the linearly independent set of functions ϕ_1 and ϕ_2 and wish to obtain the orthogonal set u_1 and u_2. First we set $u_1 = \phi_1$ and $u_2 = c_1\phi_1 + c_2\phi_2$. The task is then to determine the constants c_1 and c_2 so that:

$$\int u_2^* u_1 d\tau = \langle u_2|u_1\rangle = 0$$

and

$$\int u_2^* u_2 d\tau = \langle u_2|u_2\rangle = 1$$

We adopt the shorthand (Dirac) notation for convenience. Expanding these integrals we find:

$$\langle u_1|u_2\rangle = \langle\phi_1|c_1\phi_1 + c_2\phi_2\rangle = c_1\langle\phi_1|\phi_1\rangle + c_2\langle\phi_1|\phi_2\rangle = 0$$

$$c_1 = -c_2\langle\phi_1|\phi_2\rangle \quad \text{since} \quad \langle\phi_1|\phi_1\rangle = 1 \tag{1}$$

$$\langle u_2|u_2\rangle = \langle c_1\phi_1 + c_2\phi_2|c_1\phi_1 + c_2\phi_2\rangle = 1$$

$$= c_1^2 + c_2^2 + c_1c_2\langle\phi_1|\phi_2\rangle + c_2c_1\langle\phi_2|\phi_1\rangle \tag{2}$$

Substituting (1) into (2)

$$1 = c_2^2\langle\phi_1|\phi_2\rangle^2 + c_2^2 - c_2^2\langle\phi_1|\phi_2\rangle^2 - c_2^2\langle\phi_2|\phi_1\rangle\langle\phi_1|\phi_2\rangle$$

$$1 = c_2^2(1 - |\langle\phi_1|\phi_2\rangle|^2)$$

or

$$c_2 = (1 - |\langle\phi_1|\phi_2\rangle|^2)^{-1/2}$$

where we have made use of the fact that

$$\langle\phi_1|\phi_2\rangle^* = \langle\phi_2|\phi_1\rangle$$

and

$$\langle\phi_1|\phi_2\rangle\langle\phi_1|\phi_2\rangle^* = |\langle\phi_1|\phi_2\rangle|^2$$

The orthogonal, normalized set (orthonormal) is:

$$\boxed{\begin{aligned} u_1 &= \phi_1 \\ u_2 &= (1 - |\langle u_1|\phi_2\rangle|^2)^{-1/2}(-\langle u_1|\phi_2\rangle u_1 + \phi_2) \end{aligned}}$$

The quantity $\langle\phi_2|u_1\rangle$ may be interpreted as the projection of the function ϕ_2 on the function u_1.

The procedure described here is known as <u>Schmidt Orthogonalization</u>. It may be continued for a set of k functions to give the general term:

$$u_n = (1 - \sum_{j=1}^{n-1} |\langle u_j|\phi_n\rangle|^2)^{-1/2}(\phi_n - \sum_{j=1}^{n-1} \langle u_j|\phi_n\rangle u_j)$$

3.8 Suppose that $\{u_i\}$ constitutes a complete set of eigenfunctions for the two linear operators $\hat{R}$ and $\hat{P}$. Show that $[\hat{R}, \hat{P}] = 0$.

Solution: Ref. LE1,157

We are given

$$\hat{R}u_i = r_i u_i, \quad \hat{P}u_i = p_i u_i \quad (i = 1, \cdots n)$$

where r_i and p_i are the eigenvalues of $\hat{R}$ and $\hat{P}$, respectively. We must prove

$$[\hat{R}, \hat{P}] = 0$$

or

$$[\hat{R}\hat{P} - \hat{P}\hat{R}]f = 0$$

Expanding f in the complete set of functions we then have

$$[\hat{R}\hat{P} - \hat{P}\hat{R}]\sum_i c_i u_i = \sum_i c_i(\hat{R}\hat{P} - \hat{P}\hat{R})u_i = \sum_i c_i(\hat{R}p_i - \hat{P}r_i)u_i$$

$$= \sum c_i(r_i p_i - p_i r_i)u_i = 0$$

Since $f = \sum_i c_i u_i \neq 0$, we must have

$$[\hat{R}\hat{P} - \hat{P}\hat{R}] = [\hat{R}, \hat{P}] = 0$$

Note that the existence of a common eigenfunction of $\hat{P}$ and $\hat{R}$ is not sufficient to conclude that $[\hat{R}, \hat{P}] = 0$.

3.9 $\hat{A}$ and $\hat{B}$ are operators which commute, and $u_{A'}$ is an eigenfunction of $\hat{A}$ which has the eigenvalue A'.

(a) Show that $u_{A'}$ is also an eigenfunction of $\hat{B}$ in the case that $u_{A'}$ is nondegenerate.

(b) If $u_{A'i}$ (i = 1, n) are linearly independent functions having the same eigenvalue A', show that linear combinations of these functions can be chosen which are simultaneously eigenfunctions of $\hat{B}$. (Hint: Expand $\hat{B}u_{A'i}$ in terms of the eigenfunctions of $\hat{A}$.)

Solution: Ref. ME,157; DI,49; RA,301

(a) It is given that

$$\hat{A}u_{A'} = A'u_{A'}$$

and

$$[\hat{A}, \hat{B}] = 0$$

Therefore

$$\hat{B}\hat{A}u_{A'} = \hat{A}\hat{B}u_{A'} = A'\hat{B}u_{A'}$$

and $\hat{B}u_{A'}$ is an eigenfunction of $\hat{A}$ having the eigenvalue A'. Since there is no degeneracy, $\hat{B}u_{A'}$ must be the same state as $u_{A'}$. This implies that $\hat{B}u_{A'} = B'u_{A'}$ where B' is a number.

(b) Proceeding as in (a) we find that $\hat{B}u_{A'i}$ is an eigenfunction of $\hat{A}$ corresponding to the eigenvalue A'. Then we must be able to expand $\hat{B}u_{A'i}$ as follows:

$$\hat{B}u_{A'i} = \sum_{j=1}^{n} c_{ji} u_{A'j}$$

This does not prove that $u_{A'i}$ is an eigenfunction of $\hat{B}$; however, a new function $\phi_{A'B'i}$ can be constructed as follows:

$$\phi_{A'B'i} = \sum_{j=1}^{n} d_{ji} u_{A'j}$$

which is still an eigenfunction of $\hat{A}$ and is also an eigenfunction of $\hat{B}$ having the eigenvalue B'_i.

$$\hat{B}\phi_{A'B'i} = B'_i \phi_{A'B'i} = B'_i \sum_{j=1}^{n} d_{ji} u_{A'j} \tag{1}$$

but

$$\hat{B}\phi_{A'B'i} = \sum_{j=1}^{n} d_{ji} \hat{B}u_{A'j} = \sum_{j=1}^{n} d_{ji} \left(\sum_{k=1}^{n} c_{kj} u_{A'k} \right)$$

$$= B'_i \sum_{j=1}^{n} d_{ji} u_{A'k} \delta_{kj}$$

to be consistent with (1), therefore

$$\sum_{j=1}^{n} d_{ji} (c_{kj} - B'_i \delta_{kj}) = 0 \tag{2}$$

Non-trivial solutions to this equation exist when

$$\det |c_{kj} - B'_i \delta_{kj}| = 0 \tag{3}$$

The roots of this secular equation are the eigenvalues $B'_1, B'_2, \ldots, B'_n$ of $\hat{B}$ and equation (2) can be solved for the coefficients d_{ji}.

3.10 A certain system is in a state that is specified by the time independent state function Ψ. According to Postulate II a measurement of the physical quantity q can only give one of the eigenvalues of the operator $\hat{Q}$ that is associated with q. What is the probability of obtaining the eigenvalue q_n in a given measurement? (Hint: Expand ψ in the eigenfunctions of $\hat{Q}$.)

Solution: Ref. ME,153

The eigenfunctions u_n of the Hermitian operator $\hat{Q}$ can be chosen to form a complete orthonormal set. Therefore we can expand Ψ as follows

$$\Psi = \sum_{n=1}^{\infty} c_n u_n$$

According to Postulate III the expectation value of q is

$$\langle q\rangle = \int \Psi^* \hat{Q}\Psi d\tau = \sum_n \sum_m c_n^* c_m \int u_n^* \hat{Q} u_m d\tau$$

but

$$\hat{Q}u_n = q_n u_n$$

and

$$\int u_n^* u_m d\tau = \delta_{nm}$$

Therefore

$$\langle q\rangle = \sum_{n=1}^{\infty} |c_n|^2 q_n$$

In this average $|c_n|^2$ is the weighting factor for the eigenvalue q_n. If measurements are carried out on a large number of identical systems each in a state described by Ψ, $|c_n|^2$ will be the fraction of measurements that give q_n. Recall that Ψ is normalized so that

$$\int \Psi^* \Psi d\tau = \sum_n \sum_m c_n^* c_m \int u_n^* u_m d\tau = \sum_n |c_n|^2 = 1$$

The normalized probability of obtaining q_n in a measurement is $|c_n|^2$.

3.11 For a system with N degrees of freedom the Poisson Bracket of the quantities r and s is defined as:

$$\{r, s\}_{PB} = \sum_{n=1}^{N} \left\{ \frac{\partial r}{\partial q_n} \frac{\partial s}{\partial p_n} - \frac{\partial r}{\partial p_n} \frac{\partial s}{\partial q_n} \right\}$$

where q_n and p_n are the canonical coordinates and momenta, respectively. Verify the following properties of the Poisson Brackets:

(a) $\{u, v\}_{PB} = -\{v, u\}_{PB}$

(b) $\{u, v_1 + v_2\}_{PB} = \{u, v_1\}_{PB} + \{u, v_2\}_{PB}$

(c) $\{u_1 u_2, v\}_{PB} = u_1\{u_2, v\}_{PB} + \{u_1, v\}_{PB} u_2$

(d) $\{u, v_1 v_2\}_{PB} = \{u, v_1\}_{PB} v_2 + v_1\{u, v_2\}_{PB}$

Solution: Ref. DI,84

For convenience we carry out these proofs for one degree of freedom. The summations can easily be added at the end if needed.

(a) By definition:

$$\{u, v\}_{PB} = \frac{\partial u}{\partial q} \frac{\partial v}{\partial p} - \frac{\partial u}{\partial p} \frac{\partial v}{\partial q} = -\left[\frac{\partial v}{\partial q} \frac{\partial u}{\partial p} - \frac{\partial v}{\partial p} \frac{\partial u}{\partial q} \right]$$

$$= -\{v, u\}_{PB}$$

(b) $$\{u, v_1 + v_2\}_{PB} = \frac{\partial u}{\partial q}\frac{\partial}{\partial p}(v_1 + v_2) - \frac{\partial u}{\partial p}\frac{\partial}{\partial q}(v_1 + v_2)$$

$$= \frac{\partial u}{\partial q}\left(\frac{\partial v_1}{\partial p} + \frac{\partial v_2}{\partial p}\right) - \frac{\partial u}{\partial p}\left(\frac{\partial v_1}{\partial q} + \frac{\partial v_2}{\partial q}\right)$$

$$= \{u, v_1\}_{PB} + \{u, v_2\}_{PB}$$

(c) $$\{u_1u_2, v\}_{PB} = \frac{\partial}{\partial q}(u_1u_2)\frac{\partial v}{\partial p} - \frac{\partial}{\partial p}(u_1u_2)\frac{\partial v}{\partial q}$$

but

$$\frac{\partial}{\partial q}(u_1u_2) = \left(\frac{\partial u_1}{\partial q}\right)u_2 + u_1\left(\frac{\partial u_2}{\partial q}\right), \text{ etc.}$$

Therefore:

$$\{u_1u_2, v\}_{PB} = \left\{\left[\left(\frac{\partial u_1}{\partial q}\right)u_2 + u_1\left(\frac{\partial u_2}{\partial q}\right)\right]\frac{\partial v}{\partial p} - \left[\left(\frac{\partial u_1}{\partial p}\right)u_2 + u_1\left(\frac{\partial u_2}{\partial p}\right)\right]\frac{\partial v}{\partial q}\right\}$$

$$= \left[\left(\frac{\partial u_1}{\partial q}\frac{\partial v}{\partial p} - \frac{\partial u_1}{\partial p}\frac{\partial v}{\partial q}\right)u_2 + u_1\left(\frac{\partial u_2}{\partial q}\frac{\partial v}{\partial p} - \frac{\partial u_2}{\partial p}\frac{\partial v}{\partial q}\right)\right]$$

$$= \{u_1, v\}_{PB}u_2 + u_1\{u_2, v\}_{PB}$$

(d) $$\{u, v_1v_2\}_{PB} = \frac{\partial u}{\partial q}\frac{\partial}{\partial p}(v_1v_2) - \frac{\partial u}{\partial p}\frac{\partial}{\partial q}(v_1v_2)$$

As before

$$\{u, v_1v_2\}_{PB} = \left\{\frac{\partial u}{\partial q}\left[\left(\frac{\partial v_1}{\partial p}\right)v_2 + v_1\left(\frac{\partial v_2}{\partial p}\right)\right] - \frac{\partial u}{\partial p}\left[\left(\frac{\partial v_1}{\partial q}\right)v_2 + v_1\left(\frac{\partial v_2}{\partial q}\right)\right]\right\}$$

$$= \left[\left(\frac{\partial u}{\partial q}\frac{\partial v_1}{\partial p} - \frac{\partial u}{\partial p}\frac{\partial v_1}{\partial q}\right)v_2 + v_1\left(\frac{\partial u}{\partial q}\frac{\partial v_2}{\partial p} - \frac{\partial u}{\partial p}\frac{\partial v_2}{\partial q}\right)\right]$$

$$= \{u, v_1\}_{PB}v_2 + v_1\{u, v_2\}_{PB}$$

3.12 Verify the following identities:

(a) $\{u_1u_2, v_1v_2\}_{PB} = \{u_1, v_1\}_{PB}v_2u_2 + v_1\{u_1, v_2\}_{PB}u_2 + u_1\{u_2, v_1\}_{PB}v_2 + u_1v_1\{u_2, v_2\}_{PB}$

(b) $\{u_1u_2, v_1v_2\}_{PB} = \{u_1, v_1\}_{PB}u_2v_2 + u_1\{u_2, v_1\}_{PB}v_2 + v_1\{u_1, v_2\}_{PB}u_2 + v_1u_1\{u_2, v_2\}_{PB}$

(c) $[u_1, v_1]/\{u_1, v_1\}_{PB} = [u_2, v_2]/\{u_2, v_2\}_{PB}$ and hence that $[u, v] = \text{const} \times \{u, v\}_{PB}$ where $[u, v] = uv - vu$.

<u>Solution</u>: Ref. DI,86

(a) First we use the result of 3.11c to obtain:

$$\{u_1u_2, v_1v_2\}_{PB} = u_1\{u_2, v_1v_2\}_{PB} + \{u_1, v_1v_2\}_{PB}u_2$$

Then using 3.11d we find

$$\{u_1u_2, v_1v_2\}_{PB} = u_1(\{u_2, v_1\}_{PB}v_2 + v_1\{u_2, v_2\}_{PB}) + (\{u_1, v_1\}_{PB}v_2 + v_1\{u_1, v_2\}_{PB})u_2$$

which is the desired result.

(b) The expansion can also be made by using 3.11d first.

$$\{u_1u_2, v_1v_2\}_{PB} = \{u_1u_2, v_1\}_{PB}v_2 + v_1\{u_1u_2, v_2\}_{PB}$$

Then using 3.11c we find

$$\{u_1u_2, v_1v_2\}_{PB} = (u_1\{u_2, v_1\}_{PB} + \{u_1, v_1\}_{PB}u_2)v_2 + v_1(u_1\{u_2, v_2\}_{PB} + \{u_1, v_2\}_{PB}u_2)$$

(c) The expansions for $\{u_1u_2, v_1v_2\}_{PB}$ found in parts (a) and (b) must clearly be equal. Therefore,

$$\{u_1, v_1\}_{PB}(v_2u_2 - u_2v_2) + \{u_2, v_2\}_{PB}(u_1v_1 - v_1u_1) = 0$$

$$\{u_2, v_2\}_{PB}[u_1, v_1] = \{u_1, v_1\}_{PB}[u_2, v_2]$$

or

$$\frac{\{u_2, v_2\}_{PB}}{[u_2, v_2]} = \frac{\{u_1, v_1\}_{PB}}{[u_1, v_1]}$$

Since this proportionality holds for any arbitrary pair of functions, we conclude:

$$[u, v] = \text{const} \times \{u, v\}_{PB}$$

Postulate IV is thus found to be consistent. The observation of quantum effects depends on the magnitude of the proportionality constant.

3.13 A particularly useful set of operators to represent a coordinate and its conjugate momentum is the so called coordinate representation:

$$\hat{p}_r \to -i\hbar \frac{\partial}{\partial x_r}, \quad \hat{x}_r \to x_r$$

Another choice, much less widely used, is the momentum representation where:

$$\hat{p}_r \to p_r, \quad \hat{x}_r \to i\hbar \frac{\partial}{\partial p_r}$$

Verify that the operators in both the coordinate and the momentum representations satisfy the basic quantum condition (Postulate IV), namely $[\hat{x}, \hat{p}] = i\hbar\{x, p\}$.

Solution: Ref. DI,89; Problem 3.28

For convenience we assume only one degree for freedom. In the coordinate representation:

$$[\hat{p}, x]u = -i\hbar\left[\left(\frac{d}{dx}\right)x - x\left(\frac{d}{dx}\right)\right]u$$

$$= -i\hbar u$$

or

$$[\hat{p}, x] = -i\hbar$$

Also,

$$\{p, x\}_{PB} = -1$$

Therefore:

$$[\hat{p}, x] = i\hbar\{p, x\}_{PB}$$

as required.

With the momentum representation

$$[p, \hat{x}]u = i\hbar\left[p\left(\frac{d}{dp}\right) - \left(\frac{d}{dp}\right)p\right]u$$

$$= -i\hbar u$$

Or

$$[p, \hat{x}] = -i\hbar$$

Again, we have

$$\{p, x\}_{PB} = -1$$

Therefore,

$$[\hat{p}, \hat{x}] = i\hbar\{p, x\}_{PB}$$

3.14 Evaluate the following commutators:

(a) $[x, d/dx]$

(b) $[d/dx, x^2]$

(c) $[\hat{\mathcal{H}}, x]$, $\hat{\mathcal{H}}$ = Hamiltonian operator in 1 dimension.

(d) $[\hat{\mathcal{H}}, \hat{p}]$, $\hat{\mathcal{H}}$ = Hamiltonian operator in 1 dimension.

Solution:

Less confusion will result in this type of calculation if we let each commutator operate on the arbitrary function u.

(a) $$[x\frac{d}{dx} - \frac{d}{dx}x]u = x\frac{du}{dx} - u - x\frac{du}{dx} = -u$$

Therefore

$$[x, d/dx] = -1$$

(b) $$\left[\frac{d}{dx}x^2 - x^2\frac{d}{dx}\right]u = 2xu + x^2\frac{du}{dx} - x^2\frac{du}{dx} = 2xu$$

Therefore

$$[d/dx, x^2] = 2x$$

Another way to evaluate this commutation is to notice in part (a) that $[d/dx, x] = 1$. Then

$$x[d/dx, x] + [d/dx, x]x = [d/dx, x^2] = 2x$$

(c) $$\hat{\mathcal{H}} = \frac{\hat{p}^2}{2m} + V(x)$$

Therefore

$$[\hat{\mathcal{H}}, x] = \frac{1}{2m}[\hat{p}^2, x] + [V, x]$$

Using the expansion in Problem 3.26 we have

$$[\hat{\mathcal{H}}, x] = \frac{1}{2m}([\hat{p}, x]\hat{p} + \hat{p}[\hat{p}, x])$$

$[V, x] = 0$ since $V(x)$ can be expanded in a power series in x and

$$[\{V(0) + \left(\frac{dV}{dx}\right)_o x + \frac{1}{2}\left(\frac{d^2V}{dx^2}\right)_o x^2 \ldots\}, x] = 0$$

Then using Problem 3.13

$$[\hat{\mathcal{H}}, x] = \frac{1}{2m}(-i\hbar\hat{p} - i\hbar\hat{p})$$

$$= -i\hbar\frac{\hat{p}}{m} = -i\hbar\frac{m\dot{x}}{m} = -i\hbar\dot{x}$$

Rearranging we have

$$\dot{x} = \frac{i}{\hbar}[\hat{\mathcal{H}}, x] \quad (\dot{x} \equiv dx/dt)$$

This equation for the time derivative of x is identical to the equation introduced in Postulate VI. The similarity to the classical result is striking.

(d) As in part (c) we write

$$\left[\frac{\hat{p}^2}{2m} + V(x), \hat{p}\right] = \frac{1}{2m}[\hat{p}^2, \hat{p}] + [V, \hat{p}]$$

With the coordinate representation as usual we have (using $[\hat{p}^2, \hat{p}] = \hat{p}[\hat{p}, \hat{p}] = 0$)

$$[V, \hat{p}]u = \left[V\left(-i\hbar\frac{d}{dx}\right) - \left(-i\hbar\frac{d}{dx}\right)V\right]u$$

$$= -i\hbar\left(V\frac{du}{dx} - u\frac{dV}{dx} - V\frac{du}{dx}\right)$$

or

$$[\hat{\mathcal{H}}, \hat{p}] = [V, \hat{p}] = +i\hbar\frac{dV}{dx}$$

3.15 Show that the momentum eigenfunction for a <u>free particle</u> moving in one dimension is

$$u_{p_x}(x) = h^{-1/2}e^{ip_x x/\hbar}$$

(Hint: Make use of the definition of the delta function

$$\delta(\gamma) = \frac{1}{2\pi}\int_{-\infty}^{+\infty} e^{i\gamma x}dx \tag{1}$$

<u>and</u> the normalization property of continuous eigenfunctions

$$\int u_f^* u_{f'} d\tau = \delta(f - f')\;) \tag{2}$$

<u>Solution</u>: Ref. LA,42; SC1,58

The eigenvalue equation for momentum in one dimension is

$$-i\hbar\frac{\partial}{\partial x}u = p_x u$$

which gives on indefinite integration

$$u_{p_x}(x) = ce^{ip_x x/\hbar} \tag{3}$$

Then, by use of Eqs. (1) and (2) we may determine the normalization c

$$\int u_{p_x'}^*(x)\,u_{p_x}(x)\,dx = \delta(p_x - p_x') = \int_{-\infty}^{+\infty} c^2 e^{\frac{ix}{\hbar}(p_x - p_x')}\,dx$$

$$= 2\pi\hbar \cdot c^2 \left(\frac{1}{2\pi}\int_{-\infty}^{+\infty} e^{iq(p_x - p_x')}\,dq\right), \quad q = x/\hbar$$

$$= 2\pi\hbar \cdot c^2\,\delta(p_x - p_x')$$

Thus

$$c = (2\pi\hbar)^{-1/2} = h^{-1/2}$$

and

$$u_{p_x}(x) = h^{-1/2} e^{ip_x x/\hbar}$$

3.16 For a conservative (non-relativistic) system having 3N degrees of freedom and N particles of mass m the Hamiltonian function is

$$\sum_{n=1}^{3N} \frac{p_i^2}{2m} + V(x_1, y_1, z_1, \ldots, x_N, y_N, z_N) = E$$

where $p_1 = p_{x_1}$, $p_2 = p_{y_1}$, etc.

(a) Obtain the Hamiltonian operator $\hat{\mathcal{H}}$ for this system in the coordinate representation. (see Problem 3.13)

(b) Assume that $\Psi(x, t) = \phi(t)u(x)$ in the time dependent Schrödinger equation (Postulate V) and obtain the time independent Schrödinger equation for u(x). Here x represents all necessary spatial coordinates.

Solution: Ref. PW,53

The coordinate representation (sometimes called Schrödinger's representation or viewpoint) is obtained by the substitution

$$p_i \to -i\hbar \frac{\partial}{\partial q_i}$$

or in three dimensions

$$\underset{\sim}{p} \to -i\hbar \underset{\sim}{\nabla}$$

Since

$$\underset{\sim}{\nabla}\cdot\underset{\sim}{\nabla} = \nabla^2$$

substitution into the Hamiltonian function gives

$$\hat{\mathcal{H}} = \sum_{i=1}^{N} -\frac{\hbar^2}{2m}\nabla_i^2 + V(\underset{\sim}{r}_1, \underset{\sim}{r}_2, \ldots, \underset{\sim}{r}_N)$$

In cartesian coordinates:

$$\nabla^2 = \frac{\partial^2}{\partial x^2} + \frac{\partial^2}{\partial y^2} + \frac{\partial^2}{\partial z^2}$$

Other convenient forms of the Laplacian operator ∇^2 are presented in Appendix 5.

(b) We are given

$$\hat{\mathcal{H}}\,\Psi = i\hbar\,\frac{\partial \Psi}{\partial t}$$

Substituting $\Psi(x, t) = \phi(t)u(x)$ and dividing by Ψ we find

$$\frac{\hat{\mathcal{H}}\,u}{u} = \frac{i\hbar}{\phi}\,\frac{d\phi}{dt} \tag{1}$$

This is possible since $\hat{\mathcal{H}}$ operates only on the spatial coordinates. The two sides of Eq. (1) depend on separate independent variables. For the equality to hold in general each side must equal a constant. In anticipation of the fact that this constant turns out to be the energy we use the symbol E and obtain

$$\hat{\mathcal{H}}\,u = Eu$$

and

$$\frac{d\phi}{dt} = -\frac{iE}{\hbar}\,\phi$$

The separation constant E must be an eigenvalue of $\hat{\mathcal{H}}$. We must therefore solve the <u>time-independent Schrödinger equation</u>:

$$\left[\sum_{i=1}^{N} -\frac{\hbar^2}{2m}\nabla_i^2 + V(\underset{\sim}{r}_1, \ldots, \underset{\sim}{r}_N)\right] u(\underset{\sim}{r}_1, \ldots, \underset{\sim}{r}_N) = Eu(\underset{\sim}{r}_1, \ldots, \underset{\sim}{r}_N)$$

3.17 A certain system is described by the Hamiltonian operator:

$$\hat{\mathcal{H}} = \left(-\frac{d^2}{dx^2} + x^2\right)$$

(a) Show that $Ax\exp(-x^2/2)$ is an eigenfunction of $\hat{\mathcal{H}}$ and determine the associated eigenvalue.

(b) Determine A so that $Ax\exp(-x^2/2)$ is normalized.

(c) What is the expectation value of x for the state described by $u(x) = Ax\exp(-x^2/2)$?

<u>Solution</u>:

(a) The function $u = Axe^{-x^2/2}$ is an eigenfunction of $\hat{\mathcal{H}}$ if $\hat{\mathcal{H}}\,u = \lambda u$ since by inspection we see that u is well-behaved. It only remains to evaluate $\hat{\mathcal{H}}\,u$.

$$-\frac{d^2}{dx^2}\left(Axe^{-x^2/2}\right) = -A\frac{d}{dx}\left[(1 - x^2)e^{-x^2/2}\right]$$

$$= A(3 - x^2)xe^{-x^2/2}$$

Then

$$\left(-\frac{d^2}{dx^2}+x^2\right)Axe^{-x^2/2} = 3Axe^{-x^2/2}$$

and

$$\hat{\mathcal{H}}\,u = 3u$$

Therefore u is an eigenfunction of $\hat{\mathcal{H}}$ and the associated eigenvalue is 3.

(b) Normalization requires that

$$\int_{-\infty}^{+\infty} u^* u dx = 1$$

Therefore:

$$\int_{-\infty}^{+\infty} A^2x^2e^{-x^2}dx = 1; \quad A^2\frac{\sqrt{\pi}}{2} = 1$$

using integral tables and

$$A = (4/\pi)^{1/4}$$

(c) If a time independent state is described by the wavefunction $u = Axe^{-x^2/2}$ the expectation value of x is given by:

$$\langle x\rangle = \int_{-\infty}^{+\infty}|u|^2 x dx = \int_{-\infty}^{+\infty}A^2x^3e^{-x^2}dx$$

$$= \int_{-\infty}^{0}A^2x^3e^{-x^2}dx + \int_{0}^{\infty}A^2x^3e^{-x^2}dx = 0$$

This can easily be seen by replacing the dummy variable x in the first integral on the right with -x. Here $\langle x\rangle$ vanishes as a consequence of the symmetry of the Hamiltonian.

3.18 The radial part of the Hamiltonian for the isolated Hydrogen atom (in a.u. with energy in Rydbergs) is

$$\hat{\mathcal{H}} = -\frac{1}{r^2}\frac{d}{dr}r^2\frac{d}{dr} - \frac{2}{r}$$

Show that $u(r) = Ae^{-r}$ is an eigenfunction of $\hat{\mathcal{H}}$.

Solution:

$$\hat{\mathcal{H}}\,u(r) = \left(-\frac{1}{r^2}\frac{d}{dr}r^2\frac{d}{dr} - \frac{2}{r}\right)Ae^{-r}$$

$$= \left(\frac{1}{r^2}\frac{d}{dr}r^2 - \frac{2}{r}\right)Ae^{-r}$$

$$= \left[\frac{1}{r^2}\frac{d}{dr}(2r - r^2) - \frac{2}{r}\right]Ae^{-r}$$

$$= \left[\frac{2}{r} - 1 - \frac{2}{r}\right]Ae^{-r}$$

$$= -1Ae^{-r} = -1u(r)$$

u(r) turns out to be the ground state eigenfunction for the hydrogen atom associated with the eigenvalue -1 Rydberg.

3.19 Consider the particle-in-box problem in which $V(x) = 0$ for $-a/2 < x < +a/2$ and $V(x) = \infty$ otherwise. The ground state wave function is known to be:

$$u(x) = \sqrt{2/a}\,\cos(\pi x/a)$$

In this problem we define $\delta x = x - \langle x\rangle$ and $\delta p = p - \langle p\rangle$.

(a) Calculate the magnitude of the expectation value of the product $\delta x\delta p$, i.e. determine $|\langle\delta x\delta p\rangle|$, for a particle in the state described by u(x).

(b) Determine the expectation value of the product $\delta p\delta x$ and relate the result to part (a).

Solution: Ref. AN1,37

(a) By definition:

$$\langle\delta x\delta p\rangle = \int u^*(x)(\delta x\delta p)_{op}u(x)dx$$

For the symmetrical box problem $\langle x\rangle = \langle p\rangle = 0$ and $\delta x = x$, $\delta p = p$. Choosing the coordinate representation ($p \to -i\hbar d/dx$) we have

$$\langle xp\rangle = \int_{-a/2}^{+a/2} \frac{2}{a}\cos\left(\frac{\pi x}{a}\right)x(-i\hbar)\frac{d}{dx}\cos\left(\frac{\pi x}{a}\right)dx$$

$$= \frac{2i\hbar\pi}{a^2}\int_{-a/2}^{+a/2}\cos\left(\frac{\pi x}{a}\right)\sin\left(\frac{\pi x}{a}\right)xdx$$

but $\sin 2y = 2\sin y\cos y$. So

$$\langle xp\rangle = \frac{i\hbar\pi}{a^2}\int_{-a/2}^{+a/2}\sin\left(\frac{2\pi x}{a}\right)xdx$$

$$= \frac{i\hbar\pi}{a^2}\left\{\frac{\sin(2\pi x/a)}{(2\pi/a)^2} - x\frac{\cos(2\pi x/a)}{(2\pi/a)}\right\}\Bigg|_{-a/2}^{+a/2}$$

$$= \frac{i\hbar}{2} \quad \text{using tables}$$

Therefore

$$|\langle\delta x\delta p\rangle| = \hbar/2$$

(b) Direct evaluation of the integral for $\langle \delta p \delta x \rangle = \langle px \rangle$ gives $-i\hbar/2$. However, it is easier to use the commutation relation: $[\hat{p},x] = -i\hbar$. Thus

$$\langle px \rangle = \langle [\hat{p},x] \rangle + \langle xp \rangle = -i\hbar + \frac{i\hbar}{2} = -\frac{i\hbar}{2}$$

3.20 The uncertainties in x and p_x are defined as

$$\Delta x = \left[\int u^*(x - \langle x \rangle)^2 u dx \right]^{1/2}, \qquad \Delta p_x = \left[\int u^*(\hat{p}_x - \langle \hat{p}_x \rangle)^2 u dx \right]^{1/2}$$

Determine the uncertainty product $\Delta x \Delta p_x$ for a harmonic oscillator in its ground state. The ground state eigenfunction, which is derived in Problem 4.10 is $u_o(x) = (\alpha/\pi)^{1/4}\exp(-\alpha x^2/2)$.

<u>Solution</u>: Ref. ME,158

First we set $\langle x \rangle = \langle p_x \rangle = 0$ since the integrands in both cases are odd functions of x and thus integrate to zero. Straightforward evaluation of the integrals then gives:

$$(\Delta x)^2 = \int_{-\infty}^{+\infty} u_o^*(x) x^2 u_o(x) dx = \sqrt{\frac{\alpha}{\pi}} \int_{-\infty}^{+\infty} e^{-\alpha x^2/2} x^2 e^{-\alpha x^2/2} dx$$

$$= 2\sqrt{\frac{\alpha}{\pi}} \int_o^{\infty} x^2 e^{-\alpha x^2} dx = \frac{1}{2\alpha}$$

$$(\Delta p_x)^2 = \int_{-\infty}^{+\infty} u_o^*(x) \left[-i\hbar \frac{d}{dx} \right]^2 u_o(x) dx = \hbar^2 \sqrt{\frac{\alpha}{\pi}} \int_{-\infty}^{+\infty} e^{-\alpha x^2/2} \frac{d^2}{dx^2} e^{-\alpha x^2/2} dx$$

$$= 2\alpha\hbar^2 \sqrt{\frac{\alpha}{\pi}} \int_o^{+\infty} e^{-\alpha x^2/2} \{ e^{-\alpha x^2/2} - \alpha x^2 e^{-\alpha x^2/2} \} dx$$

$$= \frac{1}{2} \alpha \hbar^2 \qquad \text{(using tables)}$$

Thus $(\Delta x)^2 (\Delta p_x)^2 = \hbar^2/4$ and $(\Delta x)(\Delta p_x) = \hbar/2$. In general the <u>Uncertainty Principle</u> states that

$$\boxed{(\Delta x)(\Delta p_x) \geq \frac{\hbar}{2}}$$

The equality holds only for the special case of a Gaussian wave packet.

Actually, it is possible to derive a general and powerful uncertainty relationship for any two noncommuting operators $\hat{A}$ and $\hat{B}$. The procedure is (following Merzbacher) to note the definitions of the commutator $\hat{A}\hat{B} - \hat{B}\hat{A} = i\hat{C}$ and the variances

$$(\Delta A)^2 = \int \psi^* (\hat{A} - \langle \hat{A} \rangle)^2 \psi d\tau = \int |(\hat{A} - \langle \hat{A} \rangle)\psi|^2 d\tau$$

$$(\Delta B)^2 = \int \psi^* (\hat{B} - \langle \hat{B} \rangle)^2 \psi d\tau = \int |(\hat{B} - \langle \hat{B} \rangle)\psi|^2 d\tau$$

where the last step involves use of the Hermitian property of $\hat{A}$ and $\hat{B}$. Application of the Schwarz inequality gives

$$(\Delta A)^2(\Delta B)^2 \geq \left|\int \psi^* \underline{(\hat{A} - \langle\hat{A}\rangle)(\hat{B} - \langle\hat{B}\rangle)}\psi d\tau\right|^2$$

The underlined operator is non-Hermitian, but any non-Hermitian operator $\hat{F}$ can be decomposed into two Hermitian operators by the equation

$$\hat{F} = \frac{\hat{F} + \hat{F}^+}{2} + i\,\frac{\hat{F} - \hat{F}^+}{2i}$$

Substitution of this equation with $\hat{F} = (\hat{A} - \langle\hat{A}\rangle)(\hat{B} - \langle\hat{B}\rangle)$ leads to

$$(\Delta A)^2(\Delta B)^2 \geq \left(\frac{1}{2}\langle C\rangle\right)^2 \quad \text{or} \quad \boxed{\Delta A \Delta B \geq \left|\frac{1}{2}\langle C\rangle\right|}$$

3.21 The virial theorem of classical mechanics states that

$$\boxed{\overline{T} = \frac{1}{2}\sum_i m_i\overline{\dot{x}_i^2} = -\frac{1}{2}\sum_i \overline{x_i F_{xi}}}$$

where F_{xi} is the x component of force for the ith particle and T is the kinetic energy. The bars indicate time averages. For a single particle in one dimension in a pure energy eigenstate, the quantum mechanical version of the virial theorem states that

$$\frac{1}{2m}\langle n|\hat{p}_x^2|n\rangle = \frac{1}{2}\langle n|x\frac{d}{dx}V(x)|n\rangle$$

Derive the latter result making use of the commutators:

$$[\hat{p}_x, \hat{\mathcal{H}}] = -i\hbar\frac{dV}{dx} \quad \text{and} \quad [x, \hat{\mathcal{H}}] = \frac{\hbar^2}{m}\frac{d}{dx}$$

Solution: Ref. S3,400; EWK,355

Using the commutation relations given above we can write

$$\frac{1}{2}\langle n|x\frac{dV}{dx}|n\rangle = \frac{i}{2\hbar}\langle n|x[\hat{p}_x, \hat{\mathcal{H}}]|n\rangle$$

$$= \frac{i}{2\hbar}\langle n|x\hat{p}_x\hat{\mathcal{H}} - \hat{\mathcal{H}}x\hat{p}_x - [x, \hat{\mathcal{H}}]\hat{p}_x|n\rangle$$

$$= \frac{i}{2\hbar}\langle n|x\hat{p}_x\hat{\mathcal{H}} - \hat{\mathcal{H}}x\hat{p}_x - \frac{\hbar^2}{m}\left(\frac{d}{dx}\right)\hat{p}_x|n\rangle$$

but

$$\hat{p}_x = -i\hbar\frac{d}{dx}, \quad \text{and} \quad \hbar\frac{d}{dx} = i\hat{p}_x$$

Therefore

$$\frac{1}{2}\langle n|x\frac{dV}{dx}|n\rangle = \frac{i}{2\hbar}\langle n|x\hat{p}_x\hat{\mathcal{H}} - \hat{\mathcal{H}}x\hat{p}_x - \frac{\hbar}{m}i\hat{p}_x^2|n\rangle$$

$$= \frac{i}{2\hbar}[(E_n - E_n)\langle n|x\hat{p}_x|n\rangle - \frac{\hbar i}{m}\langle n|\hat{p}_x^2|n\rangle] = \frac{1}{2m}\langle n|\hat{p}_x^2|n\rangle$$

where we have used the fact that:

$$\langle n|\hat{\mathcal{H}}x\hat{p}_x|n\rangle = \sum_{n'} \langle n|\hat{\mathcal{H}}|n'\rangle\langle n'|x\hat{p}_x|n\rangle = E_n\langle n|x\hat{p}_x|n\rangle$$

3.22 Assume that for a system of particles the potential energy is equal to the sum of pairwise contributions each proportional to r^n, where r is an interparticle distance, and show that for this case the Virial Theorem of Problem 3.21 reduces to $2\overline{T} = n\overline{V}$.

Solution: Ref. EWK,357; CHB,68

We are given

$$V = \sum_{k<\ell} V_{k\ell} = \sum_{k<\ell} c_{k\ell} r_{k\ell}^n$$

where

$$r_{k\ell} = [(x_k - x_\ell)^2 + (y_k - y_\ell)^2 + (z_k - z_\ell)^2]^{1/2}$$

Since

$$\frac{\partial V_{k\ell}}{\partial x_\ell} = \frac{\partial V_{k\ell}}{\partial r_{k\ell}} \cdot \frac{\partial r_{k\ell}}{\partial x_\ell}$$

and

$$\frac{\partial r_{k\ell}}{\partial x_k} = \frac{(x_k - x_\ell)}{r_{k\ell}}, \qquad \frac{\partial r_{k\ell}}{\partial x_\ell} = -\frac{(x_k - x_\ell)}{r_{k\ell}}$$

we have then for a given pair potential

$$\sum_j x_j \frac{\partial V_{k\ell}}{\partial x_j} = \sum_j x_j \frac{\partial V_{k\ell}}{\partial r_{k\ell}} \cdot \frac{\partial r_{k\ell}}{\partial x_j} = \sum_j x_j c_{k\ell} \cdot n r_{k\ell}^{n-1} \cdot \frac{\partial r_{k\ell}}{\partial x_j}$$

$$= \frac{nV_{k\ell}}{r_{k\ell}} \cdot \left(x_k \frac{\partial r_{k\ell}}{\partial x_k} + x_\ell \frac{\partial r_{k\ell}}{\partial x_\ell} + y_k \frac{\partial r_{k\ell}}{\partial y_k} + \dots\ z_\ell \frac{\partial r_{k\ell}}{\partial z_\ell} \right)$$

$$= \frac{nV_{k\ell}}{r_{k\ell}} \cdot \left[\frac{x_k(x_k - x_\ell)}{r_{k\ell}} - \frac{x_\ell(x_k - x_\ell)}{r_{k\ell}} + \dots\ \frac{-z_\ell(z_k - z_\ell)}{r_{k\ell}} \right]$$

$$= \frac{nV_{k\ell}}{r^2_{k\ell}} [(x_k - x_\ell)^2 + \dots] = nV_{k\ell}$$

Thus

$$\sum_{\text{all pairs}} x_j \frac{\partial V}{\partial x_j} = n \sum_{\text{all pairs}} V_{k\ell} = nV$$

and

$$\langle T\rangle = \frac{1}{2}\left\langle \sum_j x_j \frac{\partial V}{\partial x_j} \right\rangle = \frac{1}{2} n\langle V\rangle$$

If all forces in the system are coulombic (n = -1) then

$$\langle T\rangle = -\frac{1}{2}\langle V\rangle$$

Note that the results of this problem are a consequence of Euler's theorem related to homogeneous functions i.e.

if
$$f(\lambda x_1, \lambda x_2, \ldots.) = \lambda^N f(x_1, x_2, \ldots.)$$

then
$$\sum \frac{\partial f}{\partial x_i} x_i = Nf(x_1, x_2, x_3, \ldots.)$$

where λ is an arbitrary number and $f(\underset{\sim}{x})$ is homogeneous of degree N. Then

$$\sum_j x_j \frac{\partial V}{\partial x_j} = nV$$

3.23 The state function $\Psi(x, t)$ must obey the time dependent wave equation. In the event that $|\Psi(x, t)|^2$ is independent of time we say that Ψ describes a stationary state.

(a) $\Psi(x, t)$ can be expanded as

$$\Psi(x, t) = \sum_n c_n\phi_n(t)u_n(x)$$

where the c_n's are constants and the u_n's are eigenfunctions of $\hat{\mathcal{H}}$. Determine the form of $\phi_n(t)$ by direct substitution into the wave equation.

(b) What conditions on the c_n's must be satisfied in order for Ψ to represent a stationary state.

Solution: Ref. ME,45; RA,405

(a)
$$\hat{\mathcal{H}}\,\Psi = i\hbar \frac{\partial}{\partial t}\Psi$$

$$\hat{\mathcal{H}}\sum_n c_n\phi_n(t)u_n(x) = i\hbar \frac{\partial}{\partial t}\sum_n c_n\phi_n(t)u_n(x)$$

$$\sum_n c_n\phi_n(t)\hat{\mathcal{H}}u_n(x) = i\hbar\sum_n c_n\left(\frac{d}{dt}\phi(t)\right)u_n(x)$$

Therefore

$$\hat{\mathcal{H}}\,u_n(x) = E_n u_n(x) = i\hbar\frac{d}{dt}\phi(t)$$

$$\frac{d}{dt}\phi_n(t) = -\frac{iE_n}{\hbar}\phi_n(t)$$

and

$$\phi_n(t) = e^{-iE_n t/\hbar}$$

(b) Consider the product

$$\Psi^*\Psi = \left(\sum_n c_n e^{-iE_n t/\hbar} u_n(x)\right)^* \left(\sum_m c_m e^{-iE_m t/\hbar} u_m(x)\right)$$

$$= (c_o^* e^{+iE_o t/\hbar} u_o^* + c_1^* e^{+iE_1 t/\hbar} u_1^* + \ldots)(c_o e^{-iE_o t/\hbar} u_o + c_1 e^{-iE_1 t/\hbar} u_1 + \ldots)$$

$$= |c_o|^2|u_o|^2 + |c_1|^2|u_1|^2 \ldots c_o^* c_1 e^{i(E_o - E_1)t/\hbar} u_o^* u + \ldots$$

All of the time dependence arises from cross-terms between states having different energies. In the non-degenerate case a stationary state can exist only when all c_n's save one vanish. For example let $c_n \neq 0$ and $c_m = 0$ $(n \neq m)$. Then

$$\Psi(x, t) = c_n e^{-iE_n t/\hbar} u_n(x)$$

$$|\Psi(x, t)|^2 = |c_n|^2 |u_n(x)|^2$$

But normalization requires that

$$\sum_\ell |c_\ell|^2 = 1$$

(see Problem 3.10). So

$$|\Psi(x, t)|^2 = |u_n(x)|^2$$

for this stationary state.

3.24 At time zero a linear harmonic oscillator is in a state that is described by the normalized wave function:

$$\Psi(x, 0) = \sqrt{\frac{1}{5}}\, u_o(x) + \sqrt{\frac{1}{2}}\, u_2(x) + c_3 u_3(x)$$

where $u_n(x)$ is the nth time independent eigenfunction for the oscillator.

(a) Determine the numerical value of c_3 assuming it to be real and positive.

(b) Write out the wave function at time t.

(c) What is the expectation value of the energy of the oscillator at $t = 0$? At $t = 1$ sec?

Solution:

(a) Normalization requires that:

$$\int_{-\infty}^{+\infty} \Psi^*(x, 0)\Psi(x, 0)\,dx = 1$$

or

$$\frac{1}{5}\int_{-\infty}^{+\infty} u_o^2(x)\,dx + \frac{1}{2}\int_{-\infty}^{+\infty} u_2^2(x)\,dx + c_3^2 \int_{-\infty}^{+\infty} u_3^2(x)\,dx = 1$$

$$1/5 + 1/2 + c_3^2 = 1$$

so

$$c_3 = \sqrt{3/10}$$

(b) At time t

$$\Psi(x, t) = \sum_{n=0}^{\infty} c_n e^{-iE_n t/\hbar} u_n(x), \quad E_n = (n + 1/2)h\nu_o$$

so

$$\Psi(x, t) = \sqrt{\frac{1}{5}}\, u_o(x)e^{-i\pi\nu_o t} + \sqrt{\frac{1}{2}}\, u_2(x)e^{-i\pi 5\nu_o t} + \sqrt{\frac{3}{10}}\, u_3(x)e^{-i\pi 7\nu_o t}$$

(c)

$$\langle E\rangle = \int_{-\infty}^{+\infty} \Psi^*(x, t)\hat{\mathcal{H}}\,\Psi(x, t)\,dx$$

$$= \frac{1}{5}\int_{-\infty}^{+\infty} u_o\hat{\mathcal{H}}\,u_o dx + \frac{1}{2}\int_{-\infty}^{+\infty} u_2\hat{\mathcal{H}}u_2 dx + \frac{3}{10}\int_{-\infty}^{+\infty} u_3\hat{\mathcal{H}}\,u_3 dx$$

$$= \frac{1}{5}\left(\frac{h\nu_o}{2}\right) + \frac{1}{2}\left(\frac{5}{2}\,h\nu_o\right) + \frac{3}{10}\left(\frac{7}{2}\,h\nu_o\right) = \frac{12}{5}\,h\nu_o$$

$\langle E\rangle$ is independent of time since energy must be conserved.

SUPPLEMENTARY PROBLEMS

3.25 The functions ϕ_1 and ϕ_2 are degenerate eigenfunctions of $\hat{\mathcal{H}}$ which are linearly independent and normalized but are not orthogonal. Prove that no more than two linearly independent combinations of the functions ϕ_1 and ϕ_2 can be formed. (Hint: Attempt to construct three linearly independent solutions.)
Ref. Problem 3.7; AN1,282

3.26 The <u>commutator</u> of the two dynamical variables r and s is defined as:

$$[r, s] = rs - sr$$

Verify the following properties of commutators:

(a) $[u, v] = -[v, u]$

(b) $[u, v_1 + v_2] = [u, v_1] + [u, v_2]$

(c) $[u_1u_2, v] = [u_1, v]u_2 + u_1[u_2, v]$

(d) $[u, v_1v_2] = [u, v_1]v_2 + v_1[u, v_2]$

3.27 Let F(p, q; t) represent a classical dynamical variable which depends on time explicitly and implicitly through the time dependence of the momenta p and the coordinates

x. Show that:

$$\frac{dF}{dt} = \frac{\partial F}{\partial t} + \{F, H\}_{PB}$$

where H is the Hamiltonian function for the system. (Hint: Use the canonical equations of Hamilton.)
Ref. Problem 3.11; PI1,76; DI,113; GO,217

3.28 Evaluate the following Poisson brackets:

(a) $\{x, x^2\}_{PB}$

(b) $\{x, p\}_{PB}$

(c) $\{x^2, p^2\}_{PB}$

(d) $\{xp, x\}_{PB}$

Ans:

(a) 0

(b) 1

(c) 4xp

(d) -x

3.29 Given that $\hat{\mathcal{H}} u = Eu$, what is the effect on u and E of adding a constant potential V_o to $\hat{\mathcal{H}}$.

Ans:

u is unaffected and E is shifted.

CHAPTER **4**

SIMPLE EXACTLY SOLUBLE PROBLEMS IN WAVE MECHANICS

Analytical solutions for the wave equation can be obtained in only a limited number of cases. In this chapter we have collected a number of problems which do permit exact solutions. We start with the motion of a free particle in one dimension with the potential energy constant, then introduce step discontinuities in the constant potential. These problems illustrate traveling waves, standing waves, bound states, and tunneling through barriers. We also include the simple harmonic oscillator and several problems in two and three dimensions. The systems discussed here provide simple models for many atomic and molecular systems which will be considered in later chapters.

The selection of problems for this chapter is somewhat arbitrary. In the chapters which follow additional exactly soluble problems appear which have been placed with closely related problems. For example the particle in a spherical potential box is included in the chapter on hydrogen-like atoms and the solution for a Morse potential is included with the problems on molecular spectroscopy.

PROBLEMS

4.1 A free particle of mass m is constrained to move along the x-direction. For all values of x the potential V_o is constant.

(a) Set up the time dependent Schrödinger equation for this particle and obtain the general solution for the wave function. Show that this solution can be interpreted as the sum of waves which move in the +x and -x directions.

(b) Show that the initial conditions can be chosen so that the wave function for the free particle is an eigenfunction of the operator $\hat{p}_x = -i\hbar\partial/\partial x$, i.e. that this special solution corresponds to a wave moving in one direction.

(c) Show that the initial conditions can be chosen so that the probability distribution function is independent of time. The result will be a standing wave.

(d) Show that the operator $\hat{p}_x$ commutes with $\hat{\mathcal{H}}$ for the free particle.

Solution:

(a) From Chapter 3 the time dependent wave equation is

$$\hat{\mathcal{H}}\,\Psi(x, t) = i\hbar \frac{\partial}{\partial t}\Psi(x, t)$$

These variables are separated by assuming that

$$\Psi(x, t) = u(x)\phi(t)$$

Substitution in the wave equation followed by division with Ψ gives:

$$\frac{\hat{\mathcal{H}}u(x)}{u(x)} \equiv \frac{i\hbar}{\phi(t)}\frac{d\phi(t)}{dt} = E$$

The sides of this identity depend on independent variables and, therefore, can only be equal for arbitrary values of x and t if the functions are equal to the same constant. The constant E is, in fact, the energy.

The resulting equations are

$$\hat{\mathcal{H}}u = Eu \quad \text{and} \quad i\hbar\frac{d\phi}{dt} = E\phi$$

The equation for ϕ can immediately be integrated to give

$$\int\frac{d\phi}{\phi} = \frac{-i}{\hbar}\int Edt \quad \text{or} \quad \phi(t) = e^{-iEt/\hbar}$$

The time independent wave equation for the particle is:

$$\frac{d^2u}{dx^2} + \frac{\hbar^2}{2m}(E - V_o)u = 0 \tag{1}$$

Linear differential equations with constant coefficients must have solutions of the form Ae^{Mx}. Substitution of this function into Eq. (1) leads to the general solution:

$$u(x) = Ae^{i\sqrt{2m(E - V_o)}x/\hbar} + Be^{-i\sqrt{2m(E - V_o)}x/\hbar}$$

and the time dependent solution is:

$$\Psi(x, t) = u(x)\phi(t)$$

$$\boxed{\Psi(x, t) = Ae^{(i/\hbar)[\sqrt{2m(E - V_o)}x - Et]} + Be^{(i/\hbar)[-\sqrt{2m(E - V_o)}x - Et]}}$$

A sum of classical traveling waves can be written as:

$$Ae^{2\pi i(x/\lambda - \nu t)} + Be^{2\pi i(-x/\lambda - \nu t)}$$

Inspection indicates that these equations are identical if:

$$2\pi/\lambda = \sqrt{2m(E - V_o)}/\hbar \quad \text{and} \quad E/\hbar = 2\pi\nu$$

$$1/\lambda = p_x/h \qquad\qquad E = h\nu$$

which are the wellknown equations for the momentum and energy of a free particle. The first term in the sum of waves represents a wave moving in the +x direction. This can be seen by setting the exponent equal to zero to obtain the crest of the wave and then solving for the velocity of the crest.

$$x/\lambda - \nu t = 0; \quad x/t = v_x = \nu\lambda$$

(b) The initial condition can be chosen so that

$$\Psi(x, 0) = Ae^{i\sqrt{2m(E - V_o)}x/\hbar}$$

which requires that B = 0. Then

$$\hat{p}_x\Psi(x, t) = -i\hbar \frac{\partial}{\partial x}\left(Ae^{(i/\hbar)[\sqrt{2m(E - V_o)}x - Et]}\right) = \sqrt{2m(E - V_o)}\Psi(x, t)$$

Therefore $\Psi(x, t)$ is an eigenfunction of $\hat{p}_x$ in this special case with the eigenvalue:

$$p_x = \sqrt{2m(E - V_o)} = mv_x$$

Since $p_x > 0$ the wave is moving in the +x direction. If the initial condition had been chosen so that A = 0, the sign of the momentum would be reversed.

(c) The initial condition can be chosen so that:

$$\Psi(x, 0) = A[e^{(i/\hbar)\sqrt{2m(E - V_o)}x} + e^{-(i/\hbar)\sqrt{2m(E - V_o)}x}]$$

Then

$$\Psi(x, t) = 2Ae^{-iEt/\hbar}\cos(\sqrt{2m(E - V_o)}x/\hbar)$$

The probability distribution function for this situation is

$$|\Psi(x, t)|^2 = 4|A|^2\cos^2(\sqrt{2m(E - V_o)}x/\hbar)$$

which is clearly independent of time.

(d) The commutator of $\hat{p}_x$ with $\hat{\mathcal{H}}$ is:

$$[\hat{p}_x, \hat{\mathcal{H}}] = \hat{p}_x\hat{\mathcal{H}} - \hat{\mathcal{H}}\hat{p}_x = \hat{p}_x\left(\frac{\hat{p}_x^2}{2m} + V_o\right) - \left(\frac{\hat{p}_x^2}{2m} + V_o\right)\hat{p}_x$$

$$= [\hat{p}_x, \hat{p}_x^2]/2m + [\hat{p}_x, V_o] = (\hat{p}_x^3 - \hat{p}_x^3)/2m + V_o(\hat{p}_x - \hat{p}_x) = 0$$

Since $\hat{p}_x$ commutes with $\hat{\mathcal{H}}$, these operators can have a complete set of simultaneous eigenfunctions. Part (b) gives an example of such a set. However, in general the eigenfunctions of $\hat{\mathcal{H}}$ are not necessarily eigenfunctions of $\hat{p}_x$. Part (c) gives a solution $\Psi(x, t)$ which is not an eigenfunction of $\hat{p}_x$.

4.2 The probability current density vector $\mathbf{j}$ for a particle of mass m is defined by the equation:

$$\mathbf{j}(\mathbf{r}, t) = \frac{i\hbar}{2m}(\Psi\nabla\Psi^* - \Psi^*\nabla\Psi)$$

(a) Determine the current density $\mathbf{j}$ for the plane wave:

$$\Psi(x, t) = Ae^{(i/\hbar)[\sqrt{2mE}x - Et]}$$

(b) Determine the current density $\underset{\sim}{j}$ for the wave function:

$$\Psi(x, t) = Ae^{(i/\hbar)[\sqrt{2mE}x - Et]} + Be^{(i/\hbar)[-\sqrt{2mE}x - Et]}$$

which represents free particles restricted to the x-direction.

Solution: Ref. ME,37; SC1,23; LA,57; S1,104

(a) In one dimension

$$j = \frac{i\hbar}{2m}(\Psi \frac{d}{dx}\Psi^* - \Psi^* \frac{d}{dx}\Psi)$$

Then

$$j = \frac{i\hbar}{2m}\left(\Psi\left[\frac{-i}{\hbar}\sqrt{2mE}\right]\Psi^* - \Psi^*\left[\frac{i}{\hbar}\sqrt{2mE}\right]\Psi\right) = \frac{\sqrt{2mE}}{m}\Psi\Psi^* = \frac{p_x}{m}|\Psi|^2 = v_x|\Psi|^2$$

This expression gives the number of particles passing through a unit area of a plane perpendicular to the x-axis in one second.

(b) For the general solution $\Psi(x, t)$ we find:

$$j = \frac{i\hbar}{2m}\left\{\frac{i}{\hbar}\sqrt{2mE}\left(-|A|^2 + |B|^2 + AB^*e^{(2i/\hbar)\sqrt{2mE}x} - A^*Be^{-(2i/\hbar)\sqrt{2mE}x}\right)\right.$$

$$\left.-\frac{i}{\hbar}\sqrt{2mE}\left(|A|^2 - |B|^2 - A^*Be^{-(2i/\hbar)\sqrt{2mE}x} + AB^*e^{(2i/\hbar)\sqrt{2mE}x}\right)\right\}$$

$$= \frac{\sqrt{2mE}}{m}(|A|^2 - |B|^2) = v_x(|A|^2 - |B|^2)$$

4.3 For a particle in one-dimensional space the function u(x) and its derivative du/dx must be continuous for all values of x in order for u to be an eigenfunction of $\hat{\mathcal{H}}$. Demonstrate the continuity of du/dx at a discontinuity in the potential when u satisfies the wave equation.

Solution: Ref. SC1,32; S1,62; S3,37

For this argument we replace the discontinuity in V(x) by a continuous change in the interval $x_o - \delta$ to $x_o + \delta$ shown below:

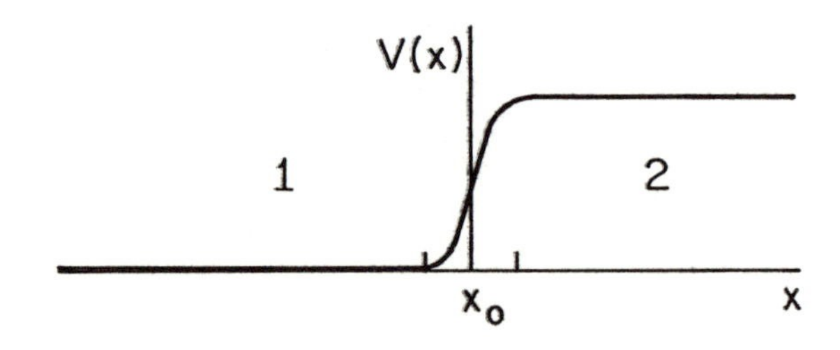

The wave equation for this problem is:

$$\frac{d^2u}{dx^2} = \frac{2m}{\hbar^2}(V - E)u$$

Integration over the interval about x_o gives:

$$\left.\frac{du}{dx}\right|_{x + \delta} - \left.\frac{du}{dx}\right|_{x - \delta} = \frac{2m}{\hbar^2}\int_{x - \delta}^{x + \delta}(V - E)u\,dx$$

In the limit that $\delta \to 0$ and in the event that $(V - E)$ remains finite the integral on the right approaches zero. Therefore du/dx is continuous at x_o. The continuity of $u(x)$ and du/dx is used in the following problems to connect solutions to the wave equation obtained on the two sides of the discontinuity in the potential. A special case of the discontinuity is treated in Problem 4.6 where the potential is assumed to be infinite on one side of the discontinuity. In such problems the wave function is taken to be zero in the region where the potential is infinite, but it should be realized that this can be justified only as the limit of the properly connected wave function (u and du/dx continuous) as the potential goes to infinity.

4.4 Consider a particle of mass m and energy E' approaching a one dimensional step potential V_o' as shown in the figure below.

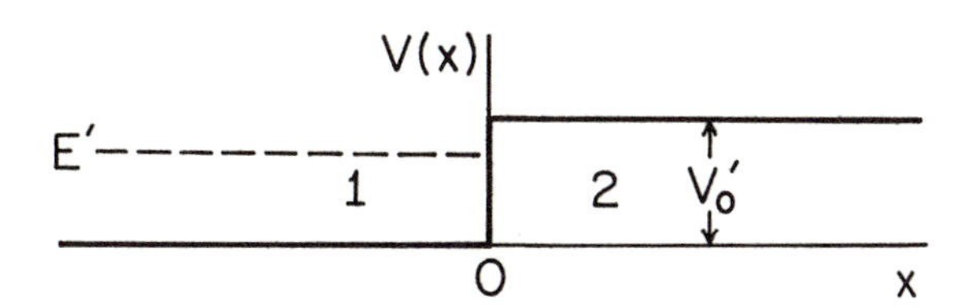

(a) For $E' < V_o'$ find

i) the probability of reflection

ii) the probability of finding the particle at $x > 0$

(b) For $E' > V_o'$ develop equations for

i) the transmission probability T

ii) the reflection probability R

iii) T + R

Solution: Ref. E1,227; DU,47

(a) i) $E' < V_o'$

First we must solve the Schrödinger equation in the two different regions. Using the reduced variables

$$E = \frac{2m}{\hbar^2} E' \quad \text{and} \quad V = \frac{2m}{\hbar^2} V_o'$$

we have

$$\text{region 1:} \quad u'' = -Eu; \quad u^{(1)} = Ae^{i\sqrt{E}\,x} + Be^{-i\sqrt{E}\,x}$$

$$\text{region 2:} \quad u'' = -(E - V_o)u; \quad u^{(2)} = Ce^{\sqrt{V_o - E}\,x} + De^{-\sqrt{V_o - E}\,x}$$

The manipulations can be simplified by defining $k_1 = \sqrt{E}$ and $k_2 = \sqrt{V_o - E}$. Furthermore, we realize that C must be zero in order to keep $u^{(2)}$ finite as x increases without limit.

The probability of reflection is

$$R = \frac{\text{Reflected current density}}{\text{Incident current density}} = \frac{v_1}{v_1}\frac{B^*B}{A^*A} = \frac{B^*B}{A^*A}$$

where v_1 is the particle velocity in region 1. To obtain R, we first note the boundary conditions

$$u^{(1)}(0) = u^{(2)}(0) \quad \text{so} \quad A + B = D \tag{1}$$

$$u'^{(1)}(0) = u'^{(2)}(0) \quad \text{so} \quad ik_1(A - B) = -k_2D \tag{2}$$

or

$$A - B = \beta D \quad \text{where} \quad \beta = ik_2/k_1$$

Thus (1) + (2) gives

$$A = \frac{D}{2}(1 + \beta) = \frac{D}{2}\left(\frac{k_1 + ik_2}{k_1}\right)$$

and (1) - (2) gives

$$B = \frac{D}{2}(1 - \beta) = \frac{D}{2}\left(\frac{k_1 - ik_2}{k_1}\right)$$

Using the definition of R we find

$$R = \frac{|B|^2}{|A|^2} = \frac{|D|^2}{|D|^2} \cdot \frac{(k_1^2 + k_2^2)}{(k_1^2 + k_2^2)} = 1$$

This result is in agreement with our classical expectations.

ii) The probability of finding the particle at $x > 0$ is given by

$$u^{(2)}u^{(2)*} = DD^*e^{-2\sqrt{V_o - E}\,x} = DD^*e^{-\frac{2}{\hbar}\sqrt{2m(V_o' - E')}\,x}$$

Thus, there is a finite probability that the particle will exist within the barrier. This is clearly a non-classical prediction.

(b) i) $E' > V_o'$

The solutions of the Schrödinger equation in regions 1 and 2 are now

$$u^{(1)} = Ae^{ik_1x} + Be^{-ik_1x}$$

$$u^{(2)} = Ce^{i\sqrt{E - V_o}\,x} + De^{-i\sqrt{E - V_o}\,x} = Ce^{ik_3x} + De^{-ik_3x}$$

where

$$k_3 = \sqrt{E - V_o} \quad \text{and} \quad k_1 = \sqrt{E}$$

Note that the solution in region 2 is now complex.

A boundary condition that we may apply immediately is that D = 0. The term De^{-ik_3x} is the wave function for a particle current in the direction of -x in region 2. This is not possible since it would imply that there is a scattering region (reflective) beyond x = 0 and with $V' > V_o'$. Such is not the case, so D = 0. Now applying the continuity boundary conditions we have

$$u^{(1)}(0) = u^{(2)}(0) \text{ yields } A + B = C \tag{1}$$

and

$$u'^{(1)}(0) = u'^{(2)}(0) \text{ yields } ik_1(A - B) = ik_3C \tag{2}$$

or

$$A - B = \gamma C \quad \text{where} \quad \gamma = k_3/k_1$$

Thus (1) + (2) gives

$$A = \frac{C}{2}(1 + \gamma) = \frac{C}{2}\left(\frac{k_1 + k_3}{k_1}\right)$$

and (1) - (2) gives

$$B = \frac{C}{2}(1 - \gamma) = \frac{C}{2}\left(\frac{k_1 - k_3}{k_1}\right)$$

Now

$$T = \frac{v_2}{v_1} \cdot \frac{|C|^2}{|A|^2} = \frac{\text{transmitted current density}}{\text{reflected current density}} = \left(\frac{E - V_o}{E}\right)^{1/2} \frac{4k_1^2}{(k_1 + k_3)^2}$$

$$= \left(\frac{E - V_o}{E}\right)^{1/2} \frac{4E}{(\sqrt{E} + \sqrt{E - V_o})^2} = \frac{4\sqrt{E} \cdot \sqrt{E - V_o}}{(\sqrt{E} + \sqrt{E - V_o})^2}$$

ii)

$$R = \frac{v_1}{v_1} \frac{|B|^2}{|A|^2} = \frac{(C^*C/4) \cdot (k_1 - k_3)^2/k_1^2}{(C^*C/4) \cdot (k_1 + k_3)^2/k_1^2}$$

$$R = \left(\frac{\sqrt{E} - \sqrt{E - V_o}}{\sqrt{E} + \sqrt{E - V_o}}\right)^2$$

iii) Finally

$$T + R = \frac{1}{(\sqrt{E} + \sqrt{E - V_o})^2} [4\sqrt{E} \cdot \sqrt{E - V_o} + (\sqrt{E} - \sqrt{E - V_o})^2]$$

$$= \frac{1}{(\sqrt{E} + \sqrt{E - V_o})^2} [(\sqrt{E} + \sqrt{E - V_o})^2] = 1$$

4.5 Given a rectangular repulsive barrier of the following shape

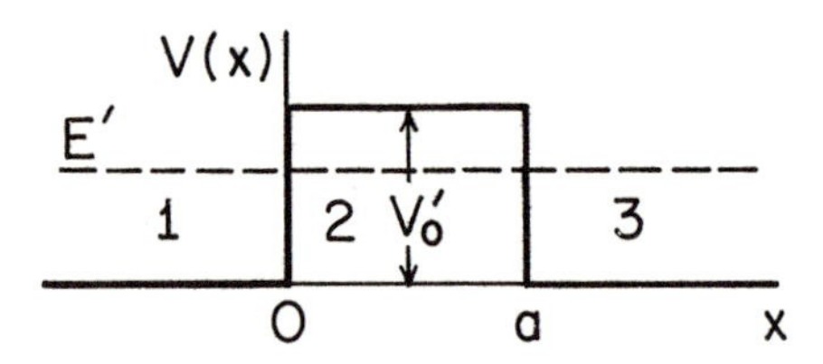

derive an equation for the probability of transmission as a function of the kinetic energy of a particle of mass m and energy E' incident from the left.

Solution: Ref. EI,231; SA,151; DU,63

Let us break the problem into three spatial regions (see figure) and into two energy cases: $E' < V_o'$ and $E' > V_o'$. First, we must solve the Schrödinger equation in each region. Using the reduced variables

$$E = \frac{2m}{\hbar^2} E' \quad \text{and} \quad V_o = \frac{2m}{\hbar^2} V_o'$$

we then have for $E' < V_o'$

region 1: $u'' = -Eu$; $\qquad u^{(1)} = Ae^{i\sqrt{E}x} + Be^{-i\sqrt{E}x}$

region 2: $u'' = -(E - V)u$; $\qquad u^{(2)} = Ce^{\sqrt{V_o - E}x} + De^{-\sqrt{V_o - E}x}$

region 3: $u'' = -Eu$; $\qquad u^{(3)} = Ge^{i\sqrt{E}x} + Fe^{-i\sqrt{E}x}$

For ease of handling these equations, we may define $k_1 = \sqrt{E}$ and $k_2 = \sqrt{V_o - E}$. Also F must equal zero since $Fe^{-i\sqrt{E}x}$ is a wave function representing a particle current in the -x direction; such a current would imply the existence of a scattering barrier in region 3. We may proceed with the solution by using the condition that the wave functions and their derivatives must be continuous at the boundaries (x = 0 and x = a). Thus

$$u^{(1)}(0) = u^{(2)}(0); \qquad A + B = C + D \tag{1}$$

$$u'^{(1)}(0) = u'^{(2)}(0); \qquad ik_1(A - B) = k_2(C - D) \tag{2}$$

$$u^{(2)}(a) = u^{(3)}(a); \qquad Ce^{k_2 a} + De^{-k_2 a} = Ge^{+ik_1 a} \tag{3}$$

$$u'^{(2)}(a) = u'^{(3)}(a); \qquad k_2(Ce^{k_2 a} - De^{-k_2 a}) = iGk_1e^{+ik_1 a} \tag{4}$$

The probability of transmission of the incident particle may be evaluated from

$$T = \frac{\text{transmitted current } (x > a)}{\text{incident current } (x < 0)} = \frac{v_3 G^* G}{v_1 A^* A} = \frac{G^* G}{A^* A}$$

since v_3 and v_1, the velocities in the respective regions, are equal. We may find C and D in terms of G from the solutions of 3 and 4. Letting $x = e^{k_2 a}$ and $y = e^{ik_1 a}$ we have

$$Cx + Dx^{-1} = Gy \quad (5) \qquad Cx - Dx^{-1} = \beta G y \quad (6)$$

where

$$\beta = ik_1/k_2$$

Now, (5) + (6) gives

$$C = \frac{G}{2}\frac{y}{x}(1 + \beta) = \frac{G}{2}\left(\frac{k_2 + ik_1}{k_2}\right)e^{(ik_1 - k_2)a} \quad (7)$$

and (5) - (6) gives

$$D = \frac{G}{2}yx(1 - \beta) = \frac{G}{2}\left(\frac{k_2 - ik_1}{k_2}\right)e^{(ik_1 + k_2)a} \quad (8)$$

Similarly for (1) and (2)

$$A + B = C + D, \qquad A - B = \gamma(C - D)$$

where

$$\gamma = k_2/ik_1$$

Therefore

$$A = \frac{1}{2}[(1 + \gamma)C + (1 - \gamma)D] = \frac{1}{2}\left[\left(\frac{k_1 - ik_2}{k_1}\right)C + \left(\frac{k_1 + ik_2}{k_1}\right)D\right]$$

and substituting (7) and (8) we have

$$A = \frac{G}{4ik_1k_2}[(k_2 + ik_1)^2 e^{(ik_1 - k_2)a} - (k_2 - ik_1)^2 e^{(ik_1 + k_2)a}]$$

and

$$T^{-1} = \frac{|A|^2}{|G|^2} = \frac{1}{16k_1^2k_2^2}[(k_2^2 + k_1^2)^2(e^{-2k_2a} + e^{+2k_2a}) - (k_2 - ik_1)^4 - (k_2 - ik_1)^4]$$

A useful trigonometric identity at this point is

$$\cosh(x) = (e^x + e^{-x})/2 = 2\sinh^2(\tfrac{1}{2}x) + 1$$

Thus

$$\frac{|A|^2}{|G|^2} = \frac{1}{16k_1^2k_2^2}[4(k_1^2 + k_2^2)^2\sinh^2(k_2a) + 2(k_2^4 + 2k_1^2k_2^2 + k_1^4) - 2k_2^4 + 12k_1^2k_2^2 - 2k_1^4]$$

and

$$\frac{AA^*}{GG^*} = \left[\frac{1}{4}\left(\frac{k_1^2 + k_2^2}{k_1k_2}\right)^2\sinh^2(k_2a) + 1\right]$$

Now for $E' < V_o'$

$$\left(\frac{k_1^2 + k_2^2}{k_1 k_2}\right)^2 = \left(\frac{E' + V_o' - E'}{\sqrt{E'} \cdot \sqrt{V_o' - E'}}\right)^2 = \frac{V_o'^2}{E'(V_o' - E')}$$

Finally

$$\boxed{T = \left[\frac{1}{4} \frac{V_o'^2}{E'(V_o' - E')} \sinh^2\left(\sqrt{2m(V_o' - E')}\ \frac{a}{\hbar}\right) + 1\right]^{-1}}$$

For $E' > V_o'$ we could rework the entire problem by changing the initial region 2 wave function to

$$u^{(2)} = Ce^{i\sqrt{E - V_o}x} + De^{-i\sqrt{E - V_o}x}$$

However, by noting that $\sinh(ix) = i\sin x$ we have

$$\sinh^2\left(\sqrt{2m(V_o' - E')}\ \frac{a}{\hbar}\right) = \sinh^2\left(i\sqrt{2m(E' - V_o')}\ \frac{a}{\hbar}\right) = -\sin^2\left(\sqrt{\frac{2m}{\hbar^2}\ (E' - V_o')}a\right)$$

so that for $E' > V_o'$ we have

$$\boxed{T = \left[\frac{1}{4} \frac{V_o'^2}{E'(E' - V_o')} \sin^2\left(\sqrt{2m(E' - V_o')}\ \frac{a}{\hbar}\right) + 1\right]^{-1}}$$

We note that when

$$\sqrt{2m(E' - V_o')}\frac{a}{\hbar} = n\pi$$

where n is an integer T will = 1. Between these integer values of π, T will fall to some value less than one, a totally nonclassical result. For example if we consider a proton impinging on a barrier of height 1 eV and width 1.5 Å we obtain the results shown in Figure 4.5 for T vs. E'/V_o'.

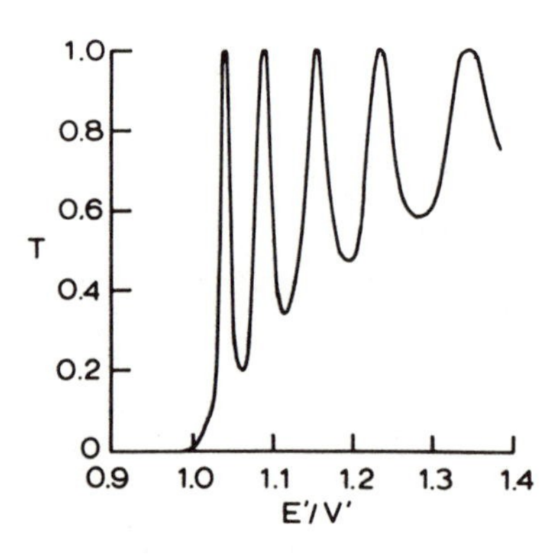

Fig. 4.5

4.6 Consider a particle of mass m confined to an infinitely deep potential well with $V = 0$ for $0 \lessgtr x \lessgtr a$ and $V = \infty$ otherwise as shown below.

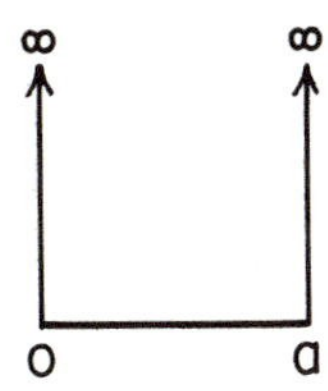

Determine the normalized wave functions and the energies for this system.

Solution: Ref. SA,147; EI,239; EWK,70; R. Seki, Am. J. Phys. 39, 929 (1971)

The Schrödinger equation for a particle of mass m is

$$-\frac{\hbar^2}{2m}\frac{d^2u}{dx^2} = E'u \quad \text{or} \quad u'' = -Eu$$

where

$$E = \frac{2m}{\hbar^2}E'$$

The solution of this equation is

$$u = A'e^{i\sqrt{E}x} + B'e^{-i\sqrt{E}x}$$

or using the relationships

$$e^{i\sqrt{E}x} = \cos(\sqrt{E}x) + i\sin(\sqrt{E}x) \quad \text{and} \quad e^{-i\sqrt{E}x} = \cos(\sqrt{E}x) - i\sin(\sqrt{E}x)$$

we can write

$$u = A\sin(\sqrt{E}x) + B\cos(\sqrt{E}x)$$

Since the potential is infinite outside the region $0 \lessgtr x \lessgtr a$, u must be zero outside the boundary. In the limit of infinite potential the wave function must approach zero at the boundary. The solution for finite potential walls is discussed in Problem 4.7. Using this limit at the boundaries we have:

$$u(0) = A\sin(\sqrt{E}\cdot 0) + B\cos(\sqrt{E}\cdot 0) = B\cdot 1 = 0$$

which implies $B = 0$. At the other boundary

$$u(a) = A\sin(\sqrt{E}a) = 0$$

which implies that

$$\sqrt{E}a = n\pi$$

Thus

$$E = \frac{n^2\pi^2}{a^2}\ ; \quad \boxed{E' = \frac{n^2h^2}{8ma^2}}$$

and

$$u(x) = A\sin\left(\frac{n\pi}{a}x\right)$$

Normalization gives

$$\int u^* u dx = \int_0^{2\pi} A^2 \sin^2\left(\frac{n\pi}{a}x\right) dx = 1; \quad A = \left[\int_0^{2\pi} \sin^2\left(\frac{n\pi}{a}x\right) dx\right]^{-1/2} = \sqrt{2/a}$$

and

$$\boxed{u(x) = \sqrt{\frac{2}{a}}\sin\left(\frac{n\pi}{a}x\right)}$$

4.7 A particle of mass m is confined to a box having finite potential walls as shown below:

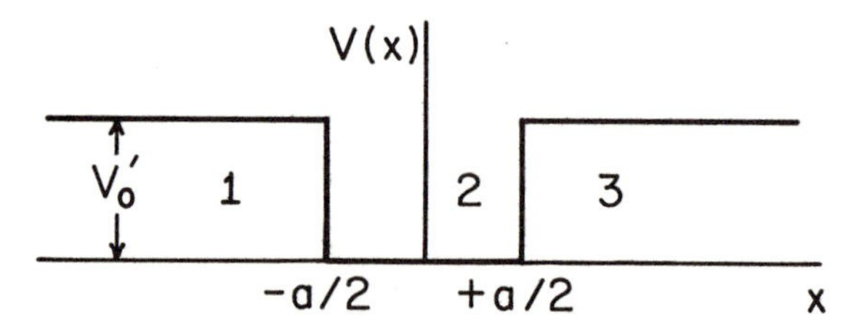

(a) Determine the energy levels for the bound states of this system, i.e. with $E' < V_o'$

(b) Derive the normalized ground state wave function for the special case in which $V_o' = 5h^2/8ma^2$

Solution: Ref. S1,65; S3,40; SC1,37; TH,1

(a) First set up the wave equations for the three regions having constant potentials.

region 1: $\dfrac{d^2u}{dx^2} = -(E - V_o)u;$ $\quad u^{(1)} = A\exp(\sqrt{V_o - E}x)$

region 2: $\dfrac{d^2u}{dx^2} = -Eu;$ $\quad u^{(2)} = B\cos(\sqrt{E}x - \alpha)$

region 3: $\dfrac{d^2u}{dx^2} = -(E - V_o)u;$ $\quad u^{(3)} = C\exp(-\sqrt{V_o - E}x)$

where $V_o = V_o'(2m/\hbar^2)$ and $E = E'(2m/\hbar^2)$. The solutions for regions 1 and 3 have been chosen so that u remains finite as $|x| \to \infty$, and in region 2 α is a phase factor to permit even and odd eigenfunctions.

To determine the energy eigenvalues it is sufficient to require that the logarithmic derivative $u^{-1}du/dx$ be continuous at $x = \pm a/2$. Therefore:

at $x = -a/2$: $\dfrac{1}{u}\dfrac{du}{dx} = \sqrt{V_o - E} = \sqrt{E}\tan(\sqrt{E}\,\frac{a}{2} + \alpha)$

at $x = +a/2$: $\frac{1}{u}\frac{du}{dx} = -\sqrt{V_o - E} = -\sqrt{E}\tan(\sqrt{E}\frac{a}{2} - \alpha)$

These conditions show that

$$\tan(\sqrt{E}\frac{a}{2} + \alpha) = \tan(\sqrt{E}\frac{a}{2} - \alpha)$$

which implies that $\alpha = 0, \pm\pi/2, \pm\pi, \ldots \pm n\pi/2$. The energy E must satisfy the condition:

$$\sqrt{V_o - E} = \sqrt{E}\tan\left(\sqrt{E}\frac{a}{2} + \frac{n\pi}{2}\right) \tag{1}$$

Various graphical methods can be used to solve Eq. (1).* For convenience we define: $k_1 = \sqrt{V_o - E}$ and $k_2 = \sqrt{E}$. Equation (1) takes the form:

$$k_1 = k_2\tan(k_2a/2) \quad \text{and} \quad k_1 = -k_2\cot(k_2a/2) \tag{2}$$

One method of solution is to recognize that $(ak_1)^2 + (ak_2)^2 = a^2V_o$, so that constant V_o is represented by a circle of radius $a\sqrt{V_o}$ in the ak_1, ak_2 plane. The intersection of the curves represented by Eqs. (1) and (2) with this circle give the allowed energies as shown below:

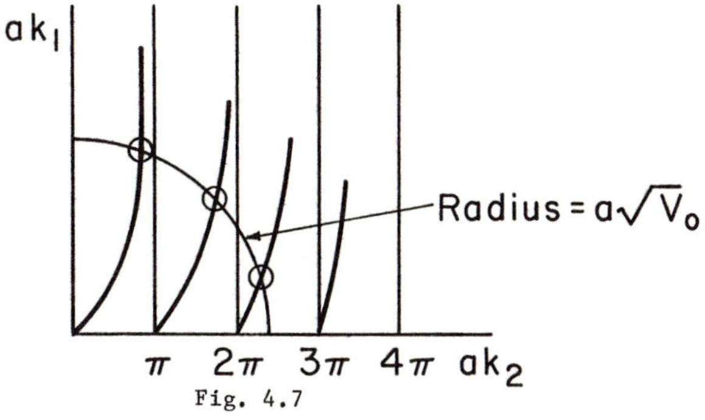

Fig. 4.7

Two conclusions can be drawn immediately from this figure. First, there will always be at least one intersection no matter how small V_o is, i.e. there will be at least one bound state. In the second place for intersections where $k_1 >> k_2$ it turns out that $ak_2 = n\pi$. At this limit the allowed energies are obtained from the equation:

$$a\sqrt{E} = n\pi \quad \text{or} \quad E' = \frac{\hbar^2}{2m}E = \frac{n^2h^2}{8ma^2}$$

As expected the solutions for a box with infinite potential walls are recovered.

(b) From Fig. 4.7 the lowest value of $\sqrt{E}$ is found to be $0.775\ \pi/a$ when $V_o = 5\pi^2/a^2$. For

---

* For another graphical method see:

C. D. Cantrell, Am. J. Phys. 39, 107 (1970).

W. C. Elmore, Am. J. Phys. 39, 976 (1971).

this energy level $\alpha = 0$, and the continuity condition gives:

$$\text{at } x = a/2: \quad C\exp(-\sqrt{V_o - E}\ a/2) = B\cos(\sqrt{E}\ a/2)$$
$$C\exp(-3.2946) = B\cos(1.217)$$
$$C/B = 9.334$$
$$u_o(x) = u_o^{(1)}(x) + u_o^{(2)}(x) + u_o^{(3)}(x)$$

Normalization requires that:

$$\int_{-\infty}^{+\infty} |u_o(x)|^2 dx = 1 = 2\int_0^{a/2} |u_o^{(2)}(x)|^2 dx + 2\int_{a/2}^{\infty} |u_o^{(3)}(x)|^2 dx$$

$$= N_o^2\{2\int_0^{a/2} \cos^2(2.434x/a)dx + 2\int_{a/2}^{\infty} (9.334)^2 e^{-13.179x/a}dx\} = N_o^2(0.6335 + 0.0045)a$$

or $\quad N_o = 1.252/\sqrt{a}$

The normalized ground state wave function is:

$$u_o^{(1)}(x) = (11.68/\sqrt{a})\exp(+6.589x/a) \qquad x < -a/2$$
$$u_o^{(2)}(x) = (1.252/\sqrt{a})\cos(+2.434x/a) \qquad -a/2 < x < +a/2$$
$$u_o^{(3)}(x) = (11.68/\sqrt{a})\exp(-6.589x/a) \qquad x > +a/2$$

4.8 A particle of mass m is confined to the double potential well shown below. The potential is zero for $-b-a/2 < x < -a/2$ and $a/2 < x < b+a/2$, and is equal to V_o' for $-a/2 \leq x \leq +a/2$. Otherwise it is infinite.

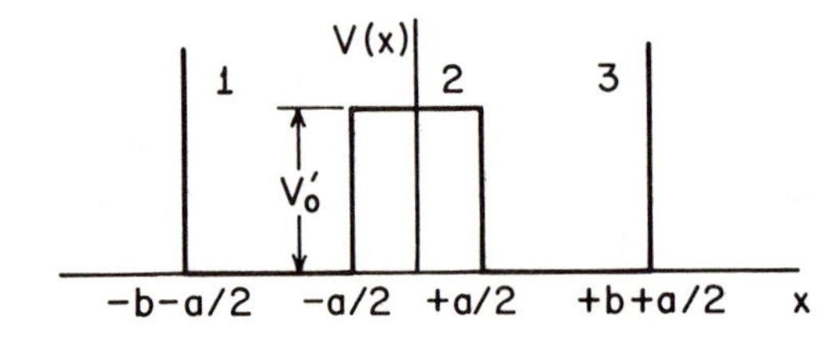

(a) Use the continuity condition for the logarithmic derivative of the wave function to obtain transcendental equations for the allowed energies of the bound states.

(b) Obtain expressions for the two lowest energies (E') in the high barrier limit, i.e., $\sqrt{2m(V_o' - E')}\ a >> \hbar$

Solution: Ref. TH,10

(a) For convenience we choose $\hbar^2/2m$ as the energy unit, i.e., $V_o = V_o'(2m/\hbar^2)$. The boundary conditions in the three regions of constant potential lead to the following solutions ($E < V_o$):

$-b-a/2 < x < -a/2;$ $u^{(1)} = D\sin(\sqrt{E}x + \alpha_1)$

$-a/2 \lessgtr x \lessgtr +a/2;$ $u^{(2)} = N\exp(\sqrt{V_o - E}x) + M\exp(-\sqrt{V_o - E}x)$

$+a/2 < x < b+a/2;$ $u^{(3)} = D\sin(\sqrt{E}x + \alpha_2)$

Since the potential function is symmetric about $x = 0$, N must equal $\pm M$.

First consider the even eigenfunctions where $N = +M$. The continuity of the logarithmic derivative of u at $x = \pm a/2$ leads to the following equations:

$$x = -a/2;\ -\sqrt{V_o - E}\tanh\left(\sqrt{V_o - E}\,\frac{a}{2}\right) = \sqrt{E}\cot\left(-\sqrt{E}\,\frac{a}{2} + \alpha_1\right)$$

$$x = +a/2;\ \sqrt{V_o - E}\tanh\left(\sqrt{V_o - E}\,\frac{a}{2}\right) = \sqrt{E}\cot\left(\sqrt{E}\,\frac{a}{2} + \alpha_2\right)$$

For the odd eigenfunctions where $N = -M$ the corresponding equations are:

$$x = -a/2;\ -\sqrt{V_o - E}\coth\left(\sqrt{V_o - E}\,\frac{a}{2}\right) = \sqrt{E}\cot\left(-\sqrt{E}\,\frac{a}{2} + \alpha_1\right)$$

$$x = +a/2;\ \sqrt{V_o - E}\coth\left(\sqrt{V_o - E}\,\frac{a}{2}\right) = \sqrt{E}\cot\left(\sqrt{E}\,\frac{a}{2} + \alpha_2\right)$$

The equations in each set are consistent if $\alpha_1 = -\alpha_2$. In addition the boundary conditions at $x = \pm(b + \frac{a}{2})$ require that the eigenfunctions vanish at these points which gives:

$$-\sqrt{E}\left(b + \frac{a}{2}\right) + \alpha_1 = n\pi; \qquad n = 0, 1, 2, \ldots$$

$$+\sqrt{E}\left(b + \frac{a}{2}\right) + \alpha_2 = m\pi; \qquad m = 0, 1, 2, \ldots$$

These conditions can then be combined to give

$$-\alpha_1 = \alpha_2 = -\sqrt{E}\left(b + \frac{a}{2}\right) + n\pi$$

The $n\pi$ term can be neglected since the cotangent is periodic in π and we conclude that:

$$\cot\left(-\sqrt{E}\,\frac{a}{2} + \alpha_1\right) = \cot(\sqrt{E}b); \quad \cot\left(\sqrt{E}\,\frac{a}{2} + \alpha_2\right) = -\cot(\sqrt{E}b)$$

The final transcendental equations turn out to be for the even eigenstates:

$$\sqrt{V_o - E}\tanh\left(\sqrt{V_o - E}\,\frac{a}{2}\right) = -\sqrt{E}\cot(\sqrt{E}b)$$

and for the odd eigenstates:

$$\sqrt{V_o - E}\coth\left(\sqrt{V_o - E}\,\frac{a}{2}\right) = -\sqrt{E}\cot(\sqrt{E}b)$$

Graphical solutions can be obtained for any desired value of V_o.

(b) To obtain the lowest eigenvalues in the high barrier limit we assume that $\sqrt{V_o - E}a \gg 1$. The following approximations are useful:

$$y \gg 1:\quad \tanh y = \frac{e^y - e^{-y}}{e^y + e^{-y}} = \frac{1 - e^{-2y}}{1 + e^{-2y}} \simeq (1 - e^{-2y})^2 \simeq 1 - 2e^{-2y}$$

and

$$\coth y \simeq 1 + 2e^{-2y}$$

For the even eigenstate where N = M we find:

$$\frac{1}{\sqrt{V_o - E}} \coth\left(\sqrt{V_o - E}\,\frac{a}{2}\right) = -\frac{1}{\sqrt{E}} \tan\left(\sqrt{E}b\right)$$

therefore

$$\tan(\sqrt{E}b) \simeq \frac{-\sqrt{E}}{\sqrt{V_o - E}} \left(1 + 2\exp[-\sqrt{V_o - E}a]\right)$$

Here we also assume that $\sqrt{E/(V_o - E)} << 1$ in the high barrier limit.

For the zeroth order solution we take the r.h.s. of the above equation to be zero. Then

$$\tan(\sqrt{E}b) = 0 \quad \text{or} \quad \sqrt{E}b = n\pi$$

The well-known solution for an isolated potential box emerges as expected:

$$E_n^{(o)} = \frac{n^2\pi^2}{b^2} \quad \text{or} \quad E_n^{(o)'} = \frac{n^2h^2}{8mb^2}$$

For the odd eigenstate with N = -M the solution differs only in the sign of the exponential term and thus the same zeroth order eigenvalues appear.

In the next order of approximation we use the fact that $\tan\alpha = \beta$ for small values of β implies that $\alpha \simeq n\pi + \beta$. It follows that:

$$\sqrt{E_n^{(1)}}\,b \cong n\pi - \frac{\sqrt{E_n^{(o)}}}{\sqrt{V_o - E_n^{(o)}}} \left(1 \pm 2\exp[-\sqrt{V_o - E_n^{(o)}}a]\right)$$

and

$$E_n^{(1)} \cong E_n^{(o)} - \frac{2E_n^{(o)}}{b\sqrt{V_o - E_n^{(o)}}} \left(1 \pm 2\exp[-\sqrt{V_o - E_n^{(o)}}a]\right)$$

where we have used the fact that $E_n^{(o)} = n^2\pi^2/b^2$ and have approximated the square of the r.h.s. by means of the equation $(1 - \varepsilon)^2 \simeq 1 - 2\varepsilon$ for $\varepsilon << 1$. In the equations above the upper signs go with the even eigenstates.

The presence of the neighboring well causes each of the energy levels of the isolated potential wells to split into two which have the separation:

$$\Delta E_n' = \frac{8E_n'^{(o)}}{b\sqrt{2m(V_o' - E_n'^{(o)})/\hbar^2}} \exp[-\sqrt{2m(V_o' - E_n'^{(o)})}\,a/\hbar]$$

The lowest eigenvalues of course occur when n = 1. In this case the form of the resulting eigenfunctions are shown below:

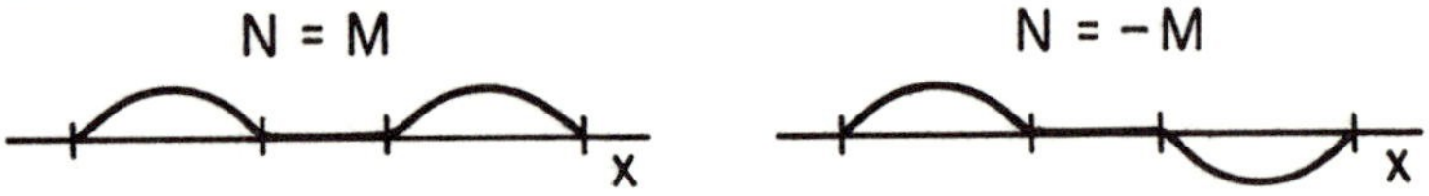

N = M

X

N = -M

X

The quantity $\nu_n = \Delta E_n/h$ is called the tunneling frequency for the nth level. The reason for this terminology will become clear when we discuss time dependent solutions for barrier problems in Problem 4.28.

4.9 A particle of mass m is acted on by a restoring force proportional to its displacement from the equilibrium position, i.e. $f = -k(x' - x'_e)$. Solve the classical equations of motion to show that $E = \frac{1}{2} kA^2$ where A is the amplitude of the vibration. Derive also an expression for the frequency of vibration.

Solution:

The restoring force f is related to a potential function V(x) by

$$f = -\frac{\partial V}{\partial x} = -kx$$

where

$$x = x' - x'_e = \text{displacement}$$

Thus

$$V = \int_0^x kx dx = kx^2/2 \quad \text{(Hooke's Law potential)}$$

and

$$E = \text{total energy} = \frac{1}{2} mv^2 + \frac{kx^2}{2} = \text{constant} \tag{1}$$

At the turning point of the vibration

$$x = A \quad \text{and} \quad v = 0$$

Thus

$$\boxed{E = \frac{1}{2} kA^2} \tag{2}$$

Classically A, the amplitude, can have any positive value and E may have any positive value. Thus from (1) and (2)

$$v^2 = \frac{k}{m}(A^2 - x^2) = \left(\frac{dx}{dt}\right)^2$$

so

$$\frac{dx}{dt} = \sqrt{\frac{k}{m}(A^2 - x^2)}$$

and

$$\int_{x_1}^{x_2} \frac{dx}{\sqrt{(A^2 - x^2)}} = \int_{\alpha}^{t+\alpha} \left(\frac{k}{m}\right)^{1/2} dt$$

This integral gives

$$\sin^{-1}\left(\frac{x_2}{A}\right) - \sin^{-1}\left(\frac{x_1}{A}\right) = \left(\frac{k}{m}\right)^{1/2} t$$

or

$$\sin^{-1}\left(\frac{x_2}{A}\right) = \left(\frac{k}{m}\right)^{1/2} t + \delta$$

Thus

$$x = A\sin\left[\left(\frac{k}{m}\right)^{1/2} t + \delta\right] = A\sin(2\pi\nu t + \delta)$$

where

$$\boxed{\nu = \frac{1}{2\pi}\sqrt{\frac{k}{m}}}$$

Another way to obtain the frequency of the oscillator is to recognize that the basic equation, $f = m\ddot{x} = -kx$, has the solution $x = A\sin(\omega t + \delta)$. Substitution of x into this equation immediately gives $\omega = \sqrt{k/m}$.

4.10 A particle of mass m is acted on by a linear restoring force so that the potential energy is given by $V = \frac{1}{2} kx^2$ where k is constant and x is the displacement from equilibrium. (a) The classical solution of this problem (see Problem 4.9) gives $\frac{1}{2} k = 2\pi^2 m\nu^2$. Use this result and the definitions $\alpha \equiv 2\pi m\nu/\hbar$, $\beta \equiv 2\pi E/\hbar^2$ and $\eta \equiv \sqrt{\alpha}x$ to obtain the Schrödinger equation in the form:

$$\left[\frac{d^2}{d\eta^2} + \left(\frac{\beta}{\alpha} - \eta^2\right)\right] u(\eta) = 0$$

(b) Find the asymptotic form of the solution to the equation in part (a) and justify the equation

$$u(\eta) = H(\eta)e^{-\eta^2/2} \quad \text{(for all } \eta\text{, } H(\eta) \text{ well-behaved)}$$

Finally show that $H(\eta)$ obeys the equation

$$H'' - 2\eta H' + (\beta/\alpha - 1)H = 0$$

(c) With the assistance of Appendix 3 on Hermite polynomials, derive the energies and unnormalized wave functions.

Solution: Ref. EWK,75; EI,256; SC1,66

Although this problem is solved in almost all elementary quantum mechanics texts, it is included in this collection because of its considerable pedagogical value.

(a) The quantum mechanical Hamiltonian is derived from the classical total energy

$$E = H = \frac{p^2}{2m} + \frac{kx^2}{2}$$

With the substitution $p = -i\hbar\partial/\partial x$, the quantum mechanical Hamiltonian becomes

$$\hat{\mathcal{H}} = -\frac{\hbar^2}{2m}\frac{d^2}{dx^2} + \frac{1}{2}kx^2$$

The Schrödinger equation is

$$\left(-\frac{\hbar^2}{2m}\frac{d^2}{dx^2} + \frac{1}{2}kx^2\right)u(x) = Eu(x) \quad \text{or} \quad \left(-\frac{\hbar^2}{2m}\frac{d^2}{dx^2} + 2\pi^2 m\nu^2 x^2\right)u(x) = Eu(x)$$

We may place this equation in a more convenient form by defining

$$\alpha \equiv 2\pi m\nu/\hbar, \qquad \beta \equiv 2mE/\hbar^2$$

so that

$$\left\{\frac{d^2}{dx^2} + \left[\frac{2mE}{\hbar^2} - \left(\frac{2\pi m\nu}{\hbar}\right)^2 x^2\right]\right\}u(x) = 0$$

$$\left\{\frac{d^2}{dx^2} + [\beta - \alpha^2x^2]\right\}u(x) = 0$$

A change of variables $\eta = \sqrt{\alpha}x$ further simplifies the equation. By the chain rule

$$\frac{du}{dx} = \frac{du}{d\eta}\cdot\frac{d\eta}{dx} = \sqrt{\alpha}\,\frac{du}{d\eta} \quad \text{and} \quad \frac{d^2u}{dx^2} = \sqrt{\alpha}\,\frac{d(du/d\eta)}{d\eta}\,\frac{d\eta}{dx} = \alpha\,\frac{d^2u}{d\eta^2}$$

thus

$$\left[\frac{d^2}{d\eta^2} + \left(\frac{\beta}{\alpha} - \eta^2\right)\right]u(\eta) = 0 \tag{1}$$

(b) The solutions $u(\eta)$ must be continuous and must be finite (to allow normalization). The asymptotic form of the solution may be studied by examining the solution of (1) as $|\eta| \to \infty$. Then β/α will be small compared to η^2 so

$$u''(\eta) = +\eta^2 u(\eta) \qquad |\eta| \to \infty \tag{2}$$

The general solution is

$$u(\eta) = Ae^{\eta^2/2} + Be^{-\eta^2/2} \tag{3}$$

(The reader should check this by substituting (3) into (2) and using the limit $|\eta| \to \infty$.) Clearly $A = 0$ since $u(\eta)$ must remain finite for all values of η. Thus

$$u(\eta) = Be^{-\eta^2/2} \qquad |\eta| \to \infty$$

We now try a solution of (1) of the form

$$u(\eta) = H(\eta)e^{-\eta^2/2} \tag{4}$$

The $H(\eta)$ must vary with η so that as $|\eta| \to \infty$, $e^{-\eta^2/2}$ dominates, i.e. so that $u(\eta) \to 0$ as $\eta \to \infty$.

We now find an equation for the $H(\eta)$ functions by substituting Eq. (4) into Eq. (1). Differentiating $u(\eta)$ we obtain:

$$u' = H'e^{-\eta^2/2} - \eta He^{-\eta^2/2}$$

and

$$u'' = H''e^{-\eta^2/2} - \eta H'e^{-\eta^2/2} - [\eta(H'e^{-\eta^2/2} - \eta He^{-\eta^2/2})] - He^{-\eta^2/2}$$

$$u'' = [-H + \eta^2 H - 2\eta H' + H'']e^{-\eta^2/2} \tag{5}$$

Then substituting Eqs. (4) and (5) into Eq. (1) we have

$$[-H + \eta^2 H - 2\eta H' + H'' + \beta/\alpha H - \eta^2 H]e^{-\eta^2/2} = 0$$

or

$$\boxed{H'' - 2\eta H' + (\beta/\alpha - 1)H = 0} \tag{6}$$

This is the much studied Hermite differential equation.

(c) Well-behaved solutions for Eq. 6 exist only for values of β/α which satisfy the condition $\beta/\alpha = 2n + 1$, $n = 0, 1, 2...$ (see Appendix 3). Thus a family of solutions denoted by $H(\eta)$ are obtained and the solutions are given by

$$\boxed{u_n(\eta) = N_n H_n(\eta)e^{-\eta^2/2}}$$

The normalization which is performed in Problem 4.12 gives $\int_{-\infty}^{+\infty} u^*(x)u(x)dx = 1$ ($\eta = \sqrt{\alpha}\, x$) and

$$N_n = \left(\frac{\sqrt{\alpha}}{\sqrt{\pi}\, 2^n n!}\right)^{1/2}$$

The first few values of $H_n(\eta)$ are

$$H_o = 1 \qquad H_3 = 12\eta - 8\eta^3$$

$$H_1 = 2\eta \qquad H_4 = 12 - 48\eta^2 + 16\eta^4$$

$$H_2 = 2 - 4\eta^2 \qquad H_5 = 120\eta - 160\eta^3 + 32\eta^5$$

The energy may be found from Eq. (7) as follows

$$\beta/\alpha = \frac{2mE}{\hbar^2} \cdot \frac{\hbar}{2\pi m\nu} = \frac{2E}{h\nu} = 2n + 1$$

or

$$\boxed{E = (n + 1/2)h\nu} \qquad n = 0, 1, 2, 3, 4$$

where

$$\nu = \frac{1}{2\pi}\sqrt{\frac{k}{m}}$$

The value of E for n = 0 gives

$$E = h\nu/2$$

which is the zero point energy.

The harmonic oscillator wave functions serve as good approximations to many problems involving potential minima since the potential energy term

$$V = \frac{1}{2}\, k(x' - x_e')^2$$

is the first non-zero term in the expansion of a function with a minimum [i.e.

$$V = V_o + \left(\frac{\partial V}{\partial x'}\right)_{x'=x'_e}(x' - x'_e) + \frac{1}{2}\left(\frac{\partial^2 V}{\partial x'^2}\right)_{x'=x'_e}(x' - x'_e)^2 + \ldots$$

$$= V_o + \frac{1}{2}\left(\frac{\partial^2 V}{\partial x'^2}\right)_{x'=x'_e}(x' - x'_e)^2 + \ldots$$

and

$$\left(\frac{\partial V}{\partial x'}\right)_{x'=x'_e} = 0 \ .]$$

4.11 Plot the probability density of a one dimensional harmonic oscillator versus the displacement for the state $v = 12$. Compare this result to the classical probability density and discuss the limit $v \to \infty$. For convenience let

$$\alpha = \frac{\sqrt{mk}}{\hbar} = 1$$

to complete the definition of the energy.

Solution: Ref. Problem 4.10; DU,122

The wave function is $u_v = N_v H_v(\eta)e^{-\eta^2/2}$ where

$$N_{12} = \left(\frac{1}{\pi^{1/2}2^{12}\cdot 12!}\right)^{1/2}$$

and since

$$\frac{\sqrt{mk}}{\hbar} = 1,$$

$x = \eta$. Thus the probability density is

$$u_v^* u_v = \left(\frac{1}{\pi^{1/2}2^{12}12!}\right) H_{12}{}^2(x)e^{-x^2}$$

where

$$H_{12}{}^2(x) = [665280 - 7983360\ x^2 + 13305600\ x^4 - 7096320\ x^6 + 1520640\ x^8 - 135168\ x^{10} + 4096\ x^{12}]^2$$

The plot of $u_v^* u_v$ is given in Fig. 4.11. Classically we may define the probability density as the fraction of a period that the oscillator spends in a small increment Δx. Then using the results of Problem 4.9

$$P(x)\Delta x \propto \frac{\Delta x}{v} = \frac{\Delta x}{\sqrt{2/m}\ (E - \frac{1}{2}kx^2)^{1/2}} = \frac{\Delta x}{\sqrt{k/m}\ (A^2 - x^2)^{1/2}}$$

Thus

$$P(x)\Delta x = N\frac{\Delta x}{\sqrt{k/m}\ (A^2 - x^2)^{1/2}}$$

since

$$\int_{-A}^{+A} P(x)dx = 1 = N\int_{-A}^{+A} \frac{dx}{\sqrt{k/m}\ (A^2 - x^2)^{1/2}}$$

we find

$$N = \frac{1}{\pi}\sqrt{k/m}$$

The classical probability density corresponding to u^*u is then given by

$$P(x) = \frac{1}{\pi\sqrt{A^2 - x^2}}$$

Since v = 12, we have

$$E = h\nu(v + 1/2) = \frac{25}{2}h\nu = \frac{25}{2}\cdot\frac{h}{2\pi}\cdot\sqrt{k/m} = \frac{k}{2}A^2$$

so that

$$A^2 = 25\hbar/\sqrt{km}$$

We are given $\sqrt{km}/\hbar = 1$ which fixes A to be 5. Thus

$$P(x) = \frac{1}{\pi\sqrt{25 - x^2}}$$

This function is plotted in Fig. 4.11 (dashed curve). It is clear that the "average" quantum mechanical result is close to that of the classical result. As $v \to \infty$ the quantum mechanical probability density will accumulate at the turning points as expected from the classical result. This, of course, is anticipated by the Bohr correspondence principle.

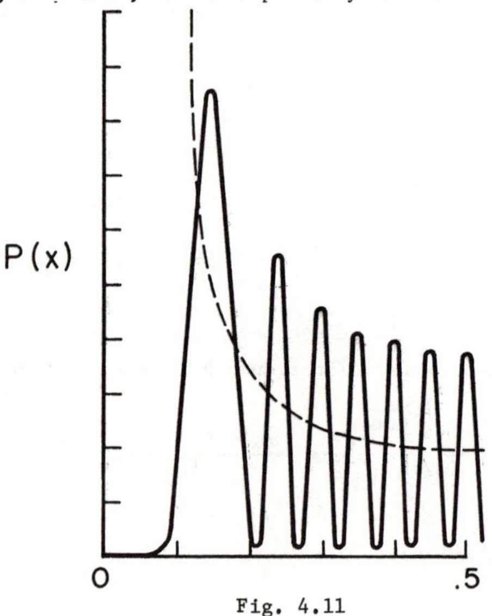

Fig. 4.11

4.12 The Hermite polynomials have several definitions. One is

$$H_n(\eta) = (-1)^n e^{\eta^2} \frac{d^n}{d\eta^n} e^{-\eta^2} \tag{1}$$

Another is

$$S(\eta, s) \equiv e^{+\eta^2-(s-\eta)^2} \equiv \sum_{n=0}^{\infty} \frac{H_n(\eta)}{n!} s^n \tag{2}$$

i.e. the Hermite polynomials are given as expansion coefficients of the exponential function.

(a) Prove that the Hermite polynomials are orthogonal and find the normalization constant of the harmonic oscillator wave function $u_n(x)$ by use of Eq. (2). Here

$$u_n(x) = N_n e^{-\eta^2/2} H_n(\eta) \qquad \eta = \sqrt{\alpha}\, x \tag{3}$$

(b) Evaluate the following integral in terms of α and n.

$$x_{nm} = \int_{-\infty}^{+\infty} u_n^*(x) x u_m(x)\, dx = \langle u_n | x | u_m \rangle$$

This is known as the nm matrix element of x for the harmonic oscillator. (Hint: In each case consider the evaluation of the integral

$$\int_{-\infty}^{+\infty} S(\eta,s) S(\eta,t) e^{-\eta^2} d\eta$$

by using both expressions for S given in Eq. (2).)

Solution: Ref. PW,77

(a)

$$1 = \langle u_n(x) | u_m(x) \rangle = \int_{-\infty}^{+\infty} u_n(x) u_m(x)\, dx = \frac{N_n N_m}{\sqrt{\alpha}} \int_{-\infty}^{+\infty} \underline{H_n(\eta) H_m(\eta) e^{-\eta^2} d\eta} \tag{4}$$

To obtain a function such as that underlined in the above equation we consider the following integral

$$\int_{-\infty}^{+\infty} \left(\sum_{n=0}^{\infty} \frac{H_n}{n!} s^n \right) \left(\sum_{m=0}^{\infty} \frac{H_m}{m!} t^m \right) e^{-\eta^2} d\eta \tag{5}$$

By use of Eq. (2), Eq. (5) is equal to

$$\int_{-\infty}^{+\infty} (e^{+\eta^2 - (s - \eta)^2})(e^{+\eta^2 - (t - \eta)^2}) e^{-\eta^2} d\eta = \int_{-\infty}^{+\infty} e^{-t^2 - s^2 + 2s\eta + 2t\eta - \eta^2} d\eta = I$$

This integral can be evaluated by completing the square since

$$(\eta - s - t)^2 = \eta^2 + s^2 + t^2 - 2\eta s - 2\eta t + 2st$$

thus

$$I = e^{2st} \int_{-\infty}^{+\infty} e^{-(\eta - s - t)^2} d(\eta - s - t) \qquad (\text{since } d\eta = d(\eta - s - t))$$

Now,

$$\int_{-\infty}^{+\infty} e^{-x^2} dx = \sqrt{\pi}$$

so

$$I = \sqrt{\pi}\, e^{2st} = \sqrt{\pi}\left(1 + \frac{2st}{1} + \frac{4s^2t^2}{2!} + \frac{8s^3t^3}{3!} + \frac{16s^4t^4}{4!} + \ldots + \frac{2^n s^n t^n}{n!} \ldots\right) \tag{6}$$

I is also equal to Eq. (5) which we rewrite after interchanging summations and integration

$$(5) = \sum_n \sum_m s^n t^m \int_{-\infty}^{+\infty} \frac{H_n H_m}{n!m!} e^{-\eta^2} d\eta = I \tag{7}$$

From Eq. (6) we see that all terms in the expansion have $n = m$. Since Eq. (5) = I this means that the coefficients of terms such as $s^n t^m$, $n \neq m$, must vanish, or

$$\int_{-\infty}^{+\infty} H_n H_m e^{-\eta^2} d\eta = 0, \quad n \neq m$$

This proves that the Hermite polynomials are orthogonal. To obtain the normalization result we note that by the use of Eqs. (4), (6) and (7)

$$\frac{\int_{-\infty}^{+\infty} H_n H_n e^{-\eta^2} d\eta}{(n!)^2} = \frac{2^n}{n!}\sqrt{\pi}$$

so

$$\frac{N_n^{\,2}}{\sqrt{\alpha}} \int_{-\infty}^{+\infty} H_n H_n e^{-\eta^2} d\eta = 1 = \frac{N_n^{\,2}}{\sqrt{\alpha}}\, 2^n n! \sqrt{\pi}$$

thus

$$\boxed{N_n = \left(\frac{\sqrt{\alpha/\pi}}{2^n n!}\right)^{1/2}} \tag{8}$$

(b)

$$x_{nm} = \langle u_n(x)|x|u_m(x)\rangle = \int_{-\infty}^{+\infty} u_n x u_m dx = \frac{N_n N_m}{\alpha} \int_{-\infty}^{+\infty} \underline{H_n H_m e^{-\eta^2} \eta d\eta} \quad (\text{using } \eta = \sqrt{\alpha}\, x) \tag{9}$$

Following the procedure used in part (a) we have to consider

$$\int_{-\infty}^{+\infty} \sum_n \left(\frac{H_n}{n!} s_n\right) \sum_m \left(\frac{H_m}{m!} t^m\right) \eta e^{-\eta^2} d\eta \tag{10}$$

$$= \sum_n \sum_m s^n t^m \int_{-\infty}^{+\infty} \frac{H_n H_m}{n!m!} \eta e^{-\eta^2} d\eta = I \tag{11}$$

and by use of the middle term in Eq. (2)

$$I = \int_{-\infty}^{+\infty} e^{+\eta^2 - (s-\eta)^2} e^{+\eta^2 - (t-\eta)^2} e^{-\eta^2} \eta d\eta = e^{2st} \int_{-\infty}^{+\infty} e^{-(\eta - s - t)^2} \eta d(\eta - s - t)$$

$$= e^{2st} \int_{-\infty}^{+\infty} e^{-(\eta - s - t)^2} [(\eta - s - t) + (s + t)] d(\eta - s - t)$$

$$= e^{2st} \left[\int_{-\infty}^{+\infty} e^{-x^2} x dx + (s + t) \int_{-\infty}^{+\infty} e^{-x^2} dx \right] \tag{12}$$

The first integral on the right vanishes and the second is equal to $\sqrt{\pi}$. Thus

$$I = \sqrt{\pi}\,(s + t) \sum_{n=0} \frac{2^n (st)^n}{n!} = \sqrt{\pi} \sum_{n=0} \frac{2^n}{n!} (s^n t^{n+1} + s^{n+1} t^n) \tag{13}$$

Comparing coefficients of Eqs. (13) and (11), we see that only if $m = n \pm 1$ is there a finite value for

$$\int_{-\infty}^{+\infty} H_n H_m \eta e^{-\eta^2} d\eta$$

Then

$$\langle u_n | x | u_{n+1} \rangle = \frac{N_n N_{n+1}}{\alpha} \sqrt{\pi}\, 2^n (n + 1)! = \left(\frac{\sqrt{\alpha/\pi}}{2^n n!} \right)^{1/2} \left(\frac{\sqrt{\alpha/\pi}}{2^{n+1} (n+1)!} \right)^{1/2} \frac{\sqrt{\pi}\, 2^n (n+1)!}{\alpha}$$

$$\boxed{x_{n,n+1} = \sqrt{\frac{n+1}{2\alpha}}}$$

and

$$\langle u_n | x | u_{n-1} \rangle = \frac{N_n N_{n-1}}{\alpha} \sqrt{\pi}\, 2^{n-1} (n - 1)!$$

or

$$\boxed{x_{n,n-1} = \sqrt{\frac{n}{2\alpha}}}$$

4.13 An electron is trapped in a well defined by the potential energy function $V(x) = -e^2/[(4\pi\varepsilon_o)x]$ for $x > 0$; $V(x) = \infty$ for $x \leq 0$.

(a) Write the Hamiltonian for this system in terms of the variable z instead of x by using the definitions:

$$x = \frac{1}{2} \alpha a_o z \text{ where } \alpha^{-2} = -(2m_e E a_o^2)/\hbar^2 \quad \text{and } a_o = \frac{\hbar^2 (4\pi\varepsilon_o)}{m_e e^2}$$

(b) Show that substitution of $u(z) = e^{-z/2}t(z)$, where $t(z)$ is an appropriately bounded function, into the transformed Schrödinger equation gives a differential equation which can be solved by the standard power series method i.e. that the solution has the form

$$t(z) = \sum_{n=0} c_n z^{n+\gamma}$$

(c) Determine the unnormalized wave function and energy.

Solution: Ref. L. K. Haines and D. H. Roberts, Am. J. Phys. 37, 1145 (1969).

(a) The Schrödinger equation is

$$\hat{\mathcal{H}}u = \left(-\frac{\hbar^2}{2m}\frac{d^2u}{dx^2} - \frac{e^2}{(4\pi\varepsilon_o)x}\right)u = Eu, \ x > 0 \tag{1}$$

The suggested transformation gives in a straightforward manner

$$\frac{d^2u}{dz^2} - \frac{1}{4}u + \frac{\alpha}{z}u = 0 \tag{2}$$

This differential equation may be identified as Kummer's equation (NBS Applied Math Series, No. 55, 1964, p. 504) which has as its general solution the confluent hypergeometric Whittaker Functions (see Whittaker and Watson, A Course in Modern Analysis, Cambridge University Press, 1969, p. 337).

(b) The bounded asymptotic solution of Eq. (2) is clearly

$$u(z) = e^{-z/2}, \ z \to \infty$$

which suggests the substitution into Eq. (2) of

$$u(z) = e^{-z/2}t(z)$$

This gives

$$\frac{d^2t}{dz^2} - \frac{dt}{dz} + \frac{\alpha}{z}t = 0 \tag{3}$$

As in the harmonic oscillator case, let us make the substitution

$$t(z) = \sum_{n=0}^{\infty} c_n z^{n+\gamma} \tag{4}$$

which gives

$$t'' - t' + \alpha/zt = c_o\gamma(\gamma - 1)z^{\gamma-2} + c_1(\gamma + 1)\gamma z^{\gamma-1} + c_2(\gamma + 2)(\gamma + 1)z^{\gamma}\ldots\ldots - c_o\gamma z^{\gamma-1}$$

$$- c_1(\gamma + 1)z^{\gamma} - c_2(\gamma + 2)z^{\gamma+1}\ldots\ldots + \alpha c_o z^{\gamma-1} + \alpha c_1 z^{\gamma} + \alpha c_2 z^{\gamma+1}\ldots = 0 \tag{5}$$

Collecting coefficients of powers of z we have

$$\begin{aligned}
z^{\gamma-2}&: \quad c_o\gamma(\gamma - 1) = 0 \\
z^{\gamma-1}&: \quad c_1(\gamma + 1)\gamma - c_o\gamma + c_o\alpha = 0 \\
z^{\gamma}&: \quad c_2(\gamma + 2)(\gamma + 1) - c_1(\gamma + 1) + \alpha c_1 = 0 \\
z^{\gamma+1}&: \quad c_3(\gamma + 2)(\gamma + 3) - c_2(\gamma + 2) + \alpha c_2 = 0
\end{aligned} \tag{6}$$

$$c_{n+1}(n + \gamma)(n + \gamma + 1) + c_n(\alpha - n - \gamma) = 0$$

thus

$$c_{n+1} = -c_n \frac{\alpha - n - \gamma}{(n + \gamma)(n + \gamma + 1)}$$

We may choose either $\gamma = 1$ or $\gamma = 0$ [both lead to the same result]. If $\gamma = 1$ then

$$c_{n+1} = -c_n \frac{\alpha - n - 1}{(n + 1)(n + 2)} = c_n \frac{n + 1 - \alpha}{(n + 1)(n + 2)} \tag{7}$$

if

$$n = 0 \quad c_1 = c_o \frac{1 - \alpha}{1 \cdot 2}$$

$$n = 1 \quad c_2 = c_1 \frac{2 - \alpha}{2 \cdot 3} = c_o \frac{(1 - \alpha)(2 - \alpha)}{2 \cdot 2 \cdot 3}$$

$$n = 2, \quad c_3 = c_2 \frac{3 - \alpha}{3 \cdot 4} = c_o \frac{(1 - \alpha)(2 - \alpha)(3 - \alpha)}{2 \cdot 2 \cdot 3 \cdot 3 \cdot 4}$$

so

$$c_{n+1} = c_o \frac{(1 - \alpha)_n}{n!(n + 1)!} \tag{8}$$

where

$$(1 - \alpha)_n = (1 - \alpha)(2 - \alpha) \ldots\ldots (n - \alpha) \text{ and } (1 - \alpha)_o \equiv 1$$

At large z the ratio of two consecutive terms in the expansion of

$$\sum_{n=0}^{\infty} \frac{(1 - \alpha)_n}{n!(n + 1)!} z^{\gamma+1} \text{ is } \frac{z}{n} \tag{9}$$

The expansion of e^z has exactly the same ratio for consecutive terms; therefore, t has the asymptotic behavior

$$t(z) \sim e^{-z/2} \times e^{+z} = e^{+z/2}$$

which goes to infinity as $z \to \infty$. This means that to have a wave function which is square integrable, the series Eq. (9) must terminate. It can do this only if $(1 - \alpha)_n$ vanishes for some n; i.e. when $\alpha = N$ where N is a positive integer.

(c) The wavefunction is then

$$u_N(z) = b_o e^{-z/2} \sum_{n=0}^{N-1} \frac{(1 - N)_n}{n!(n + 1)!} z^{n+1} \tag{10}$$

$$= b_o e^{-\frac{x}{Na_o}} \sum_{n=0}^{N-1} \frac{(1 - N)_n}{n!(n + 1)!} \left(\frac{2x}{Na_o}\right)^{n+1}$$

where b_o is a normalization constant. The energy is

$$E_N = -\frac{\hbar^2}{2m_e a_o{}^2 \alpha^2} = -\frac{m_e e^4}{2\hbar^2 (4\pi\varepsilon_o)^2 N^2} \qquad N = 1, 2, 3, \ldots\ldots\infty$$

The energy is identical to that found in the 3-dimensional H atom.

Several interesting references have extended this problem to consider the potential $V(x) = -e^2/(4\pi\varepsilon_o)|x|$. This is the one-dimensional hydrogen atom problem. The problem is then considerably more difficult because of the pole at the origin. The interested reader should consult:

R. Loudon, Am. J. Phys. 27, 649 (1959).

M. Andrews, Am. J. Phys. 34, 1194 (1966).

4.14 A particle of mass m is bouncing elastically on a smooth, flat surface in the earth's gravitational field. Determine the energy levels and wave functions for this system. (Hint: Solution of the Schrödinger equation with the independent variable z is considerably facilitated by transforming to the variable η which is defined by the equations

$$c\eta = \frac{2m^2g}{\hbar^2} z - \frac{2mE}{\hbar^2} = az - b \quad \text{where c is a constant.)}$$

Solution: Ref. TH,5; P. W. Langhoff, Am. J. Phys. 39, 954 (1971).

The potential experienced by the particle is

$$V(z) = mgz \quad z > 0, \qquad V(z) = \infty \quad z \lesseqgtr 0 \tag{1}$$

where z is the vertical distance perpendicular to the flat plane and g is the acceleration of gravity.

The boundary conditions on the wave function are

$$u(z) \to 0 \text{ as } z \to \infty, \qquad u(z) = 0 \text{ at } z = 0 \tag{2}$$

The Schrödinger equation for this system is

$$-\frac{\hbar^2}{2m}\frac{d^2u}{dz^2} + mgzu = Eu \tag{3}$$

or

$$-\frac{d^2u}{dz^2} + \left(\frac{2m^2gz}{\hbar^2} - \frac{2mE}{\hbar^2}\right)u = 0 \tag{4}$$

This form suggests the substitution

$$\frac{2m^2g}{\hbar^2} z - \frac{2mE}{\hbar^2} = c\eta = az - b \tag{5}$$

where c is a constant. Now

$$\frac{du}{dz} = \frac{du}{d\eta}\cdot\frac{d\eta}{dz} = \frac{du}{d\eta}\cdot\frac{a}{c} \quad \text{and} \quad \frac{d^2u}{dz^2} = \frac{d^2u}{d\eta^2}\cdot\frac{a^2}{c^2} \tag{6}$$

thus Eq. (5) becomes

$$-\frac{a^2}{c^2}\frac{d^2u}{d\eta^2} + c\eta u = 0 \quad \text{or} \quad u'' - (c^3/a^2)\eta u = 0$$

If we take $c^3 = a^2$ then the differential equation that we must solve takes on the very simple form

$$u'' - \eta u = 0$$

The solution of this equation is an Airy Function [see Handbook of Mathematical functions, NBS, No. 55, 1964, p. 446-447, Eqs. 10.4.1 and 10.4.32]. The Airy function, which is actually closely related to a Bessel function of fractional order, can be expressed in integral form as

$$\mathrm{Ai}[\eta] = \frac{1}{\pi}\int_0^\infty \cos\left(\frac{1}{3}t^3 + \eta t\right)dt$$

so

$$u(\eta) = B\,\mathrm{Ai}(\eta)$$

where B is a normalization constant. At the boundary where z = 0

$$u(\eta) = u(az - b)\Big|_{z=0} = 0$$

or

$$u\left[-\left(\frac{2}{\hbar^2 g^2 m}\right)^{1/3} E\right] = 0$$

Thus the eigenvalues E are proportional to the roots of the Airy function, <u>i.e.</u>

$$\mathrm{Ai}\left[-\left(\frac{2}{\hbar^2 g^2 m}\right)^{1/3} E\right] = 0$$

$$-\left(\frac{2}{\hbar^2 g^2 m}\right)^{1/3} E_n = (\mathrm{Root})_n = -R_n$$

where n = 1, 2, 3 ..... (root label) and the minus sign is chosen for convenience. Thus

$$\boxed{E_n = \left(\frac{\hbar^2 g^2 m}{2}\right)^{1/3} R_n}$$

and

$$u_n(\eta) = B_n \mathrm{Ai}\left[\left(\frac{2m^2 g^2}{\hbar^2}\right)^{1/3} z - R_n\right]$$

The normalization constant will be given by

$$B_n = \left[\int_{-R_n}^\infty u_n^2(\eta)\,d\eta\right]^{-1/2}$$

The first few roots of the Airy function which correspond to Ai(x) = 0 are: $x_1 = -2.34$, $x_2 = -4.09$, $x_3 = -5.52$, $x_4 = -6.78$. Graphs of the lowest eigenfunctions can be found in Langhoff's article.

4.15 Two masses, m_1 and m_2, which are restricted to move in a plane are connected by a massless rod of length R to form a rigid rotator.

(a) Set up the Schrödinger equation for this rotator and obtain equations for the energy eigenvalues and the normalized eigenfunctions.

(b) Obtain an equation for the eigenvalues of the operator $\hat{\ell}^2 = -\hbar^2\partial^2/\partial\phi^2$, which corresponds to the square of the angular momentum of the rotator.

(c) What is the expectation value of the angular momentum for the state of lowest energy for the rotator? For a state having the energy $E = m^2\hbar^2/2I$ with $m \neq 0$?

Solution: Ref. Problems 1.3 and 1.24

(a) The moment of inertia for this system is given by $I = \mu R^2$ where $\mu = m_1m_2/(m_1 + m_2)$ is the reduced mass. This result permits the rotator to be treated as a one particle problem. We consider the motion of a particle having mass μ around the circumference of a circle of radius R. The single degree of freedom for this problem can be chosen to be the length of the arc S or the angle ϕ. Since $S = R\phi$ we can write:

$$-\frac{\hbar^2}{2\mu}\frac{d^2\psi(\phi)}{dS^2} = E\psi(\phi)$$

or

$$-\frac{\hbar^2}{2\mu R^2}\frac{d^2\psi(\phi)}{d\phi^2} = -\frac{\hbar^2}{2I}\frac{d^2\psi(\phi)}{d\phi^2} = E\psi(\phi)$$

This is a linear differential equation which must have a solution of the form $\psi(\phi) = Ae^{M\phi}$. For convenience we write:

$$\frac{d^2\psi(\phi)}{d\phi^2} = -\frac{2IE}{\hbar^2}\psi(\phi)$$

Then by substitution

$$M^2 = -(2IE/\hbar^2)\psi(\phi) \quad \text{and} \quad \psi(\phi) = Ae^{\pm i\sqrt{2IE}\,\phi/\hbar}$$

Since $\psi(\phi)$ must be a single valued function of ϕ the boundary conditions require that $\psi(\phi) = \psi(\phi + 2\pi)$. Thus:

$$e^{i\sqrt{2IE}\,\phi/\hbar} = e^{i\sqrt{2IE}\,\phi/\hbar}e^{i\sqrt{2IE}\,2\pi/\hbar}$$

In order for the function $e^{i\sqrt{2IE}\,2\pi/\hbar}$ to be unity it is required that:

$$\sqrt{2IE}/\hbar = m, \quad m = 0, \pm1, \pm2, \ldots$$

since:

$$e^{i\alpha} = \cos\alpha + i\sin\alpha$$

Solving for E we find:

$$\boxed{E = \frac{m^2\hbar^2}{2I} = \frac{m^2h^2}{8\pi^2I}}$$

The eigenfunctions can be chosen to be

$$\psi(\phi) = Ae^{im\phi}, \quad m = 0, \pm 1, \pm 2, \ldots$$

but since the functions having the same value of $|m|$ are degenerate this choice is not unique. For example linear combinations of the degenerate pairs can be chosen to give real functions:

$$\psi(\phi) = C \sin|m|\phi, \quad \psi(\phi) = D \cos|m|\phi$$

Adopting the complex form of $\psi(\phi)$ we find the normalization constant.

$$\int_{\phi=0}^{\phi=2\pi} |\psi(\phi)|^2 d\phi = \int_0^{2\pi} A^2 d\phi = A^2 2\pi = 1 \quad \text{or} \quad A = 1/\sqrt{2\pi}$$

Finally

$$\boxed{\psi(\phi) = (1/\sqrt{2\pi})e^{im\phi}}, \quad m = 0, \pm 1, \pm 2, \ldots$$

(b) The operator for the momentum conjugate to the coordinate ϕ is $\hat{\ell} = -i\hbar\partial/\partial\phi$. For the angular momentum squared we have $\hat{\ell}^2 = -\hbar^2\partial^2/\partial\phi^2$. With either choice of eigenfunction in part (a) we find:

$$\hat{\ell}^2\psi(\phi) = -\hbar^2 \frac{d^2}{d\phi^2} \psi(\phi) = +\hbar^2 m^2 \psi(\phi)$$

The eigenvalues of the $\hat{\ell}^2$ operator are, therefore, $\hbar^2 m^2$. This result shows that the eigenfunctions $\psi(\phi)$ for the energy are also eigenfunctions of $\hat{\ell}^2$ as expected since $\hat{\mathcal{H}} = \hat{\ell}^2/2m$. In general simultaneous eigenfunctions are possible for operators which commute.

(c) The state of lowest energy has the quantum number $m = 0$ and the eigenfunction is:

$$\psi(\phi) = 1/\sqrt{2\pi}$$

$$\langle\hat{\ell}\rangle = \int_{\phi=0}^{\phi=2\pi} (1/\sqrt{2\pi})(-\hbar \frac{d}{d\phi})(1/\sqrt{2\pi})d\phi = 0$$

The eigenfunctions which correspond to the energy $m^2\hbar^2/2I$ can be written as

$$\psi(\phi) = N \frac{e^{im\phi}}{\sqrt{2\pi}} \pm \sqrt{1 - N^2} \frac{e^{-im\phi}}{\sqrt{2\pi}} \qquad |N| \lesssim 1$$

$$= Nu_+ \pm \sqrt{1 - N^2}\, u_-$$

which is properly normalized.

The expectation value of $\hat{\ell}$ is then:

$$\langle\hat{\ell}\rangle = \int_{\phi=0}^{\phi=2\pi} (Nu_+ \pm \sqrt{1 - N^2}\, u_-)^* \hat{\ell}(Nu_+ \pm \sqrt{1 - N^2}\, u_-)d\phi = \hbar[mN^2 - m(1 - N^2)] = \hbar m(2N^2 - 1)$$

The expectation value of $\hat{\ell}$ can range from $+m$ to $-m$ in units of $\hbar$ depending on the choice of eigenfunction. This illustrates that $\psi(\phi)$, which is an eigenfunction of the Hamiltonian $\hat{\mathcal{H}}$

and of the squared angular momentum $\hat{\ell}^2$, is not necessarily an eigenfunction of $\hat{\ell}$. Since $\hat{\ell}$ commutes with $\hat{\mathcal{H}}$, it is possible to find simultaneous eigenfunctions for these operators.

4.16 A particle of mass m is confined to a cylindrical potential box as shown below. The potential V_o' is zero for $0 \lessapprox z \lessapprox H$ and $(x^2 + y^2) \lessapprox \rho$, otherwise it is infinite.

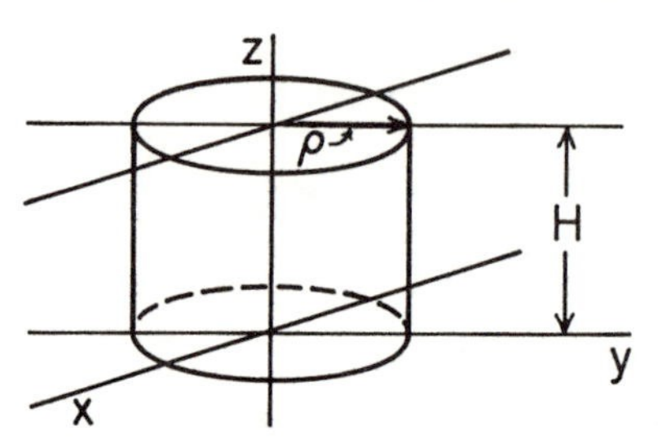

(a) Separate the wave equation for this system into three equations which depend separately on the cylindrical coordinates r, ϕ, and z.

(b) Obtain equations for the eigenfunctions of the ϕ, z, and r equations of part (a).

(c) Obtain an equation for the allowed energies of this system in terms of the separation constants, and explicitly write out the energy equation for the ground state.

Solution:

(a) The Laplacian in cylindrical coordinates is: (see Appendix 5)

$$\nabla^2 = \frac{1}{r}\frac{\partial}{\partial r}\left(r\frac{\partial}{\partial r}\right) + \frac{1}{r^2}\frac{\partial^2}{\partial\phi^2} + \frac{\partial^2}{\partial z^2}$$

Therefore the wave equation for the cylindrical potential box is:

$$-\frac{\hbar^2}{2m}\left[\frac{1}{r}\frac{\partial}{\partial r}\left(r\frac{\partial u}{\partial r}\right) + \frac{1}{r^2}\frac{\partial^2 u}{\partial\phi^2} + \frac{\partial^2 u}{\partial z^2}\right] = E'u$$

For separation of the variables we assume that $u(r,\phi,z) = R(r)\Phi(\phi)Z(z)$ and then divide through the wave equation with $u(r,\phi,z)$. The result is:

$$\frac{1}{rR}\frac{\partial}{\partial r}\left(r\frac{\partial R}{\partial r}\right) + \frac{1}{\Phi r^2}\frac{\partial^2\Phi}{\partial\phi^2} + \frac{1}{Z}\frac{\partial^2 Z}{\partial z^2} = -E$$

where E is in units of $\hbar^2/2m$.

The constants used in the separation are not unique. One choice is as follows:

$$\frac{1}{rR}\frac{\partial}{\partial r}\left(r\frac{\partial R}{\partial r}\right) + \frac{1}{\Phi r^2}\frac{\partial^2\Phi}{\partial\phi^2} \equiv -\frac{1}{Z}\frac{\partial^2 Z}{\partial z^2} - E = -\lambda^2$$

$$\frac{r}{R}\frac{\partial}{\partial r}\left(r\frac{\partial R}{\partial r}\right) + \frac{1}{\Phi}\frac{\partial^2\Phi}{\partial\phi^2} = -\lambda^2 r^2$$

$$\frac{r}{R}\frac{d}{dr}\left(r\frac{dR}{dr}\right) + \lambda^2 r^2 \equiv -\frac{1}{\Phi}\frac{\partial^2\Phi}{\partial\phi^2} = m^2$$

When two sides of an equation are identically equal in spite of the fact that they depend on different independent variables, they must equal the same constant. We are left with the following differential equations:

$$\Phi: \quad \frac{d^2\Phi}{d\phi^2} + m\Phi = 0$$

$$Z: \quad \frac{d^2Z}{dz^2} + (E - \lambda^2)Z = 0$$

$$R: \quad \frac{1}{r}\frac{d}{dr}\left(r\frac{dR}{dr}\right)+\left(\lambda^2 - \frac{m^2}{r^2}\right)R = 0$$

(b) The solutions proceed as follows:

Φ: Linear equation with constant coefficients:

$$\Phi(\phi) = \frac{1}{\sqrt{2\pi}} e^{im\phi}, \quad m = 0, \pm 1, \pm 2, \ldots$$

Z: The general solution is:

$$Z(z) = c_1 e^{i\sqrt{E - \lambda^2}\, z} + c_2 e^{-i\sqrt{E - \lambda^2}\, z}$$

The boundary conditions require that: $Z(0) = c_1 + c_2 = 0$ or $c_1 = -c_2$. Therefore:

$$Z(z) = c \sin\sqrt{E - \lambda^2}\, z$$

Also:

$$Z(H) = c \sin\sqrt{E - \lambda^2}\, H = 0$$

which implies that

$$\sqrt{E - \lambda^2}\, H = \ell\pi, \quad \ell = 1, 2, 3, \ldots$$

$$Z(z) = c \sin\left(\frac{\ell\pi}{H} z\right)$$

Normalization gives:

$$\int_{z=0}^{Z=H} Z^2(z)dz = c^2\int_o^H \sin^2(\ell\pi z/h)dz = c^2H/2 = 1$$

Therefore:

$$Z(r) = \sqrt{\frac{2}{H}} \sin(\ell\pi z/H)$$

R: This equation can be brought into the form of Bessel's equation by the change of variable: $v = \lambda r$.

$$v^2 \frac{d^2R}{dv^2} + v\frac{dR}{dv} + (v^2 - m^2)R = 0$$

Therefore, $R_m(v) = J_m(\lambda r)$ where $J_m(\lambda r)$ is a Bessel function. The boundary conditions require that:

$$J_m(\lambda\rho) = 0$$

The roots of this equation then give the values of λ in terms of ρ.

(c) From part (b) the energy is given by:

$$E = \frac{\ell^2\pi^2}{H^2} + \lambda^2, \quad \ell = 1, 2, 3, \ldots$$

The ground state energy is obtained when $m = 0$, $\ell = 1$, and λ is adjusted so that $J_o(\lambda\rho) = 0$. From tables the first zero for $J_o(\lambda\rho)$ occurs when $\lambda\rho = 2.405$. Therefore:

$$E' = \frac{\hbar^2}{2m} E = \frac{\hbar^2}{2m}\left[\frac{\pi^2}{H^2} + \frac{(2.405)^2}{\rho^2}\right]$$

4.17 The momentum eigenfunction for a particle moving in one dimension was shown in Problem 3.15 to be

$$\phi_p(x) = h^{-1/2}e^{ipx/\hbar}$$

The energy eigenfunction for a particle in a one dimensional box of length a is

$$u(x) = \sqrt{\frac{2}{a}}\sin(n\pi x/a)$$

If $u(x)$ is expanded in terms of $\phi_p(x)$, the expansion coefficient may be interpreted as the momentum probability amplitude; its square gives the probability distribution function for momentum. Determine the momentum probability distribution for the given function $u(x)$.

Solution: Ref. SC1,58; LS,44

The general expansion of any function $u(x)$ in terms of $\phi_p(x)$ is

$$u(x) = \sum_p A(p)\phi_p(x)$$

The coefficients $A(p)$ may be found by multiplying both sides by $\phi_p^*(x)$ and integrating over x. Thus

$$A(p) = \int \phi_p^*(x)u(x)dx$$

The probability distribution that we seek is

$$P(p) = |A(p)|^2$$

where

$$A(p) = \int_0^a \sqrt{\frac{2}{a}}\, h^{-1/2}e^{-ipx/\hbar}\sin(n\pi x/a)dx$$

This integral can be evaluated by expressing the sine function as

$$\sin(n\pi x/a) = (e^{in\pi x/a} - e^{-in\pi x/a})/2i$$

to obtain elementary integrals or by using a standard integral table. The result is

$$A(p) = \sqrt{\frac{2}{ah}}\frac{(n\pi a)[1 - (-1)^n e^{-ipa/\hbar}]}{[(pa/\hbar)^2 - (n\pi)^2]} \tag{1}$$

and the probability is given by

$$P(p) = A^*(p)A(p) = \frac{4n^2\pi^2 a}{h} \frac{[1 - (-1)^n \cos(pa/\hbar)]}{[(pa/\hbar)^2 - (n\pi)^2]^2}$$

Alternate expressions can be obtained from Eq. (1) by factoring out $\exp(-ipa/2\hbar)$. Thus

$$\text{n-even:} \qquad A(p) = 2i\sqrt{\frac{2}{ah}} \frac{(n\pi a)\sin(pa/2\hbar)}{[(pa/\hbar)^2 - (n\pi)^2]}$$

$$\text{n-odd:} \qquad A(p) = 2\sqrt{\frac{2}{ah}} \frac{(n\pi a)\cos(pa/2\hbar)}{[(pa/\hbar)^2 - (n\pi)^2]}$$

The function A(p) may be regarded as the wave function in the momentum representation. $|A(p)|^2 dp$ is then the probability that the momentum has a value between p and p + dp. Plots of P(p) vs. p and a discussion of this function can be found in:

F. L. Markley, Am. J. Phys. 40, 1545 (1972).

4.18 A particle of mass m is confined to a one-dimensional potential box as discussed in Problem 4.6. The potential V(x) = 0 for $0 \leq x \leq a$ and is ∞ otherwise. Suppose that at time t = 0 the state function is given by:

$\Psi(x, 0) = (2b/a)x; \qquad 0 \leq x \leq a/2$

$\Psi(x, 0) = 2b(1 - x/a); \quad a/2 \leq x \leq a$

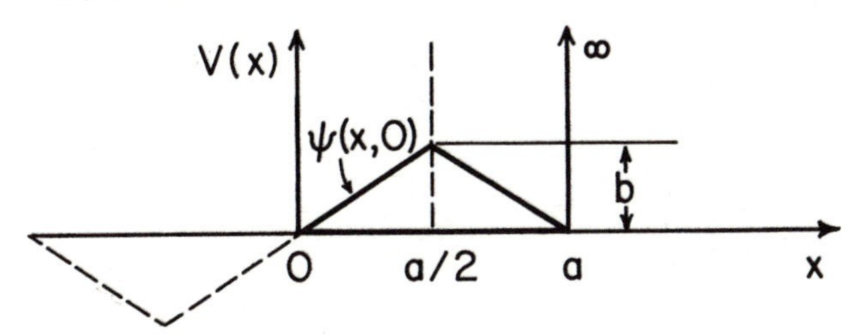

(a) Determine b so that $\Psi(x, 0)$ is normalized.

(b) Obtain the time-dependent solution $\Psi(x, t)$.

(c) Determine the expectation value of the energy $\langle E \rangle$.

Solution:

(a) Normalization requires that:

$$\int_0^a |\Psi(x, 0)|^2 dx = 1$$

or

$$\int_0^{a/2} \frac{4b^2}{a^2} x^2 dx + \int_{a/2}^{a} 4b^2(1 - \frac{2x}{a} + \frac{x^2}{a^2}) dx = 1$$

$$\frac{b^2 a}{6} + \frac{b^2 a}{6} = 1 \quad \text{or} \quad b = \sqrt{\frac{3}{a}}$$

(b) The general time dependent solution to the wave equation is given by:

$$\Psi(x, t) = \sum_{n=1}^{\infty} c_n e^{-iE_n t/\hbar} u_n(x)$$

Here the c_n's are constants, which are chosen to satisfy the initial condition:

$$\Psi(x, 0) = \sum_{n=1}^{\infty} c_n u_n(x) \tag{1}$$

Since $u_n(x) = \sqrt{2/a}\,\sin(n\pi x/a)$, this amounts to a Fourier series expansion of $\Psi(x, 0)$. The coefficients c_n can be obtained immediately by multiplying through Eq. (1) with $u_n(x)$ and integrating over x. Thus

$$c_n = \int_0^a u_n(x)\Psi(x, 0)\,dx = \sqrt{\frac{2}{a}}\left[\int_0^{a/2} \frac{2b}{a} \sin\left(\frac{n\pi x}{a}\right) dx + \int_{a/2}^{a} 2b(1 - \frac{x}{a}) \sin\left(\frac{n\pi x}{a}\right) dx\right]$$

$$= \sqrt{\frac{2}{a}}\left(\frac{4ab}{n\pi^2}\right) \sin\left(\frac{n\pi}{2}\right) \qquad n = 1, 2, 3, \ldots$$

$$= \sqrt{2a}\left(\frac{4b}{n^2\pi^2}\right) (-1)^{(n-1)/2} \qquad n = 1, 3, 5, \ldots$$

Therefore

$$\Psi(x, 0) = \sum_{n=1,3\ldots} \left(\frac{4\sqrt{6}}{\pi^2}\right) \frac{(-1)^{(n-1)/2}}{n^2} u_n(x)$$

so that

$$\Psi(x, t) = \sum_{n=1,3,5} \left(\frac{4\sqrt{6}}{\pi^2}\right) \frac{(-1)^{(n-1)/2}}{n^2} e^{-iE_n t/\hbar} u_n(x)$$

(c) The expectation value of E is given by:

$$\langle E\rangle = \int_0^a \Psi^*(x, t)\mathcal{H}_{op}\Psi(x, t)\,dx = \sum_n \sum_m c_n^* c_m E_m \langle u_n | u_m\rangle = \sum_n |c_n|^2 E_n$$

where

$$E_n = \frac{n^2 h^2}{8ma^2}$$

From part (b) we have:

$$\langle E\rangle = \sum_{n=1,3,5,\ldots} \frac{12h^2}{\pi^4 ma^2}\left(\frac{1}{n^2}\right) = \frac{12h^2}{\pi^4 ma^2}\left(\frac{\pi^2}{8}\right) = \frac{12}{\pi^2} E_1$$

4.19 The time-dependent state function for a particle in a one dimensional potential box can be expressed as

$$\Psi(x,\ t) = \sum_{n=1}^{\infty} c_n e^{-iE_n t/\hbar} u_n(x)$$

Derive an equation for the expectation value $<x>$ for the special case that $c_1 = c_2 = 1/\sqrt{2}$ and $c_n = 0$ for $n > 2$.

Solution: Ref. Problems 4.6 and 4.18

For this special case the state function becomes

$$\Psi(x,\ t) = \frac{1}{\sqrt{2}}\,[u_1(x)e^{-iE_1t/\hbar} + u_2(x)e^{-iE_2t/\hbar}]$$

$$= \frac{1}{\sqrt{2}}\left\{\sqrt{\frac{2}{a}}\sin\left(\frac{\pi x}{a}\right)e^{-iE_1t/\hbar} + \sqrt{\frac{2}{a}}\sin\left(\frac{2\pi x}{a}\right)e^{-iE_2t/\hbar}\right\}$$

and the expectation value of x is given by:

$$<x> = \int_0^a \Psi^*(x,\ t)x\Psi(x,\ t)dx$$

$$= \frac{1}{a}\int_0^a \left[\sin\left(\frac{\pi x}{a}\right)e^{+iE_1t/\hbar} + \sin\left(\frac{2\pi x}{a}\right)e^{+iE_2t/\hbar}\right]x$$

$$\left[\sin\left(\frac{\pi x}{a}\right)e^{-iE_1t/\hbar} + \sin\left(\frac{2\pi x}{a}\right)e^{-iE_2t/\hbar}\right]dx$$

$$= \frac{1}{a}\left\{I_1 + I_2 + 2I_3\cos\left(\frac{\Delta Et}{\hbar}\right)\right\}\ ; \qquad \Delta E = E_2 - E_1$$

where:

$$I_1 = \int_0^a x\sin^2(\pi x/a)dx = a^2/4, \quad I_2 = \int_0^a x\sin^2(2\pi x/a)dx = a^2/4$$

$$I_3 = \int_0^a \sin(\pi x/a)\sin(2\pi x/a)x dx = \frac{1}{2}\int_0^a [x\cos(\pi x/a) - x\cos(3\pi x/a)]dx = -\frac{8}{9}\frac{a^2}{\pi^2}$$

Here we have made use of the identity:

$$\sin A\sin B = \frac{1}{2}\,[\cos(A - B) - \cos(A + B)]$$

Finally we conclude that:

$$<x> = \left[\frac{a}{2} - \frac{16a}{9\pi^2}\cos\left(\frac{\Delta Et}{\hbar}\right)\right]$$

The expectation value of the position coordinate is thus found to oscillate with the frequency $2\pi\nu = \omega = \Delta E/\hbar$. If the particle has the charge q, then the electric moment has the form $q(x - x_o)$. Thus the dipole also oscillates with the frequency ν.

SUPPLEMENTARY PROBLEMS

4.20 A particle of mass m moves in the direction $\underset{\sim}{r}$ under the influence of a constant potential.

(a) Set up the time independent Schrödinger equation for this particle, separate variables, and obtain the general plane wave solution.

(b) Derive the probability current j for the particle in three dimensions.

Ans: Ref. EI,212; SC1,22,26

(a) $u(x, y, z) = A\exp(i\underset{\sim}{p} \cdot \underset{\sim}{r}/\hbar)$

(b) $\underset{\sim}{j}(\underset{\sim}{r}, t) = \underset{\sim}{v}\psi^*\psi$

4.21 Given a barrier of the following shape

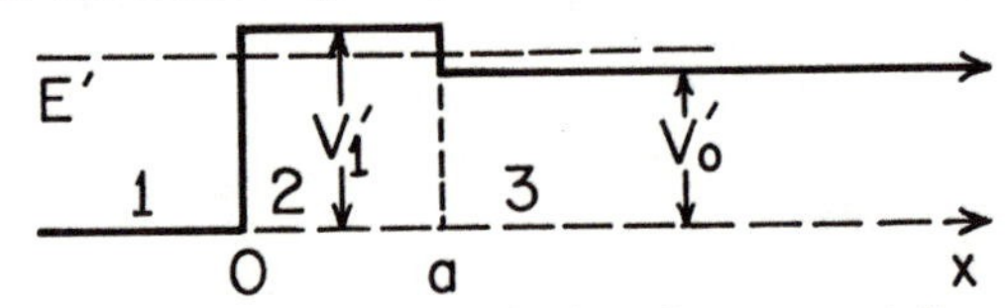

derive an equation for the probability of transmission for a particle of mass m and energy E' incident from the left. Consider the two cases (a) $V_1' > E' > V_o'$ and (b) $E' > V_1'$.

Ans: Ref. E1,231; SA,151; DU,63

(a) For $V_1' > E' > V_o'$

$$T = \left(\frac{E' - V_o'}{E'}\right)^{1/2} \left\{\frac{1}{4}\left[\frac{V_1'}{E'} \left(\frac{V_1' - V_o'}{V_1' - E'}\right) \sinh^2 \left(\sqrt{2m(V_1' - E')}\ \frac{a}{\hbar}\right) + \left(\frac{2E' - V_o'}{E'}\right)^2\right]\right\}^{-1}$$

(b) For $E' > V_1'$

$$T = \left(\frac{E' - V_o'}{E'}\right)^{1/2} \left\{\frac{1}{4}\left[\frac{V_1'}{E'}\left(\frac{V_1' - V_o'}{E' - V_1'}\right) \sin^2\left(\sqrt{2m(E' - V_1')}\ \frac{a}{\hbar}\right) + \left(\frac{2E' - V_o'}{E'}\right)^2\right]\right\}^{-1}$$

4.22 A particle of mass m is trapped in a one dimensional box defined by the potential $V = 0$ for $|x'| \leq a/2$, $V = \infty$ otherwise. Determine the energies and wave functions for this system.

Ans: Ref. Problem 4.6

$$E_n = \frac{n^2h^2}{8ma^2}$$

n even: $u_n(x') = \sqrt{\frac{2}{a}} \sin\left(\frac{n\pi}{a} x'\right)$, $n = 2, 4, 6, \ldots$

n odd: $u_n(x') = \sqrt{\frac{2}{a}} \cos\left(\frac{n\pi}{a} x'\right)$, $n = 1, 3, 5, \ldots$

4.23 A particle of mass m is confined to a box having an infinite potential barrier at x = 0 and a finite barrier at x = b as shown below:

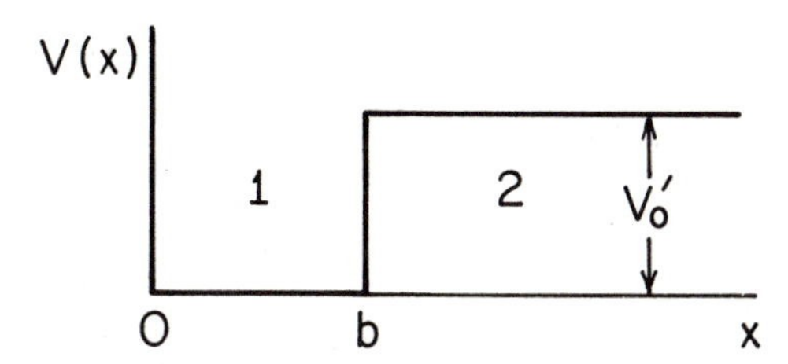

Determine the energy levels for the bound states of this system, i.e. with $E < V_o$.

Ans: Ref. Problem 4.7

The condition $\sqrt{E}\cot(\sqrt{E}\,b) = -\sqrt{V_o - E}$ leads to the same solutions that were encountered in Problem 4.7 for odd values of n. This is made evident by using the identity:

$\cot\sqrt{E}\,b = -\tan(\sqrt{E}\,b + n\pi/2)$, n odd

4.24 A particle of mass m moves in a rectangular box having dimensions a, b, c. Inside the box the particle experiences zero potential; outside however the potential is infinite:

$V = 0 \quad 0 \lesseqgtr x \lesseqgtr a, \quad 0 \lesseqgtr y \lesseqgtr b, \quad 0 \lesseqgtr z \lesseqgtr c$

$V = \infty \quad$ otherwise

(a) Determine the allowed energies for this system.

(b) Write down the first few levels for the case of a = b = c.

Ans: Ref. PW,95; AT1,53; EWK,70; SC3,25; MKT,10; H. D. Schreiber and J. N. Spencer, J. Chem. Ed. 48, 185 (1971)

(a) $E = \left(\frac{n_x^2}{a^2} + \frac{n_y^2}{b^2} + \frac{n_z^2}{c^2}\right)\frac{h^2}{8m}$

4.25 An electron is confined to a cubical potential box of length 1 cm. How many energy states will have energies less than 1 eV for this system?

Ans:

$(n_x^2 + n_y^2 + n_z^2)_{max} = 2.66\text{x}10^{14}$

and using Fig. 4.25 we find that Number = $2.27\text{x}10^{21}$.

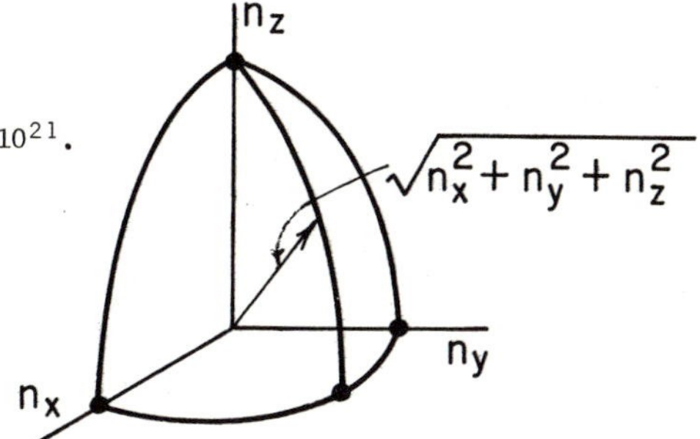

Fig. 4.25

4.26 Derive the equations for the energies and wave functions for the anisotropic and the isotropic three-dimensional harmonic oscillators.

Ans: Ref. SC3,51; KG,53; Problem 4.10

For the general oscillator

$$E = h[(n_x + 1/2)\nu_x + (n_y + 1/2)\nu_y + (n_z + 1/2)\nu_z]$$

For the isotropic oscillator

$$E = (n_x + n_y + n_z + 3/2)h\nu_o$$

4.27 The wave function for a planar rigid rotor at t = 0 is given by $\Psi(\phi, 0) = A\cos^2\phi$. Determine the normalized time dependent wave function and find the times at which the rotor will pass through its initial state. (Hint: $\Psi(\phi, t) = \sum_m c_m\psi_m(\phi, t)$)

Ans:

$$\Psi(\phi, t) = \frac{1}{\sqrt{3\pi}} [1 + \cos 2\phi e^{-i2\hbar t/I}]$$

The initial function recurrs each time $e^{-i2\hbar t/I} = 1$ or $t = \pi In/\hbar$ where n is an integer.

4.28 The double potential well in Problem 4.8 has two eigenfunctions when n = 1. These are denoted by u_1^+ and u_1^- and their energy separation is: $(V_o' >> E_1')$

$$\Delta E_1' = \frac{8E_1'^{(0)}}{b\sqrt{2m(V_o' - E_1'^{(0)})/\hbar^2}} \exp[-\sqrt{2m(V_o' - E_1'^{(0)})}\ a/\hbar]$$

Consider the special case in which $\Psi(x, 0) = (1/\sqrt{2})(u_1^+ + u_1^-)$ so that the probability density is concentrated in the well on the left.

(a) How long will be required for the probability density to become maximized in the well on the right?

(b) What is the frequency of tunneling?

Ans: Ref. P. A. Deutchman, Am. J. Phys. 39, 952 (1971).

(a) $t' = \pi\hbar/\Delta E_1$ (b) $1/t'$

4.29 The Hamiltonian function for a simple harmonic oscillator can be written as

$$\mathcal{H} = \frac{\hbar\omega}{2} (P^2 + Q^2)$$

where $P = (m\hbar\omega)^{-1/2}p$ and $Q = \sqrt{m\omega/\hbar}\ q$.

(a) Show that in the coordinate representation $\hat{P} = -i\partial/\partial Q$.

(b) Show that $[\hat{Q},\hat{P}] = i$.

Ref. Problem 3.13

4.30 It is convenient to introduce the operators

$$\hat{a} = \frac{1}{\sqrt{2}}(\hat{Q} + i\hat{P}) \tag{1}$$

$$\hat{a}^{+} = \frac{1}{\sqrt{2}}(\hat{Q} - i\hat{P}) \tag{2}$$

for the harmonic oscillator where P and Q are defined in Problem 4.29. Show that:

(a) $[\hat{a},\hat{a}^{+}] = 1$ (b) $\hat{\mathcal{H}} = \frac{\hbar\omega}{2}(\hat{a}\hat{a}^{+} + \hat{a}^{+}\hat{a})$

Ref. AN3,191

4.31 For the harmonic oscillator we define the nth eigenvalue E_n by the equation $\hat{\mathcal{H}}u_n = E_n u_n$ where $\hat{\mathcal{H}}$ is defined in Problem 4.29.

(a) Show that $\hat{\mathcal{H}}(\hat{a}u_n) = (E_n - \hbar\omega)(\hat{a}u_n)$ and thus demonstrate that $u_{n-1} \propto \hat{a}u_n$.

(b) Use the result of part (a) to determine the lowest energy eigenvalue and establish the general equation:

$$E_n = (n + 1/2)\hbar\omega$$

(Hint: Use the fact that the eigenvalues must be positive. It is interesting to note that $\langle p^2 + q^2 \rangle \geq 0$ is required <u>if the wave function is well behaved</u>.)

Ref. AN3,195; R. J. Swenson and J. C. Hermanson, Am. J. Phys. 40, 1258 (1972)

4.32 If u_o is the eigenfunction associated with the lowest energy state of the harmonic oscillator, then $\hat{a}u_o = 0$. Use this equation to determine the form of u_o, *i.e.* obtain u_o as a function of Q.

Ans:

$$u_o = N_o e^{-Q^2/2}$$

4.33 Show that for the harmonic oscillator $\hat{a}^{+}u_n \propto u_{n+1}$, and use this result with u_o from Problem 4.32 to derive the unnormalized functions u_1, u_2, and u_3.

Ans:

$N_1 Q e^{-Q^2/2}$, $N_2(-1 + 2Q^2)e^{-Q^2/2}$, $N_3(2Q^3 - 3Q)e^{-Q^2/2}$

CHAPTER 5

ANGULAR MOMENTUM

In this chapter we introduce the quantum theory of angular momentum. We start with the mechanical definition of angular momentum and then introduce the differential operators for angular momentum. Considerable attention is devoted to the transformation from cartesian to spherical polar coordinates. The properties of orbital angular momentum operators and their eigenfunctions are presented to provide a basis for the study of hydrogen-like atoms and other spherical systems which we undertake in Chapter 7. The later problems in this chapter illustrate transformation theory, rotation operators, and vector coupling theory. This material can profitably be studied before many-electron atoms are approached.

We have adopted the convention of using $\underset{\sim}{L}$ for orbital angular momentum, $\underset{\sim}{S}$ for spin angular momentum, and $\underset{\sim}{J}$ for the resultant $\underset{\sim}{L} + \underset{\sim}{S}$. In general derivations which do not depend on differential operators or functions of spatial coordinates, $\underset{\sim}{J}$ will be used. We denote the simultaneous eigenfunctions of $\hat{J}_z$ and $\hat{J}^2$ by u_{jm}, but at times it is convenient to use the ket $|j,m\rangle$. The terms eigenket, eigenvector, and eigenfunction are used interchangeably. We continue the practice of using a wavy underline $\sim$ to denote a vector <u>or</u> matrix and a hat ^ to denote an operator <u>or</u> a unit vector.

General references: RO3; BS; CS

PROBLEMS

5.1 For a particle having mass m, linear momentum $\underset{\sim}{p}$, and position $\underset{\sim}{r}$ the <u>angular momentum</u> is defined as

$$\boxed{\underset{\sim}{\mathcal{L}} = \underset{\sim}{r} \times \underset{\sim}{p}}$$

(a) Use this definition to obtain equations for $\mathcal{L}_x$, $\mathcal{L}_y$, and $\mathcal{L}_z$.

(b) What are the differential operators for $\mathcal{L}_x$, $\mathcal{L}_y$, and $\mathcal{L}_z$ in the x-representation of quantum mechanics, <u>i.e.</u> in the Schrödinger representation?

Solution:

(a) From the definition

$$\mathcal{L} = \begin{vmatrix} \hat{i} & \hat{j} & \hat{k} \\ x & y & z \\ p_x & p_y & p_z \end{vmatrix} = \hat{i}(yp_z - zp_y) + \hat{j}(zp_x - xp_z) + \hat{k}(xp_y - yp_x)$$

Therefore

$$\mathcal{L}_x = yp_z - zp_y, \quad \mathcal{L}_y = zp_x - xp_z, \quad \mathcal{L}_z = xp_y - yp_x$$

(b) In the x-representation $\hat{x} \to x$ and $\hat{p}_x \to -i\hbar\partial/\partial x$. Thus

$$\hat{\mathcal{L}}_x = \hbar\hat{L}_x = -i\hbar(y\partial/\partial z - z\partial/\partial y)$$
$$\hat{\mathcal{L}}_y = \hbar\hat{L}_y = -i\hbar(z\partial/\partial x - x\partial/\partial z)$$
$$\hat{\mathcal{L}}_z = \hbar\hat{L}_z = -i\hbar(x\partial/\partial y - y\partial/\partial x)$$

5.2 This problem illustrates the relation between <u>torque</u> and <u>angular momentum</u>. Consider a mass m located at the point (x, y) and subjected to the force $\underset{\sim}{F}$ in the plane (see Fig. 5.2). If the mass is rotated through the small angle $\Delta\theta$, the work done will equal the component of the force parallel to the displacement multiplied by the magnitude of the displacement. Thus

$$\Delta W \equiv (\underset{\sim}{F}\cdot\hat{\theta})r\Delta\theta \equiv (\underset{\sim}{T}\cdot\hat{k})\Delta\theta$$

where $\hat{\theta}$ is a unit vector in the xy plane that is perpendicular to $\underset{\sim}{r}$, and $\underset{\sim}{T}$ is defined as the torque.

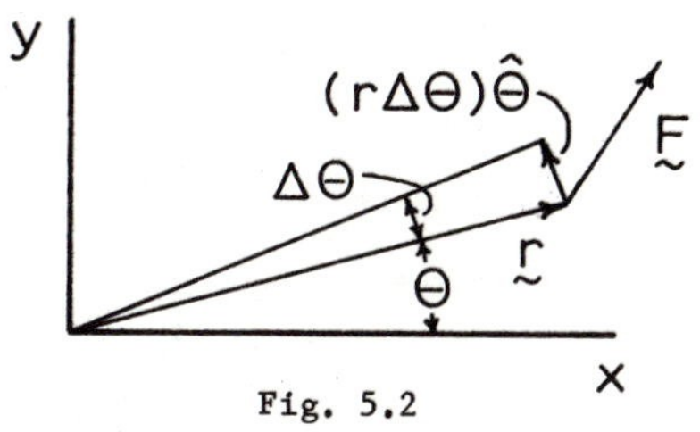

Fig. 5.2

(a) Show that $T_z = xF_y - yF_x$ and in general that $\underset{\sim}{T} = \underset{\sim}{r}\times\underset{\sim}{F}$.

(b) Show that $d\underset{\sim}{\mathcal{L}}/dt = \underset{\sim}{T}$.

Solution: Ref. FE1,18-2

(a) From the work equation

$$(\underset{\sim}{T}\cdot\hat{k}) = (\underset{\sim}{F}\cdot\hat{\theta})r$$

$$T_z = (F_x\hat{i} + F_y\hat{j})\cdot(-\sin\theta\hat{i} + \cos\theta\hat{j})r = -r\sin\theta F_x + r\cos\theta F_y = xF_y - yF_x$$

Rotations about x and y axes, respectively, lead to the equations:

$$T_x = yF_z - zF_y; \quad T_y = zF_x - xF_z$$

Taken together the equations show that $\underset{\sim}{T} = \underset{\sim}{r} x \underset{\sim}{F}$.

(b) From Problem 5.1 we have $\underset{\sim}{\mathcal{L}} = \underset{\sim}{r} x \underset{\sim}{p}$. Then

$$\frac{d\mathcal{L}_z}{dt} = \frac{d}{dt}(xp_y - yp_x) = m\frac{d}{dt}\left(x\frac{dy}{dt} - y\frac{dx}{dt}\right) = x\left(m\frac{d^2y}{dt^2}\right) - y\left(m\frac{d^2x}{dt^2}\right) = xF_y - yF_x$$

Similarly for $\mathcal{L}_x$ and $\mathcal{L}_y$ we find

$$\frac{d\mathcal{L}_x}{dt} = yF_z - zF_y; \quad \frac{d\mathcal{L}_y}{dt} = zF_x - xF_z$$

and finally

$$\frac{d\underset{\sim}{\mathcal{L}}}{dt} \equiv \underset{\sim}{r} x \underset{\sim}{F} \equiv \underset{\sim}{T}$$

5.3 Obtain expressions for the differential operators $\hat{L}_x$, $\hat{L}_y$, and $\hat{L}_z$ in spherical polar coordinates.

Solution: Ref. EWK,40

From Problem 5.1 we have

$$\begin{aligned} \hat{\mathcal{L}}_x &= -i\hbar(y\partial/\partial z - z\partial/\partial y) = \hbar\hat{L}_x \\ \hat{\mathcal{L}}_y &= -i\hbar(z\partial/\partial x - x\partial/\partial z) = \hbar\hat{L}_y \\ \hat{\mathcal{L}}_z &= -i\hbar(x\partial/\partial y - y\partial/\partial x) = \hbar\hat{L}_z \end{aligned} \tag{1}$$

Spherical polar coordinates are defined by the transformations

$$x = r\sin\theta\cos\phi; \quad y = r\sin\theta\sin\phi; \quad z = r\cos\theta \tag{2}$$

where $r^2 = x^2 + y^2 + z^2$, $\cos\theta = z/r$, $\tan\phi = y/x$. The partial differential operators required in Eq. (1) can be obtained from the chain rule

$$\partial/\partial x_i = (\partial r/\partial x_i)\partial/\partial r + (\partial\theta/\partial x_i)\partial/\partial\theta + (\partial\theta/\partial x_i)\partial/\partial\phi \tag{3}$$

where x_i represents x, y, or z. Evaluation of the derivatives proceeds as follows

$$\frac{\partial r}{\partial x} = \partial(x^2 + y^2 + z^2)^{1/2}/\partial x = x/r = \sin\theta\cos\phi$$

$$\frac{\partial\theta}{\partial x} = \partial[\cos^{-1}(z/r)]/\partial x = \frac{1}{\sin\theta}\left(\frac{z}{r}\right)\frac{\sin\theta\cos\phi}{r} = \frac{\cos\theta\cos\phi}{r}$$

$$\frac{\partial\phi}{\partial x} = \partial[\tan^{-1}(y/x)]/\partial x = \frac{-\tan\phi}{(1 + \tan^2\phi)r\sin\theta\cos\phi} = -\frac{\sin\phi}{r\sin\theta}$$

In a similar manner we find

$$\frac{\partial r}{\partial y} = \sin\theta\sin\phi; \quad \frac{\partial r}{\partial z} = \cos\theta$$

$$\frac{\partial\theta}{\partial y} = \frac{\cos\theta\sin\phi}{r}; \quad \frac{\partial\theta}{\partial z} = \frac{-\sin\theta}{r}$$

$$\frac{\partial\phi}{\partial y} = \frac{\cos\phi}{r\sin\theta}; \quad \frac{\partial\phi}{\partial z} = 0$$

Equation (3) then gives

$$\frac{\partial}{\partial x} = \sin\theta\cos\phi \frac{\partial}{\partial r} + \frac{\cos\theta\cos\phi}{r}\frac{\partial}{\partial\theta} - \frac{\sin\phi}{r\sin\theta}\frac{\partial}{\partial\phi}$$

$$\frac{\partial}{\partial y} = \sin\theta\sin\phi \frac{\partial}{\partial r} + \frac{\cos\theta\sin\phi}{r}\frac{\partial}{\partial\theta} + \frac{\cos\phi}{r\sin\theta}\frac{\partial}{\partial\phi} \tag{4}$$

$$\frac{\partial}{\partial z} = \cos\theta \frac{\partial}{\partial r} - \frac{\sin\theta}{r}\frac{\partial}{\partial\theta}$$

Substitution of Eqs. (4) into Eqs. (1) yields

$$\hat{L}_x = -i\left[r\sin\theta\sin\phi\left(\cos\theta\frac{\partial}{\partial r} - \frac{\sin\theta}{r}\frac{\partial}{\partial\theta}\right) - r\cos\theta\left(\sin\theta\sin\phi\frac{\partial}{\partial r} + \frac{\cos\theta\sin\phi}{r}\frac{\partial}{\partial\theta} + \frac{\cos\phi}{r\sin\theta}\frac{\partial}{\partial\phi}\right)\right]$$

$$= -i\left[(-\sin^2\theta\sin\phi - \cos^2\theta\sin\phi)\frac{\partial}{\partial\theta} + \left(-\frac{\cos\phi\cos\theta}{\sin\theta}\frac{\partial}{\partial\phi}\right)\right]$$

$$\hat{L}_x = -i\left\{-\sin\phi\frac{\partial}{\partial\theta} - \cos\phi\cot\theta\frac{\partial}{\partial\phi}\right\}$$

and similarly

$$\hat{L}_y = -i\left\{\cos\phi\frac{\partial}{\partial\theta} - \sin\phi\cot\theta\frac{\partial}{\partial\phi}\right\}$$

$$\hat{L}_z = -i\frac{\partial}{\partial\phi}$$

5.4 Derive equations for the differential operators $\hat{L}_+$, $\hat{L}_-$, and $\hat{L}^2$ in spherical polar coordinates. By definition $\hat{L}_\pm = \hat{L}_x \pm i\hat{L}_y$.

Solution:

Using $\hat{L}_x$ and $\hat{L}_y$ from Problem 5.3 we can write

$$\hat{L}_x + i\hat{L}_y = \left(+i\sin\phi\frac{\partial}{\partial\theta} + i\cos\phi\cot\theta\frac{\partial}{\partial\phi}\right) + \left(\cos\phi\frac{\partial}{\partial\theta} - \sin\phi\cot\theta\frac{\partial}{\partial\phi}\right)$$

$$= (\cos\phi + i\sin\phi)\left[\frac{\partial}{\partial\theta} + i\cot\theta\frac{\partial}{\partial\phi}\right]$$

Therefore

$$\hat{L}_+ = e^{i\phi}\left[\frac{\partial}{\partial\theta} + i\cot\theta\frac{\partial}{\partial\phi}\right]$$

Also we have

$$\hat{L}_x - i\hat{L}_y = \left(i\sin\phi\frac{\partial}{\partial\theta} + i\cos\phi\cot\theta\frac{\partial}{\partial\phi}\right) - \left(\cos\phi\frac{\partial}{\partial\theta} - \sin\phi\cot\theta\frac{\partial}{\partial\phi}\right)$$

$$= (\cos\phi - i\sin\phi)\left[-\frac{\partial}{\partial\theta} + i\cot\theta\frac{\partial}{\partial\phi}\right]$$

and

$$\hat{L}_- = e^{-i\phi}\left[-\frac{\partial}{\partial\theta} + i\cot\theta\frac{\partial}{\partial\phi}\right]$$

To obtain $\hat{L}^2$ we can use the identity $\hat{L}^2 = \hat{L}_x^2 + \hat{L}_y^2 + \hat{L}_z^2$; however, it is more convenient to use $\hat{L}^2 = \hat{L}_+\hat{L}_- + \hat{L}_z^2 - \hat{L}_z$. The latter follows by inspection of the product $\hat{L}_+\hat{L}_-$. Thus

$$\hat{L}_+\hat{L}_- f = e^{i\phi}\left[\frac{\partial}{\partial\theta} + i\cot\theta\,\frac{\partial}{\partial\phi}\right]e^{-i\phi}\left[-\frac{\partial}{\partial\theta} + i\cot\theta\,\frac{\partial}{\partial\phi}\right]f = -\left[\frac{\partial^2 f}{\partial\theta^2} + \cot\theta\,\frac{\partial f}{\partial\theta} + \cot^2\theta\,\frac{\partial^2 f}{\partial\phi^2} + i\,\frac{\partial f}{\partial\phi}\right]$$

$$\hat{L}_z f = -i\,\frac{\partial f}{\partial\phi}\ ; \quad \hat{L}_z^2 f = -\frac{\partial^2 f}{\partial\phi^2}$$

Combining these equations we obtain

$$\hat{L}^2 = \hat{L}_+\hat{L}_- + \hat{L}_z^2 - \hat{L}_z = -\left[\frac{\partial^2}{\partial\theta^2} + \cot\theta\,\frac{\partial}{\partial\theta} + \cot^2\theta\,\frac{\partial^2}{\partial\phi^2} + \frac{\partial^2}{\partial\phi^2}\right]$$

The identities

$$\frac{1}{\sin\theta}\,\frac{\partial}{\partial\theta}\,\sin\theta\,\frac{\partial f}{\partial\theta} = \cot\theta\,\frac{\partial f}{\partial\theta} + \frac{\partial^2 f}{\partial\theta^2} \quad \text{and} \quad \cot^2\theta + 1 = (\sin\theta)^{-2}$$

permit $\hat{L}^2$ to be written in the form

$$\boxed{\hat{L}^2 = -\left[\frac{1}{\sin\theta}\,\frac{\partial}{\partial\theta}\,\sin\theta\,\frac{\partial}{\partial\theta} + \frac{1}{\sin^2\theta}\,\frac{\partial^2}{\partial\phi^2}\right]}$$

5.5 Show that the Laplacian

$$\nabla^2 = \frac{\partial^2}{\partial x^2} + \frac{\partial^2}{\partial y^2} + \frac{\partial^2}{\partial z^2}$$

when transformed to spherical polar coordinates becomes

$$\nabla^2 = \frac{1}{r^2}\,\frac{\partial}{\partial r}\,r^2\,\frac{\partial}{\partial r} + \frac{1}{r^2\sin\theta}\,\frac{\partial}{\partial\theta}\,\sin\theta\,\frac{\partial}{\partial\theta} + \frac{1}{r^2\sin^2\theta}\,\frac{\partial^2}{\partial\phi^2} \tag{1}$$

and show that

$$\nabla^2 = \frac{1}{r^2}\left[\frac{\partial}{\partial r}\,r^2\,\frac{\partial}{\partial r} - \hat{L}^2\right]$$

where the angular momentum operator $\hat{L}^2$ is defined in Problem 5.4.

<u>Solution</u>: Ref. Problems 5.3 and 5.4

For the x component we have

$$\frac{\partial f}{\partial x} = \frac{\partial f}{\partial r}\,\frac{\partial r}{\partial x} + \frac{\partial f}{\partial\theta}\,\frac{\partial\theta}{\partial x} + \frac{\partial f}{\partial\phi}\,\frac{\partial\phi}{\partial x} = f_r\,\frac{\partial r}{\partial x} + f_\theta\,\frac{\partial\theta}{\partial x} + f_\phi\,\frac{\partial\phi}{\partial x}$$

and

$$\frac{\partial^2 f}{\partial x^2} = \left(\frac{\partial f_r}{\partial r}\,\frac{\partial r}{\partial x} + \frac{\partial f_r}{\partial\theta}\,\frac{\partial\theta}{\partial x} + \frac{\partial f_r}{\partial\phi}\,\frac{\partial\phi}{\partial x}\right)\frac{\partial r}{\partial x} + f_r\,\frac{\partial^2 r}{\partial x^2} + \left(\frac{\partial f_\theta}{\partial r}\,\frac{\partial r}{\partial x} + \frac{\partial f_\theta}{\partial\theta}\,\frac{\partial\theta}{\partial x} + \frac{\partial f_\theta}{\partial\phi}\,\frac{\partial\phi}{\partial x}\right)\frac{\partial\theta}{\partial x} + f_\theta\,\frac{\partial^2\theta}{\partial x^2}$$

$$+ \left(\frac{\partial f_\phi}{\partial r}\,\frac{\partial r}{\partial x} + \frac{\partial f_\phi}{\partial\theta}\,\frac{\partial\theta}{\partial x} + \frac{\partial f_\phi}{\partial\phi}\,\frac{\partial\phi}{\partial x}\right)\frac{\partial\phi}{\partial x} + f_\phi\,\frac{\partial^2\phi}{\partial x^2}$$

$$= \frac{\partial^2 f}{\partial r^2}\left(\frac{\partial r}{\partial x}\right)^2 + \frac{\partial^2 f}{\partial\theta\partial r}\frac{\partial\theta}{\partial x}\frac{\partial r}{\partial x} + \frac{\partial^2 f}{\partial\phi\partial r}\frac{\partial\phi}{\partial x}\frac{\partial r}{\partial x} + \frac{\partial f}{\partial r}\frac{\partial^2 r}{\partial x^2} + \frac{\partial^2 f}{\partial r\partial\theta}\frac{\partial r}{\partial x}\frac{\partial\theta}{\partial x} + \frac{\partial^2 f}{\partial\theta^2}\left(\frac{\partial\theta}{\partial x}\right)^2$$

$$+ \frac{\partial^2 f}{\partial\phi\partial\theta}\frac{\partial\phi}{\partial x}\frac{\partial\theta}{\partial x} + \frac{\partial f}{\partial\theta}\frac{\partial^2\theta}{\partial x^2} + \frac{\partial^2 f}{\partial r\partial\phi}\frac{\partial r}{\partial x}\frac{\partial\phi}{\partial x} + \frac{\partial^2 f}{\partial\theta\partial\phi}\frac{\partial\theta}{\partial x}\frac{\partial\phi}{\partial x} + \frac{\partial^2 f}{\partial\phi^2}\left(\frac{\partial\phi}{\partial x}\right)^2 + \frac{\partial f}{\partial\phi}\frac{\partial^2\phi}{\partial x^2} \qquad (2)$$

Also

$$\frac{\partial^2 f}{\partial y^2} \quad \text{and} \quad \frac{\partial^2 f}{\partial z^2}$$

are analogous in form so that

$$\nabla^2 f = \frac{\partial^2 f}{\partial r^2}\sum_i\left(\frac{\partial r}{\partial x_i}\right)^2 + \frac{\partial^2 f}{\partial\theta^2}\sum_i\left(\frac{\partial\theta}{\partial x_i}\right)^2 + \frac{\partial^2 f}{\partial\phi^2}\sum_i\left(\frac{\partial\phi}{\partial x_i}\right)^2 + 2\left[\frac{\partial^2 f}{\partial\theta\partial r}\sum_i\frac{\partial\theta}{\partial x_i}\frac{\partial r}{\partial x_i}\right.$$

$$\left. + \frac{\partial^2 f}{\partial\phi\partial r}\sum_i\frac{\partial\phi}{\partial x_i}\frac{\partial r}{\partial x_i} + \frac{\partial^2 f}{\partial\phi\partial\theta}\sum_i\frac{\partial\phi}{\partial x_i}\frac{\partial\theta}{\partial x_i}\right] + \frac{\partial f}{\partial r}\sum_i\frac{\partial^2 r}{\partial x_i^2} + \frac{\partial f}{\partial\theta}\sum_i\frac{\partial^2\theta}{\partial x_i^2} + \frac{\partial f}{\partial\phi}\sum_i\frac{\partial^2\phi}{\partial x_i^2} \qquad (3)$$

where $x_i = x, y, z$ for $i = 1, 2, 3$ respectively. The necessary derivatives are found with the aid of the transformations (see Problem 5.3)

$$\left.\begin{aligned} x &= r\sin\theta\cos\phi \\ y &= r\sin\theta\sin\phi \\ z &= r\cos\theta \end{aligned}\right\} \qquad \begin{aligned} r &= \sqrt{x^2 + y^2 + z^2} \\ \theta &= \cos^{-1}\left(\frac{z}{r}\right) \\ \phi &= \tan^{-1}\frac{y}{x} \end{aligned} \qquad (4)$$

	$\frac{\partial r}{\partial x_i}$	$\frac{\partial^2 r}{\partial x_i^2}$	$\frac{\partial\theta}{\partial x_i}$	$\frac{\partial^2\theta}{\partial x_i^2}$	$\frac{\partial\phi}{\partial x_i}$	$\frac{\partial^2\phi}{\partial x_i^2}$
$i = 1$	$\frac{x}{r}$	$\frac{y^2+z^2}{r^3}$	$\frac{1}{r^2}\frac{xz}{\sqrt{x^2+y^2}}$	$\frac{z[y^4-2x^4-y^2(x^2-z^2)]}{r^4(x^2+y^2)^{3/2}}$	$-\frac{y}{x^2+y^2}$	$\frac{2xy}{(x^2+y^2)^2}$
$i = 2$	$\frac{y}{r}$	$\frac{x^2+z^2}{r^3}$	$\frac{1}{r^2}\frac{yz}{\sqrt{x^2+y^2}}$	$\frac{z[x^4-2y^4-x^2(y^2-z^2)]}{r^4(x^2+y^2)^{3/2}}$	$\frac{x}{x^2+y^2}$	$\frac{-2xy}{(x^2+y^2)^2}$
$i = 3$	$\frac{z}{r}$	$\frac{x^2+y^2}{r^3}$	$-\frac{1}{r^2}\sqrt{x^2+y^2}$	$\frac{2z\sqrt{x^2+y^2}}{r^4}$	0	0

(5)

Substitution of Eq. (5) into Eq. (3) gives

$$\nabla^2 f = \frac{\partial^2 f}{\partial r^2} + \frac{2}{r}\frac{\partial f}{\partial r} + \frac{1}{r^2}\frac{\partial^2 f}{\partial\theta^2} + \frac{1}{r^2}\cot\theta\frac{\partial f}{\partial\theta} + \frac{1}{r^2\sin^2\theta}\frac{\partial^2 f}{\partial\phi^2} \qquad (6)$$

and noting that

$$\frac{\partial}{\partial r} r^2 \frac{\partial}{\partial r} = 2r \frac{\partial}{\partial r} + r^2 \frac{\partial^2}{\partial r^2} \quad \text{and} \quad \frac{\partial}{\partial \theta} \sin\theta \frac{\partial}{\partial \theta} = \cos\theta \frac{\partial}{\partial \theta} + \sin\theta \frac{\partial^2}{\partial \theta^2} \tag{7}$$

we have

$$\boxed{\nabla^2 = \frac{1}{r^2} \frac{\partial}{\partial r} r^2 \frac{\partial}{\partial r} + \frac{1}{r^2 \sin\theta} \frac{\partial}{\partial \theta} \sin\theta \frac{\partial}{\partial \theta} + \frac{1}{r^2 \sin^2\theta} \frac{\partial^2}{\partial \phi^2}} \tag{8}$$

Since

$$\hat{L}^2 = -\left[\frac{1}{\sin\theta} \frac{\partial}{\partial \theta} \sin\theta \frac{\partial}{\partial \theta} + \frac{1}{\sin^2\theta} \frac{\partial^2}{\partial \phi^2}\right] \tag{9}$$

as shown in Problem 5.4, we can then write

$$\boxed{\nabla^2 f = \frac{1}{r^2}\left[\frac{\partial}{\partial r} r^2 \frac{\partial}{\partial r} - \hat{L}^2\right] f}$$

For an alternative derivation see F. M. Chen, Am. J. Phys. 40, 553 (1972).

5.6 Obtain $\underset{\sim}{\nabla}$, $\underset{\sim}{\hat{L}}$ and ∇^2 in spherical polar coordinates by the use of the theory of curvilinear coordinates.

Solution: Ref. Appendix 5

First we establish the metric tensor g:

$$\underset{\sim}{r} = r\sin\theta\cos\phi \, \hat{i} + r\sin\theta\sin\phi \, \hat{j} + r\cos\theta \, \hat{k}$$

so

$$\frac{\partial \underset{\sim}{r}}{\partial r} = \sin\theta\cos\phi \, \hat{i} + \sin\theta\sin\phi \, \hat{j} + \cos\theta \, \hat{k} \, ; \qquad \frac{\partial \underset{\sim}{r}}{\partial \theta} = r\cos\theta\cos\phi \, \hat{i} + r\cos\theta\sin\phi \, \hat{j} - r\sin\theta \, \hat{k}$$

$$\frac{\partial \underset{\sim}{r}}{\partial \phi} = -r\sin\theta\sin\phi \, \hat{i} + r\sin\theta\cos\phi \, \hat{j}$$

and

$$\left|\frac{\partial \underset{\sim}{r}}{\partial r}\right| = 1, \quad \left|\frac{\partial \underset{\sim}{r}}{\partial \theta}\right| = r, \quad \left|\frac{\partial \underset{\sim}{r}}{\partial \phi}\right| = r\sin\theta$$

Then

$$\hat{e}_1 = \frac{\partial \underset{\sim}{r}/\partial r}{|\partial \underset{\sim}{r}/\partial r|} = \sin\theta\cos\phi \, \hat{i} + \sin\theta\sin\phi \, \hat{j} + \cos\theta \, \hat{k}$$

$$\hat{e}_2 = \frac{\partial \underset{\sim}{r}/\partial \theta}{|\partial \underset{\sim}{r}/\partial \theta|} = \cos\theta\cos\phi \, \hat{i} + \cos\theta\sin\phi \, \hat{j} - \sin\theta \, \hat{k}$$

and

$$\hat{e}_3 = \frac{\partial \underset{\sim}{r}/\partial\phi}{|\partial \underset{\sim}{r}/\partial\phi|} = -\sin\phi\ \hat{i} + \cos\phi\ \hat{j}$$

It is seen immediately that $\hat{e}_i \cdot \hat{e}_j = 0$, thus g is diagonal. The diagonal elements are given by

$$\left\{\left|\frac{\partial \underset{\sim}{r}}{\partial u_i}\right|^2 ;\ u_i = r,\ \theta,\ \phi\right\}: \quad g_{11} = 1,\ g_{22} = r^2,\ g_{33} = r^2\sin^2\theta$$

$\underset{\sim}{\nabla}$: $$\underset{\sim}{\nabla} = \sum_{j=1}^{3} \frac{1}{\sqrt{g_{jj}}}\hat{e}_j \frac{\partial}{\partial u_j} = \frac{\partial}{\partial r}\hat{e}_1 + \frac{1}{r}\frac{\partial}{\partial\theta}\hat{e}_2 + \frac{1}{r\sin\theta}\frac{\partial}{\partial\phi}\hat{e}_3$$

$\underset{\sim}{L}$: Since $\hat{r} = r\,\hat{e}_1$ then

$$\underset{\sim}{r} \times \underset{\sim}{\nabla} = r\,\hat{e}_1 \times \left(\hat{e}_1 \frac{\partial}{\partial r} + \hat{e}_2 \frac{1}{r}\frac{\partial}{\partial\theta} + \hat{e}_3 \frac{1}{r\sin\theta}\frac{\partial}{\partial\phi}\right)$$

Now

$$\hat{e}_j \times \hat{e}_j = 0,\quad \hat{e}_1 \times \hat{e}_2 = \hat{e}_3,\quad \hat{e}_2 \times \hat{e}_3 = \hat{e}_1 \quad \text{and} \quad \hat{e}_1 \times \hat{e}_3 = -\hat{e}_2$$

Thus

$$\hat{\underset{\sim}{L}} = -i\hbar\ \underset{\sim}{r} \times \underset{\sim}{\nabla} = -i\hbar\left[-\frac{1}{\sin\theta}\frac{\partial}{\partial\phi}\hat{e}_2 + \frac{\partial}{\partial\theta}\hat{e}_3\right]$$

To obtain the components along the x, y, z axes, we simply take the dot product with the unit vectors of the (x, y, z) set so that

$$\hat{L}_x = \hat{i}\cdot\hat{\underset{\sim}{L}};\quad \hat{L}_y = \hat{j}\cdot\hat{\underset{\sim}{L}};\quad \hat{L}_z = \hat{k}\cdot\hat{\underset{\sim}{L}}$$

The required terms are

$$\hat{i}\cdot\hat{e}_1 = \sin\theta\cos\phi,\quad \hat{j}\cdot\hat{e}_1 = \sin\theta\sin\phi,\quad \hat{k}\cdot\hat{e}_1 = \cos\theta$$

$$\hat{i}\cdot\hat{e}_2 = \cos\theta\cos\phi,\quad \hat{j}\cdot\hat{e}_2 = \cos\theta\sin\phi,\quad \hat{k}\cdot\hat{e}_2 = -\sin\theta$$

$$\hat{i}\cdot\hat{e}_3 = -\sin\phi,\quad \hat{j}\cdot\hat{e}_3 = \cos\phi,\quad \hat{k}\cdot\hat{e}_3 = 0$$

Thus the components of $\underset{\sim}{L}$ are

$$\hat{L}_x = -i\hbar\left[-\frac{1}{\sin\theta}\cos\theta\cos\phi\frac{\partial}{\partial\phi} - \sin\phi\frac{\partial}{\partial\theta}\right] = i\hbar\left[\cot\theta\cos\phi\frac{\partial}{\partial\phi} + \sin\phi\frac{\partial}{\partial\theta}\right]$$

$$\hat{L}_y = -i\hbar\left[-\frac{1}{\sin\theta}\cos\theta\sin\phi\frac{\partial}{\partial\phi} + \cos\phi\frac{\partial}{\partial\theta}\right] = i\hbar\left[\cot\theta\sin\phi\frac{\partial}{\partial\phi} - \cos\phi\frac{\partial}{\partial\theta}\right]$$

$$\hat{L}_z = -i\hbar\left[+\frac{\sin\theta}{\sin\theta}\frac{\partial}{\partial\phi}\right] = -i\hbar\frac{\partial}{\partial\phi}$$

∇^2: By use of Appendix 5 we have

$$\nabla^2 = \frac{1}{r^2\sin\theta}\left[\frac{\partial}{\partial r}\frac{r^2\sin\theta}{1}\frac{\partial}{\partial r} + \frac{\partial}{\partial\theta}\frac{r^2\sin\theta}{r^2}\frac{\partial}{\partial\theta} + \frac{\partial}{\partial\phi}\frac{r^2\sin\theta}{r^2\sin^2\theta}\frac{\partial}{\partial\phi}\right]$$

or

$$\nabla^2 = \frac{1}{r^2}\frac{\partial}{\partial r}r^2\frac{\partial}{\partial r} + \frac{1}{r^2\sin\theta}\frac{\partial}{\partial\theta}\sin\theta\frac{\partial}{\partial\theta} + \frac{1}{r^2\sin^2\theta}\frac{\partial^2}{\partial\phi^2}$$

5.7 Show that for orbital angular momentum $[\hat{L}_x,\hat{L}_y] = i\hat{L}_z$, $[\hat{L}_y,\hat{L}_z] = i\hat{L}_x$, and $[\hat{L}_z,\hat{L}_x] = i\hat{L}_y$. Use these results to show that $\hat{\tilde{L}}x\hat{\tilde{L}} = i\hat{\tilde{L}}$. (Hint: Make use of the commutation relations for linear momentum, i.e. $[x_i,\hat{p}_j] = i\hbar\delta_{ij}$)

Solution: Ref. Problem 5.1 and AN3,252

We start with the equations:

$$\hbar\hat{L}_x = y\hat{p}_z - z\hat{p}_y; \quad \hbar\hat{L}_y = z\hat{p}_x - x\hat{p}_z; \quad \hbar\hat{L}_z = x\hat{p}_y - y\hat{p}_x$$

Therefore:

$$\hbar^2[\hat{L}_x,\hat{L}_y] = [(y\hat{p}_z - z\hat{p}_y), (z\hat{p}_x - x\hat{p}_z)] = [y\hat{p}_z,z\hat{p}_x] + [z\hat{p}_y,x\hat{p}_z] - [y\hat{p}_z,x\hat{p}_z] - [z\hat{p}_y,z\hat{p}_x]$$

where

$$[y\hat{p}_z,z\hat{p}_x] = y\hat{p}_z z\hat{p}_x - z\hat{p}_x y\hat{p}_z = y\hat{p}_x[\hat{p}_z,z] = -i\hbar y\hat{p}_x$$

$$[z\hat{p}_y,x\hat{p}_z] = z\hat{p}_y x\hat{p}_z - x\hat{p}_z z\hat{p}_y = \hat{p}_y x[z,\hat{p}_z] = +i\hbar x\hat{p}_y$$

$$[y\hat{p}_z,x\hat{p}_z] = [z\hat{p}_y,z\hat{p}_x] = 0$$

Therefore

$$[\hat{L}_x,\hat{L}_y] = i\hbar^{-1}(-y\hat{p}_x + x\hat{p}_y) = i\hat{L}_z$$

In a similar manner we find: $[\hat{L}_y,\hat{L}_z] = i\hat{L}_x$ and $[\hat{L}_z,\hat{L}_x] = i\hat{L}_y$. These commutators can be obtained with roughly the same amount of work by starting with $\hat{L}_x$, $\hat{L}_y$, and $\hat{L}_z$ in terms of differential operators.

The cross product $\tilde{L}x\tilde{L}$ is just:

$$\hat{\tilde{L}}x\hat{\tilde{L}} = i(\hat{L}_y\hat{L}_z - \hat{L}_z\hat{L}_y) + j(\hat{L}_z\hat{L}_x - \hat{L}_x\hat{L}_z) + k(\hat{L}_x\hat{L}_y - \hat{L}_y\hat{L}_x) = i[\hat{L}_y,\hat{L}_z] + j[\hat{L}_z,\hat{L}_x] + k[\hat{L}_x,\hat{L}_y]$$

$$= i(i\hat{L}_x + j\hat{L}_y + k\hat{L}_z) = i\hat{\tilde{L}}$$

5.8 Show that $[\hat{J}^2,\hat{J}_z] = 0$ by considering the commutators of $\hat{J}_z$ with $\hat{J}_x^2$, $\hat{J}_y^2$, and $\hat{J}_z^2$ separately. You may assume the commutation relation $\hat{\tilde{J}}x\hat{\tilde{J}} = i\hat{\tilde{J}}$.

Solution: Ref. Problem 3.26

Since $\hat{J}^2 = \hat{J}_x^2 + \hat{J}_y^2 + \hat{J}_z^2$ we can write

$$[\hat{J}^2,\hat{J}_z] = [\hat{J}_x^2,\hat{J}_z] + [\hat{J}_y^2,\hat{J}_z] + [\hat{J}_z^2,\hat{J}_z] \tag{1}$$

where

$$[\hat{J}_x^2,\hat{J}_z] = \hat{J}_x\hat{J}_x\hat{J}_z - \hat{J}_z\hat{J}_x\hat{J}_x + (\hat{J}_x\hat{J}_z\hat{J}_x - \hat{J}_x\hat{J}_z\hat{J}_x)$$

$$= \hat{J}_x(\hat{J}_x\hat{J}_z - \hat{J}_z\hat{J}_x) + (\hat{J}_x\hat{J}_z - \hat{J}_z\hat{J}_x)\hat{J}_x = -i(\hat{J}_x\hat{J}_y + \hat{J}_y\hat{J}_x) \tag{2}$$

$$[\hat{J}_y^2,\hat{J}_z] = \hat{J}_y\hat{J}_y\hat{J}_z - \hat{J}_z\hat{J}_y\hat{J}_y + (\hat{J}_y\hat{J}_z\hat{J}_y - \hat{J}_y\hat{J}_z\hat{J}_y)$$

$$= \hat{J}_y(\hat{J}_y\hat{J}_z - \hat{J}_z\hat{J}_y) + (\hat{J}_y\hat{J}_z - \hat{J}_z\hat{J}_y)\hat{J}_y = +i(\hat{J}_y\hat{J}_x + \hat{J}_x\hat{J}_y) \tag{3}$$

$$[\hat{J}_z^2,\hat{J}_z] = \hat{J}_z^3 - \hat{J}_z^3 = 0 \tag{4}$$

Inserting Eqs. (2)-(4) into (1) immediately gives

$$[\hat{J}^2,\hat{J}_z] = 0$$

It can also be shown that $[\hat{J}^2,\hat{J}_y] = [\hat{J}^2,\hat{J}_x] = 0$. This implies that a complete set of simultaneous eigenfunctions can be found for $\hat{J}^2$ and any one of the components of $\hat{J}$ (see Problem 3.9). It is important to remember that $\hat{J}\times\hat{J} = i\hat{J}$ so that no two components of $\hat{J}$ can have a complete set of simultaneous eigenfunctions.

5.9 In many applications it is convenient to define the operators $\hat{J}_+ = \hat{J}_x + i\hat{J}_y$ and $\hat{J}_- = \hat{J}_x - i\hat{J}_y$. These are called raising and lowering operators or ladder operators.

(a) Show that $[\hat{J}_z,\hat{J}_\pm] = \pm\hat{J}_\pm$. (b) Show that $[\hat{J}^2,\hat{J}_\pm] = 0$.

(c) Show that $\hat{J}^2 = \hat{J}_-\hat{J}_+ + \hat{J}_z + \hat{J}_z^2 = \hat{J}_+\hat{J}_- - \hat{J}_z + \hat{J}_z^2$

Solution:

(a) $$[\hat{J}_z,\hat{J}_\pm] = [\hat{J}_z,\hat{J}_x \pm i\hat{J}_y] = [\hat{J}_z,\hat{J}_x] \pm i[\hat{J}_z,\hat{J}_y]$$
$$= i\hat{J}_y \pm i(-i\hat{J}_x) = i\hat{J}_y \pm \hat{J}_x = \pm(\hat{J}_x \pm i\hat{J}_y) = \pm\hat{J}_\pm$$

(b) $$[\hat{J}^2,\hat{J}_\pm] = [\hat{J}^2,\hat{J}_x \pm i\hat{J}_y] = [\hat{J}^2,\hat{J}_x] \pm i[\hat{J}^2,\hat{J}_y] = 0$$

(c) We start with the equation
$$\hat{J}^2 = \hat{J}_x^2 + \hat{J}_y^2 + \hat{J}_z^2$$

But

$$\hat{J}_+\hat{J}_- = (\hat{J}_x + i\hat{J}_y)(\hat{J}_x - i\hat{J}_y) = \hat{J}_x^2 + \hat{J}_y^2 - i[\hat{J}_x,\hat{J}_y] = \hat{J}_x^2 + \hat{J}_y^2 + \hat{J}_z$$
$$\hat{J}_-\hat{J}_+ = (\hat{J}_x - i\hat{J}_y)(\hat{J}_x + i\hat{J}_y) = \hat{J}_x^2 + \hat{J}_y^2 + i[\hat{J}_x,\hat{J}_y] = \hat{J}_x^2 + \hat{J}_y^2 - \hat{J}_z$$

Therefore:

$$\hat{J}^2 = \hat{J}_+\hat{J}_- - \hat{J}_z + \hat{J}_z^2 \quad \text{and} \quad \hat{J}^2 = \hat{J}_-\hat{J}_+ + \hat{J}_z + \hat{J}_z^2$$

The last equations are particularly convenient to use when $\hat{J}^2$ is operating on eigenfunctions of $\hat{J}_z$.

5.10 Since $\hat{J}^2$ and $\hat{J}_z$ commute, they can have a complete set of simultaneous eigenfunctions. We denote such functions by u_{jm} and write

$$\hat{J}^2 u_{jm} = a_j u_{jm}, \quad \hat{J}_z u_{jm} = m u_{jm}$$

where a_j and m are constants which must be determined.

(a) Show that $\hat{J}^2(\hat{J}_\pm u_{jm}) = a_j(\hat{J}_\pm u_{jm})$ (b) Show that $\hat{J}_z(\hat{J}_\pm u_{jm}) = (m \pm 1)(\hat{J}_\pm u_{jm})$

(Hint: Since $[\hat{J}_\pm,\hat{J}^2] = 0$, it can be added or substracted arbitrarily.)

Solution: Ref. R03,24

(a) $$\hat{J}^2(\hat{J}_\pm u_{jm}) = \{\hat{J}^2\hat{J}_\pm + [\hat{J}_\pm,\hat{J}^2]\}u_{jm} = \{\hat{J}^2\hat{J}_\pm + \hat{J}_\pm\hat{J}^2 - \hat{J}^2\hat{J}_\pm\}u_{jm} = \hat{J}_\pm\hat{J}^2 u_{jm} = a_j(\hat{J}_\pm u_{jm})$$

(b) Here we make use of the trick of adding the substracting $\hat{J}_\pm \hat{J}_z$.

$$\hat{J}_z(\hat{J}_\pm u_{jm}) = \{\hat{J}_z\hat{J}_\pm + \hat{J}_\pm\hat{J}_z - \hat{J}_\pm\hat{J}_z\}u_{jm} = \{[\hat{J}_z,\hat{J}_\pm] + \hat{J}_\pm\hat{J}_z\}u_{jm} = \{\pm\hat{J}_\pm + \hat{J}_\pm\hat{J}_z\}u_{jm} = (m \pm 1)\hat{J}_\pm u_{jm}$$

Therefore, $\hat{J}_\pm u_{jm}$ is a new eigenfunction of $\hat{J}^2$ and $\hat{J}_z$ which has the eigenvalues a_j and $(m \pm 1)$, respectively. The operator $\hat{J}_\pm$ creates the new function $\hat{J}_\pm u_{jm}$ from the function u_{jm} by raising or lowering the eigenvalue m. This result can be written as

$$\hat{J}_\pm u_{jm} = C_\pm u_{j,m\pm 1}$$

where $C_\pm$ is a constant which must be determined.

5.11 The effects of the ladder operators $\hat{J}_\pm$ were shown in Problem 5.10 to be:

$$\hat{J}_\pm u_{jm} = C_\pm u_{j,m\pm 1}$$

where the $C_\pm$ are constants which must be determined. Here u_{jm} is an eigenfunction of $\hat{J}^2$ and $\hat{J}_z$. Determine $C_\pm$ so that $u_{j,m\pm 1}$ is normalized.

Solution:

Since $\{\hat{J}_\pm u_{jm}\}$ is a set of eigenfunctions, we can determine the "overlap."

$$\langle \hat{J}_\pm u_{jm} | \hat{J}_\pm u_{jm} \rangle = \langle C_\pm u_{j,m\pm 1} | C_\pm u_{j,m\pm 1} \rangle = |C_\pm|^2 \langle u_{j,m\pm 1} | u_{j,m\pm 1} \rangle$$

Then

$$|C_\pm|^2 = \langle \hat{J}_\pm u_{j,m} | \hat{J}_\pm u_{j,m} \rangle$$

This can be rewritten using the fact that $\hat{J}_\pm^+ = \hat{J}_\mp$.

$$|C_\pm|^2 = \langle j,m | \hat{J}_\mp \hat{J}_\pm | j,m \rangle = \langle j,m | \hat{J}^2 - \hat{J}_z^2 \mp \hat{J}_z | j,m \rangle$$

where the last step follows from Problem 5.9c. Finally,

$$|C_\pm|^2 = \langle j,m | j(j+1) - m^2 \mp m | j,m \rangle = j(j+1) - m^2 \mp m$$

We choose the phase factor to be unity and write:

$$C_\pm = [j(j+1) - m(m \pm 1)]^{1/2} = [(j \mp m)(j \pm m + 1)]^{1/2}$$

and

$$\boxed{\hat{J}_\pm u_{jm} = \sqrt{(j \mp m)(j \pm m + 1)}\; u_{j,m\pm 1}}$$

5.12 Write out matrices to represent the orbital angular momentum operators $\hat{L}^2$, $\hat{L}_z$, $\hat{L}_\pm$, $\hat{L}_x$, and $\hat{L}_y$ using the simultaneous eigenfunctions of $\hat{L}^2$ and $\hat{L}_z$ (i.e. the spherical harmonics) as basis functions. Show the matrix elements from L = 0 through L = 2 and order the basis functions as follows:

$$u_{0,0},\ u_{1,1},\ u_{1,0},\ u_{1,-1},\ u_{2,2},\ u_{2,1},\ u_{2,0},\ u_{2,-1},\ u_{2,-2},\ \ldots$$

<u>Solution</u>: Ref. DW,189; RO3,27

The necessary matrix elements are

$$\langle \ell', m' | \hat{L}^2 | \ell, m \rangle = \ell(\ell+1)\delta_{\ell',\ell}\delta_{m',m} \tag{1}$$

$$\langle \ell', m' | \hat{L}_z | \ell, m \rangle = m\delta_{\ell',\ell}\delta_{m',m} \tag{2}$$

$$\langle \ell', m' | \hat{L}_\pm | \ell, m \rangle = \sqrt{(\ell \mp m)(\ell \pm m + 1)}\; \delta_{\ell,\ell'}\delta_{m',m\pm 1} \tag{3}$$

Therefore:

$$\underset{\sim}{L}^2 = \left(\begin{array}{c|ccc|ccccc} 0 & & 0 & & & & 0 & & \\ \hline & 2 & 0 & 0 & & & & & \\ 0 & 0 & 2 & 0 & & & 0 & & \\ & 0 & 0 & 2 & & & & & \\ \hline & & & & 6 & 0 & 0 & 0 & 0 \\ & & & & 0 & 6 & 0 & 0 & 0 \\ 0 & & 0 & & 0 & 0 & 6 & 0 & 0 \\ & & & & 0 & 0 & 0 & 6 & 0 \\ & & & & 0 & 0 & 0 & 0 & 6 \end{array}\right) \qquad \underset{\sim}{L}_z = \left(\begin{array}{c|ccc|ccccc} 0 & & 0 & & & & 0 & & \\ \hline & 1 & 0 & 0 & & & & & \\ 0 & 0 & 0 & 0 & & & 0 & & \\ & 0 & 0 & -1 & & & & & \\ \hline & & & & 2 & 0 & 0 & 0 & 0 \\ & & & & 0 & 1 & 0 & 0 & 0 \\ 0 & & 0 & & 0 & 0 & 0 & 0 & 0 \\ & & & & 0 & 0 & 0 & -1 & 0 \\ & & & & 0 & 0 & 0 & 0 & -2 \end{array}\right)$$

From Eq. (3) we find

$$\underset{\sim}{L}_+ = \left(\begin{array}{c|ccc|ccccc} 0 & & 0 & & & & 0 & & \\ \hline & 0 & \sqrt{2} & 0 & & & & & \\ 0 & 0 & 0 & \sqrt{2} & & & 0 & & \\ & 0 & 0 & 0 & & & & & \\ \hline & & & & 0 & 2 & 0 & 0 & 0 \\ & & & & 0 & 0 & \sqrt{6} & 0 & 0 \\ 0 & & 0 & & 0 & 0 & 0 & \sqrt{6} & 0 \\ & & & & 0 & 0 & 0 & 0 & 2 \\ & & & & 0 & 0 & 0 & 0 & 0 \end{array}\right) \qquad \underset{\sim}{L}_- = \left(\begin{array}{c|ccc|ccccc} 0 & & 0 & & & & 0 & & \\ \hline & 0 & 0 & 0 & & & & & \\ 0 & \sqrt{2} & 0 & 0 & & & 0 & & \\ & 0 & \sqrt{2} & 0 & & & & & \\ \hline & & & & 0 & 0 & 0 & 0 & 0 \\ & & & & 2 & 0 & 0 & 0 & 0 \\ 0 & & 0 & & 0 & \sqrt{6} & 0 & 0 & 0 \\ & & & & 0 & 0 & \sqrt{6} & 0 & 0 \\ & & & & 0 & 0 & 0 & 2 & 0 \end{array}\right)$$

The matrices for $\underset{\sim}{L}_x$ and $\underset{\sim}{L}_y$ are obtained using the equations

$$\underset{\sim}{L}_x = \frac{1}{2}(\underset{\sim}{L}_+ + \underset{\sim}{L}_-), \qquad \underset{\sim}{L}_y = -\frac{i}{2}(\underset{\sim}{L}_+ - \underset{\sim}{L}_-)$$

Thus

$$
\underset{\sim}{L}_x = \frac{1}{2}\left(\begin{array}{c|ccc|ccccc}
0 & 0 & 0 & 0 & 0 & 0 & 0 & 0 & 0 \\
\hline
0 & 0 & \sqrt{2} & 0 & 0 & 0 & 0 & 0 & 0 \\
0 & \sqrt{2} & 0 & \sqrt{2} & 0 & 0 & 0 & 0 & 0 \\
0 & 0 & \sqrt{2} & 0 & 0 & 0 & 0 & 0 & 0 \\
\hline
0 & 0 & 0 & 0 & 0 & \sqrt{2} & 0 & 0 & 0 \\
0 & 0 & 0 & 0 & \sqrt{2} & 0 & \sqrt{6} & 0 & 0 \\
0 & 0 & 0 & 0 & 0 & \sqrt{6} & 0 & \sqrt{6} & 0 \\
0 & 0 & 0 & 0 & 0 & 0 & \sqrt{6} & 0 & \sqrt{2} \\
0 & 0 & 0 & 0 & 0 & 0 & 0 & \sqrt{2} & 0
\end{array}\right)
\qquad
\underset{\sim}{L}_y = \frac{i}{2}\left(\begin{array}{c|ccc|ccccc}
0 & 0 & 0 & 0 & 0 & 0 & 0 & 0 & 0 \\
\hline
0 & 0 & -\sqrt{2} & 0 & 0 & 0 & 0 & 0 & 0 \\
0 & \sqrt{2} & 0 & -\sqrt{2} & 0 & 0 & 0 & 0 & 0 \\
0 & 0 & \sqrt{2} & 0 & 0 & 0 & 0 & 0 & 0 \\
\hline
0 & 0 & 0 & 0 & 0 & -2 & 0 & 0 & 0 \\
0 & 0 & 0 & 0 & 2 & 0 & -\sqrt{6} & 0 & 0 \\
0 & 0 & 0 & 0 & 0 & \sqrt{6} & 0 & -\sqrt{6} & 0 \\
0 & 0 & 0 & 0 & 0 & 0 & \sqrt{6} & 0 & -2 \\
0 & 0 & 0 & 0 & 0 & 0 & 0 & 2 & 0
\end{array}\right)
$$

5.13 (a) Write out the matricies for $\hat{J}_+$ and $\hat{J}_-$ for a system having $j = 1/2$ in the representation where $\hat{J}^2$ and $\hat{J}_z$ are diagonal, i.e. use $|j,m\rangle$ as the basis functions.
(b) Derive matricies for $\hat{J}_x$, $\hat{J}_y$, $\hat{J}_z$, and $\hat{J}^2$ for the system in part (a).

Solution:

(a) From Problem 5.11 we have

$$\langle j',m'|\hat{J}_\pm|j,m\rangle = \sqrt{(j \mp m)(j \pm m + 1)}\ \delta_{j',j}\delta_{m',m\pm 1}$$

By convention the array of elements is arranged as follows:

$$\langle j,m|\ |j,m'\rangle \to \begin{pmatrix} \langle j,j|\ |j,j\rangle & \langle j,j|\ |j,j-1\rangle & \cdots \\ \langle j,j-1|\ |j,j\rangle & & \\ \vdots & & \end{pmatrix}$$

Therefore, for $j = 1/2$ we obtain:

$$\underset{\sim}{J}_+ = \begin{pmatrix} 0 & 1 \\ 0 & 0 \end{pmatrix}; \qquad \underset{\sim}{J}_- = \begin{pmatrix} 0 & 0 \\ 1 & 0 \end{pmatrix}$$

(b) Since $\hat{J}_x = (\hat{J}_+ + \hat{J}_-)/2$ and $\hat{J}_y = (\hat{J}_+ - \hat{J}_-)/2i$, we have

$$\underset{\sim}{J}_x = \begin{pmatrix} 0 & 1/2 \\ 1/2 & 0 \end{pmatrix}; \qquad \underset{\sim}{J}_y = \begin{pmatrix} 0 & -i/2 \\ i/2 & 0 \end{pmatrix}$$

Also, $\langle j',m'|\hat{J}_z|j,m\rangle = m\delta_{j',j}\delta_{m',m}$ so

$$\underset{\sim}{J}_z = \begin{pmatrix} 1/2 & 0 \\ 0 & 1/2 \end{pmatrix}$$

The matrix for $\hat{J}^2$ is obtained from the relation

$$\hat{J}^2 = \hat{J}_x^2 + \hat{J}_y^2 + \hat{J}_z^2$$

$$\underset{\sim}{J}^2 = \begin{pmatrix} 0 & 1/2 \\ 1/2 & 0 \end{pmatrix}^2 + \begin{pmatrix} 0 & -i/2 \\ i/2 & 0 \end{pmatrix}^2 + \begin{pmatrix} 1/2 & 0 \\ 0 & 1/2 \end{pmatrix}^2$$

and

$$\underset{\sim}{J}^2 = \begin{pmatrix} 3/4 & 0 \\ 0 & 3/4 \end{pmatrix}$$

5.14 (a) Given the matrix for $\hat{J}_x$ in the $\hat{J}^2$, $\hat{J}_z$ representation with $j = 1/2$ (see Problem 5.13), obtain the unitary matrix $\underset{\sim}{S}$ which diagonalizes $\underset{\sim}{J}_x$ by the transformation $\underset{\sim}{S}^{-1}\underset{\sim}{J}_x\underset{\sim}{S}$. Compare the $\underset{\sim}{S}$ matrix with the matrix which represents a simple rotation around the y direction.

(b) Write out the matrices for $\hat{J}_y$, $\hat{J}_z$, and $\hat{J}^2$ in the representation where $\underset{\sim}{J}_x$ is diagonal.

Solution: Ref. Appendix 6

(a) The transformation matrix $\underset{\sim}{S}$ can be constructed from the eigenvectors of $\hat{J}_x$. We first diagonalize $\hat{J}_x$ (see Appendix 6) by requiring that the secular determinant vanish:

$$\begin{vmatrix} -\lambda & 1/2 \\ 1/2 & -\lambda \end{vmatrix} = \lambda^2 - \frac{1}{4} = 0$$

The eigenvalues are found to be $\lambda_1 = 1/2$, $\lambda_2 = -1/2$.

The elements of the eigenvector for $\lambda_1 = 1/2$ obey the equation

$$(0 - \lambda_1)S_{11} + (1/2)S_{21} = 0; \quad (-1/2)S_{11} + (1/2)S_{21} = 0 \quad \text{or} \quad S_{21} = +S_{11}$$

Normalization then gives:

$$S_{21}^2 + S_{11}^2 = 1, \quad S_{11} = \pm 1/\sqrt{2}$$

We choose the phase so that:

$$|1> = \frac{1}{\sqrt{2}}\,|j=1/2,m=1/2> + \frac{1}{\sqrt{2}}\,|j=1/2,m=-1/2>$$

For the eigenvalue $\lambda_2 = -1/2$ the elements of the corresponding eigenvector obey the equation:

$$(0 - \lambda_2)S_{12} + (1/2)S_{22} = 0; \quad (1/2)S_{12} = (1/2)S_{22} \quad \text{or} \quad S_{12} = -S_{22}$$

Again the normalization requires that $S_{12} = \pm 1/\sqrt{2}$ and we write the eigenvector as:

$$|2> = \frac{1}{\sqrt{2}}\,|j=1/2,m=1/2> - \frac{1}{\sqrt{2}}\,|j=1/2,m=-1/2>$$

From the eigenvectors the $\underset{\sim}{S}$ matrix is found to have the form:

$$\underset{\sim}{S} = \frac{1}{\sqrt{2}}\begin{pmatrix} 1 & -1 \\ 1 & 1 \end{pmatrix}, \quad \underset{\sim}{S}^{-1} = \frac{1}{\sqrt{2}}\begin{pmatrix} 1 & 1 \\ -1 & 1 \end{pmatrix}$$

As a check we multiply out the product $\underset{\sim}{S}^{-1}\underset{\sim}{J}_x\underset{\sim}{S}$:

$$\begin{pmatrix} 1/\sqrt{2} & 1/\sqrt{2} \\ -1/\sqrt{2} & 1/\sqrt{2} \end{pmatrix} \begin{pmatrix} 0 & 1/2 \\ 1/2 & 0 \end{pmatrix} \begin{pmatrix} 1/\sqrt{2} & -1/\sqrt{2} \\ 1/\sqrt{2} & 1/\sqrt{2} \end{pmatrix} = \begin{pmatrix} 1/2 & 0 \\ 0 & -1/2 \end{pmatrix}$$

It should be noted here that the form of $\underset{\sim}{S}$ depends on the choice of phases for the eigenvectors. For example the transformation matrices

$$\underset{\sim}{S}' = (\underset{\sim}{S}^{-1})' = \frac{1}{\sqrt{2}}\begin{pmatrix} 1 & 1 \\ 1 & -1 \end{pmatrix}$$

would also diagonalize $\underset{\sim}{J}_x$.

The matrix which represents a rotation of the coordinate system through an angle θ around the y direction is shown in the transformation below

$$\begin{pmatrix} x' \\ z' \end{pmatrix} = \begin{pmatrix} \cos\theta & -\sin\theta \\ \sin\theta & \cos\theta \end{pmatrix}\begin{pmatrix} x \\ z \end{pmatrix}$$

where (x',y') gives the coordinates of a point (x,y) as seen in the rotated coordinate system. By comparison the $\underset{\sim}{S}$ matrix for this transformation is found to be equivalent to the rotation matrix for $\theta = \pi/4$. (See Problem 5.18)

(b) The matrices for $\hat{J}_y$, $\hat{J}_z$, and $\hat{J}^2$ from part (a) can be transformed into the representation where $\hat{J}_x$ is diagonal by means of the $\underset{\sim}{S}$ matrix from part (a).

$$\underset{\sim}{J}'_y = \begin{pmatrix} 1/\sqrt{2} & 1/\sqrt{2} \\ -1/\sqrt{2} & 1/\sqrt{2} \end{pmatrix}\begin{pmatrix} 0 & -i/2 \\ i/2 & 0 \end{pmatrix}\begin{pmatrix} 1/\sqrt{2} & -1/\sqrt{2} \\ 1/\sqrt{2} & 1/\sqrt{2} \end{pmatrix} = \begin{pmatrix} 0 & -i/2 \\ i/2 & 0 \end{pmatrix}$$

$$\underset{\sim}{J}'_z = \begin{pmatrix} 1/\sqrt{2} & 1/\sqrt{2} \\ -1/\sqrt{2} & 1/\sqrt{2} \end{pmatrix}\begin{pmatrix} 1/2 & 0 \\ 0 & -1/2 \end{pmatrix}\begin{pmatrix} 1/\sqrt{2} & -1/\sqrt{2} \\ 1/\sqrt{2} & 1/\sqrt{2} \end{pmatrix} = \begin{pmatrix} 0 & -1/2 \\ -1/2 & 0 \end{pmatrix}$$

$$\underset{\sim}{J}^2 = \begin{pmatrix} 1/\sqrt{2} & 1/\sqrt{2} \\ -1/\sqrt{2} & 1/\sqrt{2} \end{pmatrix}\begin{pmatrix} 3/4 & 0 \\ 0 & 3/4 \end{pmatrix}\begin{pmatrix} 1/\sqrt{2} & -1/\sqrt{2} \\ 1/\sqrt{2} & 1/\sqrt{2} \end{pmatrix} = \begin{pmatrix} 3/4 & 0 \\ 0 & 3/4 \end{pmatrix}$$

5.15 The spherical harmonics $Y_{\ell m}(\theta,\phi)$ are simultaneous eigenfunctions of the orbital angular momentum operators $\hat{L}^2$ and $\hat{L}_z$. Thus

$$\hat{L}^2 Y_{\ell m} = \ell(\ell + 1)Y_{\ell m}; \quad \hat{L}_z Y_{\ell m} = mY_{\ell m}$$

(a) Separate the variables in the $\hat{L}^2$ equation to obtain equations for $\Phi(\phi)$ and $P_{\ell m}(\theta)$ by assuming that $Y_{\ell m}(\theta,\phi) = \Phi(\phi)P_{\ell m}(\theta)$.

(b) Determine the normalized function $\Phi(\phi)$.

Solution: Ref. PW,115

(a) We use $\mathcal{L}^2$ from Problem 5.4 to obtain

$$-\left[\frac{1}{\sin\theta}\frac{\partial}{\partial\theta}\sin\theta\frac{\partial}{\partial\theta} + \frac{1}{\sin^2\theta}\frac{\partial^2}{\partial\phi^2}\right]Y_{\ell m}(\theta,\phi) = \ell(\ell + 1)Y_{\ell m}(\theta,\phi)$$

Now making the substitution $Y_{\ell m}(\theta,\phi) = \Phi(\phi)P_{\ell m}(\theta)$ and dividing by $Y_{\ell m}(\theta,\phi)$ gives:

$$-\left[\frac{1}{P_{\ell m}\sin\theta}\frac{d}{d\theta}\sin\theta\frac{dP_{\ell m}}{d\theta} + \frac{1}{\Phi\sin^2\theta}\frac{d^2\Phi}{d\phi^2}\right] = \ell(\ell + 1)$$

$$\sin^2\theta\left[\frac{1}{P_{\ell m}\sin\theta}\frac{d}{d\theta}\sin\theta\frac{dP_{\ell m}}{d\theta} + \ell(\ell + 1)\right] = -\frac{1}{\Phi}\frac{d^2\Phi}{d\phi^2}$$

The two sides of this equation depend on different independent variables. They can be identically equal for all values of θ and ϕ only if both sides are equal to the same constant. We set this separation constant equal to m^2 to be consistent with the $\hat{L}_z$ equation above and write

$$\left[\frac{1}{\sin\theta}\frac{d}{d\theta}\sin\theta\frac{d}{d\theta} - \frac{m^2}{\sin^2\theta}\right]P_{\ell m}(\theta) = -\ell(\ell+1)P_{\ell m}(\theta)$$

which is the differential equation for the Associated Legendre Functions (see Appendix 4). Also,

$$\frac{d^2\Phi(\phi)}{d\phi^2} = -m^2\Phi(\phi) \quad \text{or} \quad \hat{L}_z^2\Phi(\phi) = m^2\Phi(\phi)$$

(b) The $\hat{L}_z$ equation (see Problem 5.3) is

$$-i\frac{d}{d\phi}\Phi = m\Phi$$

Integration gives the solution: $\Phi(\phi) = Ae^{im\phi}$ where the boundary condition $\Phi(\phi + 2\pi) = \Phi(\phi)$ requires that $m = 0, \pm1, \pm2, \cdots$.

Normalization gives

$$\int_0^{2\pi}\Phi^*(\phi)\Phi(\phi)d\phi = 2\pi|A|^2 = 1$$

We choose the phase so that $A = 1/\sqrt{2\pi}$ and write:

$$\boxed{\Phi(\phi) = \frac{1}{\sqrt{2\pi}}e^{im\phi}, \quad m = 0, \pm1, \pm2, \cdots}$$

5.16 The unitary operator $\hat{R}(\hat{n},\theta)$ can be used to relate the wave function ψ' after a rotation of the coordinate system to the wave function ψ before the rotation. Hence

$$\psi' = \hat{R}(\hat{n},\theta)\psi$$

where $\hat{n}$ defines the axis of rotation and θ is the rotation angle. The angular momentum $\hat{\underset{\sim}{J}}$ is defined as follows:

$$\hat{R}\psi = e^{-i\theta(\hat{n}\cdot\hat{\underset{\sim}{J}})}\psi$$

(a) Show that the infinitesimal rotations $\exp(-id\theta_y\hat{J}_y)$ and $\exp(-id\theta_x\hat{J}_x)$ do not commute by expanding each to second order.

(b) If $\hat{R}_1 = \exp(-id\theta_y\hat{J}_y)\exp(-id\theta_x\hat{J}_x)$ and $\hat{R}_2 = \exp(-id\theta_x\hat{J}_x)\exp(-id\theta_y\hat{J}_y)$, the difference $\hat{R}_2 - \hat{R}_1$ is found to correspond to a rotation of magnitude $d\theta_x d\theta_y$ about the z-direction (see Figure 5.16), i.e. $(\hat{R}_2 - \hat{R}_1)\psi = (\hat{R} - 1)\psi$ where $\hat{R} = \exp(-id\theta_x d\theta_y\hat{J}_z)$. Use this result with the conclusion of part (a) to show that $[\hat{J}_x,\hat{J}_y] = i\hat{J}_z$.

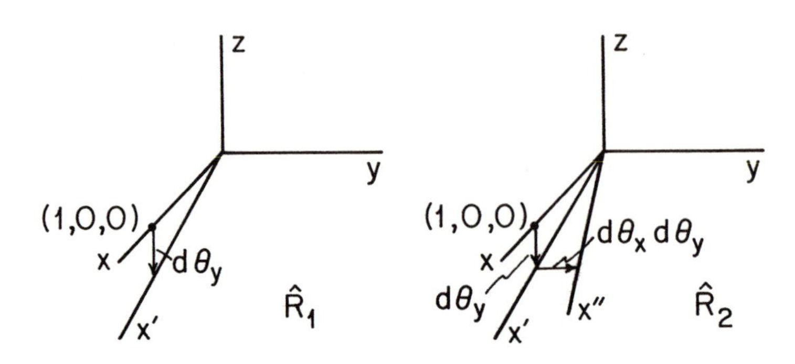

Fig. 5.16

Solution: Ref. RO3,20

(a) The commutator of the rotations $\hat{R}_x$ and $\hat{R}_y$ is:

$$[e^{-id\theta_x\hat{J}_x}, e^{-id\theta_y\hat{J}_y}] \cong [1 - id\theta_x\hat{J}_x - \left(\frac{d\theta_x}{2}\right)^2 \hat{J}_x^2][1 - id\theta_y\hat{J}_y - \left(\frac{d\theta_y}{2}\right)^2 \hat{J}_y^2]$$

$$- [1 - id\theta_y\hat{J}_y - \left(\frac{d\theta_y}{2}\right)^2 \hat{J}_y^2][1 - id\theta_x\hat{J}_x - \left(\frac{d\theta_x}{2}\right)^2 \hat{J}_x^2] \cong -d\theta_x d\theta_y[\hat{J}_x,\hat{J}_y]$$

The linear terms cancel but terms of second and higher degree in dθ do not.

(b) We are given:

$$\hat{R}_2 - \hat{R}_1 = e^{-id\theta_x d\theta_y \hat{J}_z} - 1$$

Expanding the right side to second order terms and using the results of part (a) gives

$$-d\theta_x d\theta_y[\hat{J}_x,\hat{J}_y] = 1 - id\theta_x d\theta_y\hat{J}_z - 1 \quad \text{or} \quad [\hat{J}_x,\hat{J}_y] = i\hat{J}_z$$

The definition of angular momentum given here reveals that the components of $\hat{\underset{\sim}{J}}$ can be determined by investigating the transformation properties of a system. This conclusion is not restricted to orbital angular momenta.

5.17 Consider the special case in which the Hamiltonian $\hat{\mathcal{H}}$ for a system is invariant to any rotation $\hat{R}(\hat{n},\theta)$, i.e. $\hat{R}^{-1}\hat{\mathcal{H}}\hat{R} = \hat{\mathcal{H}}$. Show that $[\hat{\mathcal{H}},\hat{J}_\alpha^n] = [\hat{\mathcal{H}},\hat{J}^2] = 0$ where $\hat{J}_\alpha = \hat{J}\cdot\hat{n}$. Here θ is the angle of rotation and $\hat{n}$ specifies the rotation axis. (Hint: Expand $\hat{R}(\hat{n},\theta)$ in a power series.)

Solution: Ref. RO3,23

The invariance of $\hat{\mathcal{H}}$ implies that

$$e^{-i\hat{J}_\alpha\theta}\hat{\mathcal{H}}\, e^{+i\hat{J}_\alpha\theta} = \hat{\mathcal{H}} \quad \text{or} \quad \hat{\mathcal{H}}\, e^{+i\hat{J}_\alpha\theta} - e^{+i\hat{J}_\alpha\theta}\hat{\mathcal{H}} = 0$$

Expanding the exponential terms gives:

$$\hat{\mathcal{H}}[1 + i\hat{J}_\alpha\theta + (i\hat{J}_\alpha)^2 \frac{\theta^2}{2!} + \cdots] - [1 + i\hat{J}_\alpha\theta + (i\hat{J}_\alpha)^2 \frac{\theta^2}{2!} + \cdots]\,\hat{\mathcal{H}} = 0$$

For this equation to hold for all values of θ the coefficient of each power of θ must vanish. Therefore,

$$[\hat{\mathcal{H}},\hat{J}_\alpha] = [\hat{\mathcal{H}},\hat{J}_\alpha^2] = \cdots = [\hat{\mathcal{H}},\hat{J}_\alpha^n] = 0$$

Since $[\hat{\mathcal{H}},\hat{J}_\alpha^2] = 0$ we have

$$[\hat{\mathcal{H}},\hat{J}_x^2 + \hat{J}_y^2 + \hat{J}_z^2] = 0 \quad \text{or} \quad [\hat{\mathcal{H}},\hat{J}^2] = 0$$

For such a system $\hat{\mathcal{H}}$, $\hat{J}^2$, and one component of $\hat{\underset{\sim}{J}}$ can have a complete set of simultaneous eigenfunctions.

5.18 From the definition of angular momentum the operator

$$\hat{R}(\hat{j},\beta) = e^{-i\beta\hat{J}_y}$$

corresponds to a rotation of the coordinate system through an angle β around the y direction. Show that the matrix for $\hat{R}$ in the representation where $\hat{J}_z$ and $\hat{J}^2$ are diagonal with $j = 1/2$ is

$$\begin{pmatrix} \cos\beta/2 & -\sin\beta/2 \\ \sin\beta/2 & \cos\beta/2 \end{pmatrix}$$

In conventional notation this is the $d^{1/2}_{mm'}(\beta)$ matrix. (Hint: Write $\hat{J}_y$ in matrix form and expand the exponential term.)

Solution: Ref. BS,25

As suggested

$$\underset{\sim}{R} = \exp\left[-i\beta\begin{pmatrix} 0 & -i/2 \\ i/2 & 0 \end{pmatrix}\right] = \exp\left[\frac{\beta}{2}\begin{pmatrix} 0 & -1 \\ 1 & 0 \end{pmatrix}\right]$$

and

$$\underset{\sim}{R} = \underset{\sim}{1} + \frac{\beta}{2}\begin{pmatrix} 0 & -1 \\ 1 & 0 \end{pmatrix} + \left(\frac{\beta}{2}\right)^2\frac{1}{2!}\begin{pmatrix} 0 & -1 \\ 1 & 0 \end{pmatrix}^2 + \cdots \left(\frac{\beta}{2}\right)^n\frac{1}{n!}\begin{pmatrix} 0 & -1 \\ 1 & 0 \end{pmatrix}^n \cdots$$

But

$$\begin{pmatrix} 0 & -1 \\ 1 & 0 \end{pmatrix}^2 = -\begin{pmatrix} 1 & 0 \\ 0 & 1 \end{pmatrix} \quad \text{and} \quad \begin{pmatrix} 0 & -1 \\ 1 & 0 \end{pmatrix}^{2n} = (-1)^n\begin{pmatrix} 1 & 0 \\ 0 & 1 \end{pmatrix}$$

Also

$$\begin{pmatrix} 0 & -1 \\ 1 & 0 \end{pmatrix}^{2n+1} = (-1)^n\begin{pmatrix} 0 & -1 \\ 1 & 0 \end{pmatrix}$$

To take advantage of these relations we write $\underset{\sim}{R}$ in the form:

$$R = \begin{pmatrix} 1 & 0 \\ 0 & 1 \end{pmatrix} + \sum_{n=0}^{\infty}\left(\frac{\beta}{2}\right)^{2n+1}\frac{1}{(2n+1)!}\begin{pmatrix} 0 & -1 \\ 1 & 0 \end{pmatrix}^{2n+1} + \sum_{n=1}^{\infty}\left(\frac{\beta}{2}\right)^{2n}\frac{1}{(2n)!}\begin{pmatrix} 0 & -1 \\ 1 & 0 \end{pmatrix}^{2n}$$

$$= \begin{pmatrix} 1 & 0 \\ 0 & 1 \end{pmatrix} + \begin{pmatrix} 0 & -1 \\ 1 & 0 \end{pmatrix}\sum_{n=0}^{\infty}\left(\frac{\beta}{2}\right)^{2n+1}\frac{(-1)^n}{(2n+1)!} + \begin{pmatrix} 1 & 0 \\ 0 & 1 \end{pmatrix}\sum_{n=1}^{\infty}\left(\frac{\beta}{2}\right)^{2n}\frac{(-1)^n}{(2n)!}$$

The elements of the matrix $\underset{\sim}{R}$ are then:

$$R_{11} = R_{22} = \sum_{n=0}^{\infty} \left(\frac{\beta}{2}\right)^{2n} \frac{(-1)^n}{(2n)!} = \cos\beta/2 \qquad (0! \equiv 1)$$

$$R_{21} = -R_{12} = \sum_{n=0}^{\infty} \left(\frac{\beta}{2}\right)^{2n+1} \frac{(-1)^n}{(2n+1)!} = \sin\beta/2$$

and as required we find

$$\underset{\sim}{R} = \begin{pmatrix} \cos\beta/2 & -\sin\beta/2 \\ \sin\beta/2 & \cos\beta/2 \end{pmatrix}$$

5.19 A particularly simple and useful representation for s = 1/2 results if we make the formal definitions:

$$\underset{\sim x}{\sigma} = 2\underset{\sim x}{s} = \begin{pmatrix} 0 & 1 \\ 1 & 0 \end{pmatrix} \qquad \underset{\sim y}{\sigma} = 2\underset{\sim y}{s} = \begin{pmatrix} 0 & -i \\ i & 0 \end{pmatrix} \qquad \underset{\sim z}{\sigma} = 2\underset{\sim z}{s} = \begin{pmatrix} 1 & 0 \\ 0 & -1 \end{pmatrix}$$

The $\underset{\sim\alpha}{\sigma}$ matrices are called the <u>Pauli Spin matrices</u>. Show that:

(a) The Pauli matrices anticommute, <u>i.e.</u> that

$$\sigma_\alpha\sigma_\beta + \sigma_\beta\sigma_\alpha = 0, \quad \alpha \neq \beta$$

(b) $\sigma_\alpha\sigma_\beta = i\sigma_\gamma, \quad \alpha \neq \beta \neq \gamma$

(c) The Pauli matrices are unitary.

(d) $\underset{\sim}{\sigma} \times \underset{\sim}{\sigma} = 2i\underset{\sim}{\sigma}$.

<u>Solution:</u>

(a)
$$\sigma_x\sigma_y + \sigma_y\sigma_x = \begin{pmatrix} 0 & 1 \\ 1 & 0 \end{pmatrix}\begin{pmatrix} 0 & -i \\ i & 0 \end{pmatrix} + \begin{pmatrix} 0 & -i \\ i & 0 \end{pmatrix}\begin{pmatrix} 0 & 1 \\ 1 & 0 \end{pmatrix} = \begin{pmatrix} i & 0 \\ 0 & -i \end{pmatrix} + \begin{pmatrix} -i & 0 \\ 0 & i \end{pmatrix} = 0$$

$$\sigma_x\sigma_z + \sigma_z\sigma_x = \begin{pmatrix} 0 & 1 \\ 1 & 0 \end{pmatrix}\begin{pmatrix} 1 & 0 \\ 0 & -1 \end{pmatrix} + \begin{pmatrix} 1 & 0 \\ 0 & -1 \end{pmatrix}\begin{pmatrix} 0 & 1 \\ 1 & 0 \end{pmatrix} = \begin{pmatrix} 0 & -1 \\ 1 & 0 \end{pmatrix} + \begin{pmatrix} 0 & 1 \\ -1 & 0 \end{pmatrix} = 0$$

$$\sigma_y\sigma_z + \sigma_z\sigma_y = \begin{pmatrix} 0 & -i \\ i & 0 \end{pmatrix}\begin{pmatrix} 1 & 0 \\ 0 & -1 \end{pmatrix} + \begin{pmatrix} 1 & 0 \\ 0 & -1 \end{pmatrix}\begin{pmatrix} 0 & -i \\ i & 0 \end{pmatrix} = \begin{pmatrix} 0 & i \\ i & 0 \end{pmatrix} + \begin{pmatrix} 0 & -i \\ -i & 0 \end{pmatrix} = 0$$

(b)
$$\sigma_x\sigma_y = \begin{pmatrix} 0 & 1 \\ 1 & 0 \end{pmatrix}\begin{pmatrix} 0 & -i \\ i & 0 \end{pmatrix} = \begin{pmatrix} i & 0 \\ 0 & -i \end{pmatrix} = i\begin{pmatrix} 1 & 0 \\ 0 & -1 \end{pmatrix} = i\sigma_z$$

$$\sigma_y\sigma_z = \begin{pmatrix} 0 & -i \\ i & 0 \end{pmatrix}\begin{pmatrix} 1 & 0 \\ 0 & -1 \end{pmatrix} = \begin{pmatrix} 0 & i \\ i & 0 \end{pmatrix} = i\begin{pmatrix} 0 & 1 \\ 1 & 0 \end{pmatrix} = i\sigma_x$$

$$\sigma_z\sigma_x = \begin{pmatrix} 1 & 0 \\ 0 & -1 \end{pmatrix}\begin{pmatrix} 0 & 1 \\ 1 & 0 \end{pmatrix} = \begin{pmatrix} 0 & 1 \\ -1 & 0 \end{pmatrix} = i\begin{pmatrix} 0 & -i \\ i & 0 \end{pmatrix} = i\sigma_y$$

(c) If a matrix is unitary $\underset{\sim}{A}^+ = \underset{\sim}{A}^{-1}$, <u>i.e.</u> the transpose of the complex conjugate of $\underset{\sim}{A}$ equals the inverse of $\underset{\sim}{A}$. By direct multiplication we can show that

$$\sigma_\alpha^+\sigma_\alpha = 1 \quad (\alpha = x, y, z)$$

thus

$$\underset{\sim}{\sigma}_{\alpha}^{+} = \underset{\sim}{\sigma}_{\alpha}^{-1}$$

so

$$\underset{\sim}{\sigma}_{x}^{+} = \begin{pmatrix} 0 & 1 \\ 1 & 0 \end{pmatrix} = \underset{\sim}{\sigma}_{x}^{-1} \qquad \underset{\sim}{\sigma}_{y}^{+} = \begin{pmatrix} 0 & -i \\ i & 0 \end{pmatrix} = \underset{\sim}{\sigma}_{y}^{-1} \quad \text{and} \quad \underset{\sim}{\sigma}_{z}^{+} = \begin{pmatrix} 1 & 0 \\ 0 & -1 \end{pmatrix} = \underset{\sim}{\sigma}_{z}^{-1}$$

so that the σ_{α}'s are unitary.

(d) $\underset{\sim}{\sigma} \times \underset{\sim}{\sigma} = \hat{i}[\underset{\sim}{\sigma}_y\underset{\sim}{\sigma}_z - \underset{\sim}{\sigma}_z\underset{\sim}{\sigma}_y] + \hat{j}[\underset{\sim}{\sigma}_z\underset{\sim}{\sigma}_x - \underset{\sim}{\sigma}_x\underset{\sim}{\sigma}_z] + \hat{k}[\underset{\sim}{\sigma}_x\underset{\sim}{\sigma}_y - \underset{\sim}{\sigma}_y\underset{\sim}{\sigma}_x]$

From part (a) we have

$$\underset{\sim}{\sigma}_{\alpha}\underset{\sim}{\sigma}_{\beta} = -\underset{\sim}{\sigma}_{\beta}\underset{\sim}{\sigma}_{\alpha}$$

so

$$\underset{\sim}{\sigma} \times \underset{\sim}{\sigma} = 2[\hat{i}(\underset{\sim}{\sigma}_y\underset{\sim}{\sigma}_z) + \hat{j}(\underset{\sim}{\sigma}_z\underset{\sim}{\sigma}_x) + \hat{k}(\underset{\sim}{\sigma}_x\underset{\sim}{\sigma}_y)]$$

and from part (b)

$$\underset{\sim}{\sigma}_{\alpha}\underset{\sim}{\sigma}_{\beta} = i\underset{\sim}{\sigma}_{\gamma}$$

Therefore:

$$\underset{\sim}{\sigma} \times \underset{\sim}{\sigma} = 2i[\hat{i}(\underset{\sim}{\sigma}_x) + \hat{j}(\underset{\sim}{\sigma}_y) + \hat{k}(\underset{\sim}{\sigma}_z)] = 2i\underset{\sim}{\sigma}$$

5.20 Let $\hat{\underset{\sim}{J}}$ be defined by the equation

$$\hat{\underset{\sim}{J}} = \hat{\underset{\sim}{J}}_1 + \hat{\underset{\sim}{J}}_2$$

where $\hat{\underset{\sim}{J}}_1$ and $\hat{\underset{\sim}{J}}_2$ are angular momentum operators. Show that $\hat{\underset{\sim}{J}}$ is also an angular momentum operator by investigating the commutation relations of the components of $\hat{\underset{\sim}{J}}$.

Solution: Ref. R03,33

Consider the commutator

$$\begin{aligned}[\hat{J}_x,\hat{J}_y] &= [\hat{J}_{1x} + \hat{J}_{2x},\hat{J}_{1y} + \hat{J}_{2y}] = [\hat{J}_{1x},\hat{J}_{1y}] + [\hat{J}_{1x},\hat{J}_{2y}] + [\hat{J}_{2x},\hat{J}_{1y}] + [\hat{J}_{2x},\hat{J}_{2y}] \\ &= i\hat{J}_{1z} + i\hat{J}_{2z} = i\hat{J}_z\end{aligned}$$

Terms involving different spaces such as $[\hat{J}_{1x},\hat{J}_{2y}]$ vanish. Similarly

$$[\hat{J}_{1y} + \hat{J}_{2y},\hat{J}_{1z} + \hat{J}_{2z}] = [\hat{J}_{1y},\hat{J}_{1z}] + [\hat{J}_{2y},\hat{J}_{2z}] = i\hat{J}_{1x} + i\hat{J}_{2x} = i\hat{J}_x$$

$$[\hat{J}_{1z} + \hat{J}_{2z},\hat{J}_{1x} + \hat{J}_{2x}] = [\hat{J}_{1z},\hat{J}_{1x}] + [\hat{J}_{2z},\hat{J}_{2x}] = i\hat{J}_{1y} + i\hat{J}_{2y} = i\hat{J}_y$$

Therefore

$$\hat{\underset{\sim}{J}} \times \hat{\underset{\sim}{J}} = i\hat{\underset{\sim}{J}}$$

and $\hat{\underset{\sim}{J}}$ is identified as an angular momentum vector operator.

5.21 The angular momentum eigenfunctions are defined as follows:

$$\hat{J}^2 u_{jm} = j(j+1)u_{jm}; \quad \hat{J}_z u_{jm} = m u_{jm}$$

$$\hat{J}_1^2 u_{j_1m_1} = j_1(j_1+1)u_{j_1m_1}; \quad J_{1z}u_{j_1m_1} = m_1 u_{j_1m_1}$$

$$\hat{J}_2^2 u_{j_2m_2} = j_2(j_2 + 1)u_{j_2m_2}; \quad J_{2z}u_{j_2m_2} = m_2 u_{j_2m_2}$$

where $\hat{\underset{\sim}{J}} = \hat{\underset{\sim}{J}}_1 + \hat{\underset{\sim}{J}}_2$. Since the functions $u_{j_1m_1}u_{j_2m_2}$ "span the space" of the operators $\hat{J}^2$ and $\hat{J}_z$, u_{jm} can be expressed as

$$u_{jm} = \sum_{m_1m_2} C(j_1j_2j;m_1m_2m)u_{j_1m_1}u_{j_2m_2} \tag{1}$$

where the constants $C(j_1j_2j;m_1m_2m)$ are called Clebsch-Gordon coefficients. Show that the coefficients vanish except when $m_1 + m_2 = m$. (Hint: Operate on the equation for u_{jm} with $\hat{J}_z = \hat{J}_{1z} + \hat{J}_{2z}$.)

Solution: Ref. RO3,33

From the definition above we can write

$$\hat{J}_z u_{jm} = \sum_{m_1m_2} C(j_1j_2j;m_1m_2m)(\hat{J}_{1z} + \hat{J}_{2z})u_{j_1m_1}u_{j_2m_2}$$

$$mu_{jm} = \sum_{m_1m_2} (m_1 + m_2)C(j_1j_2j;m_1m_2m)u_{j_1m_1}u_{j_2m_2}$$

Then expanding u_{jm} by use of Eq. (1)

$$\sum_{m_1m_2} (m - m_1 - m_2)C(j_1j_2j;m_1m_2m)u_{j_1m_1}u_{j_2m_2} = 0$$

Since the functions $u_{j_1m_1}u_{j_2m_2}$ are linearly independent, each of the coefficients must vanish:

$$(m - m_1 - m_2)C(j_1j_2j;m_1m_2m) = 0$$

We conclude that $C(j_1j_2j;m_1m_2m) = 0$ unless $m = m_1 + m_2$. The expansion can therefore be simplified by replacing m_2 with $m - m_1$:

$$u_{jm} = \sum_{m_1} C(j_1j_2j;m_1,m - m_1)u_{j_1m_1}u_{j_2m-m_1}$$

In Dirac notation the expansion of the ket $|j_1j_2;jm\rangle$ is given by

$$|j_1j_2;jm\rangle = \sum_{m_1m_2} |j_1m_1\rangle|j_2m_2\rangle\langle j_2m_2|\langle j_1m_1|j_1j_2;jm\rangle = \sum_{m_1m_2} \langle j_1j_2m_1m_2|j_1j_2jm\rangle|j_1j_2m_1m_2\rangle$$

where

$$|j_1j_2m_1m_2\rangle = |j_1m_1\rangle|j_2m_2\rangle \quad \text{and} \quad \langle j_1j_2m_1m_2|j_1j_2jm\rangle = C(j_1j_2j;m_1m_2m)$$

5.22 Construct a complete set of eigenfunctions of $\hat{J}^2$ and $\hat{J}_z$ for two independent spin 1/2 particles by making linear combinations of simple product functions of the type $|j_1m_1\rangle|j_2m_2\rangle$. Here $\hat{\underset{\sim}{J}} = \hat{\underset{\sim}{J}}_1 + \hat{\underset{\sim}{J}}_2$ and $j_1 = j_2 = 1/2$. (Hint: Let $|j = 1, m = 1\rangle = |m_1 = 1/2\rangle|m_2 = 1/2\rangle$ and operate on both sides with $\hat{J}_-$. The function $|j = 0, m = 0\rangle$ can be obtained by use of the orthogonality relation.)

Solution:

An eigenket of $\hat{J}^2$ and $\hat{J}_z$ can be written as

$$|j,m\rangle = \sum_{m_1 m_2} |m_1 m_2\rangle\langle m_1 m_2|jm\rangle = \sum_{m_1} C(j_1 j_2 j, m_1, m - m_1)|m_1 m_2\rangle$$

where the j_1 and j_2 indices have been surpressed. It is clear that

$$m_{max} = (m_1 + m_2)_{max} = j_1 + j_2 \quad \text{and} \quad j_{max} = j_1 + j_2$$

Accordingly,

$$|j, j_1 + j_2\rangle = C(j_1 j_2 j, j_1 j_2)|m_1 m_2\rangle$$

or

$$|j = 1, m = 1\rangle = C(1/2,1/2,1;1/2,1/2)|m_1 = 1/2\rangle|m_2 = 1/2\rangle$$

Normalization requires that $|C(1/2,1/2,1;1/2,1/2)|^2 = 1$ and by convention we choose this coefficient to be +1.

The coefficients for $|j = 1, m = 0\rangle$ can be determined as follows:

$$\hat{J}_-|j = 1, m = 1\rangle = \sqrt{(j + m)(j - m + 1)}\,|j = 1, m = 0\rangle = \sqrt{2}\,|j = 1, m = 0\rangle$$

$$= (J_{1-} + J_{2-})|m_1 = 1/2\rangle|m_2 = 1/2\rangle = |m_1 = -1/2\rangle|m_2 = 1/2\rangle + |m_1 = 1/2\rangle|m_2 = -1/2\rangle$$

$$|1,0\rangle = \frac{1}{\sqrt{2}}|-1/2\rangle|+1/2\rangle + \frac{1}{\sqrt{2}}|+1/2\rangle|-1/2\rangle$$

In the following, the integer indices refer to j and m and the half integers correspond to m_1 and m_2 in that order.

The ket $|1,-1\rangle$ can be determined by again operating with $\hat{J}_-$.

$$\hat{J}_-|1,0\rangle = \frac{1}{\sqrt{2}}|-1/2\rangle|-1/2\rangle + \frac{1}{\sqrt{2}}|-1/2\rangle|-1/2\rangle$$

$$\sqrt{2}\,|1,-1\rangle = \sqrt{2}\,|-1/2\rangle|-1/2\rangle \qquad |1,-1\rangle = |-1/2\rangle|-1/2\rangle$$

We have now determined three functions which correspond to $m = j_1 + j_2$, $j_1 + j_2 - 1$, and $j_1 + j_2 - 2$. However, there are $(2j_1 + 1)(2j_2 + 1) = 4$ functions of the type $|m_1\rangle|m_2\rangle$ and we require the same number of transformed functions. The missing function corresponds to $j = 0$ which is a special case of the general rule:

$$j_{min} = |j_1 - j_2|$$

and m must be zero. The rule $m = m_1 + m_2$ requires that

$$|j = 0, m = 0\rangle = a|1/2\rangle|-1/2\rangle + b|-1/2\rangle|+1/2\rangle$$

where the coefficients a and b can be determined by the orthogonality requirement:

$$\langle j = 1, m = 0|j = 0, m = 0\rangle = \frac{1}{\sqrt{2}}(\langle -1/2|\langle +1/2| + \langle +1/2|\langle -1/2|) \times$$

$$(a|+1/2\rangle|-1/2\rangle + b|-1/2\rangle|+1/2\rangle) = \frac{1}{\sqrt{2}}(a + b) = 0$$

Therefore

$$|j = 0, m = 0\rangle = \frac{1}{\sqrt{2}}|+1/2\rangle|-1/2\rangle - \frac{1}{\sqrt{2}}|-1/2\rangle|+1/2\rangle$$

All of the coefficients $C(1/2,1/2,j;m_1m_2m)$ can be found in tables of vector coupling coefficients.

5.23 When the angular momentum vectors $\hat{\mathbf{J}}_1$ and $\hat{\mathbf{J}}_2$ couple to give the resultant $\hat{\mathbf{J}}$ we require that $m = m_1 + m_2$. Therefore, $m_{max} = (m_1 + m_2)_{max} = j_1 + j_2$ and we see that $j_{max} = j_1 + j_2$. What is the minimum value of j? (Hint: Use the fact that the total number of states is $(2j_1 + 1)(2j_2 + 1)$.)

Solution:

For each value of j there are $2j + 1$ states. Thus

$$\sum_{j_{min}}^{j_{max}} (2j + 1) = (2j_1 + 1)(2j_2 + 1) \tag{1}$$

to conserve the number of states in the two representations. The unknown in this equation is j_{min}. We now simply count the states:

$$\sum_{j=a}^{b} j = \frac{(b - a + 1)(b + a)}{2}$$

and

$$\sum_{j=a}^{b} (2j + 1) = (b - a + 1)(b + a) + (b - a + 1) = (b + 1)^2 - a^2 \tag{2}$$

Since $b = j_{max} = j_1 + j_2$ Eqs. (1) and (2) can be solved for a.

$$(j_1 + j_2 + 1)^2 - a^2 = (2j_1 + 1)(2j_2 + 1)$$
$$a^2 = j_1^2 + j_2^2 - 2j_1j_2 = (j_1 - j_2)^2$$

Since $j \geq 0$, we require that $a = j_{min} = |j_1 - j_2|$.

The allowed values of j are then:

$j = j_1 + j_2,\ \ j_1 + j_2 - 1,\ \ \cdots,\ \ |j_1 - j_2|$

5.24 In constructing eigenfunctions of $\hat{S}^2$ it is convenient to define the unnormalized projection operator

$$\hat{O}_{S_j} = \prod_{i \neq j} [\hat{S}^2 - S_i(S_i + 1)]$$

which has the effect of "annihilating" those eigenfunctions of $\hat{S}^2$ which have the eigenvalues $S_i(S_i + 1)$ with $i \neq j$.

(a) Show that $\hat{O}_{S=1}[\alpha(1)\beta(2) + \beta(1)\alpha(2)] = \text{constant} \times [\alpha(1)\beta(2) + \beta(1)\alpha(2)]$ and $\hat{O}_{S=1}[\alpha(1)\beta(2) - \beta(1)\alpha(2)] = 0$

(b) Show that $\hat{O}_{S=0}[\alpha(1)\beta(2) + \beta(1)\alpha(2)] = 0$ and $\hat{O}_{S=0}[\alpha(1)\beta(2) - \beta(1)\alpha(2)] = \text{constant} \times [\alpha(1)\beta(2) - \beta(1)\alpha(2)]$

(c) Demonstrate that $\alpha(1)\beta(2)$ contains eigenfunctions of $\hat{S}^2$ having both $S = 1$ and $S = 0$ by operating on this function with $\hat{O}_{S=1}$ and $\hat{O}_{S=0}$.

Solution: Ref. S2,75

(a) By the definition we have

$$\hat{O}_{S=1} = \hat{S}^2$$

since only $S = 0, 1$ is allowed for two spin 1/2 particles. It is convenient to use the identity

$$\hat{S}^2 = \hat{S}_+\hat{S}_- + \hat{S}_z^2 - \hat{S}_z$$

and the following relations

$$\hat{S}_-\alpha(1)\beta(2) = (\hat{S}_{1-} + \hat{S}_{2-})\alpha(1)\beta(2) = \beta(1)\beta(2)$$

$$\hat{S}_+\beta(1)\beta(2) = (\hat{S}_{1+} + \hat{S}_{2+})\beta(1)\beta(2) = \alpha(1)\beta(2) + \beta(1)\alpha(2)$$

$$\hat{S}_-\beta(1)\alpha(2) = \beta(1)\beta(2), \quad \hat{S}_z\alpha(1)\beta(2) = \hat{S}_z\alpha(2)\beta(1) = 0$$

Then

$$\hat{O}_{S=1}[\alpha(1)\beta(2) + \beta(1)\alpha(2)] = \hat{S}_+\hat{S}_-[\alpha(1)\beta(2) + \beta(1)\alpha(2)] = 2[\alpha(1)\beta(2) + \beta(1)\alpha(2)]$$

and

$$\hat{O}_{S=1}[\alpha(1)\beta(2) - \beta(1)\alpha(2)] = \hat{S}_+\hat{S}_-[\alpha(1)\beta(2) - \beta(1)\alpha(2)] = \hat{S}_+[\beta(1)\beta(2) - \beta(1)\beta(2)] = 0$$

(b) Here the projection operator is

$$\hat{O}_{S=0} = \hat{S}^2 - 1(1 + 1) = \hat{S}_+\hat{S}_- + \hat{S}_z^2 - \hat{S}_z - 2$$

Therefore

$$\hat{O}_{S=0}[\alpha(1)\beta(2) + \beta(1)\alpha(2)] = (\hat{S}_+\hat{S}_- - 2)[\alpha(1)\beta(2) + \beta(1)\alpha(2)] = 0$$

and

$$\hat{O}_{S=0}[\alpha(1)\beta(2) - \beta(1)\alpha(2)] = (\hat{S}_+\hat{S}_- - 2)[\alpha(1)\beta(2) - \beta(1)\alpha(2)] = -2[\alpha(1)\beta(2) - \beta(1)\alpha(2)]$$

(c) $\hat{O}_{S=1}\alpha(1)\beta(2) = \hat{S}_+\hat{S}_-\alpha(1)\beta(2) = \alpha(1)\beta(2) + \beta(1)\alpha(2)$

$$\hat{O}_{S=0}\alpha(1)\beta(2) = (\hat{S}_+\hat{S}_- - 2)\alpha(1)\beta(2) = \alpha(1)\beta(2) - \beta(1)\alpha(2)$$

The conclusion is that $u_1 = \alpha(1)\beta(2) + \beta(1)\alpha(2)$ corresponds to $S = 1$ and $u_o = \alpha(1)\beta(2) - \beta(1)\alpha(2)$ corresponds to $S = 0$. The initial function $\alpha(1)\beta(2)$ can clearly be expressed as a linear combination of u_1 and u_o. Thus

$$\alpha(1)\beta(2) = \frac{1}{2}(u_1 + u_o)$$

where

$$\hat{O}_{S=1}u_o = 0 \quad \text{and} \quad \hat{O}_{S=0}u_1 = 0$$

5.25 Consider the spin functions for a three electron system:

(a) Demonstrate that the function $\alpha(1)\alpha(2)\alpha(3)$ is an eigenfunction of both $\hat{S}^2$ and $\hat{S}_z$ and determine the corresponding eigenvalues. (Recall that $\hat{S}_{1z}\alpha(1) = (1/2)\alpha(1)$.)

(b) Use the ladder operator $\hat{S}_- = (\hat{S}_{1-} + \hat{S}_{2-} + \hat{S}_{3-})$ to generate all $2S + 1$ of the eigenfunctions for $S = 3/2$.

(c) Determine the eigenfunctions for a three electron system having $S = 1/2$. How many sets of such functions must there be? (Hint: Use the projection operator $\hat{O}_{S=1/2}$ to determine the first function.)

Solution:

(a) We use the identity $\hat{S}^2 = \hat{S}_+\hat{S}_- + \hat{S}_z^2 - \hat{S}_z$ where

$$\hat{S}_\pm = \hat{S}_{1\pm} + \hat{S}_{2\pm} + \hat{S}_{3\pm} \quad \text{and} \quad \hat{S}_z = \hat{S}_{1z} + \hat{S}_{2z} + \hat{S}_{3z}$$

Then

$$\hat{S}_-\alpha\alpha\alpha = \beta\alpha\alpha + \alpha\beta\alpha + \alpha\alpha\beta \qquad \hat{S}_+\hat{S}_-\alpha\alpha\alpha = \alpha\alpha\alpha + \alpha\alpha\alpha + \alpha\alpha\alpha$$

$$\hat{S}_z\alpha\alpha\alpha = (1/2 + 1/2 + 1/2)\alpha\alpha\alpha = (3/2)\alpha\alpha\alpha \qquad \hat{S}_z^2\alpha\alpha\alpha = (9/4)\alpha\alpha\alpha$$

and we find

$$\hat{S}^2\alpha\alpha\alpha = 3\alpha\alpha\alpha + (9/4)\alpha\alpha\alpha - (3/2)\alpha\alpha\alpha = (15/4)\alpha\alpha\alpha = \frac{3}{2}\left(\frac{3}{2} + 1\right)\alpha\alpha\alpha$$

The eigenvalues of $\hat{S}^2$ and $\hat{S}_z$ are 15/4 and 3/2, respectively.

(b)

$$\boxed{|S = 3/2, M_S = 3/2> = \alpha\alpha\alpha}$$

$$\hat{S}_-|3/2,3/2> = \sqrt{(3/2 + 3/2)(3/2 - 3/2 + 1)}\,|3/2,1/2> = \sqrt{3}\,|3/2,1/2>$$

$$= (\hat{S}_{1-} + \hat{S}_{2-} + \hat{S}_{3-})\alpha\alpha\alpha = \beta\alpha\alpha + \alpha\beta\alpha + \alpha\alpha\beta$$

$$\boxed{|3/2,1/2> = (1/\sqrt{3})(\beta\alpha\alpha + \alpha\beta\alpha + \alpha\alpha\beta)}$$

$$\hat{S}_-|3/2,1/2> = \sqrt{(3/2 + 1/2)(3/2 - 1/2 + 1)}\,|3/2,-1/2> = 2|3/2,-1/2>$$

$$= (1/\sqrt{3})[(\beta\beta\alpha + \beta\alpha\beta) + (\beta\beta\alpha + \alpha\beta\beta) + (\beta\alpha\beta + \alpha\beta\beta)]$$

$$\boxed{|3/2,-1/2> = (1/\sqrt{3})(\beta\beta\alpha + \beta\alpha\beta + \alpha\beta\beta)}$$

$$\hat{S}_-|3/2,-1/2> = \sqrt{(3/2 - 1/2)(3/2 + 1/2 + 1)}\,|3/2,-3/2> = \sqrt{3}\,|3/2,-3/2>$$

$$= (1/\sqrt{3})[\beta\beta\beta + \beta\beta\beta + \beta\beta\beta]$$

$$\boxed{|3/2,-3/2> = \beta\beta\beta}$$

(c) To obtain the functions for S = 1/2 we use the projection operator

$$\hat{O}_{S=1/2} = \hat{S}^2 - 3/2(5/2) = \hat{S}_+\hat{S}_- + \hat{S}_z^2 - \hat{S}_z - 15/4$$

The function $\beta\alpha\alpha$ should contain $|S = 1/2, M_S = 1/2>$. The necessary relations are

$$\hat{S}_z\beta\alpha\alpha = (\hat{S}_{1z} + \hat{S}_{2z} + \hat{S}_{3z})\beta\alpha\alpha = (-1/2 + 1/2 + 1/2)\beta\alpha\alpha = (1/2)\beta\alpha\alpha$$

$$\hat{S}_z^2\beta\alpha\alpha = (1/4)\beta\alpha\alpha \qquad \hat{S}_-\beta\alpha\alpha = \beta\beta\alpha + \beta\alpha\beta$$

$$\hat{S}_+\hat{S}_-\beta\alpha\alpha = \alpha\beta\alpha + 2\beta\alpha\alpha + \alpha\alpha\beta$$

Therefore

$$\left(\hat{S}^2 - \frac{15}{4}\right)\beta\alpha\alpha = \alpha\beta\alpha + 2\beta\alpha\alpha + \alpha\alpha\beta + \frac{1}{4}\beta\alpha\alpha - \frac{1}{2}\beta\alpha\alpha - \frac{15}{4}\beta\alpha\alpha = \alpha\beta\alpha - 2\beta\alpha\alpha + \alpha\alpha\beta$$

or

$$|1/2,1/2\rangle_1 = \frac{1}{\sqrt{6}} (\alpha\beta\alpha - 2\beta\alpha\alpha + \alpha\alpha\beta)$$

The function $|1/2,-1/2\rangle_1$ is readily obtained with the lowering operator:

$$\hat{S}_-|1/2,1/2\rangle_1 = |1/2,-1/2\rangle_1 = \frac{1}{\sqrt{6}} [(\beta\beta\alpha + \alpha\beta\beta) - 2(\beta\beta\alpha + \beta\alpha\beta) + (\beta\alpha\beta + \alpha\beta\beta)]$$

$$|1/2,-1/2\rangle_1 = \frac{1}{\sqrt{6}} (-\beta\beta\alpha + 2\alpha\beta\beta - \beta\alpha\beta)$$

For three spin 1/2 particles there are a total of $[2(1/2) + 1]^3 = 8$ spin functions. Since only four are associated with $S = 3/2$, four must also be associated with $S = 1/2$. We need to find the other set of functions $|1/2,\pm1/2\rangle_2$. The function $|1/2,1/2\rangle_2$ is orthogonal to both $|3/2,1/2\rangle$ and $|1/2,1/2\rangle_1$ and this fact can be used for its determination. We write

$$|1/2,1/2\rangle_2 = c_1\alpha\beta\alpha + c_2\beta\alpha\alpha + c_3\alpha\alpha\beta \qquad \langle 3/2,1/2|1/2,1/2\rangle_2 = c_1 + c_2 + c_3 = 0$$

$$_1\langle 1/2,1/2|1/2,1/2\rangle_2 = c_1 - 2c_2 + c_3 = 0$$

Then $c_2 = 0$ and $c_1 = -c_3$. We conclude that

$$|1/2,1/2\rangle_2 = \frac{1}{\sqrt{2}} (\alpha\beta\alpha - \alpha\alpha\beta)$$

Another application of the lowering operator gives

$$\hat{S}_-|1/2,1/2\rangle_2 = |1/2,-1/2\rangle_2 = \frac{1}{\sqrt{2}} [(\beta\beta\alpha + \alpha\beta\beta) - (\beta\alpha\beta + \alpha\beta\beta)]$$

$$|1/2,-1/2\rangle_2 = \frac{1}{\sqrt{2}} (\beta\beta\alpha - \beta\alpha\beta)$$

5.26 Diagonalize the rotation matrix

$$\underset{\sim}{R} = \begin{pmatrix} \cos\frac{\beta}{2} & -\sin\frac{\beta}{2} \\ \sin\frac{\beta}{2} & \cos\frac{\beta}{2} \end{pmatrix}$$

to obtain the eigenvalues and eigenvectors.

Solution: Ref. Appendix 6

We require that the secular equation for $\underset{\sim}{R}$ vanish. Thus

$$\begin{vmatrix} (\cos\frac{\beta}{2} - \lambda) & -\sin\frac{\beta}{2} \\ \sin\frac{\beta}{2} & (\cos\frac{\beta}{2} - \lambda) \end{vmatrix} = \lambda^2 - 2\lambda\cos\frac{\beta}{2} + 1 = 0$$

or

$$\lambda = \cos\frac{\beta}{2} \pm \sqrt{\cos^2\frac{\beta}{2} - 1} = \cos\frac{\beta}{2} \pm i\sin\frac{\beta}{2}$$

The eigenvalues are then

$$\lambda_1 = \cos\frac{\beta}{2} + i\sin\frac{\beta}{2} = e^{i\beta/2} \qquad\qquad \lambda_2 = \cos\frac{\beta}{2} - i\sin\frac{\beta}{2} = e^{-i\beta/2}$$

The corresponding eigenvectors are obtained from the equations:

$$\lambda_1: \left[\cos\frac{\beta}{2} - \left(\cos\frac{\beta}{2} + i\sin\frac{\beta}{2}\right)\right] S_{11} - \left(\sin\frac{\beta}{2}\right) S_{21} = 0 \qquad\qquad -iS_{11} = S_{21}$$

If $|1\rangle$ and $|2\rangle$ are taken to be the basis vectors for $\underset{\sim}{R}$, we can write:

$$|e^{i\beta/2}\rangle = S_{11}|1\rangle - iS_{11}|2\rangle = \frac{1}{\sqrt{2}}(|1\rangle - i|2\rangle)$$

$$\lambda_2: \left[\cos\frac{\beta}{2} - \left(\cos\frac{\beta}{2} - i\sin\frac{\beta}{2}\right)\right] S_{12} - \left(\sin\frac{\beta}{2}\right) S_{22} = 0 \qquad\qquad iS_{12} = S_{22}$$

The eigenvector is thus:

$$|e^{-i\beta/2}\rangle = -iS_{22}|1\rangle + S_{22}|2\rangle = \frac{1}{\sqrt{2}}(-i|1\rangle + |2\rangle)$$

The matrix $\underset{\sim}{R}$ corresponds to a rotation through the angle $\beta/2$ about a fixed axis. Since the eigenvectors do not depend on the angle of rotation, we see that all rotations about the same axis have the same eigenvectors. These rotations thus commute.

Problem 5.16 indicates that $\hat{R}(\hat{n},\beta) = \exp(-i\beta\hat{n}\cdot\hat{J})$ where $\hat{n}$ specifies the axis of the rotation. By diagonalizing $\underset{\sim}{R}$ we have simply obtained a new representation in which the operator $(\hat{n}\cdot\hat{\underset{\sim}{J}})$ is diagonal.

5.27 The Clebsch-Gordan coefficients $\langle j_1j_2m_1m_2|j_1j_2jm\rangle = C(j_1j_2j;m_1m_2m)$ for coupling $j = 1$ or $j = 1/2$ to any other vector are given in Tables 5.27a and 5.27b. By explicit use of these tables and the defining equation

$$|jm\rangle_2 = \sum_{m_1,m_2} |j_1m_1\rangle_1|j_2m_2\rangle_1\langle j_1j_2m_1m_2|j_1j_2jm\rangle \tag{1}$$

write out all of the kets $|jm\rangle_2$ for the cases

(a) $j_1 = 1,\ j_2 = 1$; ($j = 0,\ m = 0$ and $j = 2,\ m = 0$)

(b) $j_1 = 1,\ j_2 = 1/2$; ($j = 3/2,\ m = 1/2$)

Solution: Ref. Problem 5.33

(a) The restrictions on j and m are

$$|j_1 - j_2| \leqslant j \leqslant j_1 + j_2; \qquad m_1 + m_2 = m$$

Thus, as in Problem 5.33, the possible values of j are 2, 1, 0 so that we have the possible coupled kets $|jm\rangle_2$:

$$\begin{aligned} j = 2:&\quad |2,2\rangle_2,\ |2,1\rangle_2,\ |2,0\rangle_2,\ |2,-1\rangle_2,\ |2,-2\rangle_2 \\ j = 1:&\quad |1,1\rangle_2,\ |1,0\rangle_2,\ |1,-1\rangle_2 \\ j = 0:&\quad |0,0\rangle_2 \end{aligned} \tag{2}$$

Table 5.27a then reduces to

	$m_2 = 1$	$m_2 = 0$	$m_2 = -1$
$j = j_1 + 1 = 1 + 1 = 2$	$\left[\frac{(1+m)(2+m)}{3\cdot4}\right]^{1/2}$	$\left[\frac{(2-m)(2+m)}{3\cdot2}\right]^{1/2}$	$\left[\frac{(1-m)(2-m)}{3\cdot4}\right]^{1/2}$
$j = j_1 = 1$	$-\left[\frac{(1+m)(2-m)}{2\cdot2}\right]^{1/2}$	$\left[\frac{m^2}{1\cdot2}\right]^{1/2}$	$\left[\frac{(1-m)(2+m)}{2\cdot2}\right]^{1/2}$
$j = j_1 - 1 = 1 - 1 = 0$	$\left[\frac{(1-m)(2-m)}{2\cdot3}\right]^{1/2}$	$-\left[\frac{(1-m)(1+m)}{1\cdot3}\right]$	$\left[\frac{(2+m)(1+m)}{2\cdot3}\right]^{1/2}$

Thus, for $|0,0\rangle_2$, we read from the bottom row of the reduced table (condensing $|j_1m_1\rangle|j_2m_2\rangle$ to $|m_1m_2\rangle$):

$$|0,0\rangle_2 = \left[\frac{(1-0)(2-0)}{2\cdot3}\right]^{1/2}|-1,1\rangle_1 - \left[\frac{(1-0)(1+0)}{1\cdot3}\right]^{1/2}|0,0\rangle_1 + \left[\frac{(2)(1)}{2\cdot3}\right]^{1/2}|1,-1\rangle_1$$

$$|0,0\rangle_2 = \frac{1}{\sqrt{3}}\,[|-1,1\rangle_1 - |0,0\rangle_1 + |1,-1\rangle_1]$$

For the ket $|2,0\rangle_2$, we read from the top row of the reduced table

$$|2,0\rangle_2 = \left[\frac{1\cdot2}{12}\right]^{1/2}|-1,1\rangle_1 + [4/6]^{1/2}|0,0\rangle_1 + [2/12]^{1/2}|1,-1\rangle_1$$

$$= \frac{1}{\sqrt{6}}\,[|-1,1\rangle_1 + 2|0,0\rangle_1 + |1,-1\rangle_1]$$

These results may be checked against those in Problem 5.33 where the same problem is solved with the raising and lowering operator technique.

(b) For this case, Table 5.27b reduces to

	$m_2 = 1/2$	$m_2 = -1/2$
$j = j_1 + 1/2 = 3/2$	$\left[\frac{3/2 + m}{3}\right]^{1/2}$	$\left[\frac{3/2 - m}{3}\right]^{1/2}$
$j = j_1 - 1/2 = 1/2$	$-\left[\frac{3/2 - m}{3}\right]^{1/2}$	$\left[\frac{3/2 + m}{3}\right]^{1/2}$

Then for $|3/2,1/2\rangle_2$ we read from the top row

$$|3/2,1/2\rangle_2 = \left[\frac{3/2 + 1/2}{3}\right]^{1/2}|0,1/2\rangle_1 + \left[\frac{3/2 - 1/2}{3}\right]^{1/2}|1,-1/2\rangle_1$$

$$= +\sqrt{2/3}\,|0,1/2\rangle_1 + \sqrt{1/3}\,|1,-1/2\rangle_1 = 1/\sqrt{3}(\sqrt{2}|0,1/2\rangle_1 + |1,-1/2\rangle_1)$$

Note: The possible eigenkets for this problem are

$|jm>$: $|3/2,3/2>$, $|3/2,1/2>$, $|3/2,-1/2>$, $|3/2,-3/2>$, $|1/2,1/2>$, $|1/2,-1/2>$

where, by following the outlined procedure,

$|3/2,3/2>_2 = |1,1/2>_1$ $\qquad$ $|3/2,-1/2>_2 = 1/\sqrt{3}(|-1,1/2>_1 + \sqrt{2}|0,-1/2>_1)$

$|3/2,-3/2>_2 = |-1,-1/2>_1$ $\qquad$ $|1/2,1/2>_2 = 1/\sqrt{3}(|0,1/2>_1 - \sqrt{2}|1,-1/2>_1)$

$|1/2,-1/2>_2 = 1/\sqrt{3}(\sqrt{2}|-1,1/2>_1 - |0,-1/2>_1)$

One example to which the above is applicable is the coupling of spin and orbital angular momenta for a p electron.

Table 5.27a

$j_2 = 1$

	$m_2 = 1$	$m_2 = 0$	$m_2 = -1$
$j = j_1 + 1$	$\left[\frac{(j_1+m)(j_1+m+1)}{(2j_1+1)(2j_1+2)}\right]^{1/2}$	$\left[\frac{(j_1-m+1)(j_1+m+1)}{(2j_1+1)(j_1+1)}\right]^{1/2}$	$\left[\frac{(j_1-m)(j_1-m+1)}{(2j_1+1)(2j_1+2)}\right]^{1/2}$
$j = j_1$	$-\left[\frac{(j_1+m)(j_1-m+1)}{2j_1(j+1)}\right]^{1/2}$	$\left[\frac{m^2}{j_1(j_1+1)}\right]^{1/2}$	$\left[\frac{(j_1-m)(j_1+m+1)}{2j_1(j_1+1)}\right]^{1/2}$
$j = j_1 - 1$	$\left[\frac{(j_1-m)(j_1-m+1)}{2j_1(2j_1+1)}\right]^{1/2}$	$-\left[\frac{(j_1-m)(j_1+m)}{j_1(2j_1+1)}\right]^{1/2}$	$\left[\frac{(j_1+m+1)(j_1+m)}{2j_1(2j_1+1)}\right]^{1/2}$

Table 5.27b

$j_2 = 1/2$

	$m_2 = 1/2$	$m_2 = -1/2$
$j = j_1 + 1/2$	$\left[\frac{j_1+m+1/2}{2j_1+1}\right]^{1/2}$	$\left[\frac{j_1-m+1/2}{2j_1+1}\right]^{1/2}$
$j = j_1 - 1/2$	$-\left[\frac{j_1-m+1/2}{2j_1+1}\right]^{1/2}$	$\left[\frac{j_1+m+1/2}{2j_1+1}\right]^{1/2}$

SUPPLEMENTARY PROBLEMS

5.28 Use the properties of the ladder operators $\hat{J}_\pm$ (Problem 5.11) to show that

$$<j,m|\hat{\underset{\sim}{J}}|j,m \pm 1> = \frac{1}{2}(\hat{i} \pm i\hat{j})\sqrt{(j \mp m)(j \pm m + 1)}$$

5.29 Given the spherical harmonic

$$Y_{2,-2}(\theta,\phi) = \frac{\sqrt{15}}{4}\sin^2\theta\,\frac{e^{-i2\phi}}{\sqrt{2\pi}}$$

apply the ladder operator $\hat{L}_+$ to generate the functions

(a) $Y_{2,-1}$ (b) $Y_{2,0}$ (c) $Y_{2,+1}$ (d) $Y_{2,+2}$

Ans:

(a) $Y_{2,-1}(\theta,\phi) = \frac{\sqrt{15}}{2}\sin\theta\cos\theta\,\frac{e^{-i\phi}}{\sqrt{2\pi}}$ (b) $Y_{2,0}(\theta,\phi) = \frac{\sqrt{10}}{4}(3\cos^2\theta - 1)\cdot\frac{1}{\sqrt{2\pi}}$

(c) $Y_{2,+1}(\theta,\phi) = -\frac{\sqrt{15}}{2}\sin\theta\cos\theta\,\frac{e^{i\phi}}{\sqrt{2\pi}}$ (d) $Y_{2,+2}(\theta,\phi) = \frac{\sqrt{15}}{4}\sin^2\theta\,\frac{e^{2i\phi}}{\sqrt{2\pi}}$

5.30 Apply the ladder operator $\hat{L}_-$ to the spherical harmonic $Y_{\ell,-\ell}(\phi,\theta)$ to obtain an equation for $P_{\ell,-\ell}(\theta)$ and solve the equation to determine the form of $P_{\ell,-\ell}(\theta)$.

Ans: Ref. RO3,237

$$P_{\ell,-\ell}(\theta) = C_-(\sin\theta)^{\ell}$$

5.31 The general equation for generating the spherical harmonics is

$$Y_{\ell m} = \left[\frac{2\ell + 1}{4\pi}\cdot\frac{(\ell - m)!}{(\ell + m)!}\right]^{1/2}\frac{1}{2^{\ell}\ell!}\,e^{im\phi}(-\sin\theta)^m\left[\frac{d}{d(\cos\theta)}\right]^{\ell+m}(\cos^2\theta - 1)^{\ell}$$

(a) Write out and simplify the equations for $Y_{5,5}$ and $Y_{5,-5}$.

(b) Combine $Y_{5,5}$ and $Y_{5,-5}$ so as to generate a pair of real, orthonormal functions.

Ans: Ref. RO3,240

(a)

$$Y_{5,5} = \left[\frac{2\cdot5 + 1}{4\pi}\,\frac{0!}{10!}\right]^{1/2}\frac{1}{2^5 5!}\,e^{i5\phi}(-\sin\theta)^5\left[\frac{d}{d(\cos\theta)}\right]^{10}(\cos^2\theta - 1)^5$$

$$= \left[\frac{11}{4\cdot10!\pi}\right]^{1/2}\frac{1}{2^5 5!}\,e^{i5\phi}(-\sin\theta)^5 10! = -\left[\frac{11!}{4\pi}\right]^{1/2}\frac{1}{2^5 5!}\sin^5\theta e^{i5\phi}$$

$$= -\frac{3}{2^5}\sqrt{\frac{77}{\pi}}\sin^5\theta e^{i5\phi}$$

$$Y_{5,-5} = \left[\frac{2\cdot5 + 1}{4\pi}\,\frac{10!}{0!}\right]^{1/2}\frac{1}{2^5 5!}\,e^{-i5\phi}(-\sin\theta)^{-5}\left[\frac{d}{d(\cos\theta)}\right]^{0}(\cos^2\theta - 1)^5$$

$$= \left[\frac{11!}{4\pi}\right]^{1/2}\frac{1}{2^5 5!}\,e^{-i5\phi}\sin^5\theta = +\frac{3}{2^5}\sqrt{\frac{77}{\pi}}\sin^5\theta e^{-i5\phi}$$

(b) $\psi_1 = \frac{3}{2^4}\sqrt{\frac{77}{2\pi}}\sin^5\theta\cos5\phi$, $\psi_2 = \frac{3}{2^4}\sqrt{\frac{77}{2\pi}}\sin^5\theta\sin5\phi$

5.32 Prove that the operator $\hat{S} = \exp(i\hat{\mathcal{H}})$ is unitary (i.e. that $\hat{S}^+ = \hat{S}^{-1}$) if $\hat{\mathcal{H}}$ is Hermitian. (Hint: Use the fact that $\exp(i\hat{\mathcal{H}})$ is defined by its series expansion.)
Ref. RO3,12

5.33 Work out the vector coupling coefficients for the case $j_1 = 1$, $j_2 = 1$, i.e. construct the eigenfunctions $|j,m>_2$ from the set $|m_1m_2>_1$. Keep in mind that $m = m_1 + m_2$ for all nonvanishing coefficients.

Ans: Ref. Problem 5.25

$$|2,2>_2 = |1,1>_1 \qquad |2,1>_2 = \frac{1}{\sqrt{2}}(|0,1>_1 + |1,0>_1)$$

$$|2,0>_2 = \frac{1}{\sqrt{6}}(|-1,1>_1 + 2|0,0>_1 + |1,-1>_1) \qquad |2,-1>_2 = \frac{1}{\sqrt{2}}(|0,-1>_1 + |-1,0>_1)$$

$$|2,-2>_2 = |-1,-1>_1 \qquad |1,1>_2 = \frac{1}{\sqrt{2}}(|0,1>_1 - |1,0>_1)$$

$$|1,0>_2 = \frac{1}{\sqrt{2}}(|-1,1>_1 - |1,-1>_1) \qquad |1,-1>_2 = \frac{1}{\sqrt{2}}(|-1,0>_1 - |0,-1>_1)$$

$$|0,0>_2 = \frac{1}{\sqrt{3}}(|1,-1>_1 + |-1,1>_1 - |0,0>_1)$$

5.34 Obtain the eigenvalues and eigenvectors for the rotation matrix:

$$\underset{\sim}{R} = \begin{pmatrix} \cos\theta & -\sin\theta & 0 \\ \sin\theta & \cos\theta & 0 \\ 0 & 0 & 1 \end{pmatrix}$$

Ans: Ref. HO,107; Appendix 6

$$\lambda_1 = e^{i\theta} \qquad |e^{i\theta}>_1 = \frac{1}{\sqrt{2}}(|1> - i|2>)$$

$$\lambda_2 = e^{-i\theta} \qquad |e^{-i\theta}>_2 = \frac{1}{\sqrt{2}}(-i|1> + |2>)$$

$$\lambda_3 = 1 \qquad |1>_2 = |3>$$

5.35 If $\underset{\sim}{A}$ and $\underset{\sim}{B}$ commute with each of the Pauli matrices but not necessarily with each other, show that

$$(\underset{\sim}{\sigma} \cdot \underset{\sim}{A})(\underset{\sim}{\sigma} \cdot \underset{\sim}{B}) = \underset{\sim}{A} \cdot \underset{\sim}{B} + i\underset{\sim}{\sigma} \cdot (\underset{\sim}{A} \times \underset{\sim}{B})$$

where

$$\underset{\sim}{\sigma} = \underset{\sim}{\sigma}_x\hat{i} + \underset{\sim}{\sigma}_y\hat{j} + \underset{\sim}{\sigma}_z\hat{k}$$

Ref. Problem 5.19

5.36 Let

$$\underset{\sim}{A} = \begin{pmatrix} 0 & -i \\ i & 0 \end{pmatrix}$$

and solve the equation

$$\underset{\sim}{A}\underset{\sim}{u} = \lambda \underset{\sim}{u}$$

for the eigenvalues and eigenvectors.

CHAPTER 6

PERTURBATION AND VARIATION THEORY

Since exact solutions are only available for a limited number of systems, methods for obtaining approximate solutions are of great importance in quantum mechanics. These methods fall into two general types: perturbation methods and variation methods. Here we list the basic equations for non-degenerate and degenerate perturbation theory and the variation theory. The problems illustrate the theories and their applications.

1. Perturbation Theory (discrete non-degenerate eigenvalues)

Let $\hat{\mathcal{H}}^o$ be the Hamiltonian for an unperturbed system having the eigenvalues E_n^o and the eigenfunctions u_n^o, i.e. $\hat{\mathcal{H}}^o u_n^o = E_n^o u_n^o$. We desire solutions for the eigenvalue problem

$$\hat{\mathcal{H}} u_n = E_n u_n \tag{1}$$

where $\hat{\mathcal{H}} = \hat{\mathcal{H}}^o + \hat{\mathcal{H}}'$. Perturbation theory is appropriate when $\hat{\mathcal{H}}^o >> \hat{\mathcal{H}}'$. We write:

$$E_n = E_n^o + E_n^{(1)} + E_n^{(2)} + \cdots \tag{2}$$

$$u_n = u_n^o + u_n^{(1)} + u_n^{(2)} + \cdots \tag{3}$$

where the superscripts indicate the order of the correction, i.e. the power of $\hat{\mathcal{H}}'$ on which the term depends. Substitution of Eqs. (2) and (3) into Eq. (1) and separation of the terms according to order yield the following equations:

Order

$$
\begin{aligned}
&0: && (\hat{\mathcal{H}}^o - E^o)u^o = 0 \\
&1: && (\hat{\mathcal{H}}^o - E^o)u^{(1)} + (\hat{\mathcal{H}}' - E^{(1)})u^o = 0 \\
&2: && (\hat{\mathcal{H}}^o - E^o)u^{(2)} + (\hat{\mathcal{H}}' - E^{(1)})u^{(1)} = E^{(2)}u^o \\
&\vdots \\
&k: && (\hat{\mathcal{H}}^o - E^o)u^{(k)} + (\hat{\mathcal{H}}' - E^{(1)})u^{(k-1)} = \sum_{j=2}^{k} E^{(j)}u^{(k-j)}
\end{aligned}
\tag{4}
$$

where the subscript n denoting the state has been surpressed for simplicity. The normalization convention for the wave function is: $<u_n|u_n> = 1$ and $<u_n^o|u_n^o> = 1$. Thus, for the various orders of correction we have:

$$1: \quad <u^o|u^{(1)}> + <u^{(1)}|u^o> = 0$$

2: $\langle u^o|u^{(2)}\rangle + \langle u^{(1)}|u^{(1)}\rangle + \langle u^{(2)}|u^o\rangle = 0$ (5)

$\vdots$

k: $\sum_{j=0}^{k} \langle u^{(j)}|u^{(k-j)}\rangle = 0$

The equations for the energy corrections can be obtained by multiplying each equation in set (4) by $(u^o)^*$ and integrating. This process yields:

0: $E^o = \langle u^o|\hat{\mathcal{H}}^o|u^o\rangle$

1: $E^{(1)} = \langle u^o|\hat{\mathcal{H}}'|u^o\rangle$

2: $E^{(2)} = \langle u^o|\hat{\mathcal{H}}' - E^{(1)}|u^{(1)}\rangle$ (6)

$\vdots$

k: $E^{(k)} = \langle u^o|\hat{\mathcal{H}}' - E^{(1)}|u^{(k-1)}\rangle - \sum_{j=1}^{k-2} E^{(k-j)}\langle u^o|u^{(j)}\rangle$

Terms such as $\langle u^o|\hat{\mathcal{H}}^o - E^o|u^{(k)}\rangle$ vanish because of the Hermitian property of $\hat{\mathcal{H}}^o$. It is possible to obtain solutions to Eqs. (6) in closed form for special cases, but in general formal solutions based on series expansions for the wave function corrections are used. For example, in Rayleigh-Schrödinger perturbation theory the expansions

$$u_n^{(k)} = \sum_j c_{jn}^{(k)} u_j^o$$

lead to the energy corrections:

$$E_n^{(1)} = \int u_n^{o*}\hat{\mathcal{H}}' u_n^o d\tau = \langle n|\hat{\mathcal{H}}'|n\rangle \quad \text{(First order)}$$

$$E_n^{(2)} = \sum_{k\neq n} \frac{|\langle n|\hat{\mathcal{H}}'|k\rangle|^2}{E_n^o - E_k^o} \quad \text{(Second order)}$$

and the corrections to the zeroth order eigenfunctions are

$$u_n^{(1)} = \sum_{k\neq n} \frac{\langle k|\hat{\mathcal{H}}'|n\rangle}{E_n^o - E_k^o} u_k^o \quad \text{(First order)}$$

$$u_n^{(2)} = \sum_{\ell\neq n} \left\{ \sum_{k\neq n} \frac{\langle \ell|\hat{\mathcal{H}}'|k\rangle\langle k|\hat{\mathcal{H}}'|n\rangle}{(E_\ell^o - E_n^o)(E_k^o - E_n^o)} - \frac{\langle \ell|\hat{\mathcal{H}}'|n\rangle\langle n|\hat{\mathcal{H}}'|n\rangle}{(E_\ell^o - E_n^o)^2} \right\} u_\ell^o \quad \text{(Second order)}$$

The non-degenerate theory is valid only if $|\langle k|\hat{\mathcal{H}}'|n\rangle| \ll |E_n^o - E_k^o|$.

2. Perturbation Theory (discrete degenerate eigenvalues)

We suppose now that the eigenfunctions $u_{n\alpha}^o$ all correspond to the same energy E_n^o. Here $\alpha = 1, 2, \cdots, k$ and the energy level is said to be k-fold degenerate. In the eigenvalue equation $\hat{\mathcal{H}}u_n = E_n u_n$ where $\hat{\mathcal{H}} = \hat{\mathcal{H}}^o + \hat{\mathcal{H}}'$ the perturbation $\hat{\mathcal{H}}'$ may lift the degeneracy. The k eigenvalues and eigenfunctions resulting from the nth degenerate set can be obtained by solving the secular equation

$$\begin{vmatrix} \mathcal{H}'_{11} - E'_n S_{11} & \mathcal{H}'_{12} - E'_n S_{12} & \cdots & \mathcal{H}'_{1k} - E'_n S_{1k} \\ \vdots & & & \\ \mathcal{H}'_{k1} - E'_n S_{k1} & \cdots & & \mathcal{H}'_{kk} - E'_n S_{kk} \end{vmatrix} = 0$$

where

$$\mathcal{H}_{ij} = \langle ni|\hat{\mathcal{H}}'|nj\rangle = \int u_{ni}^{o*}\hat{\mathcal{H}}' u_{nj}^{o} d\tau$$

$$S_{ij} = \langle ni|nj\rangle = \int u_{ni}^{o*} u_{nj}^{o} d\tau$$

This gives first order corrections since contributions from states having energies different from E_n^o are neglected. The essence of this method is that it permits the correct, orthogonal linear combination of the degenerate function $u_{n\alpha}^o$ to be associated with each of the k energies when the degeneracy is lifted.

3. Variation Theory

According to the Variation Principle

$$E_o \leqslant \frac{\int \psi^* \hat{\mathcal{H}} \psi d\tau}{\int \psi^* \psi d\tau} = W$$

where E_o is the lowest eigenvalue of $\hat{\mathcal{H}}$ and ψ is an arbitrary trial function. The equality holds only when ψ is identically equal to the eigenfunction of $\hat{\mathcal{H}}$ associated with E_o. If $\psi = \psi(a,b,c,\cdots)$, then the "best" function ψ is obtained when the parameters a,b,c,$\cdots$ are adjusted so that

$$\frac{\partial W}{\partial a} = \frac{\partial W}{\partial b} = \frac{\partial W}{\partial c} = \cdots = 0$$

In the Linear Variation Theory a set of functions $\{\phi_i\}$ is selected and the trial function has the form

$$\psi = c_1\phi_1 + c_2\phi_2 + \cdots = \sum_{n=1}^{N} c_n\phi_n$$

It can be shown for such a function that W is minimized when the following secular equation is satisfied

$$\begin{vmatrix} \mathcal{H}_{11} - S_{11}W & \mathcal{H}_{12} - S_{12}W & \cdots & \mathcal{H}_{1N} - S_{1N}W \\ \vdots & & & \vdots \\ \mathcal{H}_{N1} - S_{N1}W & \cdots & & \mathcal{H}_{NN} - S_{NN}W \end{vmatrix} = 0$$

where

$$\mathcal{H}_{ij} = \langle i|\hat{\mathcal{H}}|j\rangle = \int \phi_i^* \hat{\mathcal{H}} \phi_j d\tau$$

$$S_{ij} = \langle i|j\rangle = \int \phi_i^* \phi_j d\tau$$

It can be shown that the roots W_o, W_1, $\cdots$ W_N are upper bounds to E_o, E_1, $\cdots$ E_N, respectively, where the E_n are the lowest N eigenvalues of $\hat{\mathcal{H}}$.

General references: PW,151; ME,413; PI1,234, HBE

PROBLEMS

6.1 Obtain an expression for the coefficients in the expansion

$$u_n^{(1)} = \sum_j c_{jn}^{(1)} u_j^o$$

and thus derive the equation for $u_n^{(1)}$ in Rayleigh-Schrödinger perturbation theory as given in the Introduction. (Hint: Use Eq. (4) of the Introduction.)

Solution:

From the Introduction we have:

$$(\hat{\mathcal{H}}^o - E_n^o)u_n^{(1)} + (\hat{\mathcal{H}}' - E^{(1)})u_n^o = 0$$

Thus

$$(\hat{\mathcal{H}}^o - E_n^o) \sum_j c_{jn}^{(1)} u_j^o + (\hat{\mathcal{H}}' - E^{(1)})u_n^o = 0$$

Multiplying through by $(u_j^o)^*$ and integrating gives:

$$(E_j^o - E_n^o)c_{jn}^{(1)} = - \langle u_j^o|\hat{\mathcal{H}}'|u_n^o\rangle$$

Then

$$u_n^{(1)} = \sum_{j\neq n} \frac{\langle u_j^o|\hat{\mathcal{H}}'|u_n^o\rangle}{(E_n^o - E_j^o)} u_j^o$$

The coefficient $c_{nn}^{(1)}$ is zero because of the normalization convention:

$$\langle u^o|u^{(1)}\rangle + \langle u^{(1)}|u^o\rangle = 0$$

6.2 Obtain the second order correction to the energy of the nth state in Rayleigh-Schrödinger perturbation theory as given in the Introduction by making use of the result of Problem 6.1.

Solution:

From the Introduction $E_n^{(2)} = \langle u^o|\hat{\mathcal{H}}' - E^{(1)}|u^{(1)}\rangle$. Thus:

$$E_n^{(2)} = \int (u_n^o)^* (\hat{\mathcal{H}}' - E^{(1)}) \sum_{j\neq n} \frac{\langle u_j^o|\hat{\mathcal{H}}'|u_n^o\rangle}{(E_n^o - E_j^o)} u_j^o d\tau = \sum_{j\neq n} \frac{\langle n|\hat{\mathcal{H}}'|j\rangle\langle j|\hat{\mathcal{H}}'|n\rangle}{(E_n^o - E_j^o)} = \sum_{j\neq n} \frac{|\langle n|\hat{\mathcal{H}}'|j\rangle|^2}{(E_n^o - E_j^o)}$$

6.3 A system described by the Hamiltonian $\hat{\mathcal{H}}$ having only two allowed states can be described by a wave function of the form

$$\psi = c_1\phi_1 + c_2\phi_2$$

(a) Derive equations for the exact energy levels of this system in terms of the matrix elements of $\hat{\mathcal{H}}$. Assume that $\hat{\mathcal{H}}$ and the coefficients c_n are real and that ϕ_1 and ϕ_2 form a complete orthonormal set.

(b) Let $\hat{\mathcal{H}} = \hat{\mathcal{H}}^o + \hat{\mathcal{H}}'$ with $\hat{\mathcal{H}}^o\phi_n = E_n^o\phi_n$ and compare the results with nondegenerate and degenerate perturbation theory.

Solution: Ref. S3,99

(a)
$$\begin{vmatrix} H_{11} - E & H_{12} \\ H_{12} & H_{22} - E \end{vmatrix} = (H_{11} - E)(H_{22} - E) - H_{12}^2 = 0 \tag{1}$$

or rearranging

$$E^2 - (H_{11} + H_{22})E + H_{11}H_{22} - H_{12}^2 = 0 \tag{2}$$

Thus

$$E = \frac{H_{11} + H_{22} \pm \sqrt{(H_{11} + H_{22})^2 + 4(H_{12}^2 - H_{11}H_{22})}}{2} \tag{3a}$$

$$= \frac{H_{11} + H_{22}}{2} \pm \sqrt{\left(\frac{H_{22} - H_{11}}{2}\right)^2 + H_{12}^2} \tag{3b}$$

$$= \frac{H_{11} + H_{22}}{2} \pm \left(\frac{H_{22} - H_{11}}{2}\right)\sqrt{1 + \left[\frac{H_{12}}{(H_{22} - H_{11})/2}\right]^2} \tag{3c}$$

If $[2H_{12}/(H_{22} - H_{11})]^2 < 1$ we may then expand the radical in a Taylor's series

$$(1 + \theta)^{1/2} = 1 + \frac{1}{2}\theta - \frac{(1/2)(1/2)\theta^2}{2!} + \cdots \tag{4}$$

so that

$$E = \frac{H_{11} + H_{22}}{2} \pm \frac{(H_{22} - H_{11})}{2}\left[1 + 2\left(\frac{H_{12}}{H_{22} - H_{11}}\right)^2 - 2\left(\frac{H_{12}}{H_{22} - H_{11}}\right)^4 \cdots\right] \tag{5}$$

Thus the two roots may be written

$$E_1 = H_{11} - \frac{H_{12}^2}{(H_{22} - H_{11})} + \frac{H_{12}^4}{(H_{22} - H_{11})^3} - \cdots \tag{6}$$

$$E_2 = H_{22} + \frac{H_{12}^2}{(H_{22} - H_{11})} - \frac{H_{12}^4}{(H_{22} - H_{11})^3} + \cdots \tag{7}$$

(b) When Eqs. (6) and (7) are written in terms of $\hat{\mathcal{H}}'$ for $E_1^o \neq E_2^o$, we find

$$E_1 = E_1^o + H_{11}' - \frac{H_{12}'^2}{(H_{22} - H_{11})} + \frac{H_{12}'^4}{(H_{22} - H_{11})^3} + \cdots \tag{8}$$

$$E_2 = E_2^o + H'_{22} + \frac{H'^2_{12}}{(H_{22} - H_{11})} - \frac{H'^4_{12}}{(H_{22} - H_{11})^3} + \cdots \quad (9)$$

To good approximation, $H_{22} - H_{11}$ may be replaced by $E_2^o - E_1^o$, thus yielding a result in agreement with perturbation theory.

If ϕ_1 and ϕ_2 are degenerate, then

$$\hat{\mathcal{H}}^o\phi_1 = E_1^o\phi_1 \quad \text{and} \quad \hat{\mathcal{H}}^o\phi_2 = E_1^o\phi_2$$

so that the expansion of Eq. (5) is obviously invalid. However, we may return to Eq. (3b) and write

$$E = \frac{H_{11} + H_{22}}{2} \pm \sqrt{\left(\frac{H_{22} - H_{11}}{2}\right)^2 + H_{12}^2} = E_1^o \pm H_{12}$$

Substituting $\hat{\mathcal{H}} = \hat{\mathcal{H}}^o + \hat{\mathcal{H}}'$ we have

$$E = E_1^o \pm \langle\phi_1|\hat{\mathcal{H}}^o + \hat{\mathcal{H}}'|\phi_2\rangle = E_1^o \pm \langle\phi_1|\hat{\mathcal{H}}'|\phi_2\rangle = E_1^o \pm H'_{12}$$

This result is exactly that predicted by degenerate perturbation theory.

6.4 Consider an electron in a potential box having the length a. When an electric field $\mathcal{E}$ is turned on in the x-direction, the electron experiences a force equal to $-e\mathcal{E}$ and the potential function has added to it the term $+e\mathcal{E}x$. The potential then has the form shown in Figure 6.4.

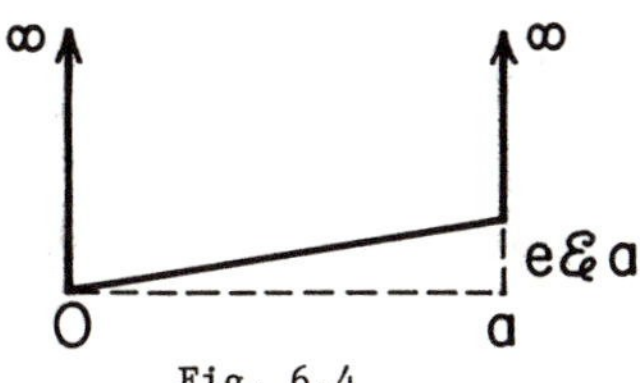

Fig. 6.4

(a) What is the lowest allowed energy (in a first order approximation) for the electron? You may assume that $e\mathcal{E}a$ is much smaller than the ground state energy in the absence of the electric field.

(b) Use first order perturbation theory to obtain an approximation to the ground state wavefunction and evaluate the first term in the correction.

<u>Solution</u>:

(a) For this system $\hat{\mathcal{H}} = \hat{\mathcal{H}}^o + \hat{\mathcal{H}}'$ with

$$\hat{\mathcal{H}}^o = -\frac{\hbar^2}{2m}\frac{d^2}{dx^2} \quad \text{and} \quad \hat{\mathcal{H}}' = e\mathcal{E}x$$

From Problem 4.6 we have the solutions for $\hat{\mathcal{H}}^o u_n^o = E_n^o u_n^o$:

$$E_n^o = \frac{n^2h^2}{8ma^2} \quad \text{and} \quad u_n^o = \sqrt{\frac{2}{a}}\sin\left(\frac{n\pi x}{a}\right)$$

and according to first order perturbation theory $E_n = E_n^o + E_n^{(1)}$ and

$$E_n^{(1)} = \int_o^a (u_n^o)^* \hat{\mathcal{H}}' u_n^o dx = e\mathcal{E}\int_o^a (u_n^o)^* x u_n^o dx = e\mathcal{E}<x> = \frac{e\mathcal{E}a}{2}$$

Therefore

$$E_1 = E_1^o + E_1^{(1)} = \frac{h^2}{8ma^2} + \frac{e\mathcal{E}a}{2}$$

(b) From first order perturbation theory we have

$$u_1 \cong u_1^o + \sum_{k \neq 1} \frac{<k|\hat{\mathcal{H}}'|1>}{E_1^o - E_k^o} u_k^o \tag{1}$$

The first term in the correction is then

$$\frac{<2|\hat{\mathcal{H}}'|1>}{E_1^o - E_2^o} = \frac{\int_o^a (u_2^o)^* e\mathcal{E}x u_1^o dx}{-3(h^2/8ma^2)} \tag{2}$$

where

$$\int_o^a (u_2^o)^* (e\mathcal{E}x) u_1^o dx = \frac{2e\mathcal{E}}{a} \int_o^a \sin\left(\frac{2\pi x}{a}\right) \sin\left(\frac{\pi x}{a}\right) x dx$$

Letting $y = \pi x/a$ the integral becomes

$$\left(\frac{a}{\pi}\right)^2 \int_o^\pi \sin y \sin 2y (y dy) = \left(\frac{a}{\pi}\right)^2 \int_o^\pi y \sin y (2 \sin y \cos y) dy = \left(\frac{a}{\pi}\right)^2 \int_o^\pi 2y(1 - \cos^2 y) \cos y dy$$

$$= 2\left(\frac{a}{\pi}\right)^2 \left[\int_o^\pi y \cos y dy - \int_o^\pi y \cos^3 y dy\right] \tag{3}$$

From tables we find

$$\int y \cos y dy = \cos y - y \sin y \qquad \int y \cos^3 y dy = \frac{y \sin 3y}{12} + \frac{\cos 3y}{36} + \frac{3}{4} y \sin y + \frac{3}{4} \cos y$$

Substitution of these formulas into Eq. (3) leads to

$$\int_o^a \sin\left(\frac{\pi x}{a}\right) \sin\left(\frac{2\pi x}{a}\right) x dx = -0.8889\left(\frac{a}{\pi}\right)^2$$

and Eq. (2) gives

$$\frac{<1|\hat{\mathcal{H}}'|2>}{E_1^o - E_2^o} = 0.889\left(\frac{2e\mathcal{E}a}{3\pi^2}\right)\left(\frac{8ma^2}{h^2}\right)$$

and finally

$$u_1 \cong u_1^o + 0.480\ e\mathcal{E}a\left(\frac{ma^2}{h^2}\right) u_2^o \cdots$$

The exact solution for this problem can be found in:

T. C. Dymski, Am. J. Phys. 36, 54 (1968).

J. N. Churchill and F. O. Arntz, Am. J. Phys. 37, 693 (1969).

6.5 An infinite potential box containing a particle of mass m is perturbed as shown in Fig. 6.5.

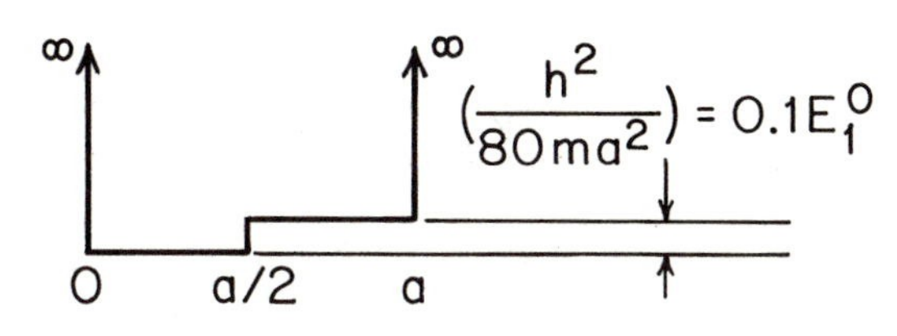

Fig. 6.5

(a) Use first order perturbation theory to calculate the correction that must be added to the "potential box" eigenvalues to obtain approximations to the energy levels for this problem.

(b) Write out the first three non-vanishing terms in the expansion of the ground state eigenfunction in terms of the unperturbed functions.

(c) Calculate the second order correction for the lowest eigenvalue.

Solution:

(a) For the unperturbed particle-in-box problem we have found:

$$E_n^o = \frac{n^2h^2}{8ma^2} \quad \text{and} \quad u_n^o = \sqrt{\frac{2}{a}}\sin\left(\frac{n\pi x}{a}\right)$$

Here $\hat{\mathcal{H}}' = (h^2/80ma^2)$ is added in the region $a/2 \lesssim x \lesssim a$ and the first order perturbation correction to the energy E_n^o is

$$E_n^{(1)} = \langle n|\hat{\mathcal{H}}|n\rangle = 0.1E_1^o \int_{a/2}^{a} (u_n^o)^* u_n dx$$

$$= 0.1E_1^o\left(\frac{1}{2}\right)\int_o^a (u_n^o)^* u_n^o dx$$

$$= \frac{0.1E_1^o}{2} = \frac{1}{2}\left(\frac{h^2}{80ma^2}\right)$$

Therefore

$$E_n \simeq E_n^{(0)} + E_n^{(1)} = \frac{h^2}{8ma^2}(n^2 + 0.05)$$

(b) The ground state wave function correct to first order is given by

$$u_1 = u_1^o + \sum_{k\neq 1} \frac{\langle 1|\hat{\mathcal{H}}'|k\rangle}{E_1^o - E_k^o} u_k^o$$

$$u_1 = u_1^o + \frac{\langle 2|\hat{\mathcal{H}}'|1\rangle}{E_1^o - E_2^o} + \frac{\langle 3|\hat{\mathcal{H}}'|1\rangle}{E_1^o - E_3^o} + \frac{\langle 4|\hat{\mathcal{H}}'|1\rangle}{E_1^o - E_4^o} + \cdots$$

The required integrals are of the form:

$$\frac{\langle n|\hat{\mathcal{H}}'|k\rangle}{|\mathcal{H}'|} = \frac{2}{a}\int_{a/2}^{a} \sin\left(\frac{n\pi x}{a}\right)\sin\left(\frac{k\pi x}{a}\right)dx = \left(\frac{2}{a}\right)\left(\frac{a}{\pi}\right)\int_{\pi/2}^{\pi} \sin(ny)\sin(ky)\,dy$$

$$= \frac{1}{\pi}\left\{\frac{\sin(n-k)y}{(n-k)} - \frac{\sin(n+k)y}{(n+k)}\right\}\Bigg|_{\pi/2}^{\pi}, \quad n^2 \neq k^2$$

where integral tables have been used in the last step. This equation gives

$$\langle 2|\hat{\mathcal{H}}'|1\rangle = 0.1E_1^o\left(-\frac{4}{3\pi}\right), \quad \langle 3|\hat{\mathcal{H}}'|1\rangle = 0, \quad \langle 4|\hat{\mathcal{H}}'|1\rangle = 0.1E_1^o\left(\frac{8}{15\pi}\right)$$

The approximate ground state wave function then becomes

$$u_1 = u_1^o + \frac{0.1E_1^o}{-3E_1^o}\left(-\frac{4}{3\pi}\right)u_2^o + \frac{0.1E_1^o}{-15E_1^o}\left(\frac{8}{15\pi}\right)u_4^o + \cdots = u_1^o + \frac{2}{45\pi}u_2^o - \frac{4}{1125\pi}u_4^o + \cdots$$

(c) The second order correction to E_1^o is given by

$$E_1^{(2)} = \sum_{k\neq 1}\frac{|\langle 1|\hat{\mathcal{H}}'|k\rangle|^2}{E_1^o - E_k^o} = \frac{|\langle 1|\hat{\mathcal{H}}'|2\rangle|^2}{E_1^o - E_2^o} + \frac{|\langle 1|\hat{\mathcal{H}}'|3\rangle|^2}{E_1^o - E_3^o} + \frac{|\langle 1|\hat{\mathcal{H}}'|4\rangle|^2}{E_1^o - E_4^o} + \cdots$$

Using the integrals from part (b) we find

$$E_1^{(2)} = -\frac{10^{-2}(E_1^o)^2}{3E_1^o}\left(\frac{4}{3\pi}\right)^2 - \frac{10^{-2}(E_1^o)^2}{15E_1^o}\left(\frac{8}{15\pi}\right)^2 - \cdots \cong -6.196 \times 10^{-4}\left(\frac{h^2}{8ma^2}\right)$$

and the eigenvalue to second order is

$$E_1 \simeq E_1^o + E_1^{(1)} + E_1^{(2)} \simeq \frac{h^2}{8ma^2}(1 + 0.05 - 6.196 \times 10^{-4})$$

The exact solution for this problem can be found in L. Melander, J. Chem. Ed. 49, 686 (1972).

6.6 A simple harmonic oscillator having the reduced mass μ and the force constant k is subjected to the quartic perturbation $\hat{\mathcal{H}}' = ax^4$. Derive an equation for the first order correction to the nth eigenvalue of this oscillator, and obtain the first nonvanishing term in the first order correction for the ground state wavefunction.

Solution:

The Hamiltonian in this problem is

$$\hat{\mathcal{H}} = \hat{\mathcal{H}}^o + \hat{\mathcal{H}}' = \left(-\frac{\hbar^2}{2\mu}\frac{d^2}{dx^2} + \frac{1}{2}kx^2\right) + ax^4$$

The unperturbed oscillator ($\hat{\mathcal{H}}^o$) was treated in Problem 4.10 where it was shown that

$$E_n^o = \left(n + \frac{1}{2}\right)h\nu \quad \text{and} \quad u_n^o(x) = \left(\frac{1}{2^n n!}\sqrt{\frac{\alpha}{\pi}}\right)^{1/2} e^{-\alpha x^2/2} H_n(\sqrt{\alpha}\, x)$$

The first order correction to the nth eigenvalue resulting from $\hat{\mathcal{H}}'$ is:

$$E_n^{(1)} = \langle n|ax^4|n\rangle = \int_{-\infty}^{+\infty} u_n^o(x)^*[ax^4]u_n^o(x)\,dx \tag{1}$$

One method for evaluating the integral is to make use of the matrix elements of x which were derived in Problem 4.12b.

$$\langle n|x|k\rangle = x_{nk} = \delta_{k,n+1}\sqrt{\frac{n+1}{2\alpha}} + \delta_{k,n-1}\sqrt{\frac{n}{2\alpha}} \tag{2}$$

The matrix element $(x^4)_{nn}$ can then be determined by matrix multiplication as follows:

$$(x^4)_{nn} = \sum_k (x^2)_{nk}(x^2)_{kn}$$

From Eq. (2) it is clear that the only nonvanishing matrix elements of x^2 are:

$$(x^2)_{n,n+2} = \sum_k x_{nk}x_{k,n+2} = x_{n,n+1}x_{n+1,n+2} = \sqrt{\frac{n+1}{2\alpha}}\sqrt{\frac{n+2}{2\alpha}} = \frac{\sqrt{(n+1)(n+2)}}{2\alpha}$$

$$(x^2)_{n,n-2} = x_{n,n-1}x_{n-1,n-2} = \frac{\sqrt{n(n-1)}}{2\alpha}$$

$$(x^2)_{nn} = x_{n,n+1}x_{n+1,n} + x_{n,n-1}x_{n-1,n} = \frac{n+1}{2\alpha} + \frac{n}{2\alpha} = \frac{(2n+1)}{2\alpha}$$

Combining these terms we find

$$(x^4)_{nn} = \frac{1}{4\alpha^2}\left[(n+1)(n+2) + n(n-1) + 4n^2 + 4n + 1\right] = \frac{3}{2\alpha^2}\left(n^2 + n + \frac{1}{2}\right)$$

Equation (1) then gives the first order correction to be

$$E_n^{(1)} = \frac{3a}{2\alpha^2}\left(n^2 + n + \frac{1}{2}\right)$$

The ground state eigenfunction corrected to first order is

$$u_o = u_o^o + \frac{\langle 1|\hat{\mathcal{H}}'|0\rangle}{E_o^o - E_1^o}u_1^o + \frac{\langle 2|\hat{\mathcal{H}}'|0\rangle}{E_o^o - E_2^o}u_2^o + \frac{\langle 3|\hat{\mathcal{H}}'|0\rangle}{E_o^o - E_3^o}u_3^o \cdots$$

where the second and fourth terms on the right vanish since the integrands are odd functions of x. In the third term we require the matrix element:

$$(x^4)_{n,n-2} = x^2_{nn}x^2_{n,n-2} + x^2_{n,n-2}x^2_{n-2,n-2}$$

$$= \frac{(2n+1)}{2\alpha}\frac{\sqrt{n(n-1)}}{2\alpha} + \frac{\sqrt{n(n-1)}}{2\alpha}\frac{(2n-3)}{2\alpha} = \frac{\sqrt{n(n-1)}\,(2n-1)}{2\alpha^2}$$

Therefore

$$u_o = u_o^o - \frac{3\sqrt{2}\,a}{4\alpha^2 h\nu} u_2^o + \cdots$$

6.7 A simple harmonic oscillator having the reduced mass μ and force constant k is subjected to a cubic perturbation $\hat{\mathcal{H}}' = ax^3$. Derive an equation for energy to second order and determine the first order correction to the wavefunction.

Solution: Ref. PW,156; AN3,339; Problem 6.6

The Hamiltonian is

$$\hat{\mathcal{H}} = \hat{\mathcal{H}}^o + \hat{\mathcal{H}}' = -\frac{\hbar^2}{2\mu}\frac{d^2}{dx^2} + \frac{1}{2}kx^2 + ax^3$$

The eigenvalue for $\hat{\mathcal{H}}^o$ is

$$E_n^o = (n + 1/2)h\nu$$

and the eigenfunction of $\hat{\mathcal{H}}^o$ is given by

$$u_n^o(x) = \left(\frac{1}{2^n n!}\sqrt{\frac{\alpha}{\pi}}\right)^{1/2} e^{-\alpha x^2/2} H_n(\alpha x)$$

(see Problem 4.10). Thus, the first order correction to the nth eigenvalue is given by

$$E_n^{(1)} = \langle n|ax^3|n\rangle = 0 \text{ (the integrand is odd)}$$

and the second order energy is

$$E_n^{(2)} = \sum_{k\neq n} \frac{|\langle n|\hat{\mathcal{H}}'|k\rangle|^2}{E_n^o - E_k^o} \tag{1}$$

$$= \frac{|\langle n|ax^3|n+3\rangle|^2}{E_n^o - E_{n+3}^o} + \frac{|\langle n|ax^3|n+1\rangle|^2}{E_n^o - E_{n+1}^o} + \frac{|\langle n|ax^3|n-1\rangle|^2}{E_n^o - E_{n-1}^o} + \frac{|\langle n|ax^3|n-3\rangle|^2}{E_n^o - E_{n-3}^o}$$

This result may be obtained with the equation

$$\langle n|ax^3|m\rangle = a\sum_k \langle n|x^2|k\rangle\langle k|x|m\rangle$$

From Problem 6.6 we see that k can be only n±2 and n; thus m can be only n±3, n±1. The necessary integrals are then:

$$a(x^3)_{n,n+3} = a(x^2)_{n,n+2}(x)_{n+2,n+3} = a\sqrt{\frac{(n+1)(n+2)(n+3)}{(2\alpha)^3}}$$

$$a(x^3)_{n,n+1} = a[(x)^2_{n,n+2}(x)_{n+2,n+1} + (x^2)_{n,n}(x)_{n,n+1}] = 3a\sqrt{\frac{(n+1)^3}{(2\alpha)^3}}$$

$$a(x^3)_{n,n-1} = a[(x)^2_{n,n-2}(x)_{n-2,n-1} + (x^2)_{n,n}(x)_{n,n-1}] = 3a\sqrt{\frac{n^3}{(2\alpha)^3}}$$

$$a(x^3)_{n,n-3} = a(x)^2{}_{n,n-2}(x)_{n-2,n-3} = a\sqrt{\frac{n(n-1)(n-2)}{(2\alpha)^3}}$$

Thus, substituting into Eq. (1), we find

$$E_n^{(2)} = -\frac{30a^2}{h\nu(2\alpha)^3}[n^2 + n + 11/30]$$

The first order wave function involves the same integrals and is found to be

$$u_n = u_n^o + \sum_{k\neq n}\frac{\langle k|\hat{\mathcal{H}}'|n\rangle}{E_n^o - E_k^o}u_k^o = u_n^o + \frac{a}{2\alpha h\nu}\left[\frac{1}{3}\sqrt{\frac{n(n-1)(n-2)}{2\alpha}}\,u_{n-3}^o + 3n\sqrt{\frac{n}{2\alpha}}\,u_{n-1}^o\right.$$
$$\left. - 3(n+1)\sqrt{\frac{(n+1)}{2\alpha}}\,u_{n+1}^o - \frac{1}{3}\sqrt{\frac{(n+1)(n+2)(n+3)}{2\alpha}}\,u_{n+3}^o\right]$$

6.8 A linear harmonic oscillator is perturbed by an electric field of strength $\mathcal{E}$. If the oscillating mass has an associated charge $-e$, the perturbing Hamiltonian becomes

$$\hat{\mathcal{H}}' = +e\mathcal{E}x$$

Determine the perturbation correction to the energy through second order.

Solution: Ref. PW,177

The energy to second order is given by

$$E_n = E_n^{\,o} + E_n^{(1)} + E_n^{(2)} = E_n^{\,o} + H'_{nn} + \sum_{k\neq n}\frac{|H'_{nk}|^2}{E_n^o - E_k^o}$$

Thus the first order correction H'_{nn} is given by

$$H'_{nn} = \langle n|\hat{\mathcal{H}}'|n\rangle = +e\mathcal{E}\langle n|x|n\rangle = 0$$

since the nth eigenfunction of $\hat{\mathcal{H}}^o$ is always either an odd or an even function and x is odd.

The matrix elements necessary for the second order correction are (Problem 4.12):

$$H'_{nk} = +e\mathcal{E}\langle n|x|k\rangle = \begin{cases} +e\mathcal{E}\sqrt{\dfrac{n+1}{2\alpha}}, & k = n+1 \\[2ex] +e\mathcal{E}\sqrt{\dfrac{n}{2\alpha}}, & k = n-1 \end{cases}$$

Here

$$\alpha = \frac{2\pi\mu\nu}{\hbar}, \quad \nu = \frac{1}{2\pi}\sqrt{\frac{k}{\mu}}$$

μ is the reduced mass of the oscillator, and k is the Hooke's Law force constant. The energy to second order is then

$$E_n = E_n^{\,o} + E_n^{(1)} + E_n^{(2)} = (n + 1/2)h\nu + 0 + \frac{|H'_{n,n+1}|^2}{E_n^{\,o} - E_{n+1}^{\,o}} + \frac{|H'_{n,n-1}|^2}{E_n^{\,o} - E_{n-1}^{\,o}}$$

$$= (n + 1/2)h\nu + (e\mathcal{E})^2\left[\frac{\left(\frac{n+1}{2\alpha}\right)}{-h\nu} + \frac{\left(\frac{n}{2\alpha}\right)}{h\nu}\right] = (n + 1/2)h\nu - \frac{e^2\mathcal{E}^2}{8\pi^2\mu\nu^2}$$

Thus the correction to second order is identical for all states. It turns out that the second order correction in this case gives the exact energy (see Problem 6.9).

6.9 Show that Problem 6.8 may be solved exactly by use of the substitution:

$$x = x' - e\mathcal{E}/k$$

Solution:

The Hamiltonian for the charged oscillator in an electric field is

$$\hat{\mathcal{H}} = -\frac{\hbar^2}{2\mu}\frac{d^2}{dx^2} + \frac{kx^2}{2} + e\mathcal{E}x$$

The suggested substitution yields

$$\hat{\mathcal{H}} = -\frac{\hbar^2}{2\mu}\frac{d^2}{dx'^2} + \frac{k}{2}(x')^2 - \frac{1}{2k}e^2\mathcal{E}^2$$

The Schrödinger equation then becomes

$$\left[-\frac{\hbar^2}{2\mu}\frac{d^2}{dx'^2} + \frac{k}{2}(x')^2\right]u(x') = \left(E' + \frac{1}{2k}e^2\mathcal{E}^2\right)u(x') = E\,u(x')$$

Clearly from previous work, $E = (n + 1/2)h\nu$ so that

$$E' = E - \frac{(e\mathcal{E})^2}{2k} \quad \text{or} \quad E' = (n + 1/2)h\nu - \frac{e^2\mathcal{E}^2}{8\pi^2\mu\nu^2}$$

6.10 A particle of mass m is constrained to move in the xy plane so that the Hamiltonian is:

$$H = \frac{1}{2m}(p_x^2 + p_y^2) + \frac{1}{2}k(x^2 + y^2) + axy$$

(a) Set up the wave equation for this problem and separate it into two equations, each depending on one independent variable, in the special case that $a = 0$.

(b) List the first few energy levels for the two dimensional harmonic oscillator ($a = 0$) and give the degeneracies.

(c) Use the degenerate perturbation theory to determine the energy splitting for the lowest degenerate states and the first order wave functions for these states when $a \neq 0$. This will involve the solution of a 2x2 secular equation. The necessary matrix elements for $(\hat{\mathcal{H}}_x^o + \hat{\mathcal{H}}_y^o + \hat{\mathcal{H}}')$ can easily be found by using the matrix elements derived for the oscillator in one dimension and the rule

$$\langle n_x n_y|xy|n_x' n_y'\rangle = \langle n_x|x|n_x'\rangle\langle n_y|y|n_y'\rangle$$

Solution:

(a) When $a = 0$ we write $\hat{\mathcal{H}}^o = \hat{\mathcal{H}}_x^o + \hat{\mathcal{H}}_y^o$ with

$$\hat{\mathcal{H}}_x^o = \frac{\hat{p}_x^2}{2m} + \frac{kx^2}{2}, \quad \hat{\mathcal{H}}_y^o = \frac{\hat{p}_y^2}{2m} + \frac{ky^2}{2}$$

The variables in the equation $\hat{\mathcal{H}}^o\psi^o = E^o\psi^o$ can be separated by assuming that $\psi^o = u^o(x)u^o(y)$ so

$$(\hat{\mathcal{H}}^o_x + \hat{\mathcal{H}}^o_y)u^o(x)u^o(y) = [E^o(x) + E^o(y)]u^o(x)u^o(y)$$

or $\hat{\mathcal{H}}^o_x u^o(x) = E^o(x)u^o(x), \quad \hat{\mathcal{H}}^o_y u^o(y) = E^o(y)u^o(y)$

(b) From Problem 4.10 we have that

$$E^o_x = (n_x + 1/2)h\nu_o \quad \text{and} \quad E^o_y = (n_y + 1/2)h\nu_o$$

Then for the two-dimensional oscillator

$$E(n_x,n_y) = (n_x + n_y + 1)h\nu_o$$

The energies and degeneracies are shown below

E	n_x	n_y	degeneracy
$h\nu_o$	0	0	1
$2h\nu_o$	0	1	2
	1	0	
$3h\nu_o$	1	1	
	2	0	3
	0	2	

(c) The degenerate functions corresponding to $2h\nu_o$ in the table are (neglecting superscripts):

$$n_x = 1, \quad n_y = 0 \rightarrow \phi_1 = u_1(x)u_o(y)$$

$$n_x = 0, \quad n_y = 1 \rightarrow \phi_2 = u_o(x)u_1(y)$$

or in Dirac notation:

$$\phi_1 \rightarrow |n_x = 1, n_y = 0\rangle = |n_x = 1\rangle|n_y = 0\rangle$$

$$\phi_2 \rightarrow |n_x = 0, n_y = 1\rangle = |n_x = 0\rangle|n_y = 1\rangle$$

When the perturbation $\hat{\mathcal{H}}' = axy$ is present the secular equation $\det|\hat{\mathcal{H}}'_{ij} - E\Delta_{ij}| = 0$ must be solved to obtain the first order corrections to the energy. Here

$$\hat{\mathcal{H}}'_{ij} = \int \phi_i^*(axy)\phi_j dv$$

We introduce the short hand notation

$$\hat{\mathcal{H}}'_{11} = \langle n_x = 1|\langle n_y = 0|axy|n_x = 1\rangle|n_y = 0\rangle = a\langle 1|x|1\rangle\langle 0|y|0\rangle, \text{ etc.}$$

and the secular equation becomes

$$\begin{vmatrix} (a\langle 1|x|1\rangle\langle 0|y|0\rangle - E_2^{(1)}) & a\langle 1|x|0\rangle\langle 0|y|1\rangle \\ a\langle 0|x|1\rangle\langle 1|y|0\rangle & (a\langle 0|x|0\rangle\langle 1|y|1\rangle - E_2^{(1)}) \end{vmatrix} = 0$$

According to Problem 4.12 the non-vanishing matrix elements are

$$\langle 1|x|0\rangle = \langle 0|x|1\rangle = \langle 1|y|0\rangle = \langle 0|y|1\rangle = 1/\sqrt{2\alpha}$$

where $\alpha = (4\pi^2 m\nu/h)$ and we obtain:

$$\begin{vmatrix} -E_2^{(1)} & a/2\alpha \\ a/2\alpha & -E_2^{(1)} \end{vmatrix} = [E_2^{(1)}]^2 - a^2/4\alpha^2 = 0$$

$$E_{2,1}^{(1)} = +a/2\alpha, \quad E_{2,2}^{(1)} = -a/2\alpha$$

The eigenfunctions are determined as follows:

$$E_{2,1} = E_2^o + E_{2,1}^{(1)} = E_2^o + a/2\alpha$$

$$-(a/2\alpha)c_1 + (a/2\alpha)c_2 = 0, \quad c_1 = c_2$$

Therefore, including normalization:

$$\psi_{2,1} = \frac{1}{\sqrt{2}}(|n_x = 0\rangle|n_y = 1\rangle + |n_x = 1\rangle|n_y = 0\rangle)$$

$$E_{2,2} = E_2^o + E_{2,2}^{(1)} = E_2^o - a/2\alpha$$

$$+(a/2\alpha)c_1 + (a/2\alpha)c_2 = 0, \quad c_1 = -c_2$$

The normalized eigenfunction is:

$$\psi_{2,2} = \frac{1}{\sqrt{2}}(|n_x = 0\rangle|n_y = 1\rangle - |n_x = 1\rangle|n_y = 0\rangle)$$

The energy levels are shown in Fig. 6.10.

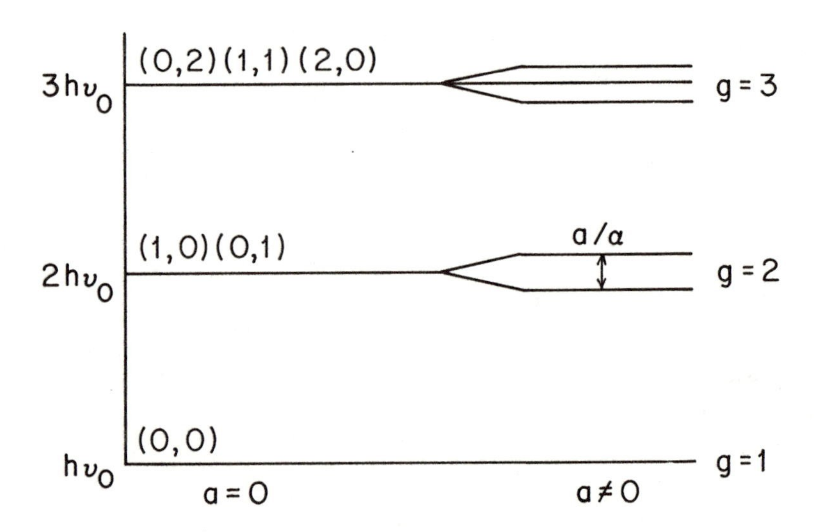

Fig. 6.10

6.11 An electron is in a cubic box of length a has the energy

$$\frac{3}{4}\frac{h^2}{ma^2}$$

(a) Determine the first order correction to the energy if an electric field of strength $\mathcal{E}$ is applied parallel to the z axis ($\perp$ to x and y). The cube is oriented so that the faces are parallel to the x, y, and z axes, and the perturbation may be taken as $\hat{\mathcal{H}}' = +e\mathcal{E}z$.

(b) Repeat the calculation for $\hat{\mathcal{H}}' = +b\mathcal{E}xy$.

Solution: Ref. PW,165; AN3,354

(a) An energy

$$\frac{3}{4}\frac{h^2}{ma^2}$$

implies $n_x^2 + n_y^2 + n_z^2 = 6$. Thus the state is triply degenerate with $(n_x, n_y, n_z) = (2,1,1)$, $(1,2,1)$ and $(1,1,2)$. The wavefunction for the electron must be a linear combination of the functions:

$$u_{2,1,1} = \sqrt{\frac{8}{a^3}}\sin\left(\frac{2\pi x}{a}\right)\sin\left(\frac{\pi y}{a}\right)\sin\left(\frac{\pi z}{a}\right)$$

$$u_{1,2,1} = \sqrt{\frac{8}{a^3}}\sin\left(\frac{\pi x}{a}\right)\sin\left(\frac{2\pi y}{a}\right)\sin\left(\frac{\pi z}{a}\right)$$

$$u_{1,1,2} = \sqrt{\frac{8}{a^3}}\sin\left(\frac{\pi x}{a}\right)\sin\left(\frac{\pi y}{a}\right)\sin\left(\frac{2\pi z}{a}\right)$$

It is easy to show that

$$\langle 2,1,1|z|2,1,1\rangle = \langle 1,2,1|z|1,2,1\rangle = \langle 1,1,2|z|1,1,2\rangle$$

$$\langle 2|2\rangle\langle 1|1\rangle\langle 1|z|1\rangle = \frac{2}{a}\int_0^a z\cdot\sin^2\left(\frac{\pi z}{a}\right)dz = \frac{a}{2}$$

In a similar manner we have

$$\langle 2,1,1|z|1,2,1\rangle = \langle 2,1,1|z|1,1,2\rangle = \langle 1,2,1|z|1,1,2\rangle = 0$$

Thus, all off-diagonal elements are zero and the energy is simply shifted by $+e\mathcal{E}a/2$. It is possible to solve without degenerate perturbation theory by simply noting that the Hamiltonian separates into (x), (y) and (z) parts where only the separated (z) part contains the perturbation. First order perturbation theory may then be applied for the z part only. It should be noticed that the energies depend on the choice of origin in this problem. If the box extends from $z = -a/2$ to $z = +a/2$, then $\langle z\rangle = 0$ and there are no shifts. The splittings are, of course, unaffected.

(b) For $\hat{\mathcal{H}}' = +b\mathcal{E}xy$ we determine the integrals necessary to construct the secular equation:

diagonal: $\langle 2,1,1|xy|2,1,1\rangle = \langle 2|x|2\rangle\langle 1|y|1\rangle\langle 1|1\rangle = \frac{a^2}{4}$

$= \langle 1,2,1|xy|1,2,1\rangle = \langle 1,1,2|xy|1,1,2\rangle$

Thus

$$H_{11} = H_{22} = H_{33} = +b\mathcal{E}a^2/4$$

off-diagonal: $\langle 2,1,1|xy|1,2,1\rangle = \langle 2|x|1\rangle\langle 1|y|2\rangle\langle 1|1\rangle = |\langle 2|x|1\rangle|^2$

$$= \left|\frac{2}{a}\int_0^a x\sin\left(\frac{2\pi x}{a}\right)\sin\left(\frac{\pi x}{a}\right)dx\right|^2 = \left|\frac{16}{9}\frac{a}{\pi^2}\right|^2 \quad \text{(see Prob. 6.4)}$$

Thus

$$H'_{12} = +b\mathcal{E}\left(\frac{16a}{9\pi^2}\right)^2 = H'_{21}$$

Similarly

$$H'_{13} = +b\mathcal{E}\langle 2,1,1|xy|1,1,2\rangle = +b\mathcal{E}\langle 2|x|1\rangle\langle 1|y|1\rangle\langle 1|2\rangle = 0 = H'_{31} = H'_{32} = H'_{23}$$

The 1st order energy secular determinant becomes

$$\begin{vmatrix} A-E & B & 0 \\ B & A-E & 0 \\ 0 & 0 & A-E \end{vmatrix} = 0$$

where

$$A = +b\mathcal{E}a^2/4 \quad \text{and} \quad B = +b\mathcal{E}\left(\frac{16}{9}\frac{a}{\pi^2}\right)^2$$

It is easily seen that this determinant yields the energies

$$E_1^{(1)} = A + B = +b\mathcal{E}a^2\left[\left(\frac{16}{9\pi^2}\right)^2 + \frac{1}{4}\right]$$

$$E_2^{(1)} = A = +b\mathcal{E}a^2/4$$

and

$$E_3^{(1)} = A - B = +b\mathcal{E}a^2\left[-\left(\frac{16}{9\pi^2}\right)^2 + \frac{1}{4}\right]$$

6.12 The following Hamiltonian matrix has been constructed using an orthonormal basis set,

$$\underset{\sim}{H} = \begin{pmatrix} 1 & 0 & 0 \\ 0 & 3 & 0 \\ 0 & 0 & -2 \end{pmatrix} + \begin{pmatrix} 0 & c & 0 \\ c & 0 & 0 \\ 0 & 0 & c \end{pmatrix}$$

Here $\underset{\sim}{\mathcal{H}} = \underset{\sim}{\mathcal{H}}^o + \underset{\sim}{\mathcal{H}}'$ and c is a constant.

(a) Find the exact eigenvalues of $\underset{\sim}{H}$.

(b) Use perturbation theory to determine the eigenvalues correct to second order.

(c) Compare the results of steps (a) and (b).

Solution: Ref. MA,86

(a) The eigenvalues are found by diagonalizing the matrix $\underset{\sim}{\mathcal{H}}$:

$$\begin{vmatrix} 1-\lambda & c & 0 \\ c & 3-\lambda & 0 \\ 0 & 0 & c-2-\lambda \end{vmatrix} = \begin{vmatrix} 1-\lambda & c \\ c & 3-\lambda \end{vmatrix} \; |c-2-\lambda| = 0$$

Thus

$$\lambda = c-2,\; 2 \pm \sqrt{1 + c^2}$$

(b) The energy, correct to second order, may be written as

$$\underset{\sim}{E} \approx \underset{\sim}{E}^{o} + \underset{\sim}{E}^{(1)} + \underset{\sim}{E}^{(2)}$$

or

$$(\underset{\sim}{E})_{ii} = (\underset{\sim}{H}^{o})_{ii} + (\underset{\sim}{H}')_{ii} + \sum_{k \neq i} \frac{H'_{ik}H'_{ki}}{(E_i^o - E_k^o)}$$

Clearly $(H^o)_{ii} = 1$, 3 and -2. And $(H')_{ii}$, the first order energy correction, is given by $H'_{11} = 0$, $H'_{22} = 0$, $H'_{33} = c$. For the second order correction

$$E_{11}^{(2)} = \frac{H'_{12}H'_{21}}{E_1^o - E_2^o} + \frac{H'_{13}H'_{31}}{E_1^o - E_3^o} = \frac{c^2}{-2} + \frac{0}{3} = -\frac{1}{2}\,c^2$$

$$E_{22}^{(2)} = \frac{H'_{21}H'_{12}}{E_2^o - E_1^o} + \frac{H'_{23}H'_{32}}{E_2^o - E_1^o} = \frac{c^2}{3-1} + \frac{0 \cdot 0}{3+2} = +\frac{1}{2}\,c^2$$

and

$$E_{33}^{(2)} = \frac{H'_{31}H'_{13}}{E_3^o - E_1^o} + \frac{H'_{32}H'_{23}}{E_3^o - E_2^o} = 0$$

Thus, in matrix notation

$$\underset{\sim}{E}^{o} = \begin{pmatrix} 1 & 0 & 0 \\ 0 & 3 & 0 \\ 0 & 0 & -2 \end{pmatrix}, \quad \underset{\sim}{E}^{(1)} = \begin{pmatrix} 0 & 0 & 0 \\ 0 & 0 & 0 \\ 0 & 0 & c \end{pmatrix}, \quad \underset{\sim}{E}^{(2)} = \begin{pmatrix} -\frac{1}{2}\,c^2 & 0 & 0 \\ 0 & +\frac{1}{2}\,c^2 & 0 \\ 0 & 0 & 0 \end{pmatrix}$$

or

$$\underset{\sim}{E} \approx \begin{pmatrix} 1 - \frac{1}{2}\,c^2 & 0 & 0 \\ 0 & 3 + \frac{1}{2}\,c^2 & 0 \\ 0 & 0 & c-2 \end{pmatrix} \qquad \text{(through second order)}$$

(c) Expanding

$$2 \pm \sqrt{1 + c^2}$$

from (a) in a binomial series

$$2 \pm \sqrt{1 + c^2} = 2 \pm (1 + \tfrac{1}{2}\,c^2 + \cdots) = 3 + \tfrac{1}{2}\,c^2,\; 1 - \tfrac{1}{2}\,c^2 \quad (c^2 \ll 1)$$

which gives the same result to second order as part (b).

6.13 Given the Hamiltonian matrix

$$\underset{\sim}{\mathcal{H}} = \underset{\sim}{\mathcal{H}}^{o} + \underset{\sim}{\mathcal{H}}' = \begin{pmatrix} 20 & 0 & 0 \\ 0 & 20 & 0 \\ 0 & 0 & 30 \end{pmatrix} + \begin{pmatrix} 0 & 1 & 0 \\ 1 & 0 & 2 \\ 0 & 2 & 0 \end{pmatrix}$$

which has been constructed using orthogonal basis functions:

(a) Determine the exact eigenvalues.

(b) Determine the eigenvalues correct to second order in the perturbation.

Solution: Ref. MA,86

(a) For the exact eigenvalues we must diagonalize the Hamiltonian matrix. The appropriate secular equation is

$$\begin{vmatrix} 20-\lambda & 1 & 0 \\ 1 & 20-\lambda & 2 \\ 0 & 2 & 30-\lambda \end{vmatrix} = 0$$

or $\quad -\lambda^3 + 70\lambda^2 - 1595\lambda + 11890 = (\lambda - 20.806)(\lambda - 18.805)(\lambda - 30.389) = 0$

The last equation was determined by a "trial and error" procedure on an electronic calculator for the first root, then using the quadratic equation for the remaining two roots.

(b) The upper 2x2 block of the secular determinant has degenerate zeroth order eigenfunctions, thus the first order corrections are given by the roots of

$$\begin{vmatrix} 20-\lambda & 1 \\ 1 & 20-\lambda \end{vmatrix} \cdot |30-\lambda| = (21 - \lambda)(19 - \lambda)(30 - \lambda) = 0$$

If we let ϕ_1, ϕ_2 and ϕ_3 be the orthonormal basis functions for which our initial Hamiltonian matrix is defined, then the eigenfunctions of the diagonalized 2x2 are given by

$$\lambda = 21: \quad x_1 = 1/\sqrt{2}\ (\phi_1 + \phi_2)$$

$$\lambda = 19: \quad x_2 = 1/\sqrt{2}\ (\phi_1 - \phi_2)$$

Then for the second order calculation, we construct a new Hamiltonian matrix based on the functions x_1, x_2 and ϕ_3.

Since

$$H'_{13} = \langle x_1|\hat{\mathcal{H}}|\phi_3\rangle = 1/\sqrt{2}\ \langle\phi_1 + \phi_2|\hat{\mathcal{H}}|\phi_3\rangle = 1/\sqrt{2}\ (H_{13} + H_{23}) = \sqrt{2} = H'_{31}$$

and similarly

$$H'_{23} = H'_{32} = -\sqrt{2}$$

we can then find the energy correct to second order from the matrix

$$\begin{pmatrix} 21 & 0 & \sqrt{2} \\ 0 & 19 & -\sqrt{2} \\ \sqrt{2} & -\sqrt{2} & 30 \end{pmatrix}$$

Thus, correct to second order, we have

$$E_1 = 21 + 0 + \frac{(\sqrt{2})^2}{21-30} = 20.777$$

$$E_2 = 19 + 0 + \frac{(\sqrt{2})^2}{19-30} = 18.818$$

and

$$E_3 = 30 + 0 + \frac{(\sqrt{2})^2}{30-21} + \frac{(\sqrt{2})^2}{30-19} = 30.404$$

In summary

root	1st order	2nd order	exact
1	21	20.777	20.806
2	19	18.818	18.805
3	30	30.404	30.389

6.14 Assume that the lowest eigenfunction of the simple harmonic oscillator is approximated by $u_o = N\exp(-cx^2)$ and use the variation method to determine c. Also, calculate the energy associated with this eigenfunction.

Solution: Ref. Problem 4.10

First we determine N from the normalization condition:

$$\int_{-\infty}^{+\infty} N^2 e^{-2cx^2} dx = N^2\sqrt{\frac{\pi}{2c}} = 1, \qquad N^2 = \sqrt{\frac{2c}{\pi}}$$

For the harmonic oscillator $\hat{\mathcal{H}} = (-\hbar^2/2m)d^2/dx^2 + kx^2/2$. The quantity W is given by:

$$W = \int_{-\infty}^{+\infty} u_o^* \hat{\mathcal{H}} u_o dx = \sqrt{\frac{2c}{\pi}} \int_{-\infty}^{+\infty} e^{-cx^2}\left[-\frac{\hbar^2}{2m}\frac{d^2}{dx^2} + \frac{k}{2}x^2\right] e^{-cx^2} dx$$

$$W = \sqrt{\frac{2c}{\pi}}\left[\left(\frac{k}{2} - \frac{2\hbar^2 c^2}{m}\right)\int_{-\infty}^{+\infty} x^2 e^{-2cx^2} dx + \frac{c\hbar^2}{m}\int_{-\infty}^{+\infty} e^{-2cx^2} dx\right]$$

since

$$\frac{d^2}{dx^2}(e^{-cx^2}) = (4c^2x^2 - 2c)e^{-cx^2}$$

From tables:

$$\int_{o}^{\infty} x^{2n} e^{-ax} dx = \frac{1\cdot 3\cdot 5\cdots(2n-1)}{2^{n+1}a^n}\sqrt{\frac{\pi}{a}}$$

Therefore

$$W = \sqrt{\frac{2c}{\pi}}\left[\left(\frac{k}{2} - \frac{2\hbar^2 c^2}{m}\right)\frac{1}{4c}\sqrt{\frac{\pi}{2c}} + \frac{c\hbar^2}{m}\sqrt{\frac{\pi}{2c}}\right] = \frac{k}{8c} + \frac{c\hbar^2}{2m} \tag{1}$$

The minimum value of W is found by setting the first derivative with respect to c equal to zero.

$$\frac{dW}{dc} = -\frac{k}{8c^2} + \frac{\hbar^2}{2m} = 0$$

Therefore, the "best" value of c^2 is

$$c^2 = \frac{km}{4\hbar^2} \qquad \text{and} \qquad c = \frac{\pi}{h}\sqrt{km} = \frac{\alpha}{2} \tag{2}$$

The variation method shows that:

$$u_o = \left(\frac{2c}{\pi}\right)^{1/4} e^{-cx^2} = \left(\frac{\alpha}{\pi}\right)^{1/4} e^{-\frac{\alpha x^2}{2}}$$

This is, of course, the <u>exact</u> ground state eigenfunction for the oscillator because of the form chosen for u_o.

Combining Eqs. (1) and (2) we find the energy associated with u_o.

$$W = \frac{k}{8\pi}\frac{h}{\sqrt{km}} + \frac{\pi}{h}\sqrt{km}\,\frac{\hbar^2}{2m} = \frac{1}{2\pi}\sqrt{\frac{k}{m}}\left(\frac{h}{2}\right) = \frac{h}{2}\left(\frac{1}{2\pi}\sqrt{\frac{k}{m}}\right) = \frac{h\nu_o}{2}$$

6.15 Suppose that the function Ae^{-cr^2} is used as an approximation to the 1s wave function for the hydrogen atom. Determine A from the normalization condition and use the variation method to find the "best" value of the parameter c. What is the minimum error that is made when the energy is calculated with this function?

<u>Solution</u>: Ref. CO,59

We choose to normalize the function $\psi = Ae^{-cr^2}$ so that:

$$\int_0^\infty A^2 e^{-2cr^2} r^2 dr = 1 \tag{1}$$

From tables we find

$$\int_0^\infty x^2 e^{-ax^2} dx = \frac{1}{4}\sqrt{\frac{\pi}{a^3}}$$

or

$$A^2\int_0^\infty e^{-2cr^2} r^2 dr = \frac{A^2}{4}\sqrt{\frac{\pi}{8c^3}} = 1, \qquad A^2 = 8\sqrt{\frac{2c^3}{\pi}}$$

In Eq. (1) the volume element $4\pi r^2 dr$ could be used if desired, but the factor 4π will cancel out in the next step of the calculation. Our convention is consistent with that of PW.

The integral which must be evaluated is:

$$W = \int_0^\infty \psi^*\left[-\frac{\hbar^2}{2\mu}\nabla^2 - \frac{e^2}{(4\pi\varepsilon_o)r}\right]\psi r^2 dr$$

Since ψ does not depend on the spherical polar angles θ and ϕ, only the r dependent part of ∇^2 must be retained. From Problem 5.5 we find

$$W = A^2\int_0^\infty e^{-cr^2}\left\{-\frac{\hbar^2}{2\mu}\left[\frac{1}{r^2}\frac{d}{dr}\left(r^2\frac{d}{dr}\right)\right] - \frac{e^2}{(4\pi\varepsilon_o)r}\right\}e^{-cr^2}r^2dr$$

After carrying out the differentiation we have

$$W = A^2\int_0^\infty e^{-cr^2}\left[-\frac{\hbar^2}{2\mu}\left(-6ce^{-cr^2} + 4c^2r^2e^{-cr^2}\right) - \frac{e^2}{(4\pi\varepsilon_o)r}\right]e^{-cr^2}r^2dr$$

$$= \frac{3A^2c\hbar^2}{\mu}\int_0^\infty e^{-2cr^2}r^2dr - \frac{2A^2c^2\hbar^2}{\mu}\int_0^\infty e^{-2cr^2}r^4dr - \frac{A^2e^2}{(4\pi\varepsilon_o)}\int_0^\infty e^{-2cr^2}rdr$$

Again resorting to tables

$$A^2\int_0^\infty e^{-2cr^2}r^4dr = 8\sqrt{\frac{2c^3}{\pi}}\left(\frac{3}{32}c\sqrt{\frac{\pi}{2c^3}}\right) = \frac{3}{4c}$$

$$A^2\int_0^\infty e^{-2cr^2}rdr = \frac{1}{4c}\left(8\sqrt{\frac{2c^3}{\pi}}\right) = \frac{2}{c}\sqrt{\frac{2c^3}{\pi}}$$

Therefore:

$$W = \frac{3c\hbar^2}{\mu} - \frac{2c^2\hbar^2}{\mu}\left(\frac{3}{4c}\right) - \frac{e^2}{(4\pi\varepsilon_o)}\left(\frac{2}{c}\sqrt{\frac{2c^3}{\pi}}\right) = +\frac{3c\hbar^2}{2\mu} - \frac{2e^2}{(4\pi\varepsilon_o)}\sqrt{\frac{2c}{\pi}} \qquad (2)$$

For a minimum in W we require that

$$\frac{dW}{dc} = \frac{3\hbar^2}{2\mu} - \frac{e^2}{(4\pi\varepsilon_o)}\sqrt{\frac{2}{\pi c}} = 0$$

and

$$c = \frac{8}{9}\left[\frac{e^4\mu^2}{(4\pi\varepsilon_o)^2\pi\hbar^4}\right]$$

Substituting this value for c into Eq. (2) we obtain:

$$W = -\frac{8}{3\pi}\left[\frac{\mu e^4}{2\hbar^2(4\pi\varepsilon_o)^2}\right] = -0.848\left[\frac{\mu e^4}{2\hbar^2(4\pi\varepsilon_o)^2}\right]$$

The exact ground state energy of the hydrogen atom is well known to be

$$-\left[\frac{\mu e^4}{2\hbar^2(4\pi\varepsilon_o)^2}\right]$$

Therefore, the variation function gives an error of +15.2%.

6.16 Apply the variation method to the particle in a box problem in one dimension. Let $V = 0$ for $-1 \le x \le +1$ and $V = \infty$ otherwise. Then use $f_1 = (1-x^2)$ and $f_2 = (1-x^4)$ to construct the trial function:

$$u = c_1 f_1 + c_2 f_2$$

(a) Calculate the energy with this function and compare with the exact solution.

(b) Determine the "best" values of c_1 and c_2.

Solution: Ref. PW,180; WI,42

(a) According to the linear variation theory the best values of c_1 and c_2 are obtained when the following secular equation is satisfied.

$$\begin{vmatrix} \mathcal{H}_{11} - S_{11}W & \mathcal{H}_{12} - S_{12}W \\ \mathcal{H}_{21} - S_{21}W & \mathcal{H}_{22} - S_{22}W \end{vmatrix} = 0$$

where

$$\mathcal{H}_{11} = \int_{-1}^{+1} (1 - x^2) \left[- \frac{\hbar^2}{2m} \frac{d^2}{dx^2} \right] (1 - x^2) dx = \frac{\hbar^2}{m} \int_{-1}^{+1} (1 - x^2) dx = \frac{4\hbar^2}{3m}$$

$$\mathcal{H}_{22} = \int_{-1}^{+1} (1 - x^4) \left[- \frac{\hbar^2}{2m} \frac{d^2}{dx^2} \right] (1 - x^4) dx = \frac{6\hbar^2}{m} \int_{-1}^{+1} (1 - x^4) x^2 dx = \frac{16\hbar^2}{7m}$$

$$\mathcal{H}_{12} = \mathcal{H}_{21} = \int_{-1}^{+1} (1 - x^4) \left[- \frac{\hbar^2}{2m} \frac{d^2}{dx^2} \right] (1 - x^2) dx = \frac{\hbar^2}{m} \int_{-1}^{+1} (1 - x^4) dx = \frac{8\hbar^2}{5m}$$

$$S_{11} = \int_{-1}^{+1} (1 - x^2)^2 dx = \frac{16}{15}, \qquad S_{22} = \int_{-1}^{+1} (1 - x^4)^2 dx = \frac{64}{45}$$

$$S_{12} = S_{21} = \int_{-1}^{+1} (1 - x^2)(1 - x^4) dx = \frac{128}{105}$$

Letting $W' = Wm/\hbar^2$, the secular equation becomes:

$$\begin{vmatrix} \left(\frac{4}{3} - \frac{16}{15} W'\right) & \left(\frac{8}{5} - \frac{128}{105} W'\right) \\ \left(\frac{8}{5} - \frac{128}{105} W'\right) & \left(\frac{16}{7} - \frac{64}{45} W'\right) \end{vmatrix} = 0 \tag{1}$$

Multiplying this out yields

$$(W')^2 - 14.00\ W' + 15.75 = 0$$

and the roots from the quadratic equation are

$$W' = 1.23,\ 12.77$$

The eigenvalues thus have the upper bounds

$$W_3 = 12.77\ \hbar^2/m; \qquad W_1 = 1.23\ \hbar^2/m$$

The exact eigenvalues from Problem 4.6 are

$$E_3 = 11.10\ \hbar^2/m; \qquad E_1 = 1.23\ \hbar^2/m$$

Note that only even eigenfunctions of x are involved in this problem since both f_1 and f_2 are even.

(b) The eigenvectors can be determined with the aid of Eq. (1). Thus

$$\left(\frac{4}{3} - \frac{16}{15} W'\right) c_{1n} + \left(\frac{8}{5} - \frac{128}{105} W'\right) c_{2n} = 0$$

$$W_1' = 1.23: \quad [1.3333 - 1.0666(1.23)]c_{11} + [1.6 - 1.2191(1.23)]c_{21} = 0$$

$$c_{21} = -0.2118\ c_{11}$$

Then

$$u_1 = c_{11}[(1 - x^2) - 0.2118(1 - x^4)]$$

The c_{11} can be determined by the normalization condition

$$\int u_1^* u_1 dv = 1 = c_{11}^2[S_{11} - 2(0.2118)S_{12} + 0.0449S_{22}] = c_{11}^2(0.6141), \quad c_{11} = 1.276$$

$$u_1 = 1.276(1 - x^2) - 0.2703(1 - x^4)$$

$$W_3' = 12.77: \quad [1.3333 - 1.0666(12.77)]c_{13} + [1.6 - 1.2191(12.77)]c_{23} = 0$$

$$c_{23} = -0.8798.c_{13}$$

$$u_3 = c_{13}[(1 - x^2) - 0.8798(1 - x^4)]$$

Normalization again gives

$$\int u_3^* u_3 dv = 1 = c_{13}^2[S_{11} - 2(0.8798)S_{12} + 0.7738S_{22}] = c_{13}^2(0.0228), \quad c_{13} = 6.6625$$

and

$$u_3 = 6.6625(1 - x^2) - 5.8285(1 - x^4)$$

These approximate eigenfunctions are shown in Fig. 6.16.

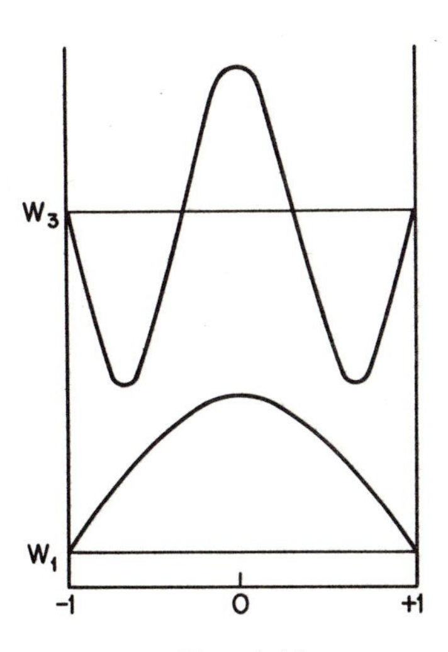

Fig. 6.16

6.17 Assume that the Hamiltonian for a system is $\hat{\mathcal{H}}$ and that the eigenfunction for the ground state can be approximated by a function of the form $c_1u_1 + c_2u_2$, where c_1 and c_2 are adjustable parameters. The functions u_1 and u_2 are normalized but not necessarily orthogonal. Use the variation principle to determine the "best" values of c_1 and c_2 and derive linear equations for these two parameters. Show that the equations lead to the well known 2x2 secular equation. You may assume that c_1 and c_2 are real.

Solution: Ref. PW,180; S3,90

By the variational principle

$$\langle\mathcal{H}\rangle = \frac{\langle c_1u_1 + c_2u_2|\hat{\mathcal{H}}|c_1u_1 + c_2u_2\rangle}{\langle c_1u_1 + c_2u_2|c_1u_1 + c_2u_2\rangle} \tag{1}$$

$$= \frac{c_1^2\langle u_1|\hat{\mathcal{H}}|u_1\rangle + 2c_1c_2\langle u_1|\hat{\mathcal{H}}|u_2\rangle + c_2^2\langle u_2|\hat{\mathcal{H}}|u_2\rangle}{c_1^2\langle u_1|u_1\rangle + 2c_1c_2\langle u_1|u_2\rangle + c_2^2\langle u_2|u_2\rangle} \tag{2}$$

$$= \frac{c_1^2H_{11} + 2c_1c_2H_{12} + c_2^2H_{22}}{c_1^2 + 2c_1c_2S_{12} + c_2^2} = \frac{N}{D} \tag{3}$$

Here we have used the fact u_1 and u_2 are normalized and that the wavefunction is real (i.e. $H_{12} = H_{21}$). Differentiating Eq. (3) with respect to c_1 yields

$$\frac{\partial\langle\mathcal{H}\rangle}{\partial c_1} = \frac{2c_1H_{11} + 2c_2H_{12}}{D} - \frac{N(2c_1 + 2c_2S_{12})}{D^2} = 0 \tag{4}$$

Identifying N/D as E we then have

$$\frac{1}{D}[2c_1H_{11} + 2c_2H_{12} - E(2c_1 + 2c_2S_{12})] = 0$$

or

$$(H_{11} - E)c_1 + (H_{12} - S_{12}E)c_2 = 0 \tag{5}$$

Differentiating Eq. (3) with respect to c_2 leads in a similar manner to

$$(H_{12} - S_{12}E)c_1 + (H_{22} - E)c_2 = 0 \tag{6}$$

For non-trivial solutions of Eqs. (5) and (6) to exist, the determinant of the coefficients must vanish:

$$\begin{vmatrix} H_{11} - E & H_{12} - S_{12}E \\ H_{12} - S_{12}E & H_{22} - E \end{vmatrix} = 0 \tag{7}$$

In a straightforward manner, the general nxn secular determinant can be derived for the general expansion

$$\psi = \sum_{i=1}^{n} c_iu_i \quad \text{(where } \psi \text{ may be imaginary)}$$

and turns out to be

$$\begin{vmatrix} H_{11} - E & H_{12} - S_{12}E & \cdots & H_{1n} - S_{1n}E \\ H_{21} - S_{21}E & H_{22} - E & \cdots & \vdots \\ \vdots & \vdots & & \vdots \\ H_{n1} - S_{n1}E & \cdots & \cdots & H_{nn} - E \end{vmatrix} = 0 \tag{8}$$

6.18 Consider a simple harmonic oscillator having the Hamiltonian

$$\hat{\mathcal{H}}^o = \frac{\hat{p}^2}{2\mu} + \frac{1}{2}kx^2$$

which is perturbed by the quartic potential ax^4 where $ax^4 << \hat{\mathcal{H}}^o$. Assume that the ground state wave function can be approximated by

$$u_o = c_o u_o^o + c_2 u_2^o$$

where u_o^o and u_2^o are the lowest even eigenfunctions for the unperturbed oscillator. Use normalization and the variation method to determine the "best" values of the constants c_o and c_2, and calculate the approximate ground state energy for the perturbed oscillator.

Solution: Ref. Problem 6.6

Since u_o^o and u_2^o are orthogonal, normalization requires

$$\int |c_o u_o^o + c_2 u_2^o|^2 dx = c_o^2 + c_2^2 = 1 \quad \text{or} \quad c_o = \pm\sqrt{1 - c_2^2}$$

where we choose c_o and c_2 to be real constants. It is convenient to define the parameter β so that $c_o = \cos\beta$ and $c_2 = \sin\beta$. Thus

$$u_o = \cos\beta u_o^o + \sin\beta u_2^o \tag{1}$$

so that normalization is automatic.

Here $\hat{\mathcal{H}} = \hat{\mathcal{H}}^o + ax^4$ and we need to calculate:

$$W = \int_{-\infty}^{+\infty} u_o \hat{\mathcal{H}} u_o dx = \{\cos\beta\langle 0| + \sin\beta\langle 2|\}\hat{\mathcal{H}}\{\cos\beta|0\rangle + \sin\beta|2\rangle\}$$

$$= \cos^2\beta\langle 0|\hat{\mathcal{H}}|0\rangle + 2\sin\beta\cos\beta\langle 0|\hat{\mathcal{H}}|2\rangle + \sin^2\beta\langle 2|\hat{\mathcal{H}}|2\rangle$$

The appropriate matrix elements of x^4 were derived in Problem 6.6.

$$\langle n|x^4|n\rangle = \frac{3}{2\alpha^2}\left(n^2 + n + \frac{1}{2}\right); \qquad \langle n|x^4|n-2\rangle = \frac{(2n-1)\sqrt{n(n-1)}}{2\alpha^2}$$

Therefore:

$$\langle 0|\hat{\mathcal{H}}|0\rangle = \frac{h\nu_o}{2} + a\langle 0|x^4|0\rangle = \frac{h\nu_o}{2} + \frac{3}{4}\frac{a}{\alpha^2}$$

$$\langle 0|\hat{\mathcal{H}}|2\rangle = a\langle 0|x^4|2\rangle = \frac{3\sqrt{2}}{2}\frac{a}{\alpha^2}$$

$$\langle 2|\hat{\mathcal{H}}|2\rangle = \frac{5}{2}h\nu_o + a\langle 2|x^4|2\rangle = \frac{5}{2}h\nu_o + \frac{39}{4}\frac{a}{\alpha^2}$$

and

$$W = \cos^2\beta\left(\frac{h\nu_o}{2} + \frac{3}{4}\frac{a}{\alpha^2}\right) + 2\sin\beta\cos\beta\left(\frac{3\sqrt{2}}{2}\frac{a}{\alpha^2}\right) + \sin^2\beta\left(\frac{5}{2}h\nu_o + \frac{39}{4}\frac{a}{\alpha^2}\right)$$

$$= \left(\frac{h\nu_o}{2} + \frac{3a}{4\alpha^2}\right) + \sin 2\beta\left(\frac{3\sqrt{2}}{2}\frac{a}{\alpha^2}\right) + \sin^2\beta\left(2h\nu_o + \frac{9a}{\alpha^2}\right) \quad (2)$$

Minimizing W with respect to β gives

$$\frac{\partial W}{\partial\beta} = 0 = 2\cos 2\beta\left(\frac{3\sqrt{2}}{2}\frac{a}{\alpha^2}\right) + 2\sin\beta\cos\beta\left(2h\nu_o + \frac{9a}{\alpha^2}\right)$$

Since $\beta \ll 1$ for sufficiently small values of a, this becomes

$$0 = \frac{3\sqrt{2}}{2}\frac{a}{\alpha^2} + \beta\left(2h\nu_o + \frac{9a}{\alpha^2}\right)$$

or

$$\beta \simeq -\frac{3\sqrt{2}}{4h\nu_o}\left(\frac{a}{\alpha^2}\right) \quad ; \quad \cos\beta \simeq 1, \quad \sin\beta \simeq \beta$$

with

$a/\alpha^2 \ll 1$

Therefore, in the limit of small a the best wave function of the form given in Eq. (1) is:

$$u_o = u_o^o - \frac{3\sqrt{2}}{4h\nu_o}\left(\frac{a}{\alpha^2}\right)u_2^o \quad (3)$$

From Eq. (2) the approximate ground state energy is

$$W = \left(\frac{h\nu_o}{2} + \frac{3}{4}\frac{a}{\alpha^2}\right) - \frac{3\sqrt{2}}{2h\nu_o}\left(\frac{a}{\alpha^2}\right)\frac{3\sqrt{2}}{2}\left(\frac{a}{\alpha^2}\right) + \left(\frac{3\sqrt{2}}{4h\nu_o}\frac{a}{\alpha^2}\right)^2\left(2h\nu_o + \frac{9a}{\alpha^2}\right)$$

$$\simeq \frac{h\nu_o}{2} + \frac{3}{4}\left(\frac{a}{\alpha^2}\right) + \frac{9}{4h\nu_o}\left(\frac{a}{\alpha^2}\right)^2 + \ldots$$

These results should be compared with the results of the first order perturbation calculation in Problem 6.6.

6.19 A system of two charges +q and -q separated by the fixed distance R constitutes an electric dipole. Show that the interaction energy of an electric dipole with an external electric field is

$W = -\mu\mathcal{E}\cos\theta$

where μ, the electric dipole moment, is qR and θ is the instantaneous angle between the direction of R and applied field.

Solution: Ref. RU,155; Problem 5.2

Figure 6.19 illustrates the problem.

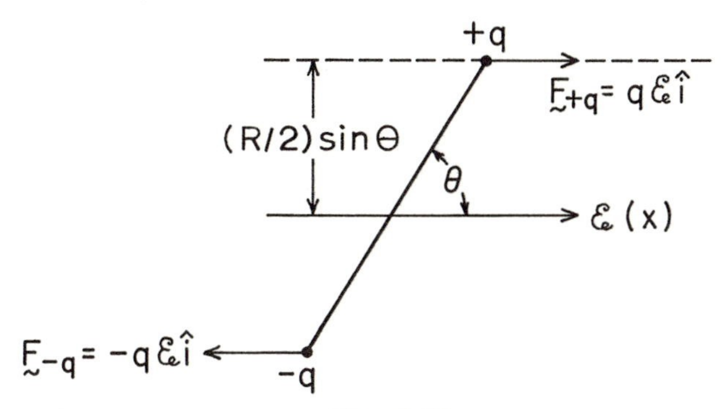

Fig. 6.19

The plus charge is acted on by a force $+q\mathcal{E}\hat{i}$ and the minus charge by a force $-q\mathcal{E}\hat{i}$. The torque is then given by

$$\tau = q\mathcal{E}\,\frac{R}{2}\sin\theta + (-q)\mathcal{E}\,\frac{R}{2}\sin\theta = q\mathcal{E}R\sin\theta$$

Thus the work done to rotate this system through the angle θ is (See Problem 5.2)

$$W = \int_0^\theta \tau d\theta = \int_0^\theta q\mathcal{E}R\sin\theta d\theta = -q\mathcal{E}R\cos\theta\Big|_0^\theta = -q\mathcal{E}R\cos\theta + q\mathcal{E}R$$

The term $q\mathcal{E}R$ merely defines a reference for the energy since it is the same for all orientations. Thus, we may neglect this term and state that the potential or interaction energy of a dipole with an electric field is

$$W = -q\mathcal{E}R\cos\theta = -\mu\mathcal{E}\cos\theta$$

Neglecting the constant term is equivalent to choosing $\theta = \pi/2$ for the zero of the potential.

A simpler method is to write

$$W = \sum_i q_i\phi_i$$

where ϕ_i is the electric potential at the position of the ith charge. Since $\phi_i = -x_i\mathcal{E}$, we immediately have

$$W = -\sum_i q_i x_i\mathcal{E} = -\mu\cos\theta\mathcal{E}$$

for any number of particles. The electric dipole moment is in general defined as

$$\underset{\sim}{\mu} = \sum_i q_i\underset{\sim}{r}_i$$

6.20 A plane rotator (see Problem 4.15) which has an electric dipole moment $\mu = \mu \underset{\sim}{R}$ is subjected to an electric field in the x-direction so that the perturbing Hamiltonian is given by $\hat{\mathcal{H}}' = -\mu \mathcal{E}_x \cos\phi$ where ϕ is the angle between $\underset{\sim}{R}$ and the direction of the field. The zeroth order wave functions are chosen to be $u_m(\phi) = (1/\sqrt{2\pi})\exp(im\phi)$, $m = 0, \pm 1, \pm 2, \cdots$. Determine the perturbation correction to the energy through second order. (Hint: If the perturbation is such that $\langle m|\hat{\mathcal{H}}'|m\rangle = 0$, then the degeneracy is not lifted in first order. However, if there exists a state $u_\ell(\phi)$ such that both $\langle m|\hat{\mathcal{H}}'|\ell\rangle$ and $\langle -m|\hat{\mathcal{H}}'|\ell\rangle$ are nonzero, then the degeneracy of these states will be lifted in second order. When the first order matrix elements vanish, the second order part of Eq. (4) in the Introduction can be used to show that:

$$\det\left| \sum_\ell{}' \frac{\langle m|\hat{\mathcal{H}}'|\ell\rangle\langle \ell|\hat{\mathcal{H}}'|n\rangle}{E_m^o - E_\ell^o} - \delta_{mn}E^{(2)} \right| = 0, \ n = m, -m \tag{1}$$

where the prime on the summation indicates that $\ell \neq m, -m$ and $E_m^o = m^2\hbar^2/2I$ is the unperturbed energy for the states $u_m(\phi)$ and $u_{-m}(\phi)$.)

Solution: Ref. SC1,248; BA2,198; Problems 4.15 and 6.13

Using $|m\rangle = |u_m(\phi)\rangle$ the first order matrix elements are

$$\langle m|\hat{\mathcal{H}}'|k\rangle = -\frac{\mu\mathcal{E}_x}{2\pi}\int_o^{2\pi} e^{i(k-m)\phi}\cos\phi d\phi = -\frac{\mu\mathcal{E}_x}{2}\delta_{k,m\pm 1} \tag{2}$$

Thus, there is no first order correction to the energy and we must resort to Eq. (1) in the Hint for the second order correction. First, we consider the case $|m| > 1$. Here there is no function $u_\ell(\phi)$ which is coupled to both $u_m(\phi)$ and $u_{-m}(\phi)$ by $\hat{\mathcal{H}}'$. The off-diagonal terms in Eq. (1) vanish and using Eq. (2) we find that

$$E_m^{(2)} = \sum_\ell{}' \frac{|\langle m|\hat{\mathcal{H}}'|\ell\rangle|^2}{E_m^o - E_\ell^o} = \frac{|H'_{m,m+1}|^2}{E_m^o - E_{m+1}^o} + \frac{|H'_{m,m-1}|^2}{E_m^o - E_{m-1}^o} = \frac{\mu^2\mathcal{E}_x^2 I}{\hbar^2(4m^2 - 1)}, \ m \neq 1 \tag{3}$$

For $|m| = 1$ the degeneracy is lifted since $\langle 1|\hat{\mathcal{H}}'|0\rangle = \langle -1|\hat{\mathcal{H}}'|0\rangle = -\mu\mathcal{E}_x/2$. Again, using Eq. (1) we have:

$$\begin{vmatrix} \dfrac{|\langle 1|\hat{\mathcal{H}}'|0\rangle|^2}{E_1^o - E_o^o} + \dfrac{|\langle 1|\hat{\mathcal{H}}'|2\rangle|^2}{E_1^o - E_2^o} - E^{(2)} & \dfrac{\langle 1|\hat{\mathcal{H}}'|0\rangle\langle 0|\hat{\mathcal{H}}'|-1\rangle}{E_1^o - E_o^o} \\ \dfrac{\langle -1|\hat{\mathcal{H}}'|0\rangle\langle 0|\hat{\mathcal{H}}'|1\rangle}{E_1^o - E_o^o} & \dfrac{|\langle -1|\hat{\mathcal{H}}'|-2\rangle|^2}{E_1^o - E_2^o} + \dfrac{|\langle -1|\hat{\mathcal{H}}'|0\rangle|^2}{E_1^o - E_o^o} - E^{(2)} \end{vmatrix} = 0$$

Thus

$$E_{1a}^{(2)} = -\frac{\alpha}{3E_1^o} = -\frac{\mu^2\mathcal{E}_x^2 I}{6\hbar^2} \quad \text{and} \quad E_{1b}^{(2)} = \frac{5\alpha}{3E_1^o} = \frac{5}{6}\frac{\mu^2\mathcal{E}_x^2 I}{\hbar^2}, \quad \text{where } \alpha = \mu^2\mathcal{E}_x^2/4$$

A detailed discussion of this problem has been given by G. L. Johnston and G. Sposito (Am. J. Phys. 44, 723 (1976)).

6.21 A knowledge of the wave function to order k is sufficient to determine the energy to order (2k + 1). Verify this general theorem for the special case that k = 3 by showing that:

$$E^{(3)} = \langle u^{(1)}|\hat{\mathcal{H}}' - E^{(1)}|u^{(1)}\rangle$$

(Hint: Use Eqs. (4), (5), and (6) of the Introduction and the Hermitian property of the operator $\hat{\mathcal{H}}'$.)

Solution: Ref. HBE,358

From Eq. (6) of the Introduction we find:

$$E^{(3)} = \langle u^o|\hat{\mathcal{H}}' - E^{(1)}|u^{(2)}\rangle - E^{(2)}\langle u^o|u^{(1)}\rangle \tag{1}$$

To remove $u^{(2)}$ from this relation we turn to the 1st and 2nd order parts of Eqs. (4) in the Introduction. Multiplying the 1st order equation by $u^{(2)*}$ and the second order equation by $u^{(1)*}$ and integrating we obtain:

$$\langle u^{(2)}|\hat{\mathcal{H}}^o - E^o|u^{(1)}\rangle + \langle u^{(2)}|\hat{\mathcal{H}}' - E^{(1)}|u^o\rangle = 0 \tag{2a}$$

$$\langle u^{(1)}|\hat{\mathcal{H}}^o - E^o|u^{(2)}\rangle + \langle u^{(1)}|\hat{\mathcal{H}}' - E^{(1)}|u^{(1)}\rangle = E^{(2)}\langle u^{(1)}|u^o\rangle \tag{2b}$$

Then taking the complex conjugate of the first of these equations, using the Hermitian property of $\hat{\mathcal{H}}^o$ and $\hat{\mathcal{H}}'$ and subtracting (2b) from (2a) gives

$$\langle u^o|\hat{\mathcal{H}}' - E^{(1)}|u^{(2)}\rangle = \langle u^{(1)}|\hat{\mathcal{H}}'|u^{(1)}\rangle - E^{(2)}\langle u^{(1)}|u^o\rangle$$

Substituting back into Eq. (1) above shows that

$$E^{(3)} = \langle u^{(1)}|\hat{\mathcal{H}}' - E^{(1)}|u^{(1)}\rangle - E^{(2)}(\langle u^o|u^{(1)}\rangle + \langle u^{(1)}|u^o\rangle)$$

This completes the derivation since $\langle u^o|u^{(1)}\rangle + \langle u^{(1)}|u^o\rangle = 0$ by the normalization convention.

SUPPLEMENTARY PROBLEMS

6.22 A particle of mass $9.1\text{x}10^{-28}$g experiences the potential shown in Fig. 6.22. Use first order perturbation theory to evaluate the correction to the ground state energy in the presence of V_1 and give the first non-vanishing correction to the wave function.

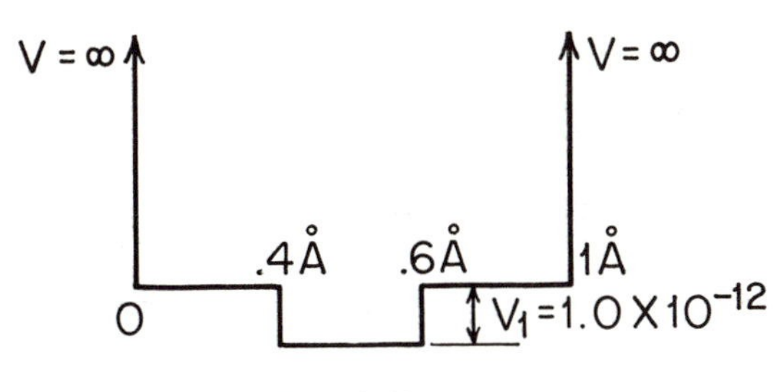

Fig. 6.22

Ans:

$E_1^{(1)} = -3.87\text{x}10^{-20}\text{J}$; $u_1 = u_1^o - 7.01\text{x}10^{-4}u_3^o + \cdots$

6.23   A particle of mass m in a potential box of length a is perturbed by the potential

$$\hat{\mathcal{H}}' = -\frac{2bx}{a} + b, \quad 0 \lesseqgtr x \lesseqgtr a/2; \qquad \hat{\mathcal{H}}' = +\frac{2bx}{a} - b, \quad a/2 \lesseqgtr x \lesseqgtr a$$

as shown in Fig. 6.23. Determine the eigenvalues of the perturbed system to first order assuming that $b << E_1^o$.

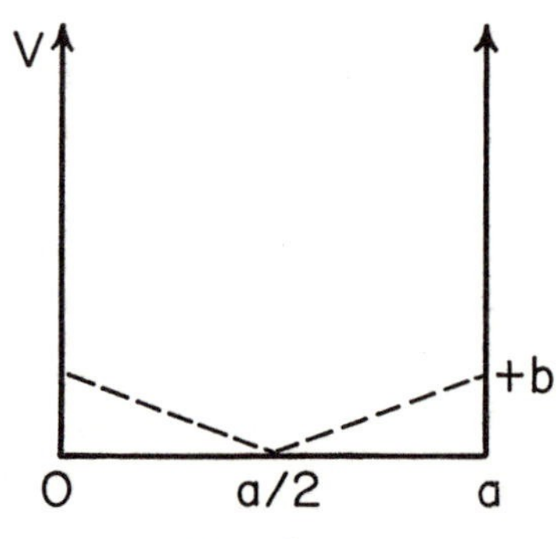

Fig. 6.23

Ans:

For n odd, $E_n^{(1)} = \frac{b}{2} - \frac{2b}{(n\pi)^2}$;   for n even, $E_n^{(1)} = \frac{b}{2}$

6.24   The trial function $u = Ax(a^2 - x^2)$ can be taken as an approximate eigenfunction for a particle of mass m in a potential box having V = 0 for $-a \lesseqgtr x \lesseqgtr +a$ and $V = \infty$ otherwise. Calculate the approximate energy using this function. Which eigenfunction does u approximate? What is the percentage error in the calculated energy?

Ans:

Approximate energy is 5.25 $\hbar^2/ma^2$. u approximates u_2 of Problem 4.6 which has the eigenvalue 4.9348 $\hbar^2/ma^2$. The error in energy is 6.39%.

6.25   A particle of mass m moves in a two dimensional potential box where V = 0 when $0 \lesseqgtr x \lesseqgtr a$ and $0 \lesseqgtr y \lesseqgtr a$, otherwise $V = \infty$.

(a) Determine the four lowest eigenvalues and display then on a diagram. Indicate the degeneracies.

(b) When the perturbation $\hat{\mathcal{H}}' = bx^2$ is applied the degeneracies are lifted. Assume that $\hat{\mathcal{H}}' << \hat{\mathcal{H}}^o$ and determine the eigenvalues of the perturbed system to first order. Illustrate the shifts on the diagram in part (a).

Ans:

n_x	n_y	E^o	E(with perturbation correction)
1	1	2A	$2A + \Delta_1$
2	1	5A	$5A + \Delta_2$
1	2	5A	$5A + \Delta_1$
2	2	8A	$8A + \Delta_2$
3	1	10A	$10A + \Delta_3$
1	3	10A	$10A + \Delta_1$

$A = h^2/8ma^2$, $\Delta_1 = 0.2827\ a^2b$, $\Delta_2 = 0.3207\ a^2b$, $\Delta_3 = 0.3277\ a^2b$

6.26 Diagonalize the matrix

$$\begin{pmatrix} 1 & \gamma \\ \gamma & .8 \end{pmatrix}$$

for γ = .01, .05, .075, .090, and .099; and compare the exact eigenvalues to the eigenvalues obtained with an expansion containing terms through 4th order.

Ans: Ref. MA31

	root 1		root 2	
γ	4th	exact	4th	exact
0.01	1.0005	1.0005	.7995	.7995
0.05	1.0117	1.0118	.7883	.7882
0.075	1.0242	1.0250	.7758	.7750
0.090	1.0323	1.0345	.7677	.7655
0.099	1.0370	1.0407	.7630	.7592

6.27 Use $u = Ae^{-cr}$ as a trial wave function for the ground state of the hydrogen atom. Apply the variation theory to determine the "best" value of the constant c. What error is made in the energy when this function is used?

Ans:

$W = -\mu e^4/[2\hbar(4\pi\varepsilon_o)^2]$. This is the exact energy for hydrogen.

6.28 Assume a variational wave function of the form $u = Are^{-br}$ for the ground state of the hydrogen atom. The variational parameter is b, A is the normalizing constant and r is the distance between the electron and the nucleus. Find the best value of b and the lowest possible value for the associated energy.

Ans:

$$b = \frac{3}{2}\left(\frac{\mu e^2}{\hbar^2(4\pi\varepsilon_o)}\right); \qquad W = \frac{3}{4}\left(-\frac{\mu e^4}{2\hbar^2(4\pi\varepsilon_o)^2}\right)$$

6.29 The harmonic oscillator in 2-dimensions with the perturbation $\hat{\mathcal{H}}' = axy$ was treated in Problem 6.10. Use degenerate state perturbation theory to calculate the splittings in the $3h\nu_o$ level which are shown in Fig. 6.10.

Ans:

$E = 3h\nu_o, \quad 3h\nu_o \pm a/\alpha$

6.30 A one-dimensional particle-in-a-box is perturbed so that (see Fig. 6.30)

$$V(x) = -A\sin\left(\frac{\pi x}{a}\right) \qquad 0 \leq x \leq a$$

$$V(x) = \infty \text{ otherwise}$$

where $A = (h^2/80\ ma^2)$. Calculate the approximate energy of the lowest state for this potential by using first order perturbation theory.

Ans:

$$E_1 \simeq E_1^o + E_1^{(1)}$$

$$= \frac{h^2}{ma^2}\left(\frac{1}{8} - \frac{1}{30\pi}\right)$$

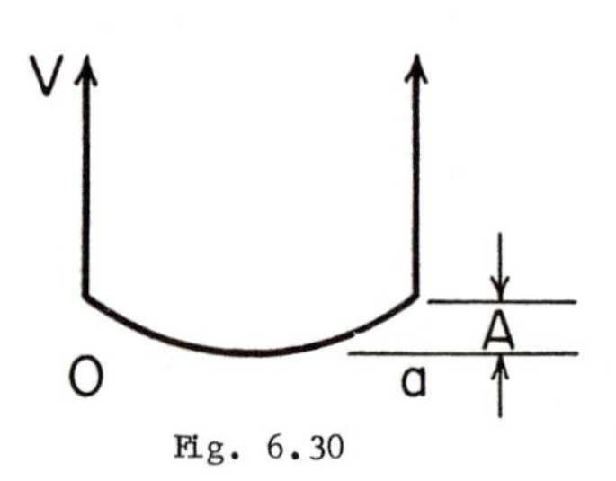

Fig. 6.30

6.31 Derive the relation:

$$E^{(4)} = \langle u^{(1)}|\hat{\mathcal{H}}' - E^{(1)}|u^{(2)}\rangle + E^{(2)}\langle u^{(2)}|u^o\rangle$$

and thus show that the fourth order energy correction can be obtained using a wave function correct through second order.

Ref. Problem 6.21

6.32 The Hamiltonian for an atom placed in the electric field $\mathcal{E}_z$ is

$$\hat{\mathcal{H}} = \hat{\mathcal{H}}^o - \mu_z\mathcal{E}_z$$

where μ_z is defined as the dipole moment operator. Show that the dipole moment for the nth state is given by:

$$(\mu_z)_n = -\frac{\partial E_n}{\partial \mathcal{E}_z}$$

(Hint: Recall that $\langle u_n|u_n\rangle = 1$.)

CHAPTER 7

HYDROGEN-LIKE ATOMS

The hydrogen-like atom falls in that small class of systems for which the wave equation can be solved exactly. It, therefore, provides a prototype for atomic calculations. In addition our qualitative ideas concerning the electronic structure of many-electron atoms are based to a large extent on the wave functions for hydrogen.

In this chapter we illustrate the quantum mechanics of one-electron atoms through numerous problems. Standard problems such as the series solution for the radial equation as well as less familiar methods such as Schrödinger's ladder operator technique are presented. Also, small corrections for spin-orbit coupling, relativistic effects, finite extent of the nucleus, etc., are considered. This chapter provides excellent opportunities for us to illustrate use of angular momentum theory and to apply the perturbation and variation theories.

General References: PW; S1; S3; BS2

PROBLEMS

7.1 A hydrogenic atom is assumed to be a system of two point charges interacting through Coulomb's law. Separate the Schrödinger equation for such a system into an equation for the motion of the center of mass and an equation for the relative motion of the two particles.

Solution: Ref. PW,112; Problem 1.2

The Schrödinger equation for the complete system is

$$\left\{-\frac{\hbar^2}{2}\left(\frac{1}{m_p}\nabla_1^2 + \frac{1}{m_e}\nabla_2^2\right) + V(r) - E_T\right\} u_T = 0 \tag{1}$$

where

$$\nabla_i^2 = \frac{\partial^2}{\partial x_i^2} + \frac{\partial^2}{\partial y_i^2} + \frac{\partial^2}{\partial z_i^2}$$

and

$$V(r) = \frac{-Ze^2/(4\pi\varepsilon_o)}{\sqrt{(x_2-x_1)^2 + (y_2-y_1)^2 + (z_2-z_1)^2}} = -\frac{Ze^2}{(4\pi\varepsilon_o)r} \tag{2}$$

The center of mass coordinates are defined by

$$\underset{\sim}{r}_c = \frac{m_p \underset{\sim}{r}_1 + m_e \underset{\sim}{r}_2}{m_p + m_e} \tag{3a}$$

and the relative coordinates are defined by

$$x_r = x_2 - x_1, \quad y_r = y_2 - y_1, \quad \text{and } z_r = z_2 - z_1 \tag{3b}$$

By the chain rule, we have

$$\frac{\partial f}{\partial x_1} = \frac{\partial x_c}{\partial x_1}\frac{\partial f}{\partial x_c} + \frac{\partial x_r}{\partial x_1}\frac{\partial f}{\partial x_r} = \frac{\partial x_c}{\partial x_1} f_{cx} + \frac{\partial x_r}{\partial x_1} f_{rx} \tag{4}$$

and

$$\frac{\partial^2 f}{\partial x_1^2} = \left(\frac{\partial f_{cx}}{\partial x_c}\frac{\partial x_c}{\partial x_1} + \frac{\partial f_{cx}}{\partial x_r}\frac{\partial x_r}{\partial x_1}\right)\frac{\partial x_c}{\partial x_1} + f_{cx}\frac{\partial^2 x_c}{\partial x_1^2} + \left(\frac{\partial f_{rx}}{\partial x_c}\frac{\partial x_c}{\partial x_1} + \frac{\partial f_{rx}}{\partial x_r}\frac{\partial x_r}{\partial x_1}\right)\frac{\partial x_r}{\partial x_1} + f_{rx}\frac{\partial^2 x_r}{\partial x_1^2}$$

Thus

$$\frac{\partial^2}{\partial x_1^2} = \left(\frac{m_p}{m_p+m_e}\frac{\partial^2}{\partial x_c^2} - \frac{\partial^2}{\partial x_c \partial x_r}\right)\frac{m_p}{m_p+m_e} - \left(\frac{m_p}{m_p+m_e}\frac{\partial^2}{\partial x_c \partial x_r} - \frac{\partial^2}{\partial x_r^2}\right)$$

$$= \left(\frac{m_p}{m_p+m_e}\right)^2\frac{\partial^2}{\partial x_c^2} - \frac{2m_p}{m_p+m_e}\frac{\partial^2}{\partial x_c \partial x_r} + \frac{\partial^2}{\partial x_r^2} \tag{5}$$

Similarly

$$\frac{\partial^2}{\partial x_2^2} = \left(\frac{m_e}{m_p+m_e}\right)^2\frac{\partial^2}{\partial x_c^2} + \frac{2m_e}{m_p+m_e}\frac{\partial^2}{\partial x_r \partial x_c} + \frac{\partial^2}{\partial x_r^2} \tag{6}$$

This procedure is then repeated to obtain similar equations for the y_1, y_2, z_1, and z_2 components. When Eq. (6) is substituted into Eq. (1), we find

$$\left\{-\frac{\hbar^2}{2}\left[\frac{1}{(m_p+m_e)}\left(\frac{\partial^2}{\partial x_c^2} + \frac{\partial^2}{\partial y_c^2} + \frac{\partial^2}{\partial z_c^2}\right) + \frac{1}{\mu}\left(\frac{\partial^2}{\partial x_r^2} + \frac{\partial^2}{\partial y_r^2} + \frac{\partial^2}{\partial z_r^2}\right)\right] - \frac{Ze^2}{(4\pi\varepsilon_o)r} - E_T\right\} u_T = 0 \tag{7}$$

where

$$\mu = \frac{m_p m_e}{m_p+m_e}$$

This equation may be further separated by choosing

$$u_T = \chi(\underset{\sim}{r}_c)u(\underset{\sim}{r}_r) \tag{8}$$

Substituting u_T into Eq. (7), dividing by u_T and rearranging we then have

$$-\frac{\hbar^2}{2\chi(\underset{\sim}{r}_c)}\left\{\frac{1}{(m_p+m_e)}\nabla_c^2\chi(\underset{\sim}{r}_c)\right\} = \frac{\hbar^2}{2u(\underset{\sim}{r}_r)}\left[\frac{1}{\mu}\nabla_r^2 + \frac{Ze^2}{(4\pi\varepsilon_o)r} + E_T\right]u(\underset{\sim}{r}_r) \tag{9}$$

For Eq. (9) to be valid for all values of $\underset{\sim}{r}_r$ and $\underset{\sim}{r}_c$, each side of the equation must equal the same constant. This separation constant is obviously the translational kinetic energy of the center of mass. Thus the separated equations are

$$\left[\frac{\hbar^2}{2(m_p+m_e)} \nabla_c^{\,2} + E_c\right]\chi(\underset{\sim}{r}_c) = 0 \tag{10a}$$

and

$$\left[\frac{\hbar^2}{2\mu} \nabla_r^{\,2} + \frac{Ze^2}{(4\pi\varepsilon_o)r} + E_r\right]u(\underset{\sim}{r}_r) = 0 \tag{10b}$$

where $E_T = E_c + E_r$. Equation (10b) gives the electronic wave function and energy.

7.2 The Schrödinger equation for the electronic energy of the hydrogen atom was shown in Problem 7.1 to be

$$\left(\frac{\hbar^2}{2\mu} \nabla^2 + \frac{Ze^2}{(4\pi\varepsilon_o)r} + E\right) u(\underset{\sim}{r}) = 0 \tag{1}$$

(we have dropped the subscripts for convenience). Assume a solution of the form

$$u(\underset{\sim}{r}) = R(r)\Theta(\theta)\Phi(\phi) \tag{2}$$

and then separate variables in Eq. (1) to obtain independent equations for $R(r)$, $\Theta(\theta)$, and $\Phi(\phi)$.

Solution: Ref. PW,115

First express ∇^2 in spherical polar coordinates (see Problem 5.5). Then substitute Eq. (2) into Eq. (1) and divide through by $u(\underset{\sim}{r})$. Upon rearrangement this gives:

$$\frac{1}{R}\left(\sin^2\theta \frac{\partial}{\partial r} r^2 \frac{\partial}{\partial r} R\right) + \frac{1}{\Theta}\left(\sin\theta \frac{\partial}{\partial \theta} \sin\theta \frac{\partial}{\partial \theta} \Theta\right) + \frac{2\mu}{\hbar^2} r^2\sin^2\theta \left(\frac{Ze^2}{(4\pi\varepsilon_o)r} + E\right) = -\frac{1}{\Phi}\frac{\partial^2\Phi}{\partial\phi^2} \tag{3}$$

Choosing the separation constant to be $+m^2$ we rewrite the left hand side of Eq. (3) to give

$$\frac{1}{R}\left(\frac{\partial}{\partial r} r^2 \frac{\partial}{\partial r} R\right) + \frac{2\mu r^2}{\hbar^2}\left(\frac{Ze^2}{(4\pi\varepsilon_o)r} + E\right) = \frac{m^2}{\sin^2\theta} - \frac{1}{\Theta}\left(\frac{1}{\sin\theta}\frac{\partial}{\partial\theta} \sin\theta \frac{\partial}{\partial\theta}\Theta\right) \tag{4}$$

Now letting the separation constant for Eq. (4) equal β, we summarize after minor rearrangement the three separate equations.

$$\frac{\partial^2\Phi}{\partial\phi^2} + m^2\Phi = 0 \tag{5a}$$

$$\frac{1}{\sin\theta}\frac{\partial}{\partial\theta} \sin\theta \frac{\partial}{\partial\theta}\Theta + \left(\beta - \frac{m^2}{\sin^2\theta}\right)\Theta = 0 \tag{5b}$$

$$\frac{1}{r^2}\frac{\partial}{\partial r} r^2 \frac{\partial}{\partial r} R + \left[\frac{2\mu}{\hbar^2}\left(\frac{Ze^2}{(4\pi\varepsilon_o)r} + E\right) - \frac{\beta}{r^2}\right] R = 0 \tag{5c}$$

Clearly m and β are to be found from boundary conditions. Comparison of Eq. (5b) with the equation for Θ in Problem 5.15 shows that $\beta = \ell(\ell + 1)$. The constants m^2 and $\ell(\ell + 1)$ can be determined after the boundary conditions are specified. It is frequently convenient to express energy in Rydbergs and distance in units of a_o, the Bohr radius. The radial equation in this system of units then has the form

$$\frac{1}{r^2}\frac{\partial}{\partial r}\left[r^2\frac{\partial R}{\partial r}\right] + \left[\frac{2Z}{r} + E - \frac{\ell(\ell + 1)}{r^2}\right]R = 0 \tag{6}$$

One Rydberg of energy is the ionization potential of the hydrogen atom. This system of units is often called atomic units. However, the reader should be aware that some authors use the Hartree (equal to two Rydbergs) as the unit of energy, and still refer to the system as atomic units.

7.3 An electron is trapped in a spherical box defined by the potential $V(r) = \infty$, $r \geq a$; $V(r) = 0$, $r < a$. The necessary radial equation is given in Problem 7.2. Determine the normalized radial solution and bound eigenvalues for the case $\ell = 0$. (Hint: The substitution $P = rR$ is convenient.)

Solution: Ref. MM2,352

The necessary separated radial equation is given in Eq. (5) of Problem 7.2.

$$\frac{1}{r^2}\frac{\partial}{\partial r}\left[r^2\frac{\partial R}{\partial r}\right] + \frac{2\mu}{\hbar^2}[E - V(r)]R = 0$$

Substituting $P = rR$ and using $V = 0$ inside a, we find

$$P'' + \frac{2\mu E}{\hbar^2}P = 0$$

This substitution is carried out in detail in Problem 7.4. The solution of this simple differential equation is

$$P = c_1\sin\left(\sqrt{\frac{2\mu E}{\hbar^2}}\,r\right) + c_2\cos\left(\sqrt{\frac{2\mu E}{\hbar^2}}\,r\right)$$

so that

$$R(r) = \frac{c_1}{r}\sin\left(\sqrt{\frac{2\mu E}{\hbar^2}}\,r\right) + \frac{c_2}{r}\cos\left(\sqrt{\frac{2\mu E}{\hbar^2}}\,r\right)$$

For $R(r)$ to be finite at the origin c_2 must vanish. $R(r)$ must vanish at $r = a$, thus

$$R(a) = 0 = \frac{c_1}{a}\sin\left(\sqrt{\frac{2\mu E}{\hbar^2}}\,a\right)$$

and

$$\sqrt{\frac{2\mu E}{\hbar^2}}\,a = n\pi \qquad n = 1, 2, 3, \cdots$$

or

$$E = \frac{n^2\pi^2\hbar^2}{2\mu a^2} = \frac{n^2h^2}{8\mu a^2}$$

c_1 is easily found by normalization. Letting

$$b = \sqrt{\frac{2\mu E}{\hbar^2}}$$

we have

$$1 = c_1^2\int_0^a \frac{1}{r^2}\sin^2(br)r^2dr = \frac{c_1^2}{b}\int_0^{ba}\sin^2x dx = \frac{c_1^2}{b}\left(\frac{x}{2} - \frac{\sin 2x}{4}\right)\Big|_0^{ba} = \frac{c_1^2 a}{2}$$

or

$$c_1 = \sqrt{\frac{2}{a}}$$

7.4 The radial equation for hydrogen-like atoms is given by Eq. (6) of Problem 7.2 with $V(r) = -2Z/r$.

(a) Transform this radial equation into a one dimensional wave equation by introducing the function $P(r) = rR(r)$.

(b) Plot the effective potential function which appears in the equation for $P(r)$ versus r for the first few values of ℓ assuming that $Z = 1$.

Solution: Ref. S1,170

(a) From Problem 7.2 we have:

$$-\frac{1}{r^2}\frac{d}{dr}\left(r^2\frac{dR}{dr}\right) + \left[\frac{\ell(\ell+1)}{r^2} + V(r)\right]R = ER \tag{1}$$

The transformation $P = rR$ can be carried out easily by working out the first term in Eq. (1).

$$\frac{1}{r^2}\frac{d}{dr}\left(r^2\frac{dR}{dr}\right) = \frac{1}{r}\left[2\frac{dR}{dr} + r\frac{d^2R}{dr^2}\right]$$

But

$$\frac{dP}{dr} = r\frac{dR}{dr} + R \quad \text{and} \quad \frac{d^2P}{dr^2} = 2\frac{dR}{dr} + r\frac{d^2R}{dr^2}$$

Therefore Eq. (1) becomes

$$-\frac{1}{r}\frac{d^2P}{dr^2} + \left[\frac{\ell(\ell+1)}{r^2} + V(r)\right]\frac{P}{r} = E\frac{P}{r}$$

$$\boxed{-\frac{d^2P}{dr^2} + \left[\frac{\ell(\ell+1)}{r^2} - \frac{2Z}{r}\right]P = EP} \tag{2}$$

(b) Equation (2) has the form of a Schrödinger equation in one dimension with the effective potential

$$V_{eff} = \frac{\ell(\ell + 1)}{r^2} - \frac{2}{r}$$

V_{eff} is plotted versus r in Fig. 7.4 for ℓ = 0, 1, and 2. The repulsive term in this potential is associated with the centrifugal force which is experienced in the rotating coordinate system where $\underset{\sim}{r}$ is fixed. Horizontal lines indicate the energy eigenvalues($E=-1/n^2$).

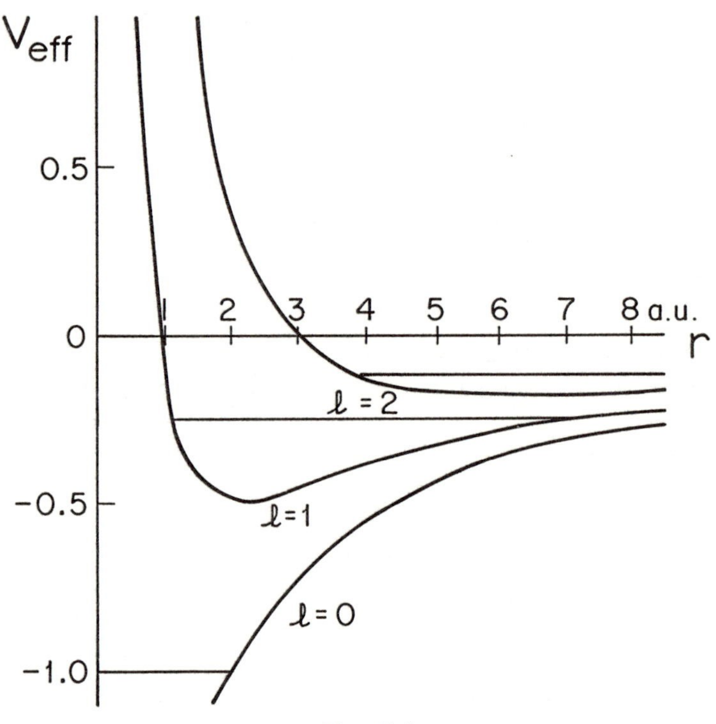

Fig. 7.4

7.5 The radial equation for hydrogen-like atoms can be written in the form:

$$-\frac{d^2P}{dr^2} + \left[\frac{\ell(\ell + 1)}{r^2} - \frac{2Z}{r}\right]P = EP \tag{1}$$

(a) Verify that $P = r^{-\ell}, r^{\ell+1}$ are solutions to Eq. (1) in the limit that $r \ll 1$.

(b) Verify that $P = \exp(\pm\sqrt{-E}\ r)$ are solutions to Eq. (1) in the limit that $r \gg 1$.

Solution: Ref. S1,172; CS,114

(a) For $r << 1$ Eq. (1) becomes

$$-\frac{d^2P}{dr^2} + \frac{\ell(\ell + 1)}{r^2} P = 0 \tag{2}$$

Using $P = r^{-\ell}$ we find

$$-(-\ell)(-\ell - 1)r^{-\ell-2} + \ell(\ell + 1)r^{-\ell-2} = 0$$

Similarly with $P = r^{\ell+1}$

$$-(\ell + 1)\ell r^{\ell-1} + \ell(\ell + 1)^{\ell-1} = 0$$

Both functions satisfy Eq. (2) but only $r^{\ell+1}$ is acceptable since $R = P/r$ must remain finite as $r \to 0$.

(b) For $r >> 1$ Eq. (1) becomes

$$+\frac{d^2P}{dr^2} + EP = 0$$

By the substitution $P = Ae^{Mr}$ we find that

$$M^2 + E = 0 \quad \text{and} \quad M = \pm\sqrt{-E}$$

The minus sign must be chosen in order to keep P finite as $r \to \infty$.

7.6 Assuming that the solution for the radial equation in Problem 7.4 has the form

$$P(r) = r^{\ell+1}e^{-\sqrt{-E}\,r}f(r) \tag{1}$$

where

$$f(r) = \sum_{k=0}^{\infty} A_k r^k$$

derive a recurrence relation for A_n.

Solution: Ref. S1,174

Using Eq. (1) for P(r) we find:

$$\frac{dP}{dr} = \left[\frac{(\ell + 1)}{r} - \sqrt{-E}\right]P + r^{\ell+1}e^{-\sqrt{-E}\,r}\frac{df}{dr}$$

$$\frac{d^2P}{dr^2} = \left[\frac{(\ell + 1)}{r} - \sqrt{-E}\right]\frac{dP}{dr} - \frac{(\ell + 1)}{r^2}P + (\ell + 1)r^{\ell}e^{-\sqrt{-E}\,r}\frac{df}{dr}$$

$$- \sqrt{-E}\,r^{\ell+1}e^{-\sqrt{-E}\,r}\frac{df}{dr} + r^{\ell+1}e^{-\sqrt{-E}\,r}\frac{d^2f}{dr^2}$$

$$= r^{\ell+1}e^{-\sqrt{-E}\,r}\left\{\frac{d^2f}{dr^2} + 2\left[\frac{(\ell + 1)}{r} - \sqrt{-E}\right]\frac{df}{dr} + \left[\frac{\ell(\ell + 1)}{r^2} - 2\sqrt{-E}\,\frac{(\ell + 1)}{r} - E\right]f\right\}$$

Substitution into Eq. (2) of Problem 7.4 then gives:

$$\frac{d^2f}{dr^2} + 2\left[\frac{(\ell + 1)}{r} - \sqrt{-E}\right]\frac{df}{dr} + \frac{2}{r}[Z - \sqrt{-E}(\ell + 1)]f = 0 \tag{2}$$

For the series solution of Eq. (2) we require:

$$f = \sum_0^\infty A_k r^k = \sum_2^\infty A_{k-2} r^{k-2}$$

$$\frac{f}{r} = \sum_0^\infty A_k r^{k-1} = \sum_1^\infty A_{k-1} r^{k-2}$$

$$\frac{df}{dr} = \sum_0^\infty A_k k r^{k-1} = \sum_2^\infty A_{k-1}(k - 1) r^{k-2}$$

$$\frac{1}{r}\frac{df}{dr} = \sum_0^\infty A_k k r^{k-2} = \sum_1^\infty A_k k r^{k-2}$$

$$\frac{d^2f}{dr^2} = \sum_0^\infty A_k k(k - 1) r^{k-2} = \sum_2^\infty A_k k(k - 1) r^{k-2}$$

Substitution of these relations into Eq. (2) then gives for the coefficient of the r^{k-2} term:

$$\{A_k k(k - 1) + A_k 2(\ell + 1)k - A_{k-1} 2\sqrt{-E}(k - 1) + A_{k-1} 2[Z - \sqrt{-E}(\ell + 1)]\} = 0$$

The coefficient of every power of r must vanish for Eq. (1) to be satisfied; therefore, in general:

$$A_k = -A_{k-1}\left\{\frac{2[Z - (\ell + k)\sqrt{-E}]}{k(k - 1) + 2(\ell + 1)k}\right\} \tag{3}$$

7.7 Use the result of Problem 7.6 to show that:

(a) For an arbitrary value of E, P(r) increases without limit as $r \to \infty$.

(b) P(r) remains finite for all values if and only if $E = -Z^2/n^2$.

(Hint: In part (a) first show that for large values of r the series

$$\sum_0^\infty A_k r^k$$

increases as $\exp(2\sqrt{-E}\, r)$.)

Solution:

(a) From Problem 7.6 we have:

$$P(r) = r^{\ell+1} e^{-\sqrt{-E}} \sum_0^\infty A_k r^k$$

where

$$\frac{A_k}{A_{k-1}} = -2\left\{\frac{Z - (\ell + k)\sqrt{-E}}{k(k - 1) + 2(\ell + 1)k}\right\} \tag{1}$$

The behavior of this series for large values of r is determined by the terms having large values of k. As k increases without limit we can write:

$$\frac{A_k}{A_{k-1}} = \frac{+2\sqrt{-E}}{k}\ ; \quad k >> \ell$$

But recall that

$$\exp[\sqrt{-E}(2r)] = \sum_{k=0}^{\infty} \frac{[\sqrt{-E}(2r)]^k}{k!}$$

Therefore, the series f(r) increases as $\exp[\sqrt{-E}(2r)]$ with increasing r and the function P(r) increases as $\exp(\sqrt{-E}\,r)$.

(b) For P(r) to remain finite as r becomes infinite, the series solution must terminate for some value of k, i.e. the ratio A_k/A_{k-1} must vanish. We require that the numerator in Eq. (1) vanish. Thus

$$Z - (\ell + k)\sqrt{-E} = 0 \quad \text{and} \quad E = -\frac{Z^2}{(\ell + k)^2}$$

We set $n = \ell + k$, where n is an integer, and obtain

$$\boxed{E = -\frac{Z^2}{n^2}}$$

as required.

7.8 The radial wave function for hydrogen-like atoms is often written as

$$P_{n\ell}(\rho) = \sqrt{\frac{(n - \ell - 1)!Z}{n^2[(n + \ell)!]^3}}\,\rho^{\ell+1}e^{-\rho/2}L_{n+\ell}^{2\ell+1}(\rho) \tag{1}$$

where $\rho = 2Zr/n$. The function $L_{n+\ell}^{2\ell+1}(\rho)$ is the associated Laguerre polynomial which is defined by the equation:

$$L_{n+\ell}^{2\ell+1}(\rho) = \sum_{k=0}^{n-\ell-1}\left\{\frac{(-1)^{k+1}[n + \ell)!]^2}{(n - \ell - 1 - k)!(2\ell + 1 + k)!k!}\right\}\rho^k$$

Show that, aside from a constant factor, the coefficients A_k derived in Problem 7.6 are equal to the coefficients in the associated Laguerre polynomial.

Solution: Ref. PW,132; S1,174

The simplest approach is to compare ratios of coefficients. Equation (1) can be written as

$$L_{n+\ell}^{2\ell+1}(\rho) = \sum_{k=0}^{n-\ell-1} B_k \rho^k = \sum_{k=0}^{n-\ell-1} C_k r^k$$

where $C_k = B_k(2Z/n)^k$. Then

$$\frac{C_k}{C_{k-1}} = -\left\{\frac{(2Z/n)(n - \ell - k)}{(\ell + k)(\ell + k + 1) - \ell(\ell + 1)}\right\} \tag{2}$$

Equation (2) of Problem 7.6 can be written as:

$$\frac{A_k}{A_{k-1}} = -\left\{\frac{2[Z - (\ell + k)\sqrt{-E}]}{(\ell + k)(\ell + k + 1) - \ell(\ell + 1)}\right\} \tag{3}$$

Equations (2) and (3) are found to be identical when we use the fact that $\sqrt{-E} = Z/n$.

7.9 The radial function for the hydrogen atom can be shown by a recursion polynomial method to be

$$R_{n\ell}(r) = N_{n\ell} e^{-\rho/2} \rho^{\ell} L_{n+\ell}^{2\ell+1}(\rho) = P_{n\ell}(r)/r \tag{1}$$

where the $L_{n+\ell}^{2\ell+1}$ are the associated Laguerre functions and $\rho = \frac{2Z}{n} r$ in atomic units. The generating function for the associated Laguerre functions of order s is given by

$$A_s(\rho,x) \equiv \sum_{i=s}^{\infty} \frac{L_i^s(\rho)}{i!} x^i = (-1)^s \frac{e^{-\frac{\rho x}{1-x}}}{(1 - x)^{s+1}} x^s \tag{2}$$

Determine the normalization constant $N_{n\ell}$ by use of the generating function. (Hint: Form the product of two generating functions and use the method of Problem 4.12.)

Solution: Ref. PW,131,451

The integral to be evaluated is

$$\int_0^{\infty} |R_{n\ell}(r)|^2 r^2 dr = 1 = N_{n\ell}^2 \left(\frac{n}{2Z}\right)^3 \int_0^{\infty} e^{-\rho} \rho^{2\ell+2} |L_{n+\ell}^{2\ell+1}(\rho)|^2 d\rho = N_{n\ell}^2 \left(\frac{n}{2Z}\right)^3 \underline{\int_0^{\infty} e^{-\rho} \rho^{s+1} |L_i^s|^2 d\rho} \tag{3}$$

where $s = 2\ell + 1$ and $i = n + \ell$. The underlined integral can be evaluated by defining a "twin" generating function to that shown in Eq. (2):

$$B_s(\rho,y) \equiv \sum_{j=s}^{\infty} \frac{L_j^s(\rho)}{j!} y^j = (-1)^s \frac{e^{-\frac{\rho y}{1-y}}}{(1 - y)^{s+1}} y^s \tag{4}$$

Then taking the product of A and B, weighting with $\rho^{s+1}e^{-\rho}$, and integrating over ρ gives:

$$\int_0^\infty e^{-\rho}\rho^{s+1}A_sB_s d\rho = \sum_{i=s}^\infty \sum_{j=s}^\infty \int_0^\infty e^{-\rho}\rho^{s+1}L_i^sL_j^s d\rho \left(\frac{x^iy^j}{i!j!}\right) \tag{5}$$

$$= \frac{(xy)^s}{(1-x)^{s+1}(1-y)^{s+1}} \int_0^\infty \rho^{s+1}e^{-\rho}\left(1 + \frac{x}{1-x} + \frac{y}{1-y}\right) d\rho$$

$$= \frac{(xy)^s(s+1)!}{(1-x)^{s+1}(1-y)^{s+1}\left(1 + \frac{x}{1-x} + \frac{y}{1-y}\right)^{s+2}}$$

$$= \frac{(xy)^s(s+1)!(1-x)(1-y)}{(1-xy)^{s+2}} \tag{6}$$

where we have used the integral:

$$\int_0^\infty z^n e^{-az} dz = \frac{n!}{a^{n+1}}$$

Application of a binomial expansion gives

$$\frac{1}{(1-xy)^{s+2}} = \sum_{k=0}^\infty \frac{(s+k+1)!}{k!(s+1)!}(xy)^k \tag{7}$$

Thus, combining Eqs. (5), (6), and (7):

$$(s+1)!(1-x-y+xy)\sum_{k=0} \frac{(s+k+1)!}{k!(s+1)!}(xy)^{k+s} = \sum_{i,j=s} \int_0^\infty e^{-\rho}\rho^{s+1}L_i^sL_j^s d\rho \frac{x^iy^j}{i!j!} \tag{8}$$

For normalization purposes $i = j$, thus equating coefficients of like powers of xy on the right and left hand sides of Eq. (8) we see that only the "1" and the "xy" of $(1 - x - y + xy)$ contribute. For the "1" term $k + s = i$ and for the "xy" term, $k + s + 1 = i$. Therefore

$$\int_0^\infty e^{-\rho}\rho^{s+1}|L_i^s|^2 d\rho = (i!)^2(s+1)!\left[\frac{(i+1)!}{(i-s)!(s+1)!} + \frac{i!}{(i-s-1)!(s+1)!}\right]$$

$$= \frac{(i!)^3(2i-s+1)}{(i-s)!} = \frac{[(n+\ell)!]^3(2n)}{(n-\ell-1)!} \tag{9}$$

and using Eq. (3) we find

$$\boxed{N_{n\ell} = \left\{\left(\frac{2Z}{n}\right)^3 \frac{(n-\ell-1)!}{[(n+\ell)!]^3 2n}\right\}^{1/2}} \tag{10}$$

7.10 Determine the average radius $\bar{r}_{n\ell m_\ell} = \langle n\ell m_\ell|r|n\ell m_\ell\rangle$ for hydrogenic orbitals. (Hint: Use the generating function approach outlined in Problem 7.9.)

Solution: Ref. PW,144; I. Waller, Z. f. Phys. 38, 635 (1926)

Following the definitions in Problem 7.9, we have

$$\overline{r^m_{n\ell m_\ell}} = \int_0^\infty |R_{n\ell}(r)|^2 r^{2+m} dr = N^2_{n\ell}\left(\frac{n}{2Z}\right)^{3+m} \int_0^\infty e^{-\rho}\rho^{(s+1+m)}|L_i^s|^2 d\rho \tag{1}$$

Again using the generating functions $A_s(x,\rho)$ and $B_s(y,\rho)$, we find

$$\int_0^\infty e^{-\rho}\rho^{(s+1+m)} A_s B_s d\rho = \sum_{i,j=s} \int_0^\infty e^{-\rho}\rho^{(s+1+m)} L_i^s L_j^s d\rho \, \frac{x^i y^j}{i!j!} \tag{2}$$

$$= \frac{(xy)^s}{(1-x)^{s+1}(1-y)^{s+1}} \int_0^\infty \rho^{(s+1+m)} e^{-\rho\left(1 + \frac{x}{1-x} + \frac{y}{1-y}\right)} d\rho$$

$$= \frac{(xy)^s (s+1+m)!}{(1-x)^{s+1}(1-y)^{s+1}\left(1 + \frac{x}{1-x} + \frac{y}{1-y}\right)^{s+2+m}}$$

$$= \frac{(xy)^s (s+1+m)!(1-x)^{m+1}(1-y)^{m+1}}{(1-xy)^{s+2+m}}$$

$$= (1-x)^{m+1}(1-y)^{m+1} \sum_{k=0} \frac{(s+m+k+1)!}{k!} (xy)^{k+s}$$

In the above use was made of

$$\int_0^\infty z^n e^{-az} dz = \frac{n!}{a^{n+1}}$$

and $(1 - xy)^{-(s+2+m)}$ was replaced by its binomial expansion. We now specialize the problem to $m = 1$ and equate coefficients of like powers of xy on the two sides of Eq. (2). Thus

$$(1 + 4xy + x^2y^2 + \cdots) \sum_{k=0} \frac{(s+k+2)!}{k!} (xy)^{k+s} = \sum_{i=s} \int_0^\infty e^{-\rho}\rho^{s+2}|L_i^s|^2 d\rho \, \frac{x^i y^i}{(i!)^2}$$

where only those terms which involve x and y to the same power have been listed. For the "1" term on the left, $k + s = i$, for the $4xy$ term, $k + s + 1 = i$, and for the x^2y^2 term, $k + s + 2 = i$. Therefore

$$\int e^{-\rho}\rho^{s+2}|L_i^s|^2 d\rho = (i!)^2 \left[\frac{(i+2)!}{(i-s)!} + 4\,\frac{(i+1)!}{(i-s-1)!} + \frac{i!}{(i-s-2)!}\right]$$

$$= \frac{(i!)^3}{(i-s)!} \left[(i+1)(i+2) + 4(i+1)(i-s) + (i-s)(i-s-1)\right] \tag{4}$$

Substituting $i = n + \ell$ and $s = 2\ell + 1$ and combining Eqs. (1) and (4) we have:

$$\overline{r_{n\ell m_\ell}} = N^2_{n\ell}\left(\frac{n}{2Z}\right)^4 \int_0^\infty e^{-\rho}\rho^{2\ell+3}|L^{2s+1}_{n+\ell}|^2 d\rho$$

$$= \left(\frac{2Z}{n}\right)^3 \frac{(n - \ell - 1)!}{[(n + \ell)!]^3 2n} \left(\frac{n}{2Z}\right)^4 \frac{[(n + \ell)!]^3}{(n - \ell - 1)!} \{6n^2 - 2\ell(\ell + 1)\}$$

or, after rearrangement,

$$\boxed{\overline{r_{n\ell m_\ell}} = \frac{n^2}{Z}\left\{1 + \frac{1}{2}\left[1 - \frac{\ell(\ell + 1)}{2n^2}\right]\right\}} \quad (5)$$

Note that Eq. (2) is general and can be used to find $\overline{r^m_{n\ell m_\ell}}$ for arbitrary values of $m > 0$.

7.11 Expand the integral

$$\int_0^\infty P^2 \frac{\partial}{\partial \ell}\left[\frac{1}{P}\left(\frac{\partial^2 P}{\partial r^2}\right)\right] dr \quad (1)$$

using the Schrödinger equation for a hydrogen-like atom to obtain an expression for

$$\frac{1}{P}\left(\frac{\partial^2 P}{\partial r^2}\right)$$

and thereby determine the expectation value $\overline{(r^{-2})_{n\ell m_\ell}} = \langle n\ell m_\ell | r^{-2} | n\ell m_\ell \rangle$. In this derivation assume that the quantum number ℓ is a continuous variable and recall that $P = rR$. (Hint: Show that the integral in (1) is equal to zero.)

Solution: Ref. KR,251; Problem 7.4

Rearranging Eq. (2) of Problem 7.4 and using the fact that $E = -Z^2/n^2$ gives:

$$\frac{1}{P}\frac{d^2P}{dr^2} = \frac{\ell(\ell + 1)}{r^2} - \frac{2Z}{r} + \frac{Z^2}{n^2} \quad (2)$$

Then treating ℓ as a continuous variable and remembering that $n = \ell$ + (an integer), we have

$$\frac{\partial}{\partial \ell}\left[\frac{1}{P}\frac{d^2P}{dr^2}\right] = \frac{2\ell + 1}{r^2} - \frac{2Z^2}{n^3}$$

Finally, integrating (1) gives

$$\int_0^\infty P^2 \frac{\partial}{\partial \ell}\left[\frac{1}{P}\frac{d^2P}{dr^2}\right] dr = \int_0^\infty P^2 \left[\frac{2\ell + 1}{r^2} - \frac{2Z^2}{n^3}\right] dr$$

$$= (2\ell + 1)\int_0^\infty P^2 r^{-2} dr - \frac{2Z^2}{n^3}\int_0^\infty P^2 dr = (2\ell + 1)\overline{r^{-2}} - \frac{2Z^2}{n^3} \quad (3)$$

since

$$\int_0^\infty P^2 dr = 1$$

The right hand side of Eq. (3) is equal to zero as can be seen by carrying out the differentiation in (1).

$$\int_0^\infty P^2 \frac{\partial}{\partial \ell}\left[\frac{1}{P}\left(\frac{\partial^2 P}{\partial r^2}\right)\right]dr = \int_0^\infty P \frac{\partial}{\partial \ell}\left(\frac{\partial^2 P}{\partial r^2}\right) dr - \int_0^\infty \frac{\partial P}{\partial \ell}\left(\frac{\partial^2 P}{\partial r^2}\right) dr = 0$$

The last step follows using integration by parts. Returning to Eq. (3) we obtain

$$\overline{r^{-2}} = + \frac{Z^2}{n^3(\ell + 1/2)}$$

7.12 A hydrogen-like wave function is shown below with r in units of a_o.

$$\psi = \frac{\sqrt{2}}{81\sqrt{\pi}} Z^{3/2}(6 - Zr)Zre^{-Zr/3}\cos\theta \tag{1}$$

(a) Determine the values of the quantum numbers n, ℓ, and m_ℓ for ψ by inspection.

(b) Generate from ψ another eigenfunction having the same values of n and ℓ but with the magnetic quantum number equal to $m_\ell + 1$.

(c) Determine the most probable value of r for an electron in the state specified by ψ when $Z = 1$.

Solution:

(a) The exponential factor in ψ is of the form $\exp(-\sqrt{-E}\ r)$ and since $E = -Z^2/n^2$ we see that $n = 3$. A much more difficult way to find n would be to carry out the operation:

$$\hat{\mathcal{H}}\psi = -(Z^2/n^2)\psi$$

The azimuthal quantum number ℓ can be determined either by recalling the factor r^ℓ which multiplies the Laguerre polynomial in hydrogen-like wave functions or by carrying out the operation:

$$\hat{L}^2\psi = \hat{L}^2 f(r)\cos\theta = f(r)\left[-\frac{1}{\sin\theta}\frac{\partial}{\partial\theta}\left(\sin\theta\frac{\partial}{\partial\theta}\cos\theta\right)\right]$$

$$= f(r)\left[\frac{1}{\sin\theta}\frac{d}{d\theta}(\sin^2\theta)\right] = 2f(r)\cos\theta = \ell(\ell + 1)\psi$$

Either way $\ell = 1$. For the magnetic quantum number we operate with $\hat{L}_z$.

$$\hat{L}_z\psi = -i\frac{\partial}{\partial\phi}[f(r)\cos\theta] = 0 = m_\ell\psi$$

Thus $m_\ell = 0$.

(b) $$\hat{L}_+\psi_{m_\ell} = \sqrt{(\ell-m_\ell)(\ell+m_\ell+1)}\ \psi_{m_\ell+1} = \sqrt{2}\ \psi_{m_\ell+1}$$

since $\ell = 1$ and $m_\ell = 0$. Using the differential operator equation from Problem 5.4 we can write:

$$\hat{L}_+ = \hat{L}_x + i\hat{L}_y = i(\sin\phi - i\cos\phi)\frac{\partial}{\partial\theta} + i(\cos\phi - i\sin\phi)\cot\theta\frac{\partial}{\partial\phi}$$

$$\hat{L}_+\psi_{m_\ell=0} = e^{i\phi}\frac{\partial}{\partial\theta} f(r)\cos\theta = -e^{i\phi}f(r)\sin\theta = \sqrt{2}\ \psi_{m_\ell+1}$$

Therefore

$$\psi_{m_\ell+1} = -\frac{1}{\sqrt{2}} f(r)\sin\theta e^{i\phi}$$

(c) The most probable value of r occurs when $(r\psi)^2$ reaches its maximum value. For convenience we find the value of r for which the magnitude of $r\psi$ is maximum. From Eq. (1) with Z = 1 we have

$$\frac{\partial(r\psi)}{\partial r} = 0 = \frac{\partial}{\partial r}(6 - r)r^2e^{-r/3} = e^{-r/3}\left(\frac{r^3}{3} - 5r^2 + 12r\right)$$

Therefore

$$r^2 - 15r + 36 = 0 \quad \text{and} \quad r = \frac{15 \pm 9}{2} = 12,\ 3$$

By evaluation of $r\psi$ we find that $r = 12a_o$ corresponds to the maximum and is, therefore, the most probable value of r.

7.13 The radial differential equation for the hydrogen atom can be solved by an operator method, which was suggested by Schrödinger. This method makes use of the fact that the second order differential operator can be decomposed into two mutually adjoint linear operators of first order if the appropriate transformation of variables is performed.

(a) First, transform the radial equation (Problem 7.2, Eq. (6)) for a given eigenfunction by means of the substitutions:

$$r = a_n y; \quad E = -1/a_n^2; \quad R_{n\ell}(r) = f_{n\ell}(y) \tag{1}$$

so that y becomes the dependent variable. Notice that the transformation depends on the particular eigenvalue chosen.

(b) Second, compare the transformed radial equation of part (a) with the postulated operator equation

$$\left(y\frac{d}{dy} + c_1y + c_2\right)\left(y\frac{d}{dy} + c_1'y + c_2'\right)f_{n\ell}(y) = c_3f_{n\ell}(y) \tag{2}$$

in order to determine the parameters c_1, c_2, c_1', c_2', and c_3.

Solution: Ref. HL,185; E. Schrödinger, Proc. Roy. Irish Acad., 46A, 9 (1940)

(a) When the eigenfunction $R_{n\ell}$ is substituted into the radial equation from Problem 7.2 the following identity results:

$$r^2\frac{d^2}{dr^2}R_{n\ell} + 2r\frac{d}{dr}R_{n\ell} + 2rR_{n\ell} + Er^2R_{n\ell} = \ell(\ell + 1)R_{n\ell} \tag{3}$$

Now

$$\frac{d}{dr} = \frac{dy}{dr}\frac{d}{dy} = \frac{1}{a_n}\frac{d}{dy} \quad \text{and} \quad \frac{d^2}{dr^2} = \frac{1}{a_n^2}\frac{d^2}{dy^2}$$

so that Eq. (3) becomes

$$y^2 \frac{d^2}{dy^2} f_{n\ell} + 2y \frac{df_{n\ell}}{dy} + 2a_n y f_{n\ell} - y^2 f_{n\ell} = \ell(\ell + 1) f_{n\ell} \tag{4}$$

(b) Expansion of Eq. (2) gives

$$[y^2 \frac{d^2}{dy^2} + y^2(c_1' + c_1) \frac{d}{dy} + y(1 + c_2' + c_2) \frac{d}{dy} + c_1 c_1' y^2 + y(c_1' + c_1 c_2' + c_2 c_1')] f_{n\ell} = (c_3 - c_2 c_2') f_{n\ell} \tag{5}$$

Comparing the coefficients of the second and fourth terms of Eq. (5) with the corresponding terms in Eq. (4) we find

$$c_1' + c_1 = 0 \quad \text{and} \quad c_1 c_1' = -1 \quad \text{or} \quad c_1 = \pm 1; \quad c_1' = \mp 1$$

Thus there are two sets of coefficients to be found. Equating the remaining coefficients, we have

$$1 + c_2' + c_2 = 2$$

$$c_1' + c_1 c_2' + c_2 c_1' = 2a_n$$

$$c_3 - c_2 c_2' = \ell(\ell + 1)$$

Simultaneous solution first for $c_1 = +1$, $c_1' = -1$, then for $c_1 = -1$, $c_1' = +1$ gives the following

(1) $c_1 = +1$, $c_1' = -1$: $c_2 = -a_n$, $c_2' = a_n + 1$, $c_3 = \ell(\ell + 1) - a_n(a_n + 1)$

(2) $c_1 = -1$, $c_1' = 1$: $c_2 = a_n$, $c_2' = 1 - a_n$, $c_3 = \ell(\ell + 1) - a_n(a_n - 1)$

Thus Eq. (5) becomes

$$(y \frac{d}{dy} + y - a_n)(y \frac{d}{dy} - y + a_n + 1) f_{n\ell} = [\ell(\ell + 1) - a_n(a_n + 1)] f_{n\ell} \tag{6a}$$

and

$$(y \frac{d}{dy} - y + a_n)(y \frac{d}{dy} + y - a_n + 1) f_{n\ell} = [\ell(\ell + 1) - a_n(a_n - 1)] f_{n\ell} \tag{6b}$$

7.14 The following operators were introduced in Problem 7.13:

$$\hat{A}_{n\ell} = y \frac{d}{dy} + y - a_n, \quad \hat{B}_{n\ell} = y \frac{d}{dy} - y + a_n$$

Schrödinger showed that these operators act as lowering and raising operators for the principal quantum n of the scaled eigenfunction $f_{n\ell}$.

(a) Show that Eqs. (6a) and (6b) of Problem 7.13 when operated on by $\hat{B}_{n\ell} + 1$ and $\hat{A}_{n\ell} + 1$, respectively, lead to the general recursion formula

$$f_{n+1,\ell} = (y \frac{d}{dy} \mp y \pm a_n + 1) f_{n\ell} \tag{1}$$

(b) Find the function $f_{n'\ell}$ which terminates the "ladder" of eigenfunctions, and determine its general form by using Eq. (1). Also, deduce the dependence of a_n on ℓ.

Solution: Ref. HL,185; Problem 7.13

(a) With the operator definitions given above, Eqs. (6a) and (6b) of Problem 7.13 become

$$\hat{A}_{n\ell}(\hat{B}_{n\ell} + 1)f_{n\ell} = [\ell(\ell + 1) - a_n(a_n + 1)]f_{n\ell} \tag{2a}$$

$$\hat{B}_{n\ell}(\hat{A}_{n\ell} + 1)f_{n\ell} = [\ell(\ell + 1) - a_n(a_n - 1)]f_{n\ell} \tag{2b}$$

The operation of $(\hat{B}_{n\ell} + 1)$ on Eq. (2a) yields

$$(\hat{B}_{n\ell} + 1)\hat{A}_{n\ell}(\hat{B}_{n\ell} + 1)f_{n\ell} = [\ell(\ell + 1) - a_n(a_n + 1)]\hat{B}_{n\ell}f_{n\ell} \tag{3}$$

Letting $n \to n + 1$ in Eq. (2b) we obtain

$$\hat{B}_{n+1,\ell}(\hat{A}_{n+1,\ell} + 1)f_{n+1,\ell} = [\ell(\ell + 1) - a_{n+1}(a_{n+1} - 1)]f_{n+1,\ell} \tag{4}$$

Comparison of Eqs. (4) and (3) shows that they are the same if

$$(\hat{B}_{n\ell} + 1)f_{n\ell} = f_{n+1,\ell}, \quad a_{n+1} = a_n + 1, \quad \hat{A}_{n\ell} = \hat{A}_{n+1,\ell} + 1, \quad \text{and} \quad \hat{B}_{n\ell} + 1 = \hat{B}_{n+1,\ell} \tag{5}$$

Similarly the operation $\hat{A}_{n\ell} + 1$ on Eq. (2b) yields

$$(\hat{A}_{n\ell} + 1)\hat{B}_{n\ell}(\hat{A}_{n\ell} + 1)f_{n\ell} = [\ell(\ell + 1) - a_n(a_n - 1)](\hat{A}_{n\ell} + 1)f_{n\ell} \tag{6}$$

Letting $n \to n - 1$ in Eq. (2a), we obtain

$$\hat{A}_{n-1,\ell}(\hat{B}_{n-1,\ell} + 1)f_{n-1,\ell} = [\ell(\ell + 1) - a_{n-1}(a_{n-1} + 1)]f_{n-1,\ell} \tag{7}$$

Equations (6) and (7) are identical if

$$(\hat{A}_{n\ell} + 1)f_{n\ell} = f_{n-1,\ell}, \quad \hat{B}_{n\ell} = B_{n-1,\ell} + 1, \quad \hat{A}_{n\ell} + 1 = \hat{A}_{n-1,\ell}, \quad \text{and} \quad a_{n-1} = a_n - 1 \tag{8}$$

Equations (5) and (8) then may be summarized

$$(y\frac{d}{dy} \mp y \pm a_n + 1)f_{n\ell} = f_{n\pm1,\ell} \tag{9}$$

(b) If there is a function $f_{n'\ell}$ which terminates the series of eigenfunctions, it can be found by solution of the equation

$$(y\frac{d}{dy} \mp y \pm a_{n'} + 1)f_{n'\ell} = 0 \tag{10}$$

i.e. for some $f_{n'\ell}$ the function $f_{n'\pm1,\ell}$ will vanish. Rearranging Eq. (10) gives

$$\frac{df_{n'\ell}}{f_{n'\ell}} = \pm dy - (1 \pm a_{n'})\frac{dy}{y}$$

This equation is easily integrated to give

$$f_{n'\ell} = c_{n'\ell}\left[\frac{e^{\pm y}}{y^{(1\pm a_{n'})}}\right] \to c_{n'\ell}y^{a_{n'}-1}e^{-y} \tag{11}$$

The upper signs are rejected since they give a function that is not well-behaved. Thus Eq. (10) can be written

$$(\hat{A}_{n'\ell} + 1)f_{n'\ell} = 0 \tag{12}$$

where $\hat{A}_{n\ell}$ may be interpreted as a "lowering" operator and $f_{n'\ell}$ is the bottom rung of the ladder.

To find $a_{n'}$ we operate on Eq. (12) with $\hat{B}_{n\ell}$, the raising operator, to give

$$\hat{B}_{n'\ell}(\hat{A}_{n'\ell} + 1)f_{n'\ell} = 0 = [\ell(\ell + 1) - a_{n'}(a_{n'} - 1)]f_{n'\ell} \quad \text{or} \quad a_{n'}(a_{n'} - 1) = \ell(\ell + 1)$$

This gives $a_{n'} = \ell + 1$ or $-\ell$. Now the allowed values of ℓ are 0, 1, 2, $\cdots$. The $-\ell$ solution, when substituted into Eq. (11) would yield non-physical singularities, thus the physical root is $a_\ell = \ell + 1 = 1, 2, 3, \cdots$. Thus we can replace a_n by n where n = 1, 2, 3, $\cdots$. From Eq. (11) we then have

$$f_{n\ell} = f_{\ell+1,\ell} = c_{\ell+1,\ell}y^\ell e^{-y} \tag{13}$$

7.15 Evaluate the functions $f_{1,0}$, $f_{2,0}$, $f_{2,1}$, $f_{3,0}$, $f_{3,1}$ and $f_{3,2}$ along with the associated energies by using the results of Problems 7.13 and 7.14.

Solution:

The energies can be found from Eq. (1) of Problem 7.13 and the discussion of Problem 7.14.

$$E_n = -\frac{1}{a_n^2} = -\frac{1}{n^2} \text{ Rydbergs}$$

The various unnormalized radial functions can be found by application of Eqs. (9) and (13) of Problem 7.14 which are repeated below for convenience.

$$f_{n\pm1,\ell} = (y\frac{d}{dy} \mp y \pm a_n + 1)f_{n\ell} \tag{9}$$

$$f_{\ell+1,\ell} = c_{\ell+1,\ell}y^\ell e^{-y} \tag{13}$$

Thus

$f_{1,0}$ = (1s): By application of Eq. (13) $f_{1,0} = c_{1,0}y^o e^{-y}$. Since $y = r/a_n = r/n$ we then have for the unnormalized solution $f_{1,0} = e^{-r}$ where r is the radial distance expressed in atomic units (i.e. units of a_o).

$f_{2,0}$ = (2s): By application of Eq. (9)

$$f_{2,0} = (y\frac{d}{dy} - y + n + 1)f_{1,0} = (y\frac{d}{dy} - y + 2)f_{1,0} = 2(1 - y)e^{-y} = 2(1 - r/2)e^{-r/2}$$

$f_{2,1}$ = (2p): From Eq. (13)

$$f_{2,1} = y^1 e^{-y} = \frac{r}{2} e^{-r/2}$$

$f_{3,0}$ = (3s): From Eq. (9)

$$f_{3,0} = (y\frac{d}{dy} - y + 3)f_{2,0} = 2[3 - 6y + 2y^2]e^{-y} = 2[3 - 2r + \frac{2}{9} r^2]e^{-r/3}$$

$f_{3,1}$ = (3p): From Eq. (9)

$$f_{3,1} = (y\frac{d}{dy} - y + 3)f_{2,1} = (4y - 2y^2)e^{-y} = \frac{2}{9}(6r - r^2)e^{-r/3}$$

$f_{3,2}$ = (3d): From Eq. (13)

$$f_{3,2} = y^2 e^{-y} = \frac{r^2}{9} e^{-r/3}$$

7.16 Unsöld's theorem states that the sum of the squares of the magnitudes of all angular functions associated with the same value of ℓ is independent of orientation. Thus

$$\sum_{m=-\ell}^{+\ell} Y_{\ell m}^{*}(\theta,\phi)Y_{\ell m}(\theta,\phi) = \text{constant}$$

Prove Unsöld's theorem. (Hint: Use the addition theorem of spherical harmonics:

$$P_{\ell}(\cos\gamma) = \sum_{m=-\ell}^{+\ell} \frac{(\ell - |m|)!}{(\ell + |m|)!} P_{\ell}^{|m|}(\cos\theta_1)P_{\ell}^{|m|}(\cos\theta_2)e^{im(\phi_1-\phi_2)} \tag{1}$$

where γ is the angle between two vectors of directions θ_1, ϕ_1 and θ_2, ϕ_2.)

Solution: Ref. S3,128; S1,182; A. Unsöld, Ann. Physik, 82, 355 (1927)

The spherical harmonics may be written as

$$Y_{\ell m}(\theta,\phi) = \frac{(-1)^{(m+|m|)/2}}{(4\pi)^{1/2}}\left[\frac{(2\ell + 1)(\ell - |m|)!}{(\ell + |m|)!}\right]^{1/2} P_{\ell}^{|m|}(\cos\theta)e^{+im\phi} \tag{2}$$

Then

$$\sum_{m_\ell=-\ell}^{+\ell} Y_{\ell m}^{*}(\theta,\phi)Y_{\ell m}(\theta,\phi) = \frac{2\ell + 1}{4\pi}\sum_{m_\ell=-\ell}^{+\ell} \frac{(\ell - |m|)!}{(\ell + |m|)!}|P_{\ell}^{|m|}(\cos\theta)|^2 \tag{3}$$

Setting $\theta_1 = \theta_2 = \theta$ and $\phi_1 = \phi_2 = \phi$ in Eq. (1), we find that

$$P_{\ell}(\cos\gamma) = P_{\ell}(0) = 1 = \sum_{m_\ell=-\ell}^{+\ell} \frac{(\ell - |m|)!}{(\ell + |m|)!}|P_{\ell}^{|m|}(\cos\theta)|^2 \tag{4}$$

Substitution of Eq. (4) into Eq. (3) then finishes the proof:

$$\sum_{-\ell}^{+\ell} Y_{\ell m}^{*}(\theta,\phi)Y_{\ell m}(\theta,\phi) = \frac{2\ell + 1}{4\pi} = \text{constant}$$

7.17 Consider a 2-dimensional hydrogen atom in which an electron is bound to the nucleus by a coulombic force and is constrained to move in a plane.

(a) Determine the eigenfunctions and eigenvalues for this system.

(b) Explain why N. Bohr was able to obtain energies in agreement with experiment even though he solved the H-atom problem assuming a planar orbit.

(Hint: Set up the Hamiltonian in cylindrical coordinates, separate variables, define $P(r) = \sqrt{r}\,R(r)$, and carry out a series solution for $P(r)$ by analogy with the 3-dimensional hydrogen atom.)

Solution: Ref. B. Zaslow and M. E. Zandler, Am. J. Phys. 35, 1118 (1967)

(a) The wave equation for this system is (in atomic units)

$$(-\nabla^2 - 2Z/r)u = Eu$$

Using cylindrical coordinates (see Appendix 5) with $x = r\cos\phi$ and $y = r\sin\phi$ we have:

$$-\left[\frac{1}{r}\frac{\partial}{\partial r}r\frac{\partial}{\partial r} + \frac{1}{r^2}\frac{\partial^2}{\partial\phi^2}\right]u - \frac{2Z}{r}u = Eu \tag{1}$$

We separate variables by assuming that $u = R(r)\Phi(\phi)$. Thus after substitution into Eq. (1) and division by u we find:

$$-\left[\frac{1}{Rr}\frac{d}{dr}\left(r\frac{dR}{dr}\right) + \frac{1}{\Phi r^2}\frac{d^2\Phi}{d\phi^2}\right] - \frac{2Z}{r} = E$$

and

$$-\frac{1}{\Phi}\frac{d^2\Phi}{d\phi^2} = \frac{r}{R}\frac{d}{dr}\left(r\frac{dR}{dr}\right) + 2Zr + Er^2 = m^2 \qquad (2)$$

where m^2 is chosen to be the separation constant.

The Φ equation

From Eq. (2) we have

$$\frac{d^2\Phi}{d\phi^2} = -m^2\Phi$$

with the normalized eigenfunctions (see Problem 4.15).

$$\Phi = \frac{1}{\sqrt{2\pi}}e^{im\phi}; \qquad m = 0, \pm 1, \pm 2, \cdots$$

The R equation

Equation (2) also yields

$$\frac{1}{r}\frac{d}{dr}\left(r\frac{dR}{dr}\right) + \left(\frac{2Z}{r} - \frac{m^2}{r^2} + E\right)R = 0 \qquad (3)$$

It is convenient to transform Eq. (3) into the form of a 1-dimensional wave equation. Recall that the radial density function in 2-dimensions is proportional to rR^2. This suggests the substitution $P = \sqrt{r}\,R$. It is easy to show that

$$\frac{1}{r}\frac{d}{dr}\left(r\frac{dR}{dr}\right) = \frac{1}{\sqrt{r}}\frac{d^2}{dr^2}(\sqrt{r}\,R) + \frac{R}{4r^2}$$

Therefore, Eq. (3) becomes:

$$-\frac{d^2P}{dr^2} + \left[\frac{(m^2 - 1/4)}{r^2} - \frac{2Z}{r}\right]P = EP \qquad (4)$$

Equation (4) can be solved using the methods of Problems 7.5 and 7.6. First consider the limits:

$$\text{As } r \to 0: \quad \frac{d^2P}{dr^2} + \frac{(1/4 - m^2)}{r^2}P = 0$$

This equation is satisfied by $r^{-(m-1/2)}$ and $r^{m+1/2}$ but only the solution $P = r^{|m|+1/2}$ gives finite values of R as $r \to 0$.

$$\text{As } r \to \infty: \quad \frac{d^2P}{dr^2} + EP = 0$$

The solution is easily found to be $P = e^{-\sqrt{-E}\,r}$.

To take into account the behavior at the limits we assume that for all r:

$$P(r) = r^{|m|+1/2} e^{-\sqrt{-E}\, r} f(r)$$

Substitution of this expression into Eq. (4) gives:

$$\frac{d^2 f}{dr^2} + 2\frac{df}{dr}\left[\frac{(|m| + 1/2)}{r} - \sqrt{-E}\right] + \frac{2}{r}[Z - (|m| + 1/2)\sqrt{-E}]f = 0$$

The calculation here is identical to that in Problem 7.6 with $\ell + 1$ replaced by $|m| + 1/2$. The series solution is carried out as before with the substitution

$$f(r) = \sum_{k=0}^{\infty} A_k r^k$$

The resulting recursion relation for A_k is

$$A_k = -A_{k-1}\left\{\frac{2[Z - (|m| - 1/2 + k)\sqrt{-E}]}{(|m| - 1/2 + k)(|m| + 1/2 + k) - (m^2 - 1/4)}\right\}$$

which should be compared with Eq. (3) of Problem 7.7.

The series solutions terminate properly to give acceptable eigenfunctions if and only if:

$$2Z - (2|m| - 1 + 2k)\sqrt{-E} = 0 \quad \text{or} \quad 2Z - n\sqrt{-E} = 0$$

Therefore

$$E = -\frac{4Z^2}{n^2} \tag{5}$$

To determine the range of n note that $|m| = 0, 1, 2, \cdots$ and that the minimum value of k is +1. Then $n_{min} = 2|m| + 1 = +1$ and $n = 1, 3, 5, 7,$ etc.

$$P(r) = r^{|m|+1/2} e^{-2Zr/n}(A_o + A_1 r + \ldots + A_{k_{max}} r^{k_{max}})$$

where

$$k_{max} = \frac{(n - 1)}{2} - |m|$$

Also,

$$P(r) = e^{-2Zr/n}(A_o r^{|m|+1/2} + \ldots + A_{k_{max}} r^{n/2})$$

and for the ground state where $n = 1$ and $|m| = 0$:

$$P(r) = A_o r^{1/2} e^{-2Zr}$$

(b) It is well-known that Bohr obtained the energy eigenvalues for the 3-dimensional hydrogen atom using (i) a planetary model for rotation of the electron in a plane and (ii) a quantum condition which amounts to the quantization of angular momentum (see Problem 1.17). Condition (i), if treated properly with wave mechanics, leads to Eq. (5). On the other hand the old quantum theory is known to lead to errors in the energy eigenvalues in certain cases.

For example the old quantum theory predicts that $E = nh\nu_o$ for the simple harmonic oscillator for which the correct result is $E = (n + 1/2)h\nu_o$.

We conclude that the errors introduced by conditions (i) and (ii) cancel and permit the correct energy equation for the hydrogen atom to be obtained. This fortunate accident was of great importance in the development of the quantum theory. It is interesting to consider the correction of Bohr's energy equation by the introduction of (n' + 1/2) in place of n. Thus

$$E = -\frac{Z^2}{(n' + 1/2)^2} \tag{6}$$

If n' = 0, 1, 2, ··· which is reasonable in view of the quantization condition, then Eqs. (5) and (6) are found to be identical.

7.18 Consider a collection of positive and negative charges at equilibrium with

$$\sum_i q_i = 0$$

i.e. the system is neutral. In the absence of an externally applied electric field the centers of positive and negative charge coincide. However, when a homogeneous electric field $\underset{\sim}{\mathcal{E}}$ is applied these centers will separate to form an induced dipole $\underset{\sim}{\mu} = \alpha\underset{\sim}{\mathcal{E}}$ where α is called the polarizability. In general α is a tensor, but for isotropic systems it is a simple constant with the units of volume in the cgs system. Show that the energy of polarization for an isotropic system is given by $E = -\alpha\mathcal{E}^2/2$. (Hint: An easy derivation makes use of the fact that at equilibrium the potential energy is a minimum. For small deviations from equilibrium the potential can be expanded about the minimum.)

Solution: Ref. BC,32; DE2,20; AT2,397

The centers of positive and negative charge act as if they are bound together by a linear restoring force for small displacements since

$$V(x) = V(0) + \left(\frac{\partial V}{\partial x}\right)_o x + \frac{1}{2}\left(\frac{\partial^2 V}{\partial x^2}\right)_o x^2 + \cdots \simeq \frac{kx^2}{2}$$

where x is the separation of the charge centers. Here V(0) can be chosen to be zero and $(\partial V/\partial x)_o$ is zero by definition at a minimum. The energy required to form the dipole is thus

$$E_1 = \int_o^{x_o} F dx = \int_o^{x_o} kx dx = \frac{kx_o^2}{2}$$

The energy of this dipole in the field $\underset{\sim}{\mathcal{E}}$ is just:

$$E_2 = -\mu\mathcal{E} = -(\alpha\mathcal{E})\mathcal{E} = -\alpha\mathcal{E}^2 \tag{1}$$

If the charge associated with the positive center is q, then $\mu = qx_o = \alpha\mathcal{E}$. Also, at equilibrium in the field the forces must be balanced. Therefore $q\mathcal{E} = kx_o$ or $x_o = q\mathcal{E}/k$.

These relations can be used to write

$$E_1 = \frac{kx_o^2}{2} = \frac{k}{2}\left(\frac{q\mathcal{E}}{k}\right)\left(\frac{\alpha\mathcal{E}}{q}\right) = \frac{\alpha\mathcal{E}^2}{2} \qquad (2)$$

The total energy of the system is given by the sum of Eqs. (1) and (2)

$$E = E_1 + E_2 = -\alpha\mathcal{E}^2 + \frac{\alpha\mathcal{E}^2}{2} = -\frac{\alpha\mathcal{E}^2}{2}$$

7.19 Consider a hydrogen atom in an electric field so that the Hamiltonian becomes:

$$\hat{\mathcal{H}} = \hat{\mathcal{H}}_o + e\mathcal{E}r\cos\theta$$

Here the nucleus is placed at the origin and the field $\mathcal{E}$ is in the z-direction. The Hamiltonian for the unperturbed hydrogen atom is $\hat{\mathcal{H}}_o$ which is assumed to be much larger than the perturbation. Use linear variation theory, i.e.

$$\psi = \sum_i c_i f_i$$

with the functions

$$f_1 = u_{1s}, \quad f_2 = u_{2p_z}$$

to determine the lowest energy state for this atom and to estimate the polarizability.

Solution: Ref. PW,180; RA,290

We are given the trial function

$$\psi = c_1 u_{1s} + c_2 u_{2p_z}$$

Minimization of the energy with respect to c_1 and c_2 leads to the secular equation

$$\begin{vmatrix} \mathcal{H}_{11} - S_{11}E & \mathcal{H}_{12} - S_{12}E \\ \mathcal{H}_{21} - S_{21}E & \mathcal{H}_{22} - S_{22}E \end{vmatrix} = 0$$

where using Appendix 7:

$$\mathcal{H}_{11} = \langle 1s|\hat{\mathcal{H}}_o|1s\rangle + e\mathcal{E}\langle 1s|z|1s\rangle = E_{1s}$$

$$\mathcal{H}_{22} = \langle 2p_z|\hat{\mathcal{H}}_o|2p_z\rangle + e\mathcal{E}\langle 2p_z|z|2p_z\rangle = E_{1s}/4$$

$$\mathcal{H}_{12} = +e\mathcal{E}\langle 1s|z|2p_z\rangle = +\frac{e\mathcal{E}a_o}{4\pi\sqrt{2}}\int_o^\infty e^{-3\sigma/2}\sigma^4 d\sigma \int_o^{2\pi} d\phi \int_o^\pi \cos^2\theta\sin\theta d\theta$$

$$= +\frac{e\mathcal{E}a_o}{\sqrt{2}}\frac{4!}{(3/2)^5}\frac{1}{3} = +\frac{e\mathcal{E}a_o}{\sqrt{2}}\frac{2^8}{3^5} = A$$

$$S_{11} = S_{22} = 1; \quad S_{12} = 0$$

Thus the secular determinant becomes

$$\begin{vmatrix} E_{1s} - E & A \\ A & \frac{E_{1s}}{4} - E \end{vmatrix} = 0$$

or

$$E^2 - \frac{5}{4} E_{1s}E + \left(\frac{E_{1s}^2}{4} - A^2\right) = 0$$

The roots are immediately given by

$$E = \frac{5}{8} E_{1s} \pm \frac{1}{2} \sqrt{\left(\frac{5}{4} E_{1s}\right)^2 - 4\left(\frac{E_{1s}^2}{4} - A^2\right)} = \frac{5}{8} E_{1s} \pm \frac{1}{2}\sqrt{\frac{9}{16} E_{1s}^2 \left(1 + \frac{16}{9E_{1s}^2} \cdot 4A^2\right)}$$

and assuming that $A^2/E_{1s}^2 << 1$ we find

$$E = + \frac{5}{8} E_{1s} \pm \frac{3}{8} E_{1s} \left(1 + \frac{32}{9} \frac{A^2}{E_{1s}^2} + \cdots\right)$$

The lowest energy is

$$E = E_{1s} + \frac{4}{3} \frac{A^2}{E_{1s}} = E_{1s} + \frac{4}{3} a_o^2 \left(\frac{2^{16}}{3^{10}}\right) \frac{e^2\mathcal{E}^2}{2E_{1s}} = E_{1s} - (2.96) \frac{\mathcal{E}^2}{2} a_o^3(4\pi\varepsilon_o) \tag{1}$$

where we have used the relation

$$E_{1s} = - \frac{e^2}{(4\pi\varepsilon_o)2a_o}$$

Since the energy of polarization is given by

$$E_{polar} = - \frac{\alpha}{2} \mathcal{E}^2$$

Eq. (1) shows that $\alpha = 2.96\ a_o^3(4\pi\varepsilon_o)$. The exact value of α for the hydrogen atom is $(9\ a_o^3/2)(4\pi\varepsilon_o)$. The units of polarizability in the SI system are Fm^2. In cgs units where the factor $(4\pi\varepsilon_o)$ is replaced by unity the polarizability has the units of volume and is the same order of magnitude as the atomic volume.

7.20 A Hamiltonian of the form $\hat{\mathcal{H}} = \hat{\mathcal{H}}^o + \zeta\hat{\underset{\sim}{L}}\cdot\hat{\underset{\sim}{S}}$ is important in atomic spectroscopy. Here ζ is a constant and $\hat{\mathcal{H}}^o$ commutes with all components of $\hat{\underset{\sim}{L}}$ and $\hat{\underset{\sim}{S}}$. The last term in $\hat{\mathcal{H}}$ arises because of spin-orbit coupling and leads to energy differences between levels with the same values of L and S but different values of J where $\hat{\underset{\sim}{J}} = \hat{\underset{\sim}{L}} + \hat{\underset{\sim}{S}}$.

(a) Show that $[\hat{\mathcal{H}}, \hat{L}^2] = [\hat{\mathcal{H}}, \hat{S}^2] = 0$ but $[\hat{\mathcal{H}}, \hat{L}_z] \neq 0$ and $[\hat{\mathcal{H}}, \hat{S}_z] \neq 0$.

(b) Show that $[\hat{\mathcal{H}}, \hat{J}_z] = [\hat{\mathcal{H}}, \hat{J}^2] = 0$.

Solution: Ref. Problems 3.26 and 5.8

(a) $$[\hat{\mathcal{H}}, \hat{L}^2] = [(\hat{\mathcal{H}}^o + \zeta\hat{\underset{\sim}{L}}\cdot\hat{\underset{\sim}{S}}), \hat{L}^2] = [\hat{\mathcal{H}}^o, \hat{L}^2] + \zeta[\hat{\underset{\sim}{L}}\cdot\hat{\underset{\sim}{S}}, \hat{L}^2] \tag{1}$$

But

$$[\hat{\mathcal{H}}^o, \hat{L}_x^2] = [\hat{\mathcal{H}}^o, \hat{L}_x]\hat{L}_x - \hat{L}_x[\hat{\mathcal{H}}^o, \hat{L}_x] = 0 \quad \text{(see Problem 3.26)}$$

since we are given that $[\hat{\mathcal{H}}^o, \hat{L}_x] = 0$. By the same argument $[\hat{\mathcal{H}}^o, \hat{L}_y^2]$ and $[\hat{\mathcal{H}}^o, \hat{L}_z^2]$ also vanish, and since $\hat{L}^2 = \hat{L}_x^2 + \hat{L}_y^2 + \hat{L}_z^2$ we find that $[\hat{\mathcal{H}}^o, \hat{L}^2] = 0$. The second commutator in Eq. (1) is:

$$[\hat{L}_x\hat{S}_x + \hat{L}_y\hat{S}_y + \hat{L}_z\hat{S}_z, \hat{L}^2]$$

This commutator also vanishes since (1) the components of $\hat{\underset{\sim}{S}}$ commute with the components of $\hat{\underset{\sim}{L}}$ which are in a different space and (2) $\hat{L}^2$ commutes with all components of $\hat{\underset{\sim}{L}}$ (see Problem 5.8). Therefore $[\hat{\mathcal{H}}, \hat{L}^2] = 0$.

The commutator $[\hat{\mathcal{H}}, \hat{L}_z]$ is straightforward.

$$[\hat{\mathcal{H}}, \hat{L}_z] = [\hat{\mathcal{H}}^o + \zeta\hat{\underset{\sim}{L}}\cdot\hat{\underset{\sim}{S}}, \hat{L}_z] = \zeta[\hat{\underset{\sim}{L}}\cdot\hat{\underset{\sim}{S}}, \hat{L}_z] = \zeta[\hat{L}_x\hat{S}_x + \hat{L}_y\hat{S}_y + \hat{L}_z\hat{S}_z, \hat{L}_z]$$
$$= \zeta\{\hat{S}_x[\hat{L}_x, \hat{L}_z] + \hat{S}_y[\hat{L}_y, \hat{L}_z] + \hat{S}_z[\hat{L}_z, \hat{L}_z]\} = \zeta\{\hat{S}_x(-i\hat{L}_y) + \hat{S}_y(i\hat{L}_x)\} \neq 0$$

By exchanging L and S in these proofs it is easily shown that $[\hat{\mathcal{H}}, \hat{S}^2] = 0$ and $[\hat{\mathcal{H}}, \hat{S}_z] \neq 0$. From these commutation relations it is evident that simultaneous eigenfunctions can be found for the operators $\hat{\mathcal{H}}$, $\hat{L}^2$, and $\hat{S}^2$, but that simultaneous eigenfunctions do not exist for $\hat{\mathcal{H}}$, $\hat{L}_z$, and $\hat{S}_z$ (see Problem 3.9).

(b) $$[\hat{\mathcal{H}}, \hat{J}_z] = [\hat{\mathcal{H}}^o, \hat{L}_z + \hat{S}_z] + \zeta[\hat{\underset{\sim}{L}}\cdot\hat{\underset{\sim}{S}}, \hat{L}_z + \hat{S}_z] = \zeta[\hat{L}_x\hat{S}_x + \hat{L}_y\hat{S}_y + \hat{L}_z\hat{S}_z, \hat{L}_z + \hat{S}_z]$$
$$= \zeta\{\hat{S}_x[\hat{L}_x, \hat{L}_z] + \hat{S}_y[\hat{L}_y, \hat{L}_z] + \hat{S}_z[\hat{L}_z, \hat{L}_z] + \hat{L}_x[\hat{S}_x, \hat{S}_z] + \hat{L}_y[\hat{S}_y, \hat{S}_z] + \hat{L}_z[\hat{S}_z, \hat{S}_z]\}$$
$$= \zeta\{\hat{S}_x(-i\hat{L}_y) + \hat{S}_y(+i\hat{L}_x) + \hat{L}_x(-i\hat{S}_y) + \hat{L}_y(+i\hat{S}_x)\} = 0$$
$$[\hat{\mathcal{H}}, \hat{J}^2] = [\hat{\mathcal{H}}, (\hat{\underset{\sim}{L}} + \hat{\underset{\sim}{S}})^2] = [\hat{\mathcal{H}}^o + \zeta\hat{\underset{\sim}{L}}\cdot\hat{\underset{\sim}{S}}, \hat{L}^2 + \hat{S}^2 + 2\hat{\underset{\sim}{L}}\cdot\hat{\underset{\sim}{S}}]$$
$$= [\hat{\mathcal{H}}, \hat{L}^2] + [\hat{\mathcal{H}}, \hat{S}^2] + 2[\hat{\mathcal{H}}^o, \hat{\underset{\sim}{L}}\cdot\hat{\underset{\sim}{S}}] + 2\zeta[\hat{\underset{\sim}{L}}\cdot\hat{\underset{\sim}{S}}, \hat{\underset{\sim}{L}}\cdot\hat{\underset{\sim}{S}}] = 2[\hat{\mathcal{H}}^o, \hat{L}_x\hat{S}_x + \hat{L}_y\hat{S}_y + \hat{L}_z\hat{S}_z]$$
$$= 2\{[\hat{\mathcal{H}}^o, \hat{L}_x]\hat{S}_x + [\hat{\mathcal{H}}^o, \hat{L}_y]\hat{S}_y + [\hat{\mathcal{H}}^o, \hat{L}_z]\hat{S}_z\} = 0$$

Both commutators in (b) confirm that $\hat{\mathcal{H}}^o$ is invariant to rotations of the coordinate system (see Problem 5.17).

7.21 The spin-orbit interaction for the electron in a hydrogen-like atom is given by:

$$\hat{\mathcal{H}}_{SO} = \xi(r)\hat{\underset{\sim}{L}}\cdot\hat{\underset{\sim}{S}} \quad \text{where} \quad \xi(r) = \frac{\hbar^2}{2m_e^2c^2}\left(\frac{1}{r}\frac{\partial V}{\partial r}\right)$$

Derive an equation for the allowed energies in such atoms in terms of the quantum numbers ℓ, s, and j, i.e. derive the Landé interval rule. (Hint: It is convenient to use unperturbed wave functions which are simultaneous eigenfunctions of $\hat{S}^2$, $\hat{L}^2$, $\hat{J}^2$, and $\hat{J}_z$.)

Solution: Ref. CS,120; S1,249

The complete Hamiltonian is $\hat{\mathcal{H}} = \hat{\mathcal{H}}^o + \hat{\mathcal{H}}_{SO}$ where $\hat{\mathcal{H}}^o|n\ell\rangle = E_o(n\ell)|n\ell\rangle$. In the representation which diagonalizes $\hat{L}^2$, $\hat{S}^2$, $\hat{J}^2$, and $\hat{J}_z$ it is easy to show that $\hat{\mathcal{H}}_{SO}$ is also diagonal. By definition $\hat{\underset{\sim}{J}} = \hat{\underset{\sim}{L}} + \hat{\underset{\sim}{S}}$ and

$$J^2 = \hat{L}^2 + \hat{S}^2 + 2\hat{\underset{\sim}{L}}\cdot\hat{\underset{\sim}{S}}$$

Therefore

$$\langle\ell, s, j, m_j|\hat{\underset{\sim}{L}}\cdot\hat{\underset{\sim}{S}}|\ell, s, j, m_j\rangle = \langle\ell, s, j, m_j|\frac{\hat{J}^2 - \hat{L}^2 - \hat{S}^2}{2}|\ell, s, j, m_j\rangle$$
$$= \frac{1}{2}[j(j+1) - \ell(\ell+1) - s(s+1)] \qquad (1)$$

Here lower case quantum numbers are used to emphasize that this is a one-electron problem.

The complete energy expression is:

$$E(n,\ \ell,\ s,\ j) = \langle n,\ \ell,\ s,\ j,\ m_j|\hat{\mathcal{H}}^o + \hat{\mathcal{H}}_{SO}|n,\ \ell,\ s,\ j,\ m_j\rangle$$

$$= E^o(n\ell) + \langle n,\ \ell|\xi(r)|n,\ \ell\rangle\langle\ell,\ s,\ j,\ m_j|\hat{\underset{\sim}{L}}\cdot\hat{\underset{\sim}{S}}|\ell,\ s,\ j,\ m_j\rangle$$

Defining the integral over the radial functions to be $\zeta(n\ell) \equiv \langle n,\ \ell|\xi(r)|n,\ \ell\rangle$ and making use of Eq. (1) we obtain:

$$\boxed{E(n,\ \ell,\ s,\ j) = E^o(n\ell) + \frac{\zeta(n\ell)}{2}\,[j(j + 1) - \ell(\ell + 1) - s(s + 1)]} \tag{2}$$

Since $s = 1/2$, Eq. (2) can be rewritten for the two special cases:

$$E(n,\ \ell,\ s,\ j) = E^o(n\ell) + \zeta(n\ell)\,\frac{\ell}{2};\quad j = \ell + 1/2$$

$$= E^o(n\ell) - \zeta(n\ell)\,\frac{(\ell + 1)}{2};\quad j = \ell - 1/2$$

For the hydrogen atom ($Z = 1$) the spin-orbit correction is the same order of magnitude as the relativistic correction so that both corrections must be used simultaneously. For heavier hydrogen-like atoms, however, the spin-orbit correction dominates. In Problem 7.22 we show that this correction is proportional to Z^4.

7.22 In Problem 7.21 we showed that the first order correction to the energy due to spin orbit coupling for hydrogen-like atoms is

$$E_{SO} = \frac{\zeta(n\ell)}{2}\,[j(j + 1) - \ell(\ell + 1) - s(s + 1)]$$

where

$$\zeta(n\ell) = \langle n\ell|\ \frac{\hbar^2}{2m_e^2c^2}\left(\frac{1}{r}\,\frac{\partial V}{\partial r}\right)|n\ell\rangle \text{ for } \ell \neq 0$$

Evaluate $\zeta(n\ell)$. (Hint: The evaluation of $\overline{r^{-3}}$ is considered in Problem 7.31.)

<u>Solution</u>:

$$\zeta(n\ell) = \frac{\hbar^2}{2m_e^2c^2}\,\langle n\ell|\left(\frac{1}{r}\,\frac{\partial V}{\partial r}\right)|n\ell\rangle = \frac{\hbar^2}{2m_e^2c^2}\,\langle n\ell|\ \frac{1}{r}\,\frac{\partial}{\partial r}\left(\frac{-Ze^2}{r(4\pi\varepsilon_o)}\right)|n\ell\rangle$$

$$= \frac{Z\hbar^2e^2}{2m_e^2c^2(4\pi\varepsilon_o)}\,\langle n\ell|\ \frac{1}{r^3}\ |n\ell\rangle = \frac{Z\hbar^2e^2}{2m_e^2c(4\pi\varepsilon_o)}\,\overline{r^{-3}}$$

From the results of Problem 7.31 we have

$$\zeta(n\ell) = \frac{Z\hbar^2e^2}{2m_e^2c^2(4\pi\varepsilon_o)}\left(\frac{Z^3}{a_o^3n^3\ell(\ell + 1/2)(\ell + 1)}\right)$$

$$= \frac{(Ze)^4m_e}{2\hbar^2(4\pi\varepsilon_o)^2}\left[\frac{e^2}{(4\pi\varepsilon_o)c\hbar}\right]^2\frac{1}{n^3\ell(\ell + 1/2)(\ell + 1)}$$

7.23 When a hydrogen-like atom is placed in a weak magnetic field $\underset{\sim}{B}$, the energy of interaction is described by the Zeeman Hamiltonian

$$\hat{\mathcal{H}}_z = \mu_B \underset{\sim}{B}\cdot\hat{\underset{\sim}{L}} + g_s\mu_B\underset{\sim}{B}\cdot\hat{\underset{\sim}{S}}$$

where g_s is approximately equal to 2. Derive an equation for the energy of this atom using first order perturbation theory. Assume that in the absence of $\underset{\sim}{B}$, the wave functions for the atom are eigenfunctions of $\hat{L}^2$, $\hat{S}^2$, $\hat{J}^2$ and $\hat{J}_z$ where $\hat{\underset{\sim}{J}} = \hat{\underset{\sim}{L}} + \hat{\underset{\sim}{S}}$. (Hint: Tables of Clebsch-Gordon coefficients can be found in Problem 5.27.)

Solution: Ref. CS,151; ME,434

The perturbing Hamiltonian may be written as:

$$\hat{\mathcal{H}}' = \mu_B B(\hat{L}_z + 2\hat{S}_z) = \mu_B B(\hat{J}_z + \hat{S}_z)$$

The first order correction to the energy is then given by:

$$E^{(1)} = \langle L,S,J,M_J|\mu_B B(\hat{J}_z + \hat{S}_z)|L,S,J,M_J\rangle = \langle \ell,1/2,\ell \pm 1/2,m|\mu_B B(\hat{J}_z + \hat{S}_z)|\ell,1/2,\ell \pm 1/2,m\rangle$$

$$= \mu_B Bm + \mu_B B\langle \ell,1/2,\ell \pm 1/2,m|\hat{S}_z|\ell,1/2,\ell \pm 1/2,m\rangle \quad (1)$$

The matrix element of $\hat{S}_z$ can be obtained by expressing the function $|\ell,1/2,\ell \pm 1/2,m\rangle$ in terms of eigenfunctions of $\hat{L}^2$, $\hat{S}^2$, $\hat{L}_z$, and $\hat{S}_z$. (See Problems 5.21 and 5.27.) The appropriate expansion is:

$$|\ell sjm\rangle = \sum_{m_\ell, m_s} |\ell s m_\ell m_s\rangle\langle \ell s m_\ell m_s|\ell sjm\rangle$$

or

$$|j,m\rangle = \sum_{m_s} |m - m_s, m_s\rangle\langle m - m_s, m_s|jm\rangle$$

where the common indices ℓ and s have been suppressed and the fact that $m = m_\ell + m_s$ has been used. Finally we write

$$|\ell \pm 1/2,m\rangle = \sum_{m_s} |m - m_s, m_s\rangle\langle m - m_s, m_s|\ell \pm 1/2,m\rangle$$

and then make use of Table 5.27b to obtain the coefficients. Thus

$$|\ell \pm 1/2,m\rangle = \pm\sqrt{\frac{\ell \pm m + 1/2}{2\ell + 1}}\,|m - 1/2,1/2\rangle + \sqrt{\frac{\ell \mp m + 1/2}{2\ell + 1}}\,|m + 1/2,-1/2\rangle$$

and the required matrix element is:

$$\langle \ell \pm 1/2,m|\hat{S}_z|\ell \pm 1/2,m\rangle = \frac{1}{2(2\ell + 1)}(\ell \pm m + 1/2 - \ell \pm m - 1/2) = \pm\frac{m}{(2\ell + 1)}$$

Equation (1) yields

$$E^{(1)} = \mu_B Bm\left[1 \pm \frac{1}{2\ell + 1}\right] = g\mu_B Bm$$

For $j = \ell \pm 1/2$ the Landé g-factor is thus found to be

$$g = 1 \pm [1/(2\ell + 1)]$$

In general it can be shown that

$$\boxed{g = 1 + \frac{J(J+1) + S(S+1) - L(L+1)}{2J(J+1)}}$$

7.24 A hydrogen-like atom is placed in a magnetic field where the Zeeman and spin-orbit interactions have approximately the same magnitudes. Obtain the energies for the states having the quantum numbers ℓ and s using the perturbing Hamiltonian:

$$\hat{\mathcal{H}}' = \mu_B B(\hat{L}_z + 2\hat{S}_z) + \zeta \hat{\underset{\sim}{L}} \cdot \hat{\underset{\sim}{S}}$$

(Hint: Use basis functions which diagonalize $\hat{L}_z$ and $\hat{S}_z$, set up the secular equation, and determine the eigenvalues.)

Solution: Ref. CS,152; TI,191

The basis functions of interest are specified by the quantum numbers ℓ, s, m_ℓ, m_s, and $m_j = m_\ell + m_s$ where s = 1/2 and the values of m_j run from $\ell + 1/2$ to $-\ell - 1/2$. Since $\hat{J}_z$ commutes with $\hat{\mathcal{H}}'$, there are no non-vanishing matrix elements between functions having different values of m_j. To see this take the matrix element of the commutator $[\hat{\mathcal{H}}', \hat{J}_z]$.

$$<m_\ell,m_s|\hat{\mathcal{H}}'\hat{J}_z - \hat{J}_z\hat{\mathcal{H}}'|m'_\ell,m'_s> = 0$$

$$<m_\ell,m_s|\hat{\mathcal{H}}'|m'_\ell,m'_s>[(m'_\ell + m'_s) - (m_\ell + m_s)] = 0$$

Therefore, either $m_j = m'_j$ or $<m_\ell,m_s|\hat{\mathcal{H}}'|m'_\ell,m'_s> = 0$. If the basis functions are arranged in descending order of m_j, it is clear that the upper left hand corner of the Hamiltonian matrix will contain a 1x1 block corresponding to $m_j = \ell + 1/2$ and the lower right hand corner will contain a 1x1 block corresponding to $m_j = -\ell - 1/2$. Each of the other values of m_j can be obtained in two ways, namely ($m_\ell = m_j - 1/2$, $m_s = +1/2$) and ($m_\ell = m_j + 1/2$, $m_s = -1/2$). The resulting 2x2 blocks for each value of m_j all have the same form and can be solved quite easily. Suppressing the ℓ and s indices we obtain the required matrix elements. First for the 1x1 blocks:

$$m_j = \ell + 1/2; \quad E = <m_\ell,m_s|\hat{\mathcal{H}}'|m_\ell,m_s> = <\ell,1/2|\hat{\mathcal{H}}'|\ell,1/2>$$

$$= \mu_B B<\ell,1/2|\hat{L}_z + 2\hat{S}_z|\ell,1/2> + \zeta<\ell,1/2|\hat{\underset{\sim}{L}}\cdot\hat{\underset{\sim}{S}}|\ell,1/2> = \mu_B B(\ell + 1) + \zeta\ell/2$$

Here we have made use of the fact that $\hat{\underset{\sim}{L}}\cdot\hat{\underset{\sim}{S}}$ can be written as

$$\hat{\underset{\sim}{L}}\cdot\hat{\underset{\sim}{S}} = \hat{L}_z\hat{S}_z + \frac{1}{2}(\hat{L}_+\hat{S}_- + \hat{L}_-\hat{S}_+) \tag{1}$$

where $\hat{L}_\pm$ and $\hat{S}_\pm$ clearly have no diagonal matrix elements.

$$m_j = -\ell - 1/2; \quad E = <-\ell,-1/2|\hat{\mathcal{H}}'|-\ell,-1/2> = -\mu_B B(\ell + 1) + \zeta\ell/2$$

For all other values of m_j four matrix elements are required. The first two are diagonal elements.

$$\langle m_j-1/2,+1/2|\hat{\mathcal{H}}'|m_j-1/2,+1/2\rangle = \mu_B B(m_j + 1/2) + \frac{\zeta}{2}(m_j - 1/2)$$

$$\langle m_j+1/2,-1/2|\hat{\mathcal{H}}'|m_j+1/2,-1/2\rangle = \mu_B B(m_j - 1/2) - \frac{\zeta}{2}(m_j + 1/2)$$

For the off-diagonal elements we find using Eq. (1):

$$\langle m_j-1/2,+1/2|\hat{\mathcal{H}}'|m_j+1/2,-1/2\rangle = \langle m_j-1/2,+1/2|\frac{\zeta}{2}(\hat{L}_+\hat{S}_- + \hat{L}_-\hat{S}_+)|m_j+1/2,-1/2\rangle$$

$$= \frac{\zeta}{2}\sqrt{(\ell + m_j + 1/2)(\ell - m_j + 1/2)}$$

$$\langle m_j+1/2,-1/2|\hat{\mathcal{H}}'|m_j-1/2,+1/2\rangle = \frac{\zeta}{2}\sqrt{(\ell - m_j + 1/2)(\ell + m_j + 1/2)}$$

Equation (3) of Problem 5.12 has also been used in this calculation.

For each value of m_j the required secular equation becomes:

$$\begin{matrix} m_s = +1/2 \\ \\ m_s = -1/2 \end{matrix}\begin{vmatrix} \mu_B B(m_j + 1/2) + \frac{\zeta}{2}(m_j - 1/2) - E & \frac{\zeta}{2}\sqrt{(\ell + m_j + 1/2)(\ell - m_j + 1/2)} \\ \frac{\zeta}{2}\sqrt{(\ell - m_j + 1/2)(\ell + m_j + 1/2)} & \mu_B B(m_j - 1/2) - \frac{\zeta}{2}(m_j + 1/2) - E \end{vmatrix} = 0$$

And the eigenvalues are easily found to be

$$E_\pm = \mu_B B m_j - \frac{\zeta}{4} \pm \frac{1}{2}\sqrt{\left(\frac{\zeta}{2} - 2\mu_B B m_j\right)^2 - \mu_B^2 B^2(4m_j^2 - 1) + 4\zeta\mu_B B m_j + \zeta^2\ell(\ell + 1)}$$

7.25 The valence electron of an alkali metal atom is excited to a p state.

(a) Into how many components is each of the levels split when a weak magnetic field B is applied?

(b) How large are the splittings in units of $\mu_B B$?

Solution: Ref. HZ1,106; Problem 7.23

(a) In the absence of the magnetic field there are two energy levels specified by ($\ell = 1$, $s = 1/2$, $j = 3/2$) and ($\ell = 1$, $s = 1/2$, $j = 1/2$). These are denoted by the term symbols $^2P_{3/2}$ and $^2P_{1/2}$, respectively. When the field B is applied the energy changes are given by (see Problem 7.23)

$$\Delta E = +g\mu_B B m_j$$

The $j = 3/2$ level ($^2P_{3/2}$) is split into four components since m_j takes on four values. The $j = 1/2$ level ($^2P_{1/2}$) is split into two components corresponding to $m_j = \pm 1/2$.

(b) The splittings depend on the value of $g(\ell, s, j)$. For the $^2P_{3/2}$ level:

$$g = \frac{2\ell + 1 + 1}{2\ell + 1} = \frac{4}{3}$$

For the $^2P_{1/2}$ level:

$$g = \frac{2\ell + 1 - 1}{2\ell + 1} = \frac{2}{3}$$

These splittings are shown schematically in Figure 7.25.

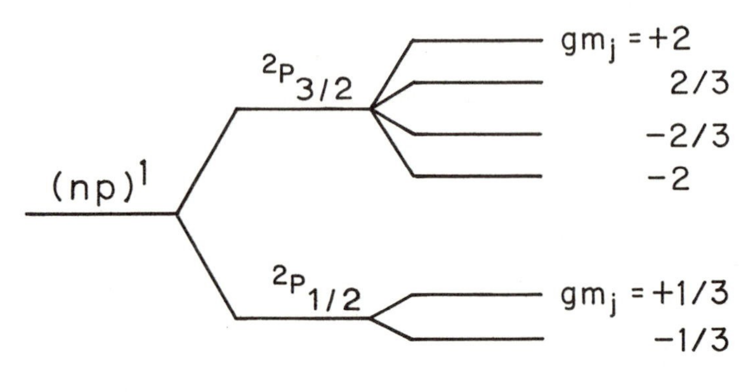

Fig. 7.25

7.26 Find the relativistic correction for the energy levels of hydrogen-like atoms using first order perturbation theory. Assume that $v << c$ and obtain the correction to the order of $(v/c)^2$. (Hint: Recall that $E^2 = p^2c^2 + m_o^2c^4$ gives the relativistic energy of the electron, determine the correction to the non-relativistic energy by expansion of E, and then apply perturbation theory.)

Solution: Ref. BS2,58; TH,249; CM2,324; Problem 7.42

Taking m_o as the rest mass of the electron we start with

$$E^2 = p^2c^2 + m_o^2c^4 \tag{1}$$

The positive energy solution of (1) permits us to write the Hamiltonian as

$$H = \sqrt{p^2c^2 + m_o^2c^4} - e\phi \tag{2}$$

where ϕ is the electrostatic potential. Since $(v/c)^2 << 1$ we expand Eq. (2) as follows:

$$H = m_oc^2\sqrt{\frac{p^2}{m_o^2c^2} + 1} - e\phi = m_oc^2\left(1 + \frac{p^2}{2m_o^2c^2} - \frac{p^4}{8m_o^4c^4} + \cdots\right) - e\phi$$

$$= m_oc^2 + \frac{p^2}{2m_o} - \frac{p^4}{8m_o^3c^2} + \cdots - e\phi$$

Choosing m_oc^2 as the reference energy we have:

$$W = (H - m_oc^2) = \left(\frac{p^2}{2m_o} - e\phi\right) - \frac{p^4}{8m_o^3c^2} = H^o + H^{(1)} \tag{3}$$

It is clear that H^o will give the non-relativistic eigenvalue problem for the hydrogen-like atom, and that $H^{(1)}$ is a small perturbation. Thus

$$\hat{\mathcal{H}}^o|n\ell m\rangle = E_n^o|n\ell m\rangle \tag{4}$$

$$E^{(1)} = \langle n\ell m|\hat{\mathcal{H}}^{(1)}|n\ell m\rangle = -\frac{1}{2m_oc^2}\langle n\ell m|\frac{\hat{p}^4}{4m_o^2}|n\ell m\rangle$$

The magnitude of $\hat{p}$ is determined primarily by H^o; therefore, using Eqs. (3) and (4) we write

$$\frac{p^2}{2m_o} = E_n^o + e\phi$$

and

$$E^{(1)} = -\frac{1}{2m_oc^2}\langle n\ell m|(E_n^o + e\phi)^2|n\ell m\rangle \tag{4}$$

Expressing both sides of Eq. (4) in atomic units (Rydbergs) gives

$$E^{(1)} = -\frac{1}{2m_oc^2}\left(\frac{2\pi^2 m_o e^4}{(4\pi\varepsilon_o)^2h^2}\right)\langle n\ell m|\left(-\frac{Z^2}{n^2}+\frac{2Z}{r}\right)^2|n\ell m\rangle$$

$$= -\frac{\alpha^2}{4}\left\{+\frac{Z^4}{n^4}-\frac{4Z^3}{n^2}\left\langle\frac{1}{r}\right\rangle + 4Z^2\left\langle\frac{1}{r^2}\right\rangle\right\}$$

where

$$\alpha = \frac{e^2}{(4\pi\varepsilon_o)\hbar c} \cong \frac{1}{137}$$

is the fine-structure constant. Finally, using the results of Problems 7.11 and 7.31 for $\langle r^{-n}\rangle$ we obtain

$$E^{(1)} = -\frac{\alpha^2}{4}\left\{+\frac{Z^4}{n^4}-\frac{4Z^4}{n^4}+\frac{4Z^4}{n^3(\ell+1/2)}\right\} = -\frac{\alpha^2Z^4}{n^3}\left\{\frac{1}{(\ell+1/2)}-\frac{3}{4n}\right\}$$

7.27 The radius of a nucleus can be approximated by the equation $R = 1.3\text{x}10^{-13}A^{1/3}$ (in cm) where A is the mass number. Use first order perturbation theory to estimate the shift in the ground state energy for a hydrogen-like atom that is expected when the finite extent of the nucleus is taken into account. In this calculation assume that the nucleus is a sphere of radius R_o having a uniform charge distribution and take $R_oZ < 0.01$ where R_o is measured in units of a_o. (Hint: Use the electrostatic potential given in Problem 1.5.)

Solution: Ref. CGT,213; CM2,207

The form of the potential changes at $r = R_o$. From Problem 1.5 we have for a nucleus having the charge +Ze

$$V_{r<R_o} = Z\left(\frac{r^2}{R_o^3}-\frac{3}{R_o}\right), \qquad V_{r>R_o} = -\frac{2Z}{r}$$

The perturbation resulting from the finite spatial extent of the nucleus is thus:

$$V' = Z\left(\frac{r^2}{R_o^3} - \frac{3}{R_o}\right) - \left(-\frac{2Z}{r}\right); \qquad 0 < r < R_o$$

and the energy correction for the ground state is

$$E^{(1)} = \int_o^{R_o} R_{100}^* V' R_{100} r^2 dr \qquad \text{where} \quad R_{n\ell m} = 2Z^{3/2} e^{-Zr}$$

Therefore

$$E^{(1)} = 4Z^3 \int_o^{R_o} e^{-2Zr}\left(\frac{r^2}{R_o^3} - \frac{3}{R_o} + \frac{2}{r}\right) r^2 dr$$

Since $R_o Z < 0.01$ the exponential factor can safely be replaced with unity and we find easily that:

$$E^{(1)} = \frac{4}{5} Z^4 R_o^2$$

7.28 In a magnetic field $\underset{\sim}{B}$ the Hamiltonian (neglecting spin) for a hydrogen atom has the form

$$\mathcal{H} = \frac{1}{2m} (\underset{\sim}{p} - e\underset{\sim}{A})^2 - \frac{e^2}{r(4\pi\varepsilon_o)} \tag{1}$$

where $\underset{\sim}{A}$ is the vector potential. Use first order perturbation theory to calculate the energy of the atom and the magnitude of its diamagnetic moment $\underset{\sim}{\mu}$ where $\mu = -\partial E/\partial B$. (Hint: Recall that $\underset{\sim}{A}$ can be chosen to be equal to $(\underset{\sim}{B}x\underset{\sim}{r})/2$ for a uniform magnetic field.)

Solution: Ref. S1,144

First we expand Eq. (1) so that the perturbation term can be extracted.

$$\hat{\mathcal{H}} = \frac{1}{2m} [\hat{p}^2 + e^2A^2 - e(\underset{\sim}{A}\cdot\hat{\underset{\sim}{p}} + \hat{\underset{\sim}{p}}\cdot\underset{\sim}{A})] - \frac{e^2}{r(4\pi\varepsilon_o)} = \hat{\mathcal{H}}_o + \hat{\mathcal{H}}_1$$

where

$$\hat{\mathcal{H}}_o = \frac{\hat{p}^2}{2m} - \frac{e^2}{r(4\pi\varepsilon_o)} \quad , \quad \hat{\mathcal{H}}_1 = \frac{e^2}{2m} A^2 - \frac{e}{2m} (\underset{\sim}{A}\cdot\hat{\underset{\sim}{p}} + \hat{\underset{\sim}{p}}\cdot\underset{\sim}{A})$$

The last term simplifies since $\underset{\sim}{\nabla}\cdot\underset{\sim}{A}f = \underset{\sim}{A}\cdot\underset{\sim}{\nabla}f + (\underset{\sim}{\nabla}\cdot\underset{\sim}{A})f$ and $\underset{\sim}{\nabla}\cdot\underset{\sim}{A} = 0$. Using $\underset{\sim}{A} = (\underset{\sim}{B}x\underset{\sim}{r})/2$, $\hat{\mathcal{H}}_1$ becomes:

$$\hat{\mathcal{H}}_1 = \frac{e^2}{8m} (\underset{\sim}{B}x\underset{\sim}{r})^2 - \frac{e}{4m} [(\underset{\sim}{B}x\underset{\sim}{r})\cdot\hat{\underset{\sim}{p}} + \hat{\underset{\sim}{p}}\cdot(\underset{\sim}{B}x\underset{\sim}{r})] = \frac{e^2}{8m} (\underset{\sim}{B}x\underset{\sim}{r})^2 - \frac{e}{2m} (\underset{\sim}{B}x\underset{\sim}{r})\cdot\underset{\sim}{p}$$

To first order the energy correction is

$$E_1^{(1)} = \frac{e^2}{8m} \langle 1s|(\underset{\sim}{B}x\underset{\sim}{r})^2|1s\rangle - \frac{e}{2m} \langle 1s|\underset{\sim}{B}\cdot\hat{\underset{\sim}{L}}|1s\rangle$$

For convenience we take $\underset{\sim}{B} = B\hat{k}$ and use the relations

$$\langle 1s|\hat{L}_z|1s\rangle = 0 \qquad \text{and} \qquad \underset{\sim}{B}x\underset{\sim}{r} = B(x\hat{j} - y\hat{i})$$

$$E_1^{(1)} = \frac{e^2B^2}{8m} \langle 1s|x^2 + y^2|1s\rangle = \frac{e^2B^2}{8m} \left(\frac{2}{3}\right)\langle 1s|r^2|1s\rangle = \frac{e^2B^2a_o^2}{4m}$$

The diamagnetic moment $\underset{\sim}{\mu}$ is defined so that

$$\mu = -\frac{\partial E_1^{(1)}}{\partial B} \quad \text{and} \quad \underset{\sim}{\mu} = -\frac{e^2 a_o^2}{2m}\underset{\sim}{B}$$

SUPPLEMENTARY PROBLEMS

7.29 Derive an equation for the <u>most probable value</u> of r for an electron in a hydrogen-like atom in the state described by the principal quantum number n and the azimuthal quantum number $\ell = n - 1$.

<u>Ans</u>: Ref. PW,132,140

$r = n^2/Z$ in units of a_o

7.30 Determine $\overline{r^{-1}_{n\ell m_\ell}}$ for hydrogenic orbitals. (Hint: Use the generating function approach outlined in Problems 7.9 and 7.10.)

<u>Ans</u>:

$$\overline{(r^{-1})}_{n\ell m_\ell} = Z/n^2$$

7.31 The following very useful recursion formula for evaluating $\overline{r^m_{n\ell m_\ell}}$ was given by Kramers:

$$Z^2\left[\frac{m+1}{n^2}\right]\overline{r^m} - (2m+1)Z\overline{r^{m-1}} + m(\ell + \frac{1}{2} + \frac{1}{2}m)(\ell + \frac{1}{2} - \frac{1}{2}m)\overline{r^{m-2}} = 0 \qquad (1)$$

Use this result to determine $\overline{r^{-3}}$ and $\overline{r^{-4}}$.

<u>Ans</u>: Ref. KR,251; Problem 7.11

$$\overline{r^{-3}} = \frac{Z\overline{r^{-2}}}{\ell(\ell+1)} = \frac{Z^3}{n^3\ell(\ell+1/2)(\ell+1)}$$

$$\overline{r^{-4}} = \frac{3Z\left[\dfrac{Z^3}{n^3\ell(\ell+1/2)(\ell+1)}\right] - \dfrac{Z^2}{n^2}\left[\dfrac{Z^2}{n^3(\ell+1/2)}\right]}{2(\ell-1/2)(\ell+3/2)} = \frac{\frac{3}{2}Z^4\left[1 - \dfrac{\ell(\ell+1)}{3n^2}\right]}{n^3(\ell-1/2)\ell(\ell+1/2)(\ell+1)(\ell+3/2)}$$

7.32 Calculate the probability that an electron in a hydrogen atom has r less than the Bohr radius when the electron is in the state described by $\psi(n = 2, \ell = 0, m_\ell = 0)$.

<u>Ans</u>:

0.176

7.33 Demonstrate that all of the 2p functions in Appendix 7 are eigenfunctions of $\hat{L}_z^2$, but that only the m = 0 function is an eigenfunction of $\hat{L}_z$.

7.34 Calculate the polarizability of the hydrogen atom by the use of perturbation theory through second order as indicated below:

(a) Obtain an expression for α by assuming that $(E_{1s}^o - E_k^o)$ can be replaced by E_{1s}^o for all $k \neq 1s$. What percentage of the total polarizability does this procedure account for?

(b) Obtain an expression for α in terms of n by using the integral formula:

$$\left[\int_o^\infty rP_{np}(r)P_{1s}(r)dr\right]^2 = a_o^2 2^8 n^7 (n - 1)^{2n-5}(n + 1)^{-2n-5}$$

from reference CS,133. What percentages of the polarizability are associated with each of the following states: 2p, 3p, 4p?

(Hint:

$$\cos\theta\Theta_{\ell,m} = \Theta_{\ell+1,m}\sqrt{\frac{(\ell + 1 - m)(\ell + 1 + m)}{(2\ell + 1)(2\ell + 3)}} + \Theta_{\ell-1,m}\sqrt{\frac{(\ell - m)(\ell + m)}{(2\ell - 1)(2\ell + 1)}}$$

see CS,53.)

<u>Ans</u>: Ref. PW,205; Problems 7.18, 7.19, and 7.41

(a) $\alpha = 4a_o^3(4\pi\varepsilon_o)$ (b) 66%, 9%, 3%

7.35 Show that $\langle 1s|z\hat{\mathcal{H}}z|1s\rangle = 0$ for the hydrogen atom where $\hat{\mathcal{H}}$ is the exact nonrelativistic Hamiltonian. (Hint: To avoid a tedious calculation try the substitution $\hat{\mathcal{H}}z = [\hat{\mathcal{H}}, z] + z\hat{\mathcal{H}}$.)

Ref. Problems 3.14 and 5.5

7.36 When a hydrogen atom is placed in a weak external electric field the wave function is distorted. A reasonable approximation to the ground state wave function is

$$u = (c_1 + c_2 z)u_{1s}$$

Use u as a variational function to obtain an estimate of the polarizability of the hydrogen atom.

<u>Ans</u>: Ref. PW,180; RA,285; Problem 6.17

$\alpha \cong 4a_o^3(4\pi\varepsilon_o)$

7.37 Use first order degenerate perturbation theory to calculate the energies of the components of the first excited energy state of the hydrogen atom in an external electric field. For this calculation take the perturbation to be $\hat{\mathcal{H}}' = +e\mathcal{E}z$.

<u>Ans</u>: Ref. PW,172

$E_2^{(1)} = 0, 0, \pm 3e\mathcal{E}a_o$

7.38 A hydrogen-like atom is placed in a very strong magnetic field where the splittings arising from the spin-orbit interaction are negligible compared to the magnetic (Zeeman) splittings. Obtain an equation for the energy of the state having the quantum numbers ℓ and s. This is called the Paschen-Back effect. (Hint: Use basis functions which diagonalize the Zeeman Hamiltonian and treat the spin-orbit interaction by first order perturbation theory.)

Ans:

$$E_Z + E_{SO} = \mu_B B(m_j + m_s) + \zeta(n\ell)m_\ell m_s$$

7.39 Consider a 2p electron in a magnetic field where the spin-orbit and Zeeman interactions have roughly the same magnitudes. Determine the six allowed energies for this electron in terms of the field strength B and the spin-orbit coupling constant ζ. (Hint: Use basis functions which are eigenfunctions of $\hat{L}_z$ and $\hat{S}_z$ to set up the secular equation.)

Ans: Ref. Problem 7.24; CS,152

$$m_j = 3/2; \quad E_1 = 2\,\mu_B B + \frac{\zeta}{2}$$

$$m_j = -3/2; \quad E_6 = -2\,\mu_B B + \frac{\zeta}{2}$$

$$m_j = +1/2; \quad E_{2,3} = \frac{\mu_B B}{2} - \frac{\zeta}{4} \pm \frac{1}{2}\sqrt{\left(\frac{\zeta}{2} - \mu_B B\right)^2 + 2(\mu_B B\zeta + \zeta^2)}$$

$$m_j = -1/2; \quad E_{4,5} = -\frac{\mu_B B}{2} - \frac{\zeta}{4} \pm \frac{1}{2}\sqrt{\left(\mu_B B + \frac{\zeta}{2}\right)^2 + 2(\zeta^2 - \mu_B B\zeta)}$$

7.40 Consider a hydrogen-like atom with Z >> 1. Use perturbation theory to derive an equation for the change in energy of each level when the perturbation $\hat{\mathcal{H}}' = -2/r$ is added. Here $\hat{\mathcal{H}}'$ is given in a.u. with energy in Rydbergs. Compare your answer with the exact result.

Ans: Ref. Problem 7.30

$$E_n^{(1)} = -\frac{2Z}{n^2}, \qquad E_n \cong -\frac{Z(Z+2)}{n^2}$$

The exact energy is $E_n = -\dfrac{(Z+1)^2}{n^2}$

7.41 The differential equation for the first order wave function in perturbation theory is

$$(\hat{\mathcal{H}}^o - E^o)u^{(1)} + (\hat{\mathcal{H}}' - E^{(1)})u^o = 0 \qquad (1)$$

(see the Introduction to Chapter 6). For a hydrogen atom in an electric field where $\hat{\mathcal{H}}' = +e\mathcal{E}r\cos\theta$, Eq. (1) can be solved exactly for $u^{(1)}$.

(a) Show that the function

$$u^{(1)} = \mathcal{E}\cos\theta e^{-r/a_o}\left(\frac{Ar}{a_o} + \frac{Br^2}{a_o}\right)$$

satisfies Eq. (1) if A and B are properly chosen.

(b) Use the results of part (a) to determine the second order energy $E^{(2)}$ and the polarizability exactly.

Ans: Ref. RA,281; Introduction to Chapter 6; Problems 7.18 and 7.19

(a) $A = 2B$, $B = a_o^2(4\pi\varepsilon_o)/(2e\sqrt{\pi a_o^3})$

(b) $E^{(2)} = -\frac{9}{4}a_o^3\mathcal{E}^2(4\pi\varepsilon_o)$, $\alpha = \frac{9}{2}a_o^3(4\pi\varepsilon_o)$

7.42 The first order relativistic correction for the ground state of the hydrogen atom is given by

$$E^{(1)} = -\frac{1}{2m_oc^2}\int_o^\infty \psi_o^*\left(\frac{\hat{p}^4}{4m_o^2}\right)\psi_o(4\pi r^2dr)$$

where $\psi_o = (\pi a_o^3)^{-1/2}\exp(-r/a_o)$. Evaluate the integral shown above by making use of the relation $\hat{p}^4 \rightarrow \hbar^4(\nabla^2)^2$. (Hint: Recall the relation $\nabla^2(r^{-1}) = -4\pi\delta(\underset{\sim}{r})$ where

$$\int_o^\varepsilon f(r)\delta(\underset{\sim}{r})r^2dr = f(0)/4\pi$$

for the spherically symmetric function f(r).)

Ans: Ref. Problem 7.26; M. Rendall and J. Arents, J. Chem. Phys. 49, 5366 (1968); PP,3

$E^{(1)} = -\frac{5}{8}\alpha^4m_oc^2$ where $\alpha = e^2/(4\pi\varepsilon_o\hbar c)$

CHAPTER 8

ELECTRONIC STRUCTURE OF ATOMS

In this chapter we consider atoms having two or more electrons. The simplest system which exhibits the repulsion between electrons is the He atom. We treat this system by means of the perturbation and variation theories. We then use the first excited manifold of states of He to illustrate the construction of antisymmetric wave functions.

The spectra of many-electron atoms can be qualitatively explained by vector coupling models for the angular momenta. The problems included here illustrate this method as well as dealing with the indistinguishability of equivalent electrons and the Pauli Exclusion Principle. Considerable attention is given to the determination of Russell-Saunders terms and to the application of the Condon-Slater rules to the determination of the energies of these terms.

1. Classification of Electronic States

An electronic configuration is specified by assigning values of n and ℓ to each electron, e.g. $(1s)^2$ or $(1s)^2(2p)^5$. We shall assume that for light atoms the spin-orbit interaction is small and that the states are characterized by the quantum numbers L, S, M_L, and M_S. This is equivalent to the assumption that the orbital angular momenta $\underset{\sim}{\ell}_i$ for the individual electrons couple tightly to give the resultant

$$\sum_i \underset{\sim}{\ell}_i = \underset{\sim}{L}$$

and that the spin angular momenta $\underset{\sim}{s}_i$ couple tightly to give the resultant

$$\sum_i \underset{\sim}{s}_i = \underset{\sim}{S}$$

while $\underset{\sim}{L}$ and $\underset{\sim}{S}$ are only weakly coupled to give the resultant $\underset{\sim}{L} + \underset{\sim}{S} = \underset{\sim}{J}$.

The electronic states of atoms having low Z are specified by term symbols of the form $^{2S+1}\mathcal{L}_J$ where S and J are quantum numbers and $\mathcal{L}$ stands for S (for L = 0), P (for L = 1), D (for L = 2), $\cdots$, e.g. L = 2, S = 1, and J = 2 implies the symbol $^3D_2$. This scheme is called LS or Russell-Saunders coupling and the symbols are often called Russell-Saunders term symbols. The L and S values define a multiplet of closely spaced energy levels having different values of J. In determining terms for a configuration the following facts are useful:

(i) A shell is specified by n and ℓ. It can accommodate $2(2\ell + 1)$ electrons.

Electrons in the same shell are equivalent.

(ii) If a shell can accommodate N electrons, the terms arising when $n < N$ electrons are present are the same as those arising when $(N - n)$ electrons are present. When $n > N/2$, the resulting multiplets are said to be inverted.

(iii) If a shell is completely filled, *i.e.* contains $2(2\ell + 1)$ electrons, it is called a closed shell. The only possible term for a closed shell is then $^1S_0$.

Hund has formulated a set of empirical rules which can be used to predict the term of lowest energy for equivalent electrons. They are:

(i) Of the terms arising for equivalent electrons, the one with the greatest multiplicity has the lowest energy. Of the terms with the maximum multiplicity the one with the greatest value of L lies lowest.

(ii) In a multiplet the term with smallest J lies lowest in energy if the shell is less than half filled (regular). The reverse is true for shells that are more than half filled (inverted).

These rules are consistent with approximate quantum mechanical calculations.

2. Matrix Elements between Determinantal Functions

In quantum chemistry we frequently must evaluate matrix elements of the form:

$$H_{k\ell} = \langle \Delta_k | \hat{\mathcal{H}} | \Delta_\ell \rangle$$

where Δ represents the Slater determinant:

$$\Delta = \frac{1}{\sqrt{N!}} \begin{vmatrix} u_1(1) & u_1(2) & \cdots & u_1(N) \\ u_2(1) & & & \\ \vdots & & & \\ u_N(1) & \cdot & \cdot \quad \cdot \quad \cdot & u_N(N) \end{vmatrix}$$

Here the functions u_i are spin-orbitals which are products of spatial orbitals and spin functions. Condon and Slater have given powerful rules for the reduction of such matrix elements to integrals over the spin-orbitals. Let $\hat{\mathcal{H}}$ be a summation of one-electron operators f_i and two-electron operators g_{ij} which are independent of spin.

$$\hat{\mathcal{H}} = \sum_i f_i + \sum_{i<j} g_{ij}$$

The determinantal functions Δ_k and Δ_ℓ are first brought into maximum coincidence by interchanging the columns of Δ_ℓ (each interchange changes the sign of Δ_ℓ). If the spin-orbitals in Δ_k are denoted by u_i and those in Δ_ℓ by u_i' the rules take the following form:

Δ_k and Δ_ℓ differ by:	$\langle\Delta_k\|\sum_{i=1} f_i\|\Delta_\ell\rangle$	$\langle\Delta_k\|\sum_{i<j} g_{ij}\|\Delta_\ell\rangle$
no spin orbitals	$\sum_i \langle i\|f\|i\rangle$	$\sum_{i>j} [\langle ij\|g\|ij\rangle - \langle ij\|g\|ji\rangle]$
one spin orbital	$\langle i\|f\|i'\rangle$	$\sum_{j\neq i} [\langle ij\|g\|i'j\rangle - \langle ij\|g\|ji'\rangle]$
two spin orbitals ($u_i \neq u_i'$, $u_j \neq u_j'$)	0	$\langle ij\|g\|i'j'\rangle - \langle ij\|g\|j'i'\rangle$
more than two spin orbitals	0	0

where:

$$\langle i|f|i\rangle = \int u_i^*(1) f_1 u_i(1) dv_1$$

$$\langle ij|g|rs\rangle = \int u_i^*(1) u_j^*(2) g_{12} u_r(1) u_s(2) dv_1 dv_2$$

It should be emphasized that u_i and u_j are spin orbitals. An integral of the type $\langle ij|g|rs\rangle$ will vanish if the spin parts of orbitals i and r differ or if the spin parts of j and s differ. The spatial parts of these orbitals are also assumed to be orthonormal.

For the evaluation of integrals involving g_{12} the reader is referred to the discussion given by Slater (S1, S3). In general these integrals must be written as a sum of radial integrals which arise from the expansion of r_{12}^{-1} (see Problem 8.1). However, in the special case that only s-electrons are involved the sum reduces to a single term. In this case the shorthand notation, e.g. $J(1s, 2s) = \langle 1s(1)2s(2)|2r_{12}^{-1}|1s(1)2s(2)\rangle$ is permissible, but it should be realized that in general such integrals depend on the m_ℓ's as well as n and m_s.

General References: HZ1, S1, S2, S3, CS

PROBLEMS

8.1 Consider two electrons at the positions specified by $\mathbf{r}_i$ and $\mathbf{r}_j$ and separated by the distance r_{ij} as shown in Figure 8.1.

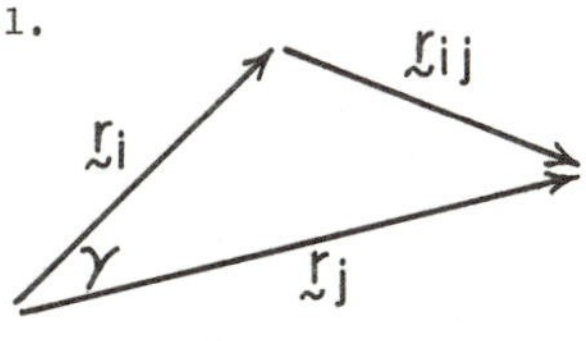

Fig. 8.1

(a) Show that $r_{ij} = r_i\sqrt{x^2 + 1 - 2x\cos\gamma}$ where $x = r_j/r_i$.

(b) Let $\xi = \cos\gamma$ and show that the coefficients $a_n(\xi)$ in the expansion

$$(x^2 + 1 - 2x\xi)^{-1/2} = \sum_{n=0}^{\infty} a_n(\xi)x^n, \quad |x| < 1 \tag{1}$$

are equal to the Legendre polynomials $P_n(\xi)$. (Hint: Differentiate Eq. (1) with respect to x and search for a recursion relation connecting a_{n+1} with a_n and a_{n-1}.)

Solution: Ref. PW,446; MM1,98; ME,183; Appendix 4

(a) From Fig. 8.1 we have $\underset{\sim}{r}_{ij} = \underset{\sim}{r}_j - \underset{\sim}{r}_i$ and thus

$$r_{ij}^2 = r_j^2 + r_i^2 - 2\underset{\sim}{r}_j \cdot \underset{\sim}{r}_i = r_j^2 + r_i^2 - 2r_j r_i \cos\gamma$$

which is known as the law of cosines. Factoring r_i^2 out of this expression and taking the square root gives

$$r_{ij} = r_i\sqrt{x^2 + 1 - 2x\cos\gamma}$$

(b) Differentiating Eq. (1) with respect to x we obtain

$$(\xi - x)(x^2 + 1 - 2x\xi)^{-3/2} = \sum_{n=0}^{\infty} na_n x^{n-1} \tag{2}$$

Then using Eq. (1) in Eq. (2) gives

$$(\xi - x)\sum_{n=0} a_n x^n = (x^2 + 1 - 2x\xi)\sum_{n=0} na_n x^{n-1} \tag{3}$$

By equating the coefficients of x^n on the two sides of Eq. (3) we find:

$$(n + 1)a_{n+1} = \xi(2n + 1)a_n - na_{n-1} \tag{4}$$

Equations (1) and (2) with $x = 0$ show that $a_o = 1$ and $a_1 = \xi$, respectively. These quantities and Eq. (4) establish all of the a_n and serve to identify $a_n(\xi)$ with $P_n(\xi)$.

An alternative approach is to expand $F(x, \xi) = (x^2 + 1 - 2x\xi)^{-1/2}$ from Eq. (1) in a Taylor's series in x about $x = 0$. Repeated differentiation of $F(x, \xi)$ with respect to x followed by setting $x = 0$ gives:

$$F(0, \xi) = 1, \quad \left(\frac{dF}{dx}\right)_o = \xi, \quad \left(\frac{d^2F}{dx^2}\right)_o = 3\xi^2 - 1, \cdots, \quad \left(\frac{d^nF}{dx^n}\right)_o = n!P_n(\xi)$$

Thus

$$F(x, \xi) = \sum_{n=0}^{\infty} \frac{1}{n!}\left(\frac{d^nF}{dx^n}\right)_o x^n = \sum_{n=0}^{\infty} P_n(\xi)x^n$$

8.2 Evaluate the two-electron integral

$I = \langle\phi_A(1)\phi_B(2)|r_{12}^{-1}|\phi_C(1)\phi_D(2)\rangle$

in terms of the atomic number Z for the special case that $\phi_A = \phi_B = \phi_C = \phi_D = u_{1s}$ by making use of the expansion for r_{12}^{-1} given in Problem 8.1. You will find it convenient to use the addition theorem for spherical harmonics which states that

$$P_\ell(\cos\gamma) = \sum_{m=-\ell}^{+\ell} \frac{(\ell - |m|)!}{(\ell + |m|)!} P_\ell^{|m|}(\cos\theta_1) P_\ell^{|m|}(\cos\theta_2) e^{im(\phi_1-\phi_2)} \tag{1}$$

The angles θ_1, ϕ_1 and θ_2, ϕ_2 are the polar coordinates of the position vectors $\underset{\sim}{r}_1$ and $\underset{\sim}{r}_2$ and γ is defined in Fig. 8.1. Here u_{1s} is the hydrogenic orbital $(Z^{3/2}/\sqrt{\pi})e^{-Zr}$.

Solution: Ref. EWK,369; TH,298; KP,174; Problem 8.1

$$I = \iint u_{1s}^2(r_1)u_{1s}^2(r_2)\left[\frac{1}{r_>}\sum_{\ell=0}^{\infty}\left(\frac{r_<}{r_>}\right)^\ell P_\ell(\cos\gamma)\right] dv_1 dv_2$$

where $r_<$ is the smaller and $r_>$ is the larger of r_1 and r_2. Introducing Eq. (1) gives:

$$I = \sum_{\ell=0}^{\infty}\sum_{m=-\ell}^{+\ell}\frac{(\ell-|m|)!}{(\ell+|m|)!}\int_0^\infty\int_0^\infty u_{1s}^2(r_1)u_{1s}^2(r_2)\frac{(r_<)^\ell}{(r_>)^{\ell+1}} r_1^2 r_2^2 dr_1 dr_2$$

$$\cdot \int_0^\pi P_\ell^{|m|}(\cos\theta_1)\sin\theta_1 d\theta_1 \int_0^\pi P_\ell^{|m|}(\cos\theta_2)\sin\theta_2 d\theta_2 \cdot \int_0^{2\pi} e^{im\phi_1}d\phi_1 \int_0^{2\pi} e^{-im\phi_2}d\phi_2 \tag{2}$$

Equation (2) can be simplified by noting that

$$\int_0^{2\pi} e^{im\phi}d\phi = 2\pi\delta_{mo}; \qquad \int_0^\pi P_\ell(\cos\theta)\sin\theta d\theta = 2\delta_{\ell o}$$

where $\delta_{\ell o} = 1$ for $\ell = 0$ and 0 otherwise. Therefore

$$I = (4\pi)^2\int_0^\infty\int_0^\infty u_{1s}^2(r_1)u_{1s}^2(r_2)\frac{1}{r_>} r_1^2 r_2^2 dr_1 dr_2$$

$$= (4\pi)^2\int_0^\infty\left[\int_0^{r_1}\frac{r_2^2}{r_1}u_{1s}^2(r_2)dr_2 + \int_{r_1}^\infty\frac{r_2^2}{r_2}u_{1s}^2(r_2)dr_2\right]u_{1s}^2(r_1)r_1^2 dr_1 \tag{3}$$

From Appendix 7 we find that

$$u_{1s}(r) = \frac{Z^{3/2}}{\sqrt{\pi}} e^{-Zr}$$

in atomic units. Therefore, the required integrals are:

$$\frac{1}{r_1}\int_0^{r_1} r^2 e^{-2Zr}dr = -\frac{e^{-2Zr_1}}{r_1}\left(\frac{r_1^2}{2Z} + \frac{r_1}{2Z^2} + \frac{1}{4Z^3}\right) + \frac{1}{4Z^3 r_1}; \quad \int_{r_1}^\infty re^{-2Zr}dr = e^{-2Zr_1}\left(\frac{r_1}{2Z} + \frac{1}{4Z^2}\right)$$

Combining these integrals and substituting into Eq. (3) gives:

$$I = 4Z^3 \int_0^\infty r_1^2 e^{-2Zr_1} \left\{ \frac{1}{r_1} [-e^{-2Zr_1}(Zr_1 + 1) + 1] \right\} dr_1 \tag{4}$$

Then using the integral formula

$$\int_0^\infty x^n e^{-ax} dx = \frac{n!}{a^{n+1}}$$

we find that $I = \frac{5}{8} Z$.

8.3 Set up Schrödinger's equation for the He atom and use the variation method to approximate the energy of the ground state. In this calculation take $u = Ae^{c(r_1+r_2)}$ for the trial function with c as a parameter to be determined by variation.

Solution: Ref. EWK,101; KP,179

The Hamiltonian is

$$\hat{\mathcal{H}} = -\nabla_1^2 - \nabla_2^2 - \frac{2Z}{r_1} - \frac{2Z}{r_2} + \frac{2}{r_{12}}$$

in atomic units and Z = 2 for He. The trial function suggested depends only on r_1 and r_2 and can be written as

$$u = \sqrt{A}\, e^{-cr_1} \sqrt{A}\, e^{-cr_2} = R(1)R(2)$$

Normalization of the radial functions gives

$$\int_0^\infty R^2(1) r_1^2 dr_1 = \int_0^\infty R^2(2) r_2^2 dr_2 = \int_0^\infty Ae^{-2cr} r^2 dr = \frac{A2!}{(2c)^3} = 1$$

or $A = 4c^3$. The integral, which must be minimized, is

$$W = \int u^* \hat{\mathcal{H}} u dv_1 dv_2 = \sum_{i=1}^{2} \int_0^\infty P^*(i) \left[-\frac{d^2}{dr_i^2} - \frac{4}{r_i} \right] P(i) dr_i$$

$$+ \int_0^\infty \int_0^\infty P^*(1)P^*(2) \left[\frac{2}{r_{12}} \right] P(1)P(2) dr_1 dr_2 = \sum_{i=1}^{2} I_i + F_{12} \tag{1}$$

Here we have introduced $P(i) = r_i R(i)$, and in the first integral on the right side of Eq. (1) have used the radial operator with $\ell = 0$ from Eq. (2) in Problem 7.4.

Evaluating I_i we obtain:

$$I_i = \int_0^\infty 4c^3 r_i e^{-cr_i} [2ce^{-cr_i} - c^2 r_i e^{-cr_i} - 4e^{-cr_i}] dr_i$$

$$= 4c^3 [(2c - 4)\left(\frac{1}{2c}\right)^2 - c^2 \left(\frac{1}{2c}\right)^3 2] = c^2 - 4c \tag{2}$$

The two-electron integral is:

$$F_{12} = 16c^6 \int_0^{\infty} e^{-2cr_1} \left(\frac{2}{r_{12}}\right) e^{-2cr_2} r_1^2 dr_1 r_2^2 dr_2$$

This integral is equal to 2 times the integral I of Eq. (3) in Problem 8.2 if Z is replaced by c. Therefore:

$$F_{12} = 5c/4 \tag{3}$$

Combining Eqs. (2) and (3) we obtain

$$W = \sum_{i=1}^{2} I_i + F_{12} = 2(c^2 - 4c) + 1.25c = 2c^2 - 6.75c$$

The minimum value of W occurs when

$$\frac{\partial W}{\partial c} = 4c - 6.75 = 0 \quad \text{or} \quad c = 1.6875$$

The inequality $\partial^2 W/\partial c^2 > 0$ confirms that this is a minimum. Therefore, $W_{min} = 2(1.6875)^2 - 6.75(1.6875) = -5.695$ and the predicted energy is equal to -77.48 eV. This is 2.68 eV lower than the result for unshielded perturbation theory and only 1.52 eV above the experimental value of -79.0 eV. The error results primarily from the fact that this calculation does not take into account the correlated motion of the electrons.

8.4 According to the first order perturbation treatment an electron in the He atom (call it electron 1) moves under the influence of a nuclear charge Z = 2 and the time average charge distribution of the other electron. The wave equation for electron 1 is:

$$[-\nabla_1^2 - 2Z(\text{eff})/r_1]u(1) = Eu(1)$$

where

$$-\frac{2Z(\text{eff})}{r_1} = -\frac{2Z}{r_1} + V(r_1)$$

(a) Obtain an expression for V(r).

(b) Plot Z(eff) with Z = 2 in the range r = 0 to 2 and discuss the results.

<u>Solution</u>: Ref. Problems 8.2, 8.25

(a) The energy of repulsion of the two electrons in He is equal to $2I = \langle 1s(1)1s(2)|2/r_{12}|1s(1)1s(2)\rangle$ according to the perturbation theory. In Eqs. (3) and (4) of Problem 8.2 we find that this integral has the form:

$$2I = \int_0^{\infty} V(r_1)u_{1s}^2(r_1)4\pi r_1^2 dr_1$$

where

$$V(r_1) = \frac{2}{r_1}[1 - e^{-2Zr_1}(1 + Zr_1)]$$

This function gives the potential of electron 1 in the field of electron 2. The total potential is $-(2Z/r_1) + V(r_1)$.

(b) By definition

$$Z(\text{eff}) = Z - \frac{r_1}{2} V(r_1) = 1 + e^{-4r_1}(1 + 2r_1)$$

This function is shown in Fig. 8.4. For very small values of r_1, electron 2 provides little shielding and electron 1 "sees" the full nuclear charge. However, for $r_1 > 1$ the shielding (or screening) is essentially complete and electron 1 experiences only one unit of charge.

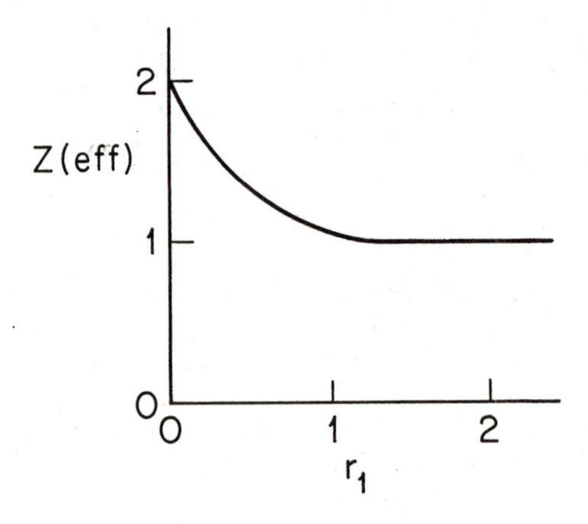

Fig. 8.4

8.5 The first manifold of excited states for the He atom is associated with the electronic transitions

$$(1s)^1(2s)^1,\ (1s)^1(2p_{x,y,z})^1 \leftarrow (1s)^2$$

Set up simple product functions using hydrogen-like orbitals to represent the eight excited states, which are degenerate for noninteracting electrons, and use perturbation theory to derive expressions for the energies of these states.

Solution: Ref. PW,210; RA,342; KP,187

The product functions which represent the eight ways that the two electrons can be assigned to the 1s, 2s and 2p orbitals in the singly excited state are:

$u_1 = 1s(1)2s(2)$ $\qquad$ $u_5 = 1s(1)2p_y(2)$

$u_2 = 1s(2)2s(1)$ $\qquad$ $u_6 = 1s(2)2p_y(1)$

$u_3 = 1s(1)2p_x(2)$ $\qquad$ $u_7 = 1s(1)2p_z(2)$

$u_4 = 1s(2)2p_x(1)$ $\qquad$ $u_8 = 1s(2)2p_z(1)$

The Hamiltonian for He is $\hat{\mathcal{H}} = -\nabla_1^2 - 4/r_1 - \nabla_2^2 - 4/r_2 + 2/r_{12} = \hat{\mathcal{H}}^o + \hat{\mathcal{H}}'$ and we find that

$$\hat{\mathcal{H}}^o u_i(1,2) = \hat{\mathcal{H}}^o \psi_{n_1\ell_1 m_{\ell_1}}(1)\psi_{n_2\ell_2 m_{\ell_2}}(2) = -4(n_1^{-2} + n_2^{-2})u_i = -5u_i$$

since $n_1 = 1$ and $n_2 = 2$ for the first excited state.

Degenerate perturbation theory requires a secular equation of the form $\det|\mathcal{H}'_{ij} - E_n' S_{ij}| = 0$. The required integrals are:

$$\left.\begin{aligned} J_s &= \langle 1s(1)2s(2)|2/r_{12}|1s(1)2s(2)\rangle \\ J_p &= \langle 1s(1)2p(2)|2/r_{12}|1s(1)2p(2)\rangle \end{aligned}\right\} \text{Coulomb integrals}$$

$$\left.\begin{aligned} K_s &= \langle 1s(1)2s(2)|2/r_{12}|1s(2)2s(1)\rangle \\ K_p &= \langle 1s(1)2p(2)|2/r_{12}|1s(2)2p(1)\rangle \end{aligned}\right\} \text{Exchange integrals}$$

where p here represents p_x, p_y, or p_z. All of the other integrals vanish because of symmetry. For example

$$\mathcal{H}'_{23} = \langle u_2|2/r_{12}|u_3\rangle = \langle 2s(1)1s(2)|2/r_{12}|1s(1)2p_x(2)\rangle = 0$$

since the integrand is an odd function of x. The secular equation is:

$$\begin{vmatrix} \begin{matrix} J_s-E' & K_s \\ K_s & J_s-E' \end{matrix} & 0 & 0 & 0 \\ 0 & \begin{matrix} J_p-E' & K_p \\ K_p & J_p-E' \end{matrix} & 0 & 0 \\ 0 & 0 & \begin{matrix} J_p-E' & K_p \\ K_p & J_p-E' \end{matrix} & 0 \\ 0 & 0 & 0 & \begin{matrix} J_p-E' & K_p \\ K_p & J_p-E' \end{matrix} \end{vmatrix} = 0$$

The J_s, K_s block gives $(J_s-E')^2 - K_s^2 = 0$; $E' = J_s \pm K_s$ and the associated eigenfunctions are found to be

$$\psi_{1,2} = \frac{1}{\sqrt{2}}[1s(1)2s(2) \pm 1s(2)2s(1)]$$

The second diagonal block gives the roots $E' = J_p \pm K_p$ with the associated eigenfunctions

$$\psi_{3,4} = \frac{1}{\sqrt{2}}[1s(1)2p_x(2) \pm 1s(2)2p_x(1)]$$

Thus, this procedure leads to 8 states, 3 pairs of which are degenerate. It turns out that $J_p > J_s$ and all of the J's and K's are positive. Figure 8.5 summarizes the calculation.

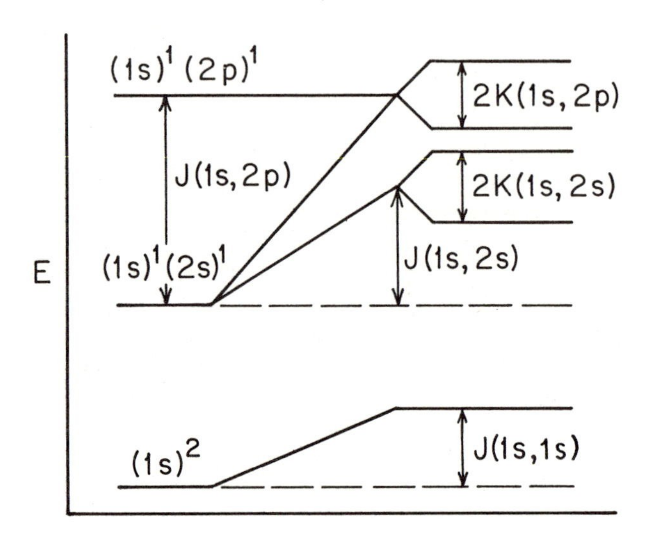

Fig. 8.5

8.6 The Pauli exclusion principle states that the wave functions of a many-electron system must be antisymmetric with respect to interchange of the coordinates of any pair of electrons.

(a) Set up antisymmetric wave functions for the 1s2s configuration of the He atom. In doing this make use of products of the spatial functions given in Problem 8.5 with the appropriate two-electron spin functions.

(b) Show that each of the antisymmetric functions of part (a) can be expressed as a linear combination of determinants of the form:

$$\frac{1}{\sqrt{2}}\begin{vmatrix} u_1(1) & u_1(2) \\ u_2(1) & u_2(2) \end{vmatrix}$$

Here $u_i(1)$ is a spin orbital, i.e. a product of a spatial orbital and a spin function. For example $u_1(1)$ might be $1s\alpha(1)$.

Solution: Ref. PW,214; KP,187; Problems 5.22 and 8.5

(a) From Problem 8.5 we have the spatial functions:

$$\psi_1 = \frac{1}{\sqrt{2}}\,[1s(1)2s(2) + 1s(2)2s(1)]$$

$$\psi_2 = \frac{1}{\sqrt{2}}\,[1s(1)2s(2) - 1s(2)2s(1)]$$

The function ψ_1 is symmetric with respect to exchange of the electrons and ψ_2 is antisymmetric. Thus, if P_{12} is an operator which exchanges the coordinates of electrons 1 and 2:

$$P_{12}\psi_1 = P_{12}\frac{1}{\sqrt{2}}[1s(1)2s(2) + 1s(2)2s(1)] = \frac{1}{\sqrt{2}}[1s(2)2s(1) + 1s(1)2s(2)] = \psi_1$$

$$P_{12}\psi_2 = \frac{1}{\sqrt{2}}[1s(2)2s(1) - 1s(1)2s(2)] = -\psi_2$$

Using α and β for the spin functions $|m_s = +1/2\rangle$ and $|m_s = -1/2\rangle$, respectively, the possible product spin functions are: $\alpha(1)\alpha(2)$, $\alpha(1)\beta(2)$, $\alpha(2)\beta(1)$, and $\beta(1)\beta(2)$. While $\alpha(1)\alpha(2)$ and $\beta(1)\beta(2)$ are symmetric with respect to electron exchange, $\alpha(1)\beta(2)$ and $\alpha(2)\beta(1)$ are neither symmetric nor antisymmetric. However, it is possible to construct normalized, symmetric and antisymmetric combinations of these functions. Thus, we adopt the set:

$$\chi_1 = \alpha(1)\alpha(2) \qquad M_s = +1$$

$$\chi_2 = \frac{1}{\sqrt{2}}[\alpha(1)\beta(2) + \alpha(2)\beta(1)] \qquad 0$$

$$\chi_3 = \frac{1}{\sqrt{2}}[\alpha(1)\beta(2) - \alpha(2)\beta(1)] \qquad 0$$

$$\chi_4 = \beta(1)\beta(2) \qquad -1$$

Comparison with Problem 5.22 shows that all of the χ's are eigenfunction of $\hat{S}^2$. The functions χ_1, χ_2, and χ_4 belong to a triplet state with $S = 1$, and χ_3 is a singlet with $S = 0$.

Antisymmetric products of the ψ's and χ's can now be constructed according to the rule:

$$\Psi_{\text{antisymm}} = \begin{cases} \psi(\text{symm})\chi(\text{antisymm}) \\ \psi(\text{antisymm})\chi(\text{symm}) \end{cases}$$

We thus obtain:

$${}^1S:\quad \Psi(S = 0, M_s = 0) = \frac{1}{\sqrt{2}}[1s(1)2s(2) + 1s(2)2s(1)]\frac{1}{\sqrt{2}}[\alpha(1)\beta(2) - \alpha(2)\beta(1)]$$

$${}^3S:\quad \Psi(S = 1, M_s = 1) = \frac{1}{\sqrt{2}}[1s(1)2s(2) - 1s(2)2s(1)]\alpha(1)\alpha(2)$$

$$\Psi(S = 1, M_s = 0) = \frac{1}{\sqrt{2}}[1s(1)2s(2) - 1s(2)2s(1)]\frac{1}{\sqrt{2}}[\alpha(1)\beta(2) + \alpha(2)\beta(1)]$$

$$\Psi(S = 1, M_s = -1) = \frac{1}{\sqrt{2}}[1s(1)2s(2) - 1s(2)2s(1)]\beta(1)\beta(2)$$

(b) Consider, for example, the ${}^1S$ state:

$$\Psi({}^1S) = \frac{1}{\sqrt{2}}[1s\alpha(1)2s\beta(2) - 1s\alpha(2)2s\beta(1)] - \frac{1}{\sqrt{2}}[1s\beta(1)2s\alpha(2) - 1s\beta(2)2s\alpha(1)]$$

$$= \frac{1}{\sqrt{2}}\begin{vmatrix} 1s\alpha(1) & 1s\alpha(2) \\ 2s\beta(1) & 2s\beta(2) \end{vmatrix} - \frac{1}{\sqrt{2}}\begin{vmatrix} 1s\beta(1) & 1s\beta(2) \\ 2s\alpha(1) & 2s\alpha(2) \end{vmatrix} = (1s\overline{2s}) - (\overline{1s}2s)$$

where the bar indicates β spin and no bar indicates α spin. By similar arguments

$$\Psi(^3S) = \begin{cases} (1s2s) \\ (1s\overline{2s}) + (\overline{1s}2s) \\ (\overline{1s}\overline{2s}) \end{cases}$$

Analogous expressions exist for the $\Psi(^1P)$ and $\Psi(^3P)$ states of the 1s2p configuration.

8.7 An antisymmetric linear combination of product functions of the type

$$\psi = \prod_{i=1}^{N} u_i(i)$$

can be constructed by operating on ψ with the "antisymmetrizer"

$$\hat{A} = \frac{1}{\sqrt{N!}} \sum_p (-1)^p \hat{P} \tag{1}$$

where $\hat{P}$ is a permutation which may be regarded as the result of p simple interchanges of pairs of orbital indices or electron labels but not both. The summation is over all permutations. Use $\hat{A}$ to generate the normalized, antisymmetric functions for the $1s\overline{1s}2s$ configuration of Li, and thus show that this function can be expressed as the Slater determinant:

$$\frac{1}{\sqrt{3!}} \begin{vmatrix} 1s\alpha(1) & 1s\alpha(2) & 1s\alpha(3) \\ 1s\beta(1) & 1s\beta(2) & 1s\beta(3) \\ 2s\alpha(1) & 2s\alpha(2) & 2s\alpha(3) \end{vmatrix} \tag{2}$$

Solution: Ref. PI1,278; KP,182

We choose the initial spin orbital to be $1s\alpha(1)1s\beta(2)2s\alpha(3)$ and then operate with $\hat{A}$.

$$\hat{A}[1s\alpha(1)1s\beta(2)2s\alpha(3)] = \frac{1}{\sqrt{3!}} \sum_p (-1)^p \hat{P}[1s\alpha(1)1s\beta(1)2s\alpha(3)]$$

$$= \frac{1}{\sqrt{3!}} [1s\alpha(1)1s\beta(2)2s\alpha(3) + 2s\alpha(1)1s\alpha(2)1s\beta(3) + 1s\beta(1)2s\alpha(2)1s\alpha(3)$$

$$- 1s\alpha(1)2s\alpha(2)1s\beta(3) - 1s\beta(1)1s\alpha(2)2s\alpha(3) - 2s\alpha(1)1s\beta(2)1s\alpha(3)] \tag{3}$$

The sum has 3! = 6 terms since that is the number of ways three distinguishable objects can be arranged in order. The right side of Eq. (3) is exactly equal to the determinant in Eq. (2) and, in fact, Eq. (1) is the definition for the expansion of a determinant.

8.8 Derive an equation for the expectation value of the energy of the $(1s)^2 2s$ configuration of Li in terms of the integrals:

$$J(1s, 2s) = \langle 1s(1)2s(2)|2r_{12}^{-1}|1s(1)2s(2)\rangle$$

$$K(1s, 2s) = \langle 1s(1)2s(2)|2r_{12}^{-1}|2s(1)1s(2)\rangle$$

In this derivation use a single Slater determinant to represent the state $(1s)^2 2s$. This determinant should be written out in full and the expectation value expression then multiplied out and simplified.

Solution: Ref. LE1,274; CS,169

We wish to evaluate the integral

$$E = \langle \hat{\mathcal{H}} \rangle = \langle \Psi | \sum_{i=1}^{3} \hat{\mathcal{H}}^{o}(i) + \hat{\mathcal{H}}' | \Psi \rangle$$

where

$$\hat{\mathcal{H}}^{o}(i) = -\nabla_i^2 - 6/r_i \quad \text{and} \quad \hat{\mathcal{H}}' = \sum_{i<j}^{3} 2/r_{ij}$$

in a.u. and

$$\Psi = \frac{1}{\sqrt{3!}} \begin{vmatrix} 1s\alpha(1) & 1s\alpha(2) & 1s\alpha(3) \\ 1s\beta(1) & 1s\beta(2) & 1s\beta(3) \\ 2s\alpha(1) & 2s\alpha(2) & 2s\alpha(3) \end{vmatrix}$$

Written out in full the one-electron integrals are:

$$\begin{aligned} \frac{1}{6} \langle & 1s\alpha(1)1s\beta(2)2s\alpha(3) + 2s\alpha(1)1s\alpha(2)1s\beta(3) + 1s\beta(1)2s\alpha(2)1s\alpha(3) - 1s\alpha(1)2s\alpha(2)1s\beta(3) \\ & - 1s\beta(1)1s\alpha(2)2s\alpha(3) - 2s\alpha(1)1s\beta(2)1s\alpha(3) | \hat{\mathcal{H}}^{o}(1) + \hat{\mathcal{H}}^{o}(2) + \hat{\mathcal{H}}^{o}(3) | 1s\alpha(1)1s\beta(2)2s\alpha(3) \\ & + 2s\alpha(1)1s\alpha(2)1s\beta(3) + 1s\beta(1)2s\alpha(2)1s\alpha(3) - 1s\alpha(1)2s\alpha(2)1s\beta(3) \\ & - 1s\beta(1)1s\alpha(2)2s\alpha(3) - 2s\alpha(1)1s\beta(2)1s\alpha(3) \rangle \end{aligned} \qquad (1)$$

Using the orthogonality conditions for the space and spin function, i.e.

$$\langle u_i \chi_k(1) | u_j \chi_\ell(1) \rangle = \langle u_i(1) | u_j(1) \rangle \langle \chi_k(1) | \chi_\ell(1) \rangle = \delta_{ij} \delta_{k\ell}$$

we can reduce (1) to:

$$\frac{1}{6} \sum_{i=1}^{3} [4\langle 1s(i) | \hat{\mathcal{H}}^{o}(i) | 1s(i) \rangle + 2\langle 2s(i) | \hat{\mathcal{H}}^{o}(i) | 2s(i) \rangle] = 2I(1s) + I(2s)$$

The two-electron integrals are more difficult. In order to simplify this calculation we factor out the common spin functions. Thus

$$\Psi = \psi_1 \alpha(1)\beta(2)\alpha(3) + \psi_2 \beta(1)\alpha(2)\alpha(3) + \psi_3 \alpha(1)\alpha(2)\beta(3) = \sum_i \psi_i S_i$$

where

$$\psi_1 = \frac{1}{\sqrt{6}} [1s(1)1s(2)2s(3) - 2s(1)1s(2)1s(3)]$$

$$\psi_2 = \frac{1}{\sqrt{6}} [1s(1)2s(2)1s(3) - 1s(1)1s(2)2s(3)]$$

$$\psi_3 = \frac{1}{\sqrt{6}} [2s(1)1s(2)1s(3) - 1s(1)2s(2)1s(3)]$$

The $\hat{\mathcal{H}}'$ integral then becomes:

$$\langle \Psi | \hat{\mathcal{H}}' | \Psi \rangle = \langle \psi_1 S_1 + \psi_2 S_2 + \psi_3 S_3 | \sum_{i<j}^{3} 2/r_{ij} | \psi_1 S_1 + \psi_2 S_2 + \psi_3 S_3 \rangle$$

The orthogonality of the spin functions gives $\langle S_i|S_j\rangle = \delta_{ij}$ and we are left with

$$\langle\Psi|\hat{\mathcal{H}}'|\Psi\rangle = \langle\psi_1|\hat{\mathcal{H}}'|\psi_1\rangle + \langle\psi_2|\hat{\mathcal{H}}'|\psi_2\rangle + \langle\psi_3|\hat{\mathcal{H}}'|\psi_3\rangle$$

Evaluating $\langle\psi_1|\hat{\mathcal{H}}'|\psi_1\rangle$ explicitly gives

$$\begin{aligned}6\langle\psi_1|\hat{\mathcal{H}}'|\psi_1\rangle &= \langle 1s(1)1s(2)2s(3) - 2s(1)1s(2)1s(3)|2/r_{12} + 2/r_{13} + 2/r_{23}| \\ &\qquad 1s(1)1s(2)2s(3) - 2s(1)1s(2)1s(3)\rangle \\ &= \langle 1s(1)1s(2)|2/r_{12}|1s(1)1s(2)\rangle + \langle 2s(1)1s(2)|2/r_{12}|2s(1)1s(2)\rangle \\ &+ \langle 1s(1)2s(3)|2/r_{13}|1s(1)2s(3)\rangle + \langle 2s(1)1s(3)|2/r_{13}|2s(1)1s(3)\rangle \\ &+ \langle 1s(2)2s(3)|2/r_{23}|1s(2)2s(3)\rangle + \langle 1s(2)1s(3)|2/r_{23}|1s(2)1s(3)\rangle \\ &- 2\langle 1s(1)2s(3)|2/r_{13}|2s(1)1s(3)\rangle\end{aligned}$$

The terms $\langle\psi_2|\hat{\mathcal{H}}'|\psi_2\rangle$ and $\langle\psi_3|\hat{\mathcal{H}}'|\psi_3\rangle$ give exactly the same results and we conclude that:

$$\langle\Psi|\hat{\mathcal{H}}'|\Psi\rangle = J(1s, 1s) + 2J(1s, 2s) - K(1s, 2s)$$

The expectation value of the energy using a single Slater determinant is thus

$$E = 2I(1s) + I(2s) + J(1s, 1s) + 2J(1s, 2s) - K(1s, 2s)$$

8.9 Assume that the ground state of Li can be described by a single Slater determinant and derive an expression for the energy in terms of the J and K integrals. You should use the Condon-Slater rules given in the Introduction in this derivation.

Solution: Ref. S3,222; Introduction

The Hamiltonian for Li is

$$\hat{\mathcal{H}} = \sum_{i=1}^{3} f(i) + \sum_{i<j}^{3} 2/r_{ij}$$

where $f(i) = -\nabla_i^2 - 2Z/r_i$ and $Z = 3$. The assumed wave function is

$$\Psi = \frac{1}{\sqrt{3!}}\begin{vmatrix} 1s\alpha(1) & 1s\alpha(2) & 1s\alpha(3) \\ 1s\beta(1) & 1s\beta(2) & 1s\beta(3) \\ 2s\alpha(1) & 2s\alpha(2) & 2s\alpha(3)\end{vmatrix} = (1s\overline{1s}2s)$$

The energy is given by:

$$E = \sum_i \langle i|f|i\rangle + \sum_{i<j} [\langle ij|g|ij\rangle - \langle ij|g|ji\rangle\delta_{m_{s_i} m_{s_j}}]$$

where f depends on the coordinates of the electron that occupies u_i and g depends on the coordinates of the two electrons which occupy the orbitals u_i and u_j. The Kronecker delta function has been included explicitly here so that space instead of spin-orbitals can be used in this expression. Thus:

$$\begin{aligned}E &= 2\langle 1s|f|1s\rangle + \langle 2s|f|2s\rangle + \langle 1s, 1s|2/r_{12}|1s, 1s\rangle + 2\langle 1s, 2s|2/r_{12}|1s, 2s\rangle \\ &- \langle 1s, 2s|2/r_{12}|2s, 1s\rangle = 2I(1s) + I(2s) + J(1s, 1s) + 2J(1s, 2s) - K(1s, 2s)\end{aligned}$$

This answer is identical to that found in Problem 8.8.

8.10 Work out the energies of the ^4S, ^2S, and $^2S'$ terms which arise for the three-electron configuration 1s2s3s in terms of the coulomb integrals J(ns, n's) and the exchange integrals K(ns, n's). The eigenfunctions of $\hat{S}^2$ for a three-electron system are given in Problem 5.25 and the Condon-Slater rules can profitably be used. (Hint: A secular equation must be solved for $\varepsilon(^2S)$ and $\varepsilon(^2S')$. The diagonal sum rule only allows $\varepsilon(^2S) + \varepsilon(^2S')$ to be determined.)

Solution: Ref. Problem 5.25

For the configuration 1s2s3s, L = 0 and the wave functions can be characterized by the spin quantum numbers: M_S = 3/2, 1/2, 1/2, 1/2, -1/2, -1/2, -1/2, -3/2. The associated determinantal wave functions constructed from simple product functions are:

	M_S		M_S
$\psi_1 = (1s2s3s)$	3/2;	$\psi_5 = (\overline{1s}\,\overline{2s}\,3s)$	-1/2
$\psi_2 = (1s2s\overline{3s})$	1/2;	$\psi_6 = (\overline{1s}\,2s\,\overline{3s})$	-1/2
$\psi_3 = (1s\overline{2s}3s)$	1/2;	$\psi_7 = (1s\,\overline{2s}\,\overline{3s})$	-1/2
$\psi_4 = (\overline{1s}2s3s)$	1/2;	$\psi_8 = (\overline{1s}\,\overline{2s}\,\overline{3s})$	-3/2

where the spin functions are α except under the bar where they are β. The function ψ_1 is uniquely associated with S = 3/2 so that for $M_L = +3/2$, $\Psi(S = 3/2, M_S = 3/2) = \psi_1$ and the energy is:

$$\varepsilon(^4S) = I(1s) + I(2s) + I(3s) + J(1s, 2s) + J(1s, 3s) + J(2s, 3s) - K(1s, 2s) - K(1s, 3s) - K(2s, 3s)$$

The $M_S = 1/2$ block of the secular equation results from functions ψ_2, ψ_3, and ψ_4 and is thus 3x3. However, in Problem 5.25 we constructed the following linear combinations of the $M_S = 1/2$ functions which are eigenfunctions of $\hat{S}^2$:

$$\Psi(^4S) = (1/\sqrt{3})(\psi_2 + \psi_3 + \psi_4); \qquad \Psi(^2S) = (1/\sqrt{6})(\psi_2 + \psi_3 - 2\psi_4)$$

$$\Psi(^2S') = (1/\sqrt{2})(\psi_3 - \psi_2)$$

The first of these functions gives $\varepsilon(^4S)$ again and thus only a 2x2 secular problem involving $\Psi(^2S)$ and $\Psi(^2S')$ must be solved. The required energies are roots of the equation:

$$\begin{vmatrix} \langle ^2S|\hat{\mathcal{H}}|^2S\rangle - E & \langle ^2S|\hat{\mathcal{H}}|^2S'\rangle \\ \langle ^2S|\hat{\mathcal{H}}|^2S'\rangle & \langle ^2S'|\hat{\mathcal{H}}|^2S'\rangle - E \end{vmatrix} = 0 \qquad (1)$$

where

$$\langle ^2S|\hat{\mathcal{H}}|^2S\rangle = (1/6)[\langle\psi_2|\hat{\mathcal{H}}|\psi_2\rangle + \langle\psi_3|\hat{\mathcal{H}}|\psi_3\rangle + 4\langle\psi_4|\hat{\mathcal{H}}|\psi_4\rangle + 2\langle\psi_2|\hat{\mathcal{H}}|\psi_3\rangle - 4\langle\psi_2|\hat{\mathcal{H}}|\psi_4\rangle - 4\langle\psi_3|\hat{\mathcal{H}}|\psi_4\rangle]$$

$$\langle ^2S'|\hat{\mathcal{H}}|^2S'\rangle = (1/2)[\langle\psi_3|\hat{\mathcal{H}}|\psi_3\rangle + \langle\psi_2|\hat{\mathcal{H}}|\psi_2\rangle - 2\langle\psi_2|\hat{\mathcal{H}}|\psi_3\rangle]$$

$$\langle ^2S|\hat{\mathcal{H}}|^2S'\rangle = (1/\sqrt{12})[-\langle\psi_2|\hat{\mathcal{H}}|\psi_2\rangle + \langle\psi_3|\hat{\mathcal{H}}|\psi_3\rangle + 2\langle\psi_2|\hat{\mathcal{H}}|\psi_4\rangle - 2\langle\psi_3|\hat{\mathcal{H}}|\psi_4\rangle]$$

Application of the Condon-Slater rules gives the individual matrix elements.

$$\langle\psi_2|\hat{\mathcal{H}}|\psi_2\rangle = I(1s) + I(2s) + I(3s) + J(1s, 2s) + J(1s, 3s) + J(2s, 3s) - K(1s, 2s)$$

$$\langle\psi_3|\hat{\mathcal{H}}|\psi_3\rangle = I(1s) + I(2s) + I(3s) + J(1s, 2s) + J(1s, 3s) + J(2s, 3s) - K(1s, 3s)$$

$$\langle\psi_4|\hat{\mathcal{H}}|\psi_4\rangle = I(1s) + I(2s) + I(3s) + J(1s, 2s) + J(1s, 3s) + J(2s, 3s) - K(2s, 3s)$$

$$\langle\psi_2|\hat{\mathcal{H}}|\psi_3\rangle = -K(2s, 3s); \quad \langle\psi_2|\hat{\mathcal{H}}|\psi_4\rangle = -K(1s, 3s); \quad \langle\psi_3|\hat{\mathcal{H}}|\psi_4\rangle = -K(1s, 2s)$$

Combining these elements we obtain:

$$\langle {}^2S|\hat{\mathcal{H}}|{}^2S\rangle = I + J - K(2s, 3s) + (1/2)[K(1s, 2s) + K(1s, 3s)]$$

$$\langle {}^2S'|\hat{\mathcal{H}}|{}^2S'\rangle = I + J + K(2s, 3s) - (1/2)[K(1s, 2s) + K(1s, 3s)]$$

$$\langle {}^2S|\hat{\mathcal{H}}|{}^2S'\rangle = (\sqrt{3}/2)[K(1s, 2s) - K(1s, 3s)]$$

where

$$I = \sum_{n=1}^{3} I(ns) \quad \text{and} \quad J = \sum_{i<j}^{3} J(is, js)$$

The roots of Eq. (1) are then:

$$E_{\pm} = I + J \pm \sqrt{K^2(1s, 2s) + K^2(1s, 3s) + K^2(2s, 3s) - K(1s, 2s)K(2s, 3s) - K(1s, 3s)K(2s, 3s) - K(1s, 2s)K(1s, 3s)}$$

8.11 Derive the Russell-Saunders terms which arise for the configuration sp by constructing a table of the possible values of m_{ℓ_1}, m_{ℓ_2}, M_L, and M_S.

Solution: Ref. HZ1,128

For orbital 1: $\ell = 0$, $m_{\ell_1} = 0$, $m_{s_1} = \pm 1/2$; orbital 2: $\ell = 1$, $m_{\ell_2} = 0, \pm 1$, $m_{s_2} = \pm 1/2$. Thus

m_{ℓ_1}	m_{ℓ_2}	M_L	M_S	Terms	Basis Functions
0	1	1	$1,0,0,\bar{1}$	${}^3P, {}^1P$	$\psi_1, \psi_2, \psi_3, \psi_4$
0	0	0	$1,0,0,\bar{1}$	${}^3P, {}^1P$	$\psi_5, \psi_6, \psi_7, \psi_8$
0	-1	-1	$1,0,0,\bar{1}$	${}^3P, {}^1P$	$\psi_9, \psi_{10}, \psi_{11}, \psi_{12}$

The maximum value $M_L = 1$ implies that $L = 1$; therefore the $M_L = 0$ and -1 states of this manifold are also present. The maximum value $M_S = 1$ similarly implies that $S = 1$ and that the $M_S = 0$ and -1 states are present. In each row $S = 1$ only accounts for 3 of the 4 values of M_S. We conclude that the state $S = 0$, $M_S = 0$ is also present. The multiplicities indicated on the term symbols are $2(1) + 1 = 3$ and $2(0) + 1 = 1$.

8.12 How many electronic states are possible for an atom having (a) two nonequivalent p-electrons (b) two equivalent p-electrons?

<u>Solution</u>: Ref. HZ1,86,120; Introduction

(a) The adiabatic change from one coupling case to another does not affect the number of states. We are, therefore, free to pick any convenient representation for the purpose of counting states. Each isolated p electron has (2S + 1)(2L + 1) = 2(3) = 6 states. The total number of possible states for two nonequivalent p electrons is thus 6x6 = 36.

(b) For equivalent electrons the number of states is limited by distinguishability and the Pauli principle. In this case (p^2) there are six possible combinations of m_ℓ and m_s which may be represented by the boxes below:

	m_ℓ = +1	0	-1
m_s = +1/2			
-1/2			

There are 6 possible locations for the first electron. This leave 5 possible locations for the second electron. Since the electrons are indistinguishable we must divide by 2. Thus the number of distinguishable ways to put 2 electrons in these boxes is:

$$(6)(5)/2 = 15$$

Another way to reach the same conclusion is to suppose that the two electrons are placed in the first two boxes. There are then 6! ways to arrange the boxes in order. One must then divide by 2! so that exchanges of the electrons are not counted and by 4! so that exchanges of the empty boxes are not counted. The result (6!)/[(2!)(4!)] is, of course, the same as that obtained above.

8.13 Derive the Russell-Saunders terms for 2 nonequivalent p-electrons.

<u>Solution</u>: Ref. HZ1,82,129

Since the electrons are not equivalent, questions of distinguishability and the Pauli principle do not arise. The vector coupling model can be used directly.

1) Couple $\underset{\sim}{\ell}_1$ and $\underset{\sim}{\ell}_2$ to obtain $\underset{\sim}{L}$.

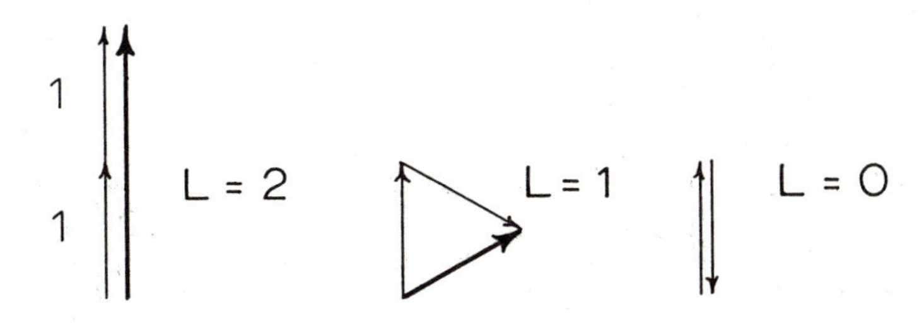

2) Couple $\underset{\sim}{s}_1$ and $\underset{\sim}{s}_2$ to obtain $\underset{\sim}{S}$.

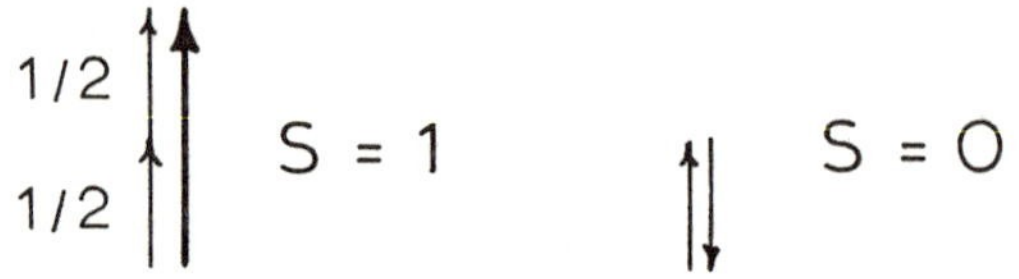

1/2
1/2
S = 1
S = 0

The terms are

$L = 2, S = 1$: ^3D $L = 1, S = 1$: ^3P $L = 0, S = 1$: ^3S

$L = 2, S = 0$: ^1D $L = 1, S = 0$: ^1P $L = 0, S = 0$: ^1S

In each case subscripts can be added to denote the values of J. For example, according to the vector model we find for the ^3D term:

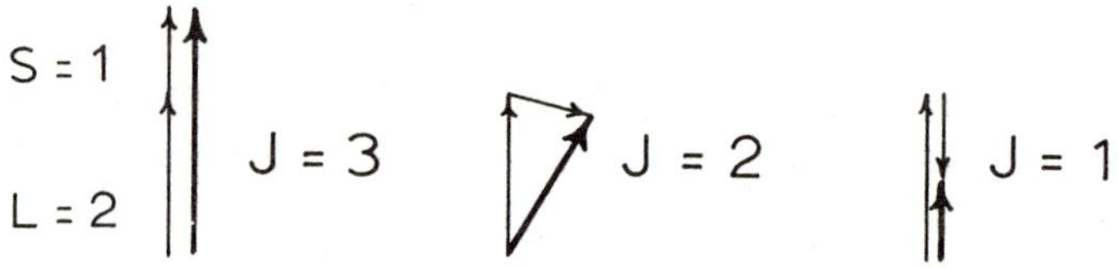

S = 1
L = 2
J = 3
J = 2
J = 1

Thus ^3D leads to $^3D_3$, $^3D_2$, and $^3D_1$.

8.14 Derive the Russell-Saunders terms for the configuration p^2, <u>i.e.</u> two equivalent p electrons, by constructing a table of the possible values of M_L and M_S.

<u>Solution</u>: Ref. HZ1,130

m_{ℓ_1}	m_{ℓ_2}	M_L	M_S	Terms	Basis Functions
1	1	2	0	^1D	ψ_1
1	0	1	$1,0,0,\bar{1}$	$^1D, {}^3P$	ψ_2-ψ_5
1	-1	0	$1,0,0,\bar{1}$	$^1D, {}^3P$	ψ_6-ψ_{10}
0	0	0	0	^1S	
-1	0	-1	$1,0,0,\bar{1}$	$^1D, {}^3P$	ψ_{11}-ψ_{14}
-1	-1	-2	0	^1D	ψ_{15}

Equivalence implies that both electrons have the same values of n and ℓ. There are two consequences of equivalence in the table above. First, the states $m_{\ell_1} = 1$, $m_{\ell_2} = 0$ and $m_{\ell_1} = 0$, $m_{\ell_2} = 1$ are indistinguishable. Second, the Pauli exclusion principle must be satisfied. Thus, when $m_{\ell_1} = m_{\ell_2}$ we require that $m_{s_1} \neq m_{s_2}$.

Since the maximum value of M_L is 2, it follows that an L = 2 term is present with 2L + 1 = 5 associated values of M_L. The only value of M_S allowed for the L = 2 states is 0; therefore the term is a singlet D. The second row of the table, which corresponds to M_L = 1, contains four states. One of these is associated with the ^1D found in the first row, and the other three imply the existence of a ^3P term. Continuing, we realize that ^1D and ^3P have four states with M_L = 0. The one remaining state has M_L = 0 and M_S = 0 and is thus ^1S. In summary the terms found from the table are ^1D, ^3P, and ^1S. For completeness the J values may be added to give: $^1D_2$, $^3P_2$, $^3P_1$, $^3P_0$, ^1S.

8.15 Derive the Russell-Saunders terms for 3 equivalent p electrons, i.e. for p^3.

Solution: Ref. HZ1,132

We determine the terms by constructing a table of allowed M_S and M_L values.

m_{ℓ_1}	m_{ℓ_2}	m_{ℓ_3}	M_L	M_S	Terms	Basis Functions
1	1	0	2	1/2,-1/2	^2D	ψ_1,ψ_2
1	1	-1	1	1/2,-1/2	$^2D,^2P$	ψ_3,ψ_4
0	0	1	1	1/2,-1/2		ψ_5,ψ_6
1	-1	0	0	3/2,1/2,1/2,1/2, -1/2,-1/2,-1/2,-3/2	$^2D,^2P,^4S$	$\psi_7-\psi_{14}$
0	0	-1	-1	1/2,-1/2		ψ_{15},ψ_{16}
-1	-1	+1	-1	1/2,-1/2		ψ_{17},ψ_{18}
-1	-1	0	-2	1/2,-1/2		ψ_{19},ψ_{20}

The total number of states must be 6!/(3!)(3!) = 20. The (2S + 1)(2L + 1) products give 10 for ^2D, 6 for ^2P, and 4 for ^4S for a total of 20 as required. It should be noticed that M_L = 3 is not allowed by the Pauli principle. No more than two equivalent electrons can have the same value of m_ℓ and then only when their m_s values differ.

8.16 Derive the Russell-Saunders terms for 9 equivalent d-electrons, i.e. for the configuration d^9.

Solution: Ref. HZ1,131

The possible combinations of m_ℓ and m_s for d electrons are represented by the boxes below. The number of ways that 9 equivalent electrons can be placed in these boxes is equal to the number of locations of the "hole." Therefore, ten states are predicted.

	m_ℓ = +2	+1	0	-1	-2
m_s = +1/2					
-1/2					

The maximum possible value for

$$\sum_{i=1}^{9} m_{\ell_i}$$

is 2 which occurs when the missing electron is assigned to $m_\ell = -2$. It is clear that the $M_L = +2$ state has $M_S = 1/2$ since there is only one unpaired electron. We conclude that the only term is ^2D.

The d^9 configuration is an example of a general rule. The terms which arise for the configuration x^r are the same as those that arise for $x^{2(2\ell+1)-r}$, e.g. d^1 and d^9 give the same terms.

8.17 List the lowest terms for the following atoms:

(a) Be, (b) C, (c) O, (d) Cl, (e) As

Solution: Ref. HZ1,135; Introduction

(a) Be$[(1s)^2(2s)^2]$: Since the shells are closed, the term is $^1S_0$.

(b) C$[(1s)^2(2s)^2(2p)^2]$: The states of maximum multiplicity have S = 1 and of these the maximum L is 1. Therefore, the lowest term is $^3P_0$. Since the shell is less than half filled, the lowest energy corresponds to the lowest value of J, i.e. the multiplet is regular.

(c) O$[(1s)^2(2s)^2(2p)^4]$: The maximum value of S is 1 and the largest L consistent with this spin state is 1. Therefore, the lowest term is $^3P_2$. Here the lowest term has the maximum value of J since the shell is more than half filled, i.e. the multiplet is inverted.

(d) Cl$[(1s)^2(2s)^2(2p)^6(3s)^2(3p)^5]$: Hund's rules for this configuration give S = 1/2 and L = 1. The highest value of J is 3/2. Therefore the lowest term is $^3P_{3/2}$.

(e) As$[(1s)^2(2s)^2(2p)^6(3s)^2(3p)^6(4s)^2(4p)^3]$: The maximum S is 3/2 and L = 0. Therefore, the lowest term is $^4S_{3/2}$.

8.18 Consider the configuration sp. In the absence of interelectronic interactions the wave functions are eigenfunctions of the operators $\hat{L}_z$ and $\hat{S}_z$. The determinantal wave functions corresponding to ψ_1 through ψ_{12} of Problem 8.11 are denoted by the symbols $(m_{\ell_1}^{m_{s1}}, m_{\ell_2}^{m_{s2}})$, e.g. $\psi_1 = (0^+, 1^+)$. Use these functions to construct the eigenfunctions of $\hat{L}^2$ and $\hat{S}^2$ which are associated with the terms ^3P and ^1P. (Hint: Start with $(0^+, 1^+)$ and use the lowering operators $\hat{S}_-$ and $\hat{L}_-$.)

Solution: Ref. S2,76; Problem 5.22, 5.24, 5.25

Since the maximum values of m_{ℓ_1} and m_{ℓ_2} are 0 and 1, respectively, we have:

$$\Psi(L = 1,\ S = 1;\ M_L = 1,\ M_S = 1) = (0^+,\ 1^+)$$

To obtain $\Psi(M_L = 1, M_S = 0)$ we operate with $\hat{S}_- = \hat{S}_{1-} + \hat{S}_{2-}$. Thus for $S = 1$:

$$\hat{S}_-\Psi(M_L = 1,\ M_S = 1) = \sqrt{(S + M_S)(S - M_S + 1)}\ \Psi(M_L = 1,\ M_S = 0) = \sqrt{2}\ \Psi(M_L = 1,\ M_S = 0)$$
$$= (\hat{S}_{1-} + \hat{S}_{2-})(0^+,\ 1^+) = (0^-,\ 1^+) + (0^+,\ 1^-)$$

and

$$\Psi(L = 1,\ S = 1;\ M_L = L,\ M_S = 0) = \frac{1}{\sqrt{2}}(0^-,\ 1^+) + \frac{1}{\sqrt{2}}(0^+,\ 1^-)$$

Operating again with $\hat{S}_-$ we obtain:

$$\hat{S}_-\Psi(M_L = 1,\ M_S = 0) = \sqrt{2}\ \Psi(M_L = 1,\ M_S = -1)$$
$$= (\hat{S}_{1-} + \hat{S}_{2-})\left[\frac{1}{\sqrt{2}}(0^-,\ 1^+) + \frac{1}{\sqrt{2}}(0^+,\ 1^-)\right] = \sqrt{2}\ (0^-,\ 1^-)$$

or

$$\Psi(L = 1,\ S = 1;\ M_L = 1,\ M_S = -1) = (0^-,\ 1^-)$$

The same procedure is used to obtain $\Psi(L = 1,\ S = 1;\ M_L = 0,\ M_S = 1)$. Applying $\hat{L}_- = \hat{L}_{1-} + \hat{L}_{2-}$ we obtain:

$$\hat{L}_-\Psi(M_L = 1,\ M_S = 1) = \sqrt{2}\ \Psi(M_L = 0,\ M_S = 1)$$
$$= \hat{L}_{1-}(0^+,\ 1^+) + \hat{L}_{2-}(0^+,\ 1^+) = 0 + \sqrt{2}\ (0^+,\ 0^+)$$

Therefore

$$\Psi(L = 1,\ S = 1;\ M_L = 0,\ M_S = 1) = (0^+,\ 0^+)$$

By continuing this procedure we obtain the following set of eigenfunctions:

^3P: $M_L = 1$;	$M_S = 1$:	$(0^+,\ 1^+)$
	$M_S = 0$:	$(2)^{-1/2}[(0^+,\ 1^-) + (0^-,\ 1^+)]$
	$M_S = -1$:	$(0^-,\ 1^-)$
$M_L = 0$;	$M_S = 1$:	$(0^+,\ 0^+)$
	$M_S = 0$:	$(2)^{-1/2}[(0^+,\ 0^-) + (0^-,\ 0^+)]$
	$M_S = -1$:	$(0^-,\ 0^-)$
$M_L = -1$;	$M_S = 1$:	$(0^+,\ -1^+)$
	$M_S = 0$:	$(2)^{-1/2}[(0^+,\ -1^-) + (0^-,\ -1^+)]$
	$M_S = -1$:	$(0^-,\ -1^-)$

The eigenfunctions for ^1P can be obtained either by the use of a projection operator on functions having $M_L = 0$ or by constructing functions having $M_L = 0$ which are orthogonal to those for the ^3P term. For example using orthogonality:

$$\int \Psi^*(L = 1, S = 1; M_L = 1, M_S = 0)\Psi(L = 1, S = 0; M_L = 1, M_S = 0)d\tau$$

$$= \int \frac{1}{\sqrt{2}} [(0^+, 1^-) + (0^-, 1^+)][a(0^+, 1^-) + b(0^-, 1^+)]d\tau_1 d\tau_2 = 0$$

Therefore

$$\frac{1}{\sqrt{2}}(a + b) = 0 \qquad \text{or} \qquad a = -b$$

Thus, for ^1P: $M_L = 1, M_S = 0$: $(2)^{-1/2}[(0^+, 1^-) - (0^-, 1^+)]$

$M_L = 0, M_S = 0$: $(2)^{-1/2}[(0^+, 0^-) - (0^-, 0^+)]$

$M_L = -1, M_S = 0$: $(2)^{-1/2}[(0^+, -1^-) - (0^-, -1^+)]$

8.19 Work out expressions for the energies associated with all of the Russell-Saunders terms for the configuration $(2p)^2$ in terms of integrals over determinantal wave functions. The basis functions for this calculation are the functions ψ_1 through ψ_{15} in Problem 8.14, e.g. $\psi_1 = (1^+, 1^-)$ and $\psi_2 = (1^+, 0^+)$. (Hint: Use the diagonal sum rule to obtain $\varepsilon(^1S)$.)

Solution: Ref. Problem 8.14

Only ψ_1 is associated with ^1D. Therefore

$$\varepsilon(^1D) = \langle\psi_1|\hat{\mathcal{H}}|\psi_1\rangle = \langle(1^+, 1^-)|\hat{\mathcal{H}}|(1^+, 1^-)\rangle$$

Similarly, ψ_2 must be associated with triplet states. Thus

$$\varepsilon(^3P) = \langle\psi_2|\hat{\mathcal{H}}|\psi_2\rangle = \langle(1^+, 0^+)|\hat{\mathcal{H}}|(1^+, 0^+)\rangle$$

An eigenfunction cannot be isolated for the ^1S state. This implies that a secular equation must be solved to obtain $\varepsilon(^1S)$. Consider the $M_L = 0$, $M_S = 0$ block. The functions $\psi_7 = (1^+, -1^-)$, $\psi_8 = (1^-, -1^+)$, and $\psi_{10} = (0^+, 0^-)$ can be used to set up a 3x3 matrix, the eigenvalues of which must equal $\varepsilon(^1D)$, $\varepsilon(^3P)$, and $\varepsilon(^1S)$. It is well known, however, that the transformation which diagonalizes this matrix cannot change the sum of the diagonal elements, i.e.

$$\langle\psi_7|\hat{\mathcal{H}}|\psi_7\rangle + \langle\psi_8|\hat{\mathcal{H}}|\psi_8\rangle + \langle\psi_{10}|\hat{\mathcal{H}}|\psi_{10}\rangle = \varepsilon(^1D) + \varepsilon(^3P) + \varepsilon(^1S)$$

We conclude then that

$$\varepsilon(^1S) = \langle\psi_7|\hat{\mathcal{H}}|\psi_7\rangle + \langle\psi_8|\hat{\mathcal{H}}|\psi_8\rangle + \langle\psi_{10}|\hat{\mathcal{H}}|\psi_{10}\rangle - \langle\psi_1|\hat{\mathcal{H}}|\psi_1\rangle - \langle\psi_2|\hat{\mathcal{H}}|\psi_2\rangle$$

8.20 An atom having the electron configuration p^3 is placed in a weak magnetic field B. Assuming LS coupling, work out the number of components into which each possible energy level is split by the field and give the magnitudes of the splittings in units of $\mu_B B$.

Solution: Ref. Problems 8.15; 7.23; 7.25; S3,204

The Russell-Saunder terms associated with p^3 were found to be ^2D, ^2P, ^4S. The associated values of J can be found by vector addition (see Problem 8.13). Thus

$$^2D \rightarrow {}^2D_{5/2},\ {}^2D_{3/2};\quad {}^2P \rightarrow {}^2P_{3/2},\ {}^2P_{1/2};\quad {}^4S \rightarrow {}^4S_{3/2}$$

When the field is applied, each level will be split into 2J + 1 components having energy shifts given by $\Delta E = g\mu_B BM_J$ where g is a function of L, S, and J. From Problem 7.23 we find:

$$g = 1 + \frac{J(J+1) + S(S+1) - L(L+1)}{2J(J+1)} \tag{1}$$

From an intuitive derivation of Eq. (1) the reader is referred to S3. Rigorous derivations of this result for many-electron systems usually make use of the Wigner-Eckart theorem to evaluate Eq. (1) of Problem 7.23, see for example ME,404. Using Eq. (1) we conclude that:

$^2D_{5/2}$: 2J + 1 = 6 components

The energy spacings are $g\mu_B B$ where

$$g = 1 + \frac{(5/2)(7/2) + (1/2)(3/2) - 2(3)}{2(5/2)(7/2)} = \frac{6}{5}$$

$^2D_{3/2}$: 2J + 1 = 4 components with g = 4/5

$^2P_{3/2}$: 2J + 1 = 4 components with g = 4/3

$^2P_{1/2}$: 2J + 1 = 2 components with g = 2/3

$^4S_{3/2}$: 2J + 1 = 4 components with g = 2

For the last term L = 0, therefore S = J, and Eq. (1) is not necessary since only spin angular momentum contributes to the magnetic moment.

8.21 The first order energy of a hydrogen atom in the magnetic field $\underset{\sim}{B}_o$ was shown in Problem 7.28 to be:

$$E_1^{(1)} = \frac{e^2B^2}{12m_e} \langle 1s|r^2|1s\rangle \tag{1}$$

The molar diamagnetic susceptibility χ_m is defined by the equation

$$\chi_m = -\frac{N_A}{\mu_o H}\left[\frac{\partial E_1^{(1)}}{\partial H}\right] \tag{2}$$

Here $B = \mu_o H$ and $\mu_o = 4\pi\times10^{-7}\ NA^{-2}$ in SI units. For many-electron atoms the susceptibilities can be approximately calculated by assuming that the electrons move independently. Estimate the diamagnetic susceptibility of He by assuming that:

(a) $\Psi_{He} = 1s(1)1s(2)$ with Z = 2.

(b) $\Psi_{He} = 1s(1)1s(2)$ with Z determined so as to minimize the energy.

Solution: Ref. AT2,414; S3,667; Problem 7.28

(a) For He using Eq. (1) and summing the contributions of the two electrons we have:

$$E_1^{(1)} = \frac{e^2B^2}{12m_e}[<1s|r_1^2|1s> + <1s|r_2^2|1s>] = \frac{e^2B^2}{6m_e}<1s|r^2|1s>$$

and

$$\frac{\partial E_1^{(1)}}{\partial H} = \frac{\partial}{\partial H}\left[\frac{e^2}{6m}<1s|r^2|1s>\ \mu_o^2H^2\right] = \frac{e^2\mu_o^2H}{3m_e}<1s|r^2|1s>$$

Therefore:

$$\chi_m = -N_A\frac{e^2\mu_o}{3m_e}<1s|r^2|1s> = -\frac{N_Ae^2\mu_oa_o^2}{4m_e} \tag{3}$$

since $<1s|r^2|1s> = 3(a_o^2/Z^2)$ and $Z = 2$. Equation (3) can also be written in the form

$$\chi_m = -4\pi\left[\frac{N_Ae^2a_o^2}{(4\pi\varepsilon_o)4m_ec^2}\right]$$

by making use of the relation $\varepsilon_o\mu_o = 1/c^2$. Evaluating Eq. (3) we obtain $\chi_m = -1.49\text{x}10^{-11}\ m^3$. When Eq. (3) is evaluated in cgs units the replacement $\mu_o = 1$ will lead to unrationalized units which appear in much of the literature, and the replacement $\mu_o = 4\pi$ will lead to rationalized units.

(b) In Problem 8.3 we found that Z(effective) = 1.688 gives the minimum energy. Thus the corresponding diamagnetic susceptibility is:

$$\chi_m = -\frac{N_Ae^2\mu_oa_o^2}{m_e(1.688)^2} = -2.09\text{x}10^{-11}\ m^3$$

The experimental value is $-2.36\text{x}10^{-11}\ m^3$.

8.22 Two hydrogen atoms having a large internuclear separation R interact through their instantaneous electric dipole moments. Show that the perturbation Hamiltonian which describes the interatomic interaction when $R >> a_o$ can be written in the form:

$$\hat{\mathcal{H}}' = \frac{e^2}{(4\pi\varepsilon_o)R^3}(x_1x_2 + y_1y_2 - 2z_1z_2) \tag{1}$$

Here $x_1y_1z_1$ and $x_2y_2z_2$ are the coordinates of electrons 1 and 2 with respect to nuclei A and B, respectively. (Hint: Expand the coulombic interactions in terms of R^{-1}.)

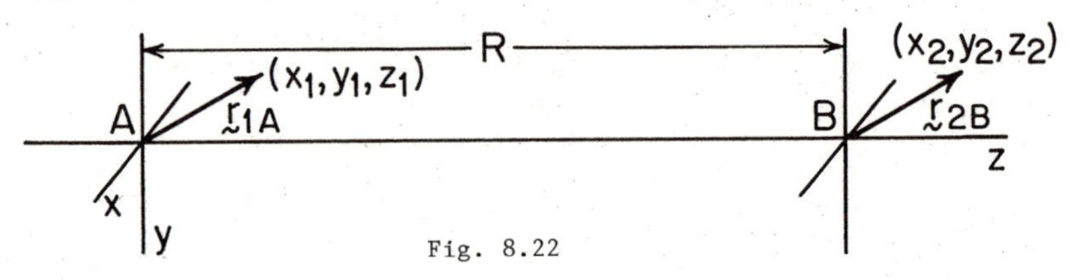

Fig. 8.22

Solution: EWK,351; AT2,456

The complete Hamiltonian for the two H-atoms is

$$\hat{\mathcal{H}} = \hat{\mathcal{H}}^o_A + \hat{\mathcal{H}}^o_B + \hat{\mathcal{H}}'$$

where

$$\hat{\mathcal{H}}' = \frac{e^2}{(4\pi\varepsilon_o)}\left[-\frac{1}{r_{1B}} - \frac{1}{r_{2A}} + \frac{1}{r_{12}} + \frac{1}{R}\right]$$

The position vectors can be written as:

$$\underset{\sim}{r}_{1A} = x_1\hat{i} + y_1\hat{j} + z_1\hat{k}; \qquad \underset{\sim}{r}_{2A} = x_2\hat{i} + y_2\hat{j} + (z_2 + R)\hat{k}$$

$$\underset{\sim}{r}_{2B} = x_2\hat{i} + y_2\hat{j} + z_2\hat{k}; \qquad \underset{\sim}{r}_{1B} = x_1\hat{i} + y_1\hat{j} + (z_1 - R)\hat{k}$$

Thus

$$\hat{\mathcal{H}}' = \frac{e^2}{(4\pi\varepsilon_o)}\left[-\frac{1}{\sqrt{x_1^2 + y_1^2 + (z_1 - R)^2}} - \frac{1}{\sqrt{x_2^2 + y_2^2 + (z_2 + R)^2}}\right.$$

$$\left. + \frac{1}{\sqrt{(x_2 - x_1)^2 + (y_2 - y_1)^2 + (z_2 - z_1 + R)^2}} + \frac{1}{R}\right]$$

Expansion of the first three terms on the right follows from the binomial series:

$$(1 + x)^{-n} = 1 - nx + \frac{n(n + 1)}{2!}x^2 + \cdots\cdot\ ; \qquad x^2 < 1$$

Therefore:

$$\hat{\mathcal{H}}' = \frac{e^2}{(4\pi\varepsilon_o)R}\left\{-\left[1 + \frac{(r_{1A}^2 - 2z_1R)}{R^2}\right]^{-1/2} - \left[1 + \frac{(r_{2B}^2 + 2z_2R)}{R^2}\right]^{-1/2}\right.$$

$$\left. + \left[1 + \frac{(r_{1A}^2 + r_{2B}^2 - 2(x_1x_2 + y_1y_2 + z_1z_2) + 2R(z_2 - z_1))}{R^2}\right]^{-1/2} + 1\right\}$$

$$\hat{\mathcal{H}}' = \frac{e^2}{(4\pi\varepsilon_o)R}\left\{+\frac{1}{R^2}(x_1x_2 + y_1y_2 + z_1z_2) + \frac{3\cdot 4R^2}{8R^4}(-2z_1z_2) + \cdots\right\}$$

Collecting terms in this equation gives Eq. (1). This equation is in the form of the interaction between two electric dipole moments. It is often written as:

$$\hat{\mathcal{H}}' = \frac{e^2}{(4\pi\varepsilon_o)R^3}(\underset{\sim}{r}_1\cdot\underset{\sim}{r}_2 - 3r_{1z}r_{2z})$$

8.23 The forces of interaction of well separated atoms are known as London forces, van der Waal's forces, or dispersion forces. Use the perturbation Hamiltonian from Problem 8.22 and perturbation theory to derive an equation for the energy of interaction of two hydrogen atoms having the internuclear separation R. You may use the approximation: $E_1 - E_n \cong E_1$ for all $n \neq 1$.

Solution: Ref. EWK,351

For the wave function of the system we use

$$\psi = 1s_A(1)\cdot 1s_B(2) \Rightarrow |1,\ 1>$$

First order perturbation theory then gives:

$$E_1^{(1)} = <1,\ 1|\ \frac{e^2}{(4\pi\varepsilon_o)R^3}\ (x_1x_2 + y_1y_2 - 2z_1z_2)|1,\ 1>$$

$$= \frac{e^2}{(4\pi\varepsilon_o)R^3}\ \{<1|x_1|1><1|x_2|1> + \cdots\} = 0 \tag{1}$$

In Eq. (1) the integrals all vanish since the integrands are all odd functions of x, y, or z.

Resorting to second order perturbation theory we find:

$$E_1^{(2)} = \sum_{n,m\neq 1} \frac{|<1,\ 1|\hat{\mathcal{H}}'|n,\ m>|^2}{E_1 - E_{n,m}}$$

$$\simeq + \frac{1}{E_1} \sum_{n,m} <1,\ 1|\hat{\mathcal{H}}'|n,\ m><n,\ m|\hat{\mathcal{H}}'|1,\ 1> \simeq + \frac{1}{E_1}\ <1,\ 1|(\hat{\mathcal{H}}')^2|1,\ 1>$$

But

$$<1|x_1^2|1> = <1|y_1^2|1> = <1|z_1^2|1> = <1|r_1^2|1>/3$$

Thus:

$$E_1^{(2)} = \frac{+6e^4}{(4\pi\varepsilon_o)^2R^6}\ \frac{<1|x_1^2|1>^2}{E_1} = \frac{+6e^4a_o^4}{(4\pi\varepsilon_o)^2R^6E_1} \tag{2}$$

In Problem 7.34 we found that:

$$\alpha = \frac{2e^2}{E_1}\ <1s|z^2|1s>$$

Equation (2) can be written as:

$$E_1^{(2)} = + \frac{3E_1\alpha^2}{2(4\pi\varepsilon_o)^2R^6} = - \frac{3I\alpha^2}{2(4\pi\varepsilon_o)^2R^6}$$

where I is the ionization potential for the hydrogen atom.

SUPPLEMENTARY PROBLEMS

8.24 The integral $I = <1s(1)1s(2)|1/r_{12}|1s(1)1s(2)>$ where

$$1s(i) = u_{1s}(r_i) = (Z^{3/2}/\sqrt{\pi})\exp(-Zr_i)$$

can be evaluated by recognizing that 2I is equal to the energy of coulombic interaction of two spherically symmetric charge distributions having the densities $u_{1s}^2(r_1)$ and $u_{1s}^2(r_2)$, respectively. Carry out this evaluation by setting up an expression for the electrostatic potential resulting from the charge distribution of one of the electrons and then finding the energy of the second electronic charge distribution in the field of the first. (Hint: The appropriate potential expressions are given in Problem 1.5 if ρ is a function of r.)

Ans: Ref. PW,446; Problems 1.5, 8.2

$I = 5Z/8$

8.25 Use first order perturbation theory to estimate the energy of the ground state of the He atom assuming that the electron-electron repulsion term is a perturbation, i.e. use $\psi_{He} = u_{1s}(1)u_{1s}(2)$ as the unperturbed wavefunction.

Ans: Ref. PW,162; Problem 8.1

$E^o + E^{(1)} = (-8 + 2.5)Rhc = -5.5$ Rhc or -74.8 eV. The experimental value is -79.0 eV.

8.26 Assume that the ground state of Be can be represented by a single Slater determinant and derive an expression for the energy of this state in terms of the J and K integrals. You should use the Condon-Slater rules given in the Introduction in this derivation.

Ans: Ref. S3,222; Problem 8.9

$E = 2I(1s) + 2I(2s) + J(1s, 1s) + J(2s, 2s) + 4J(1s, 2s) - 2K(1s, 2s)$

8.27 How many states are allowed for (a) 3 nonequivalent d electrons (b) 3 equivalent d electrons?

Ans: Ref. HZ1,128; Problem 8.12

(a) 10^3 (b) 120

8.28 Determine the Russell-Saunders terms for two equivalent d electrons.

Ans: Ref. HZ1,132

^1G, ^3F, ^1D, ^3P, ^1S

8.29 Determine the Russell-Saunders terms for two equivalent f electrons.

Ans: Ref. HZ1,132

^1I, ^3H, ^1G, ^3F, ^1D, ^3P, ^1S

8.30 Work out the eigenfunctions for the ^1D states of the configuration p^2, i.e. construct linear combinations of Slater determinants which diagonalize $\hat{L}^2$ and $\hat{S}^2$.

Ans: Ref. S2,84; Problem 8.18

$\Psi(L = 2, M_L = 2; S = 0, M_S = 0) = (1^+, 1^-)$

$\Psi(L = 2, M_L = 1; S = 0, M_S = 0) = \frac{1}{\sqrt{2}}[(1^+, 0^-) - (1^-, 0^+)]$

$$\Psi(L = 2, M_L = 0; S = 0, M_S = 0) = \frac{1}{\sqrt{6}}[(1^+, -1^-) - (1^-, -1^+) + 2(0^+, 0^-)]$$

$$\Psi(L = 2, M_L = -1; S = 0, M_S = 0) = \frac{1}{\sqrt{2}}[(0^+, -1^-) - (0^-, -1^+)]$$

$$\Psi(L = 2, M_L = -2; S = 0, M_S = 0) = (-1^+, -1^-)$$

8.31 List the lowest terms for the following atoms, i.e. give the ground states: (a) Ca, (b) N, (c) Co, (d) Al.

Ans: Ref. HZ1,138; Introduction to Chapter 8

(a) $^1S_0$, (b) $^4S_{3/2}$, (c) $^4F_{9/2}$, (d) $^2P_{1/2}$

8.32 Consider a one-dimensional system consisting of two widely separated atoms, one of which is located at $z = 0$ and the other at $z = R$. Assume that each atom has a single electron which is bound to the nucleus by a linear restoring force with the force constant k.

(a) Show that the instantaneous energy of interaction between the atoms is $\hat{\mathcal{H}}' = -2e^2z_1z_2/(4\pi\varepsilon_0)R^3$ where z_1 is the displacement of electron 1 from the origin and z_2 is the displacement of electron 2 from the point $z = R$.

(b) Use perturbation theory to show that the energy of interaction is proportional to R^{-6}.

(c) Solve for the exact energy of the system by transforming from the one electron coordinates z_1, z_2 to the coordinates $z_1 + z_2$ and $z_1 - z_2$. Expand the solution for the ground state in terms of R^{-1} for comparison with part (b).

Ans: Ref. KP,265; Problems 8.22 and 8.23

(b) $E_0^{(2)} = -\dfrac{h\nu_0 e^4}{(4\pi\varepsilon_0)^2 2k^2R^6}$ (c) $E_0 = h\nu_0 - \dfrac{h\nu_0 e^4}{(4\pi\varepsilon_0)^2 2k^2R^6} + \cdots$

8.33 Work out expressions for the energies of the Russell-Saunders terms for the configuration $(2p)^3$. Your answers should be given in terms of integrals over the determinantal functions ψ_i of Problem 8.15, e.g. $\psi_1 = (1^+, 1^-, 0^+)$. (Hint: Use the diagonal sum rule to obtain $\varepsilon(^2P)$.)

Ans: Ref. Problems 8.15, 8.19

$$\varepsilon(^2D) = \langle\psi_1|\hat{\mathcal{H}}|\psi_1\rangle$$

$$\varepsilon(^2P) = \langle\psi_3|\hat{\mathcal{H}}|\psi_3\rangle + \langle\psi_5|\hat{\mathcal{H}}|\psi_5\rangle - \langle\psi_1|\hat{\mathcal{H}}|\psi_1\rangle$$

$$\varepsilon(^4S) = \langle\psi_7|\hat{\mathcal{H}}|\psi_7\rangle$$

where $\psi_1 = (1^+, 1^-, 0^+)$, $\psi_3 = (1^+, 1^-, -1^+)$, $\psi_5 = (0^+, 0^-, 1^+)$ and $\psi_7 = (1^+, -1^+, 0^+)$.

CHAPTER **9**

ELECTRONIC STRUCTURE OF MOLECULES

The major application of quantum mechanics in chemistry has been the attempt to calculate molecular energies and geometries. The early calculations, which were restricted to H_2^+ and H_2, met with considerable success and provided a basis for the qualitative understanding of the electronic structure of molecules. For example the concept of molecular orbitals permitted chemists to apply the Aufbau principle to molecules just as Bohr's theory of the H-atom had done for atoms.

Going beyond H_2 presented formidable many-body problems, which still cannot be solved exactly. For over forty years approximate methods of solution, usually based on the variation principle, have been developed and refined. In the last decade high speed digital computers have made possible a breakthrough toward the goal of accurate energies for large molecules (~50 atom). But even so, the Born-Oppenheimer approximation must be made and rather drastic orbital approximations must be accepted. The fact that molecular orbitals are used, even though they are antisymmetrized, implies neglect of at least part of the correlation in the motion of the electrons. Beyond this, rather restricted mathematical expressions are generally used to describe the orbitals.

The details of the many approximate methods are beyond the scope of this book. Instead we have restricted our attention to rather general principles (e.g. the Born-Oppenheimer approximation), simple systems (e.g. H_2^+ and H_2), and the application of a few crude models (e.g. square-well, δ-function potential, and the Hückel method). The simple Hückel MO (HMO) method is used in a number of problems because it permits the student to actually carry through molecular calculations to their conclusion, obtaining wave functions and energies without the use of a computer. These problems also allow us to introduce the use of group theory to simplify secular equations. The defects in the Hückel method are well-known(MKT,281).

It is well-known that the HMO method is not consistent in its treatment of electron repulsions, and that the resulting energies are not very good even when the resonance integrals have been adjusted to fit experimental data. It has, in fact, been true for a decade that even routine, approximate calculations can and should be based on more sophisticated theories such as that developed by Pople. As mentioned above we have not included approximate self-consistent field calculations in this collection. To fill the gap we refer the reader to the books by Parr, Pople and Beveridge, and Dewar.

General references: PA2; PB; DE; MKT

PROBLEMS

9.1 In the Born-Oppenheimer approximation separate though coupled Schrödinger equations are solved for the electrons and nuclei in a molecule. It is assumed that the wave function for a molecule with nuclear coordinates R and electronic coordinates r can be written in the form

$$\Psi(r, R) = \psi(r; R)\chi(R) \tag{1}$$

where $\psi(r; R)$ is a solution to Eq. (2), the electronic wave equation, and $\chi(R)$ is a solution to Eq. (3), the nuclear wave equation.

$$(\hat{\mathcal{H}}_T - \hat{T}_n)\psi(r; R) = \hat{\mathcal{H}}_e\psi(r; R) = E(R)\psi(r; R) \tag{2}$$

$$[\hat{T}_n + E(R)]\chi(R) = E_a\chi(R) \tag{3}$$

Here

$\hat{\mathcal{H}}_T$ = total molecular Hamiltonian; $\hat{T}_n$ = nuclear kinetic energy terms

Substitute Eq. (1) into the full Schrödinger equation for the molecule and determine what approximations must be made in order to obtain Eqs. (2) and (3).

Solution: Ref. PI1,414; RA,349

For a molecule containing n nuclei and p electrons the Schrödinger equation is:

$$\left[-\sum_{i=1}^{n}\frac{\hbar^2}{2M_i}\nabla_i^2 - \sum_{j=1}^{p}\frac{\hbar^2}{2m_e}\nabla_j^2 + V(r, R)\right]\Psi(r, R) = E_T\Psi(r, R) \tag{4}$$

where

$$V(r, R) = \frac{e^2}{(4\pi\varepsilon_o)}\left[-\sum_{k=1}^{n}\sum_{i=1}^{p}\frac{Z_k}{r_{ik}} + \sum_{i<j}^{p}\frac{1}{r_{ij}} + \sum_{k<\ell}^{n}\frac{Z_k Z_\ell}{R_{k\ell}}\right]$$

Substitution of Eq. (1) into Eq. (4) gives:

$$-\sum_{i=1}^{n}\frac{\hbar^2}{2M_i}[\psi(r; R)\nabla_i^2\chi(R) + \chi(R)\nabla_i^2\psi(r; R) + 2\underset{\sim}{\nabla}_i\chi(R)\cdot\underset{\sim}{\nabla}_i\psi(r; R)]$$

$$-\sum_{j=1}^{p}\frac{\hbar^2}{2m_e}\chi(R)\nabla_j^2\psi(r; R) + V(r, R)\psi(r; R)\chi(R) = E_T\psi(r; R)\chi(R) \tag{5}$$

Here $\underset{\sim}{\nabla}_i\psi(r; R)$ refers to differentiation with respect to R only. If the second and third terms in the first summation of Eq. (5) are neglected and this equation is divided by

$\psi(r, R)\chi(R)$, one obtains:

$$-\sum_{j=1}^{p} \frac{\hbar^2}{2m_e} \frac{\nabla_j^2\psi(r; R)}{\psi(r, R)} + V(r, R) \cong \sum_{i=1}^{n} \frac{\hbar^2}{2M_i} \frac{\nabla_i^2\chi(R)}{\chi(R)} + E_T = E(R) \tag{6}$$

Here $E(R)$ is the eigenvalue of the operator $(\hat{\mathcal{H}}_T - \hat{T}_n)$ for a fixed value of R as shown in Eq. (2). Equation (6) also shows that

$$[\hat{T}_n + E(R)]\chi(R) = E_a\chi(R)$$

where E_a is an approximation to E_T.

The neglected terms are:

$$-\sum_{i=1}^{n} \frac{\hbar^2}{2M_i} [\chi(R)\nabla_i^2\psi(r; R) + 2\underset{\sim}{\nabla}_i\chi(R)\cdot\underset{\sim}{\nabla}_i\psi(r; R)] \tag{7}$$

The terms in (7) constitute a perturbation which can cause transitions between the electronic levels. (See: H. C. Longuet-Higgins in Advances in Spectroscopy (Interscience, New York, 1962), Vol. II, Edited by H. W. Thompson, p. 429.)

9.2 Apply the linear variation method to H_2^+ using a trial function of the form

$$\phi = c_A u_A + c_B u_B$$

where

$$u_A = \pi^{-1/2} e^{-r_A} \quad \text{and} \quad u_B = \pi^{-1/2} e^{-r_B}$$

to obtain approximate eigenfunctions and eigenvalues for the ground and first excited states. The coordinates of the problem are shown below.

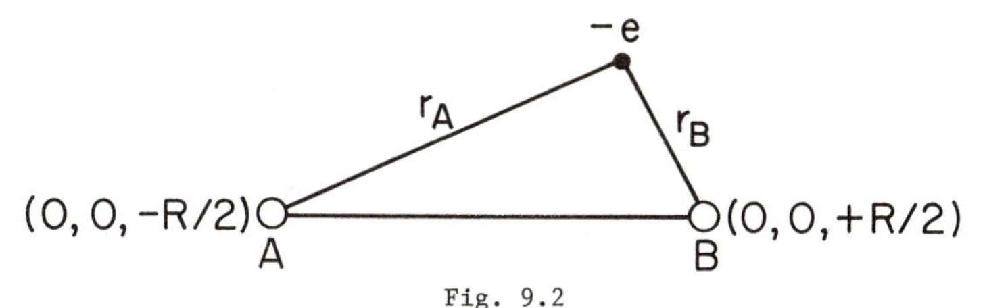

Fig. 9.2

Your answers should be given in terms of integrals over u_A and u_B except where evaluations are trivial.

Solution: Ref. Problem 6.17

The eigenvalues are readily obtained from the secular equation

$$\begin{vmatrix} (H_{AA} - E) & (H_{AB} - ES) \\ (H_{BA} - ES) & (H_{BB} - E) \end{vmatrix} = 0$$

where

$$H_{AA} = H_{BB} = \int u_A^* \hat{\mathcal{H}} u_A d\tau = \langle u_A|\hat{\mathcal{H}}|u_A\rangle; \quad H_{AB} = H_{BA} = \langle u_A|\hat{\mathcal{H}}|u_B\rangle; \quad S = \langle u_A|u_B\rangle = \langle u_B|u_A\rangle$$

Therefore:

$$(H_{AA} - E)^2 = (H_{AB} - SE)^2$$

$$H_{AA} - E = \pm(H_{AB} - SE)$$

$$E_{\pm} = \frac{H_{AA} \pm H_{AB}}{(1 \pm S)} \tag{1}$$

The coefficients are obtained from the equation

$$(H_{AA} - E)c_A + (H_{AB} - ES)c_B = 0$$

by substituting E_+ or E_- for E. For E_+ it turns out that $c_A = c_B$, and for E_- we find $c_A = - c_B$. These conclusions could have been deduced in advance by considering the symmetry of the molecular Hamiltonian. Since the electron probability distribution $|\phi|^2$ must be symmetric with respect to the center of symmetry, $c_A = \pm c_B$. If the symmetric and antisymmetric linear combinations of u_A and u_B had been used at the outset, the secular problem would have immediately given two 1x1 blocks. For the magnitudes of the coefficients we use the normalization condition:

$$\int|\phi_{\pm}|^2 d\tau = c_{\pm}^2\int(u_A^2 + u_B^2 \pm 2u_Au_B)d\tau = 1 \quad \text{or} \quad 2(1 \pm S)c_{\pm}^2 = 1 \quad \text{or} \quad c_{\pm} = [2(1 \pm S)]^{-1/2}$$

Eq. (1) can be carried a few steps further by writing out $\hat{\mathcal{H}}$ in detail. Thus

$$\langle u_A|\hat{\mathcal{H}}|u_A\rangle = \langle u_A|-\nabla^2 - \frac{2}{r_A} - \frac{2}{r_B} + \frac{2}{R}\ |u_A\rangle = E_H + \frac{2}{R} - \langle u_A|\ \frac{2}{r_B}\ |u_A\rangle \tag{2}$$

This result follows since

$$\left(-\nabla^2 - \frac{2}{r_A}\right)u_A = E_H u_A$$

Similarly for H_{AB} we find:

$$\langle u_A|\hat{\mathcal{H}}|u_B\rangle = \langle u_A|-\nabla^2 - \frac{2}{r_B} - \frac{2}{r_A} + \frac{2}{R}\ |u_B\rangle = S\left(E_H + \frac{2}{R}\right) - \langle u_A|\ \frac{2}{r_A}\ |u_B\rangle \tag{3}$$

Substitution of Eqs. (2) and (3) into Eq. (1) gives:

$$E_{\pm} = E_H + \frac{2}{R} - \left[\ \frac{\langle u_A|2r_B^{-1}|u_A\rangle \pm \langle u_A|2r_A^{-1}|u_B\rangle}{(1 \pm S)}\ \right]$$

9.3 The integrals appearing in Problem 9.2 can be evaluated without difficulty by using elliptical coordinates. The necessary relations between the coordinates are:

$$\mu = (r_A + r_B)/R \qquad 1 \lesseqgtr \mu \lesseqgtr \infty \tag{1}$$

$$\nu = (r_A - r_B)/R \qquad -1 \lesseqgtr \nu \lesseqgtr +1 \tag{2}$$

$$\phi \qquad 0 \lesseqgtr \phi \lesseqgtr 2\pi$$

and the volume element is

$$d\tau = \frac{R^3}{8}(\mu^2 - \nu^2)d\mu d\nu d\phi$$

(see Appendix 5). Obtain analytical expressions for the following integrals in terms of R:

(a) $S = \frac{1}{\pi}\int e^{-(r_A+r_B)}d\tau$ (b) $\langle u_A|2r_B^{-1}|u_A\rangle = \frac{1}{\pi}\int e^{-r_A}\frac{2}{r_B}e^{-r_A}d\tau$

Solution:

(a) Equations (1) and (2) give $\mu R = (r_A + r_B)$ and we obtain:

$$S = \frac{1}{\pi}\int_0^{2\pi}\int_{-1}^{+1}\int_{+1}^{\infty} e^{-\mu R}\frac{R^3}{8}(\mu^2 - \nu^2)d\mu d\nu d\phi = \frac{R^3}{4}\left[\int_{+1}^{\infty}\mu^2 e^{-\mu R}d\mu\int_{-1}^{+1}d\nu - \int_{+1}^{\infty}e^{-\mu R}d\mu\int_{-1}^{+1}\nu^2 d\nu\right]$$

$$= \frac{R^3}{4}\left[\frac{2e^{-R}}{R^3}\left(1 + R + \frac{R^2}{2}\right)2 - \frac{e^{-R}}{R}\frac{2}{3}\right] = e^{-R}\left(1 + R + \frac{R^2}{3}\right)$$

Integral tables were used in the last steps. A particularly convenient relation is

$$\int_b^{\infty} x^n e^{-ax}dx = \frac{n!e^{-ab}}{a^{n+1}}\left[1 + ab + \frac{1}{2!}(ab)^2 + \dots \frac{1}{n!}(ab)^n\right]$$

(b) Equations (1) and (2) give

$$r_A = \frac{R}{2}(\mu + \nu) \quad \text{and} \quad r_B = \frac{R}{2}(\mu - \nu)$$

Thus

$$\langle u_A|2r_B^{-1}|u_A\rangle = \frac{1}{\pi}\int_0^{2\pi}\int_{-1}^{+1}\int_{+1}^{\infty} e^{-R(\mu+\nu)}\frac{2\cdot 2}{R(\mu - \nu)}\cdot\frac{R^3}{8}(\mu - \nu)(\mu + \nu)d\mu d\nu d\phi$$

$$= R^2\left[\int_{+1}^{\infty}e^{-R\mu}\mu d\mu\int_{-1}^{+1}e^{-R\nu}d\nu + \int_{+1}^{\infty}e^{-R\mu}d\mu\int_{-1}^{+1}e^{-R\nu}\nu d\nu\right] = \frac{2}{R}\left[1 - e^{-2R}(1 + R)\right]$$

9.4 The two lowest electronic states for H_2 can be understood semiquantitatively in terms of simple molecular orbital theory using spatial functions of the form:

$$\phi_{\pm} = \frac{1}{\sqrt{2(1 \pm S)}}(u_A \pm u_B) \qquad (1)$$

Use Eq. (1) and the spin functions α and β to construct all possible two electron functions which are (a) antisymmetric in the exchange of the electrons and (b) eigenfunctions of the operator $\hat{S}^2$. Example: $\psi(S = 1, M_s = 1) = [\phi_+(1)\phi_-(2) - \phi_-(1)\phi_+(2)]\alpha(1)\alpha(2)/\sqrt{2}$

Solution: Ref. S5,60; M.J.S. Dewar and J. Kelemen, J. Chem. Ed. 48, 494 (1971); Problems 5.22, 5.24, 8.6

There are six possible ways to arrange the electrons in keeping with the Pauli principle:

ϕ_-	___	↑	↓	↑	↓	↑↓
ϕ_+	↑↓	↑	↑	↓	↓	___
	ψ_1	ψ_2	ψ_3	ψ_4	ψ_5	ψ_6
M_s	0	1	0	0	-1	0

The functions $\psi_1 - \psi_6$ are conveniently written as determinantal functions:

$$\psi_1 = (\phi_+\bar{\phi}_+) = \frac{1}{\sqrt{2}} \begin{vmatrix} \phi_+\alpha(1) & \phi_+\alpha(2) \\ \phi_+\beta(1) & \phi_+\beta(2) \end{vmatrix} = \frac{1}{\sqrt{2}} \phi_+(1)\phi_+(2)[\alpha(1)\beta(2) - \beta(1)\alpha(2)]$$

$$\psi_2 = (\phi_+\phi_-) = \frac{1}{\sqrt{2}} \begin{vmatrix} \phi_+\alpha(1) & \phi_+\alpha(2) \\ \phi_-\alpha(1) & \phi_-\alpha(2) \end{vmatrix} = \frac{1}{\sqrt{2}} [\phi_+(1)\phi_-(2) - \phi_-(1)\phi_+(2)]\alpha(1)\alpha(2)$$

$$\psi_3 = (\phi_+\bar{\phi}_-) = \frac{1}{\sqrt{2}} \begin{vmatrix} \phi_+\alpha(1) & \phi_+\alpha(2) \\ \phi_-\beta(1) & \phi_-\beta(2) \end{vmatrix} = \frac{1}{\sqrt{2}} [\phi_+(1)\phi_-(2)\alpha(1)\beta(2) - \phi_-(1)\phi_+(2)\alpha(2)\beta(1)]$$

$$\psi_4 = (\bar{\phi}_+\phi_-) = \frac{1}{\sqrt{2}} \begin{vmatrix} \phi_+\beta(1) & \phi_+\beta(2) \\ \phi_-\alpha(1) & \phi_-\alpha(2) \end{vmatrix} = \frac{1}{\sqrt{2}} [\phi_+(1)\phi_-(2)\beta(1)\alpha(2) - \phi_-(1)\phi_+(2)\alpha(1)\beta(2)]$$

Similarly

$$\psi_5 = \frac{1}{\sqrt{2}} [\phi_+(1)\phi_-(2) - \phi_-(1)\phi_+(2)]\beta(1)\beta(2)$$

$$\psi_6 = \frac{1}{\sqrt{2}} \phi_-(1)\phi_-(2)[\alpha(1)\beta(2) - \beta(1)\alpha(2)]$$

The functions ψ_1, ψ_2, ψ_5, and ψ_6 are eigenfunctions of $\hat{S}^2$ and $\hat{S}_z$ (see Problem 8.6) with quantum numbers as shown below:

$\psi_1 = \psi_1(S = 0, M_s = 0)$; $\qquad \psi_6 = \psi_6(S = 0, M_s = 0)$

$\psi_2 = \psi_2(S = 1, M_s = 1)$; $\qquad \psi_5 = \psi_5(S = 1, M_s = -1)$

The functions ψ_3 and ψ_4 correspond to $M_s = 0$, but are not eigenfunctions of $\hat{S}^2$. The reader should refer to Problems 5.22, 5.24, and 8.6 for the methods of constructing singlet and triplet functions. The proper combinations are:

$$\psi_3'(S = 1, M_s = 0) = \frac{1}{\sqrt{2}} (\psi_3 + \psi_4) = \frac{1}{2} [\phi_+(1)\phi_-(2) - \phi_-(1)\phi_+(2)][\alpha(1)\beta(2) + \beta(1)\alpha(2)]$$

$$\psi_4'(S = 0, M_s = 0) = \frac{1}{\sqrt{2}} (\psi_3 - \psi_4) = \frac{1}{2} [\phi_+(1)\phi_-(2) + \phi_-(1)\phi_+(2)][\alpha(1)\beta(2) - \beta(1)\alpha(2)]$$

9.5 The simplest molecular orbital wave function for the ground state of H_2 is the singlet function

$$\psi_1 = \phi_+(1)\phi_+(2) \frac{1}{\sqrt{2}} [\alpha(1)\beta(2) - \beta(1)\alpha(2)]$$

which was constructed in Problem 9.4. Use ψ_1 to derive an expression for the expectation value of the energy for H_2 in terms of one-electron integrals of the form $\langle u_A | 2r_{A1} | u_r \rangle$

and two-electron integrals of the form $\langle u_A u_B | 2r_{12}^{-1} | u_r u_s \rangle$ where r and s represent A or B. The coordinates are defined in Fig. 9.5.

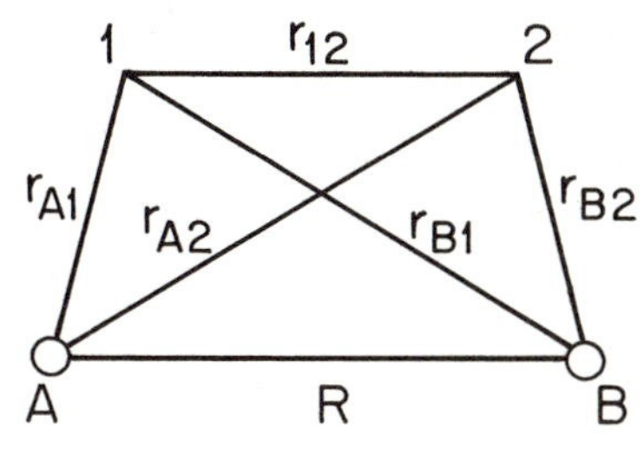

Fig. 9.5

Solution: Ref. S5,60

The Hamiltonian for H_2 is (atomic units with energy in Rydbergs):

$$\hat{\mathcal{H}} = -\nabla_1^2 - \frac{2}{r_{A1}} - \frac{2}{r_{B1}} - \nabla_2^2 - \frac{2}{r_{A2}} - \frac{2}{r_{B2}} + \frac{2}{r_{12}} + \frac{2}{R} = \hat{\mathcal{H}}(1) + \hat{\mathcal{H}}(2) + \frac{2}{r_{12}} + \frac{2}{R}$$

The expectation value of $\hat{\mathcal{H}}$ is:

$$E = \langle \psi_1 | \hat{\mathcal{H}} | \psi_1 \rangle = \langle \phi_+(1)\phi_+(2) | \hat{\mathcal{H}} | \phi_+(1)\phi_+(2) \rangle$$

since the spin function is normalized. Thus

$$E = \langle \phi_+(1) | \hat{\mathcal{H}}(1) | \phi_+(1) \rangle + \langle \phi_+(2) | \hat{\mathcal{H}}(2) | \phi_+(2) \rangle + \langle \phi_+(1)\phi_+(2) | 2r_{12}^{-1} | \phi_+(1)\phi_+(2) \rangle + 2/R \tag{1}$$

where we have used the normalization condition $\langle \phi_+(i) | \phi_+(i) \rangle = 1$. The first two terms on the r.h.s. of Eq. (1) are equal because of the symmetry of the Hamiltonian. We can thus write:

$$E = \frac{2}{2(1+S)} \int [u_A(1) + u_B(1)] \left(-\nabla_1^2 - \frac{2}{r_{A1}} - \frac{2}{r_{B1}} \right) [u_A(1) + u_B(1)] d\tau_1$$

$$+ \frac{1}{4(1+S)^2} \iint [u_A(1) + u_B(1)]^2 [u_A(2) + u_B(2)]^2 \frac{2}{r_{12}} d\tau_1 d\tau_2 + \frac{2}{R} \tag{2}$$

The terms in Eq. (2) simplify as follows:

$$\langle u_A(1) + u_B(1) | -\nabla_1^2 - \frac{2}{r_{A1}} - \frac{2}{r_{B1}} | u_A(1) + u_B(1) \rangle = \langle u_A(1) + u_B(1) |$$

$$\left(E_H - \frac{2}{r_{B1}} \right) | u_A(1) \rangle + \left(E_H - \frac{2}{r_{A1}} \right) | u_B(1) \rangle$$

$$= E_H(1 + S) - \langle u_A | \frac{2}{r_{B1}} | u_A \rangle - \langle u_B | \frac{2}{r_{B1}} | u_A \rangle + E_H(1 + S)$$

$$- \langle u_A | \frac{2}{r_{A1}} | u_B \rangle - \langle u_B | \frac{2}{r_{A1}} | u_B \rangle$$

The second terms on the r.h.s. of Eq. (2) give

$$\frac{1}{2(1+S)^2}\iint [u_A^2(1) + 2u_A(1)u_B(1) + u_B^2(1)][u_A^2(2) + 2u_A(2)u_B(2) + u_B^2(2)]r_{12}^{-1}d\tau_1 d\tau_2$$

$$= \frac{1}{2(1+S)^2}\Big[\langle u_A u_A|\frac{2}{r_{12}}|u_A u_A\rangle + 4\langle u_A u_A|\frac{2}{r_{12}}|u_A u_B\rangle + \langle u_A u_B|\frac{2}{r_{12}}|u_A u_B\rangle +$$

$$2\langle u_A u_A|\frac{2}{r_{12}}|u_B u_B\rangle\Big]$$

The natural order is always maintained for the electrons in these integrals, i.e.

$$\langle u_A u_B|2r_{12}^{-1}|u_R u_S\rangle \equiv \langle u_A(1)u_B(2)|2r_{12}^{-1}|u_R(1)u_S(2)\rangle$$

Notation varies widely in the integral evaluation literature. The orbitals are rearranged by some authors to make the charge clouds more evident. For example, Slater uses

$$[AR|BS] \equiv (AR|BS) \equiv \langle u_A u_B|2r_{12}^{-1}|u_R u_S\rangle$$

Combining the above results with Eq. (2) gives

$$E = 2E_H - 2\left[\frac{\langle u_A|2r_{B1}^{-1}|u_A\rangle + \langle u_A|2r_{B1}^{-1}|u_B\rangle}{(1+S)}\right]$$

$$+\left[\frac{\langle u_A u_A|2r_{12}^{-1}|u_A u_A\rangle + 4\langle u_A u_A|2r_{12}^{-1}|u_A u_B\rangle + \langle u_A u_B|2r_{12}^{-1}|u_A u_B\rangle + 2\langle u_A u_A|2r_{12}^{-1}|u_B u_B\rangle}{2(1+S)^2}\right] + \frac{2}{R}$$

9.6 In the molecular orbital (LCAO-MO) treatment of H_2 using atomic orbitals of the form $a(1) = \sqrt{\alpha^3/\pi}\,\exp(-\alpha r_{1a})$ with $\alpha = 1$ it is found that the energy of the ground state approaches $(-2 + \frac{5}{8})$ Rydbergs as the internuclear separation becomes infinite. This is roughly the average of the energies of nonionic and ionic H_2 for infinite separation in the AO approximation.

(a) Calculate the approximate energy of ionic H_2 in the limit of infinite internuclear separation in terms of α by calculating the energy of H^-. The expectation value of the Hamiltonian for H^- can easily be obtained in the orbital approximation. The necessary integrals have been tabulated by Slater.

$$\int a(1)(-\nabla_1^2)a(1)dv_1 = \alpha^2$$

$$\int a^2(1)(-2/r_{1a})dv_1 = -2\alpha$$

$$\int a^2(1)a^2(2)(2/r_{12})dv_1 dv_2 = 5\alpha/4$$

(b) Obtain the best value of α from the solution to part (a) and compare the minimum energy obtained with the exact energy of -1.0555 Ry for the H^- ion. To what do you attribute the difference? Here Ry = Rydberg.

Solution: Ref. S5,67; C. L. Pekeris, Phys. Rev. 112, 1649 (1958)

(a) For H^- the Hamiltonian is:

$$\hat{\mathcal{H}} = -\nabla_1^2 - \nabla_2^2 - \frac{2}{r_1} - \frac{2}{r_2} + \frac{2}{r_{12}}$$

Thus in the orbital approximation

$$\langle E\rangle = \langle a(1)a(2)|\hat{\mathcal{H}}|a(1)a(2)\rangle = \langle a(1)|-\nabla_1^2|a(1)\rangle + \langle a(2)|-\nabla_2^2|a(2)\rangle + \langle a(1)|-2/r_1|a(1)\rangle$$

$$+ \langle a(2)|-2/r_2|a(2)\rangle + \langle a(1)a(2)|2/r_{12}|a(1)a(2)\rangle = 2\alpha^2 - 4\alpha + \frac{5}{4}\alpha = 2\alpha^2 - \frac{11}{4}\alpha$$

(b) The "best" value of α corresponds to the minimum value of $\langle E\rangle$. Thus $\frac{\partial\langle E\rangle}{\partial\alpha} = 4\alpha - \frac{11}{4} = 0$ or $\alpha = \frac{11}{16}$ and the minimum value of $\langle E\rangle$ is -0.9453 Ry. This is greater than the exact energy because electron correlation has been neglected.

9.7 Assume that the total asymmetry of charge in a heteronuclear diatomic molecule is determined by the doubly occupied MO:

$$\psi = N(u_A + \lambda u_B) \tag{1}$$

where u_A and u_B are "pure" atomic orbitals, i.e. polarization and hybridization are neglected. A charge of +e is centered on each nucleus and the internuclear distance is R. Derive an expression for the electric dipole moment μ in terms of λ, R, and the overlap integral S. You may neglect the integral $\langle u_A|z|u_B\rangle$ where the bond lies along the z-direction if the origin is taken at the center point of the bond. This assumption is not always justified.

Solution: Ref. CO,107

First we obtain the normalization constant for Eq. (1)

$$\int\psi^*\psi d\tau = N^2\int(u_A^2 + \lambda^2 u_B^2 + 2\lambda u_A u_B)d\tau = N^2(1 + \lambda^2 + 2\lambda S) = 1$$

Therefore

$$N = (1 + \lambda^2 + 2\lambda S)^{-1/2}$$

In principle the dipole moment for a neutral molecule is independent of the choice of origin; however we wish to minimize the quantity $\langle u_A|z|u_B\rangle$ and so the origin is taken at the center of the bond as shown below.

```
A                 B
•--------+--------•
-R/2     0      +R/2
```

The dipole moment is given by:

$$\mu = \sum_{i=A,B} q_i z_i + \int\psi^*(-2ez)\psi d\tau = -2eN^2\int(zu_A^2 + z\lambda^2 u_B^2 + 2z\lambda u_A u_B)d\tau$$

$$= -2eN^2(\langle u_A|z|u_A\rangle + \lambda^2\langle u_B|z|u_B\rangle + 2\lambda\langle u_A|z|u_B\rangle)$$

Since u_A and u_B are pure atomic orbitals, we can replace $\langle u_A|z|u_A\rangle$ and $\langle u_B|z|u_B\rangle$ with -R/2 and +R/2, respectively. The result is

$$\mu = \frac{eR(1 - \lambda^2)}{(1 + \lambda^2 + 2\lambda S)}$$

9.8 A determinantal wave function for a molecule is denoted by

$$\Delta = \frac{1}{\sqrt{N!}} \det|\prod_j \psi_j| \quad \text{where} \quad \psi_j = \phi_j \chi_j \tag{1}$$

Here ψ_j is a molecular spin orbital having the spin part χ_j and the spatial function

$$\phi_j = \sum_i c_{ij} u_i$$

Show that the electron probability density arising from the N electron wave function Δ is given by

$$\rho(x,y,z) = \sum_{j=1}^{N} |\phi_j|^2 = \sum_{i,k} P_{ik} u_i^* u_k \tag{2}$$

where

$$P_{ik} = \sum_j c_{ij}^* c_{kj}$$

is called the density matrix. (Hint: Use the electron density operator $\hat{\rho}(\underset{\sim}{r}) = \sum_{i=1}^{N} \delta(\underset{\sim}{r}_i - \underset{\sim}{r})$)

Solution: Ref. Introduction to Chapter 8; PB,43

The Condon-Slater rules give:

$$\langle\Delta|\hat{\mathcal{H}}|\Delta\rangle = \sum_i \langle i|f|i\rangle + \sum_{i>j} [\langle ij|g|ij\rangle - \langle ij|g|ji\rangle]$$

for the matrix element of a sum of one and two electron operators. In this application we write

$$\langle\Delta|\hat{\rho}(\underset{\sim}{r})|\Delta\rangle = \sum_{i=1}^{N} \langle\phi_i(1)|\delta(\underset{\sim}{r} - \underset{\sim}{r}_1)|\phi_i(1)\rangle \tag{3}$$

Thus we obtain the function:

$$\rho(x,y,z) = \sum_{j=1}^{N} |\phi_j|^2$$

Then substituting

$$\phi_j = \sum_i c_{ij} u_i$$

we have

$$\rho(x,y,z) = \sum_{j=1}^{N} \sum_i c_{ij}^* u_i^* \sum_k c_{kj} u_k = \sum_i \sum_k (\sum_j c_{ij}^* c_{kj}) u_i^* u_k = \sum_i \sum_k P_{ik} u_i^* u_k$$

where

$$P_{ik} = \sum_j c^*_{ij} c_{kj}$$

The density matrix for closed shells is sometimes written as

$$P_{ik} = 2 \sum_j^{occ} c^*_{ij} c_{kj}$$

where the summation refers to spatial orbitals not spin orbitals.

9.9 The Slater determinant for a closed shell configuration (each spatial orbital is occupied with two electrons or none) is denoted by

$$\Delta = (\phi_1\bar{\phi}_1\phi_2\bar{\phi}_2\cdots\phi_N\bar{\phi}_N) \tag{1}$$

where the ϕ_i are molecular orbitals, the bar means β spin, and no bar means α spin. Use the Condon-Slater rules to show that the energy of this configuration is given by:

$$E = 2\sum_{i=1}^{N} \varepsilon_i^o + \sum_{i,j}^{N} (2J_{ij} - K_{ij}) \tag{2}$$

where the first term on the right side of Eq. (2) gives the expectation values of the one electron operators in the Hamiltonian,

$$J_{ij} = \langle\phi_i(1)\phi_j(2)| \frac{2}{r_{12}} |\phi_i(1)\phi_j(2)\rangle$$

and

$$K_{ij} = \langle\phi_i(1)\phi_j(2)| \frac{2}{r_{12}} |\phi_j(1)\phi_i(2)\rangle$$

(Hint: The indices in Eq. (2) refer to spatial orbitals while the Condon-Slater rules are written in terms of spin orbitals.)

Solution: Ref. DE,64; Introduction to Chapter 8; PI1,348

The Condon-Slater rules give

$$\langle\Delta|\hat{\mathcal{H}}|\Delta\rangle = \sum_i \langle i|f|i\rangle + \sum_{i<j} [\langle ij|g|ij\rangle - \langle ij|g|ji\rangle] \tag{3}$$

Here i and j refer to spin orbitals ψ_i and ψ_j and we can make a correspondence with the spin orbitals in Eq. (1) as follows: $\psi_1 = \phi_1$, $\psi_2 = \bar{\phi}_1$, $\psi_3 = \phi_2$, $\psi_4 = \bar{\phi}_2$, etc. Now

$$\langle\phi_1|f|\phi_1\rangle = \langle\phi_1(1)|-\nabla_1^2 - \sum_\ell \frac{2Z_\ell}{r_{1\ell}} |\phi_1(1)\rangle = \varepsilon_1^o = \langle 1|f|1\rangle = \langle 2|f|2\rangle$$

In converting to spatial indices the first term on the r.h.s. of Eq. (3) gives

$$\sum_i \langle i|f|i\rangle \rightarrow 2\sum_{i=1}^{N} \varepsilon_i^o$$

Now consider the electron repulsion integrals involving the spatial orbitals i and j. Integrals of the type J_{ij} arise four ways:

$$<\phi_i(1)\phi_j(2)|2r_{12}^{-1}|\phi_i(1)\phi_j(2)>, \quad <\phi_i(1)\bar{\phi}_j(2)|2r_{12}^{-1}|\phi_i(1)\bar{\phi}_j(2)>$$

$$<\bar{\phi}_i(1)\phi_j(2)|2r_{12}^{-1}|\bar{\phi}_i(1)\phi_j(2)>, \quad <\bar{\phi}_i(1)\bar{\phi}_j(2)|2r_{12}^{-1}|\bar{\phi}_i(1)\bar{\phi}_j(2)>$$

For K_{ij} there are only two nonvanishing possibilities:

$$<\phi_i(1)\phi_j(2)|2r_{12}^{-1}|\phi_j(1)\phi_i(2)>, \quad <\bar{\phi}_i(1)\bar{\phi}_j(2)|2r_{12}^{-1}|\bar{\phi}_j(1)\bar{\phi}_i(2)>$$

Finally since the orbitals i and j each contain two electrons we encounter the integrals J_{ii}, J_{jj}:

$$<\phi_i(1)\bar{\phi}_i(2)|2r_{12}^{-1}|\phi_i(1)\bar{\phi}_i(2)>, \quad <\phi_j(1)\bar{\phi}_j(2)|2r_{12}^{-1}|\phi_j(1)\bar{\phi}_j(2)>$$

Combining these results gives

$$E = 2\sum_{i=1}^{N} \varepsilon_i^o + \sum_{i=1}^{N} J_{ii} + \sum_{i<j} (4J_{ij} - 2K_{ij}) \tag{4}$$

But $K_{ii} = J_{ii}$ so that Eq. (3) can be written as:

$$E = 2\sum_{i=1}^{N} \varepsilon_i^o + \sum_{i,j} (2J_{ij} - K_{ij})$$

where the summations are over orbitals not spin orbitals.

9.10 Use the results of Problem 9.9 to derive an expression for the energy required to remove the kth electron from a molecule having a closed shell configuration. You may assume that there is no rearrangement of the remaining electrons on ionization. The ionization potential thus derived may be defined as the negative of the orbital energy of the kth electron. This is known as Koopmans' theorem.

Solution: Ref. PB,36

From Problem 9.9 we have the total energy of the N electron molecule:

$$E = 2\sum_{i=1}^{N} \varepsilon_i^o + \sum_{i,j}^{N} (2J_{ij} - K_{ij}) \tag{1}$$

The contribution to Eq. (1) from the kth electron is just:

$$\varepsilon_k = \varepsilon_k^o + \sum_{i}^{N} (2J_{ki} - K_{ki})$$

which we call the orbital energy of the kth electron. When the kth electron is removed, the energy of the singly ionized system is $E^+ = E - \varepsilon_k$. Therefore, the change in energy is $\delta E = -\varepsilon_k$.

9.11 Assuming that the Born-Oppenheimer approximation is valid, determine the conditions under which the potential energy curves for two different electronic states of a diatomic molecule can cross. (Hint: Assume that all of the wave functions are known except those associated with the two states in question. Then expand these two states in terms of the functions ψ_1 and ψ_2 which are orthogonal to each other and to all of the known wave functions.)

Solution: Ref. EWK,206

As suggested we expand the unknown states in terms of ψ_1 and ψ_2:

$$u = c_1\psi_1 + c_2\psi_2$$

The coefficients can be determined by solving the 2x2 secular determinant:

$$\begin{vmatrix} H_{11} - E & H_{12} \\ H_{12} & H_{22} - E \end{vmatrix} = 0$$

The solutions are (see Problem 6.3 and Appendix 6):

$$u = \frac{1}{\sqrt{2}}(\psi_1 \pm \psi_2)$$

and

$$E_{\pm} = \frac{H_{11} + H_{22}}{2} \pm \sqrt{\left(\frac{H_{22} - H_{11}}{2}\right)^2 + H_{12}^2} \tag{1}$$

Assuming that $[2H_{12}/(H_{22} - H_{11})]^2 < 1$ and taking $H_{22} > H_{11}$ Eq. (1) can be expanded to obtain:

$$E_1 = H_{11} - \frac{H_{12}^2}{H_{22} - H_{11}} + \cdots \tag{2}$$

$$E_2 = H_{22} + \frac{H_{12}^2}{H_{22} - H_{11}} - \cdots \tag{3}$$

In order for the levels to cross for some value of R so that $E_1 = E_2$ we must have $H_{11} - H_{22} = 0$ and $H_{12} = 0$. Otherwise there will be no simultaneous solutions for these equations. If ψ_1 and ψ_2 have different symmetry properties, $H_{12} = 0$ since $\hat{\mathcal{H}}$ is totally symmetric. In general if ψ_1 and ψ_2 have the same symmetry properties $H_{12} \neq 0$ and no crossing can occur. States having different multiplicities have different symmetry properties and thus may cross (see E. Teller, J. Phys. Chem. 41, 109 (1937)).

9.12 Apply the Hückel MO method to the π-electrons in the allyl group shown below to obtain the allowed energies in terms of the coulomb integral α and the resonance integral β. Also, obtain the normalized eigenfunctions. A direct solution is expected here without the use of symmetry in constructing the basis functions.

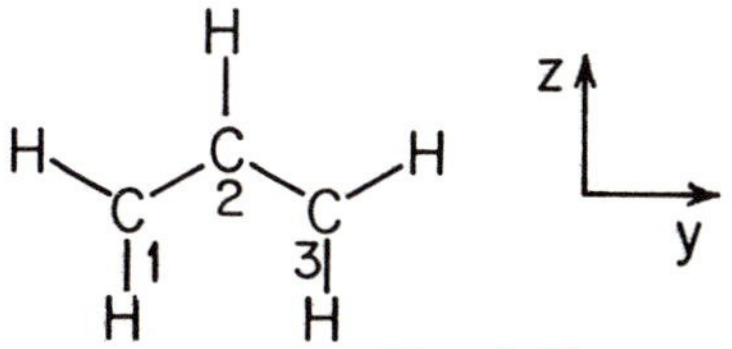

Fig. 9.12

Solution: Ref. Appendix 6

Let $\phi_i = (2p_x)_i$ and expand the μth MO as follows:

$$\psi_\mu = \sum_{i=1}^{3} c_{\mu i}\phi_i$$

In the linear variation theory the following integrals are required:

$$\mathcal{H}_{11} = \int \phi_1 \hat{\mathcal{H}} \phi_1 d\tau = \alpha_1 \qquad \mathcal{H}_{12} = \int \phi_1 \hat{\mathcal{H}} \phi_2 d\tau = \beta_{12} = \beta_{21}$$

$$\mathcal{H}_{22} = \alpha_2, \ \mathcal{H}_{33} = \alpha_3 \qquad \mathcal{H}_{23} = \beta_{23} = \beta_{32}, \ \mathcal{H}_{13} = 0$$

According to the usual Hückel assumptions $\mathcal{H}_{ii} = \alpha_i = \alpha$ for all values of i, $\mathcal{H}_{ij} = \beta_{ij} = \beta$ if atoms i and j are adjacent and $\mathcal{H}_{ij} = 0$ otherwise. The secular equation which must be solved to obtain the "best" $c_{\mu i}$'s is:

$$\begin{vmatrix} (\alpha - E) & \beta & 0 \\ \beta & (\alpha - E) & \beta \\ 0 & \beta & (\alpha - E) \end{vmatrix} = 0 \quad \text{or} \quad \text{letting } (\alpha - E)/\beta = X,$$

$$\begin{vmatrix} X & 1 & 0 \\ 1 & X & 1 \\ 0 & 1 & X \end{vmatrix} = X^3 - 2X = X(X^2 - 2) = 0$$

Thus $X = +\sqrt{2}$, 0, $-\sqrt{2}$ and $E = \alpha - \sqrt{2}\,\beta$, α, $\alpha + \sqrt{2}\,\beta$. To obtain the eigenfunctions we have for $X = -\sqrt{2}$:

$$-\sqrt{2}\,c_{11} + c_{12} = 0; \qquad c_{12} = \sqrt{2}\,c_{11}$$

$$c_{11} - \sqrt{2}\,c_{12} + c_{13} = 0$$

$$c_{11} - 2c_{11} + c_{13} = 0; \qquad c_{11} = c_{13}$$

Thus

$$\psi_1 = c_{11}(\phi_1 + \sqrt{2}\,\phi_2 + \phi_3); \qquad \sum_{i=1}^{3} c_{1i}^2 = c_{11}^2(1 + 2 + 1) = 1$$

and

$$\psi_1 = \frac{1}{2}(\phi_1 + \sqrt{2}\,\phi_2 + \phi_3); \qquad c_{11} = 1/2$$

Similar calculations for the other values of X lead to the results shown below:

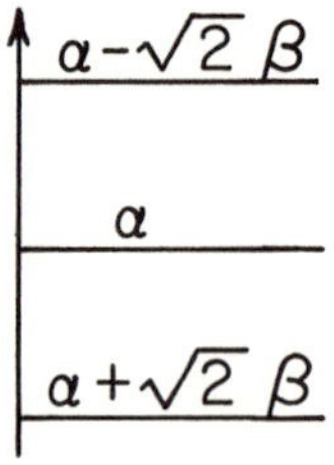

$\alpha - \sqrt{2}\,\beta$ $\qquad \psi_3 = \frac{1}{2}\phi_1 - \frac{1}{\sqrt{2}}\phi_2 + \frac{1}{2}\phi_3$

α $\qquad \psi_2 = \frac{1}{\sqrt{2}}\phi_1 - \frac{1}{\sqrt{2}}\phi_3$

$\alpha + \sqrt{2}\,\beta$ $\qquad \psi_1 = \frac{1}{2}\phi_1 + \frac{1}{\sqrt{2}}\phi_2 + \frac{1}{2}\phi_3$

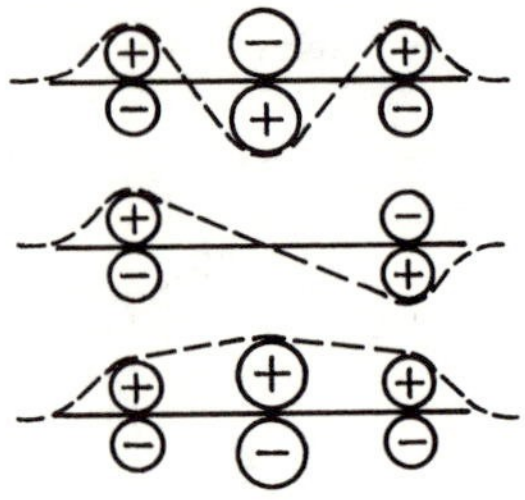

9.13 Repeat the determination of the eigenvalues and eigenfunctions for the allyl group which was carried out in Problem 9.12 but use linear combinations of the atomic $2p_x$ orbitals which are symmetric and antisymmetric with respect to reflection in the xz-mirror plane as basis functions. Thus, show that the secular problem can be simplified by the use of symmetry.

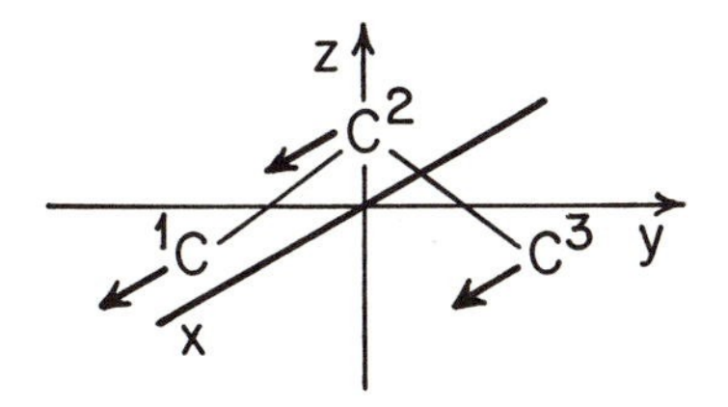

Fig. 9.13

Solution: Ref. Appendix 8

The allyl group has C_{2v} symmetry and the MO's must transform according to the irreducible representations of this point group. Even without the formalism of group theory it is clear that the eigenfunctions of the molecular Hamiltonian must be symmetric or antisymmetric with respect to reflection in the yz-plane since their squares (i.e. the charge distribution) must have the symmetry of the molecule. Accordingly, we use the basis functions

$$\left.\begin{aligned} u_1 &= \phi_2 \\ u_2 &= \frac{1}{\sqrt{2}}(\phi_1 + \phi_3) \end{aligned}\right\} \quad \text{symmetric}$$

$$u_3 = \frac{1}{\sqrt{2}}(\phi_1 - \phi_3) \qquad \text{antisymmetric}$$

where again $\phi_i = (2p_x)_i$.

The secular problem for the symmetric functions can be treated quite independently of that for the antisymmetric functions since:

$$\mathcal{H}_{13} = \int u_1 \hat{\mathcal{H}} u_3 d\tau = \int \phi_2 \hat{\mathcal{H}} \frac{1}{\sqrt{2}}(\phi_1 - \phi_3) d\tau = \frac{1}{\sqrt{2}}(\beta_{21} - \beta_{23}) = 0$$

$$\mathcal{H}_{23} = \int \frac{1}{\sqrt{2}}(\phi_1 + \phi_3) \hat{\mathcal{H}} \frac{1}{\sqrt{2}}(\phi_1 - \phi_3) d\tau = \frac{1}{2}(\alpha_{11} - \alpha_{33} + \beta_{31} - \beta_{13}) = 0$$

In general such integrals must vanish since their integrands are antisymmetric with respect to reflection in the yz-plane. The integrals thus change sign on reflection even though their values cannot change as the result of an operation that leaves the molecule and hence its Hamiltonian invariant. Accordingly, their values must be zero.

The Hamiltonian matrix has the form:

$$\begin{pmatrix} \mathcal{H}_{11} & \mathcal{H}_{12} & 0 \\ \mathcal{H}_{21} & \mathcal{H}_{22} & 0 \\ 0 & 0 & \mathcal{H}_{33} \end{pmatrix} = \left(\begin{array}{cc|c} \multicolumn{2}{c|}{\text{sym}} & 0 \\ & & 0 \\ \hline 0 & 0 & \text{asy} \end{array}\right)$$

and the secular equations for even and odd functions are completely independent.

Symmetric functions: The required integrals are:

$$\mathcal{H}_{11} = \int \phi_2 \hat{\mathcal{H}} \phi_2 d\tau = \alpha; \qquad \mathcal{H}_{22} = (1/2) \int (\phi_1 + \phi_3) \hat{\mathcal{H}} (\phi_1 + \phi_3) d\tau = \alpha$$

$$\mathcal{H}_{12} = (1/\sqrt{2}) \int \phi_2 \hat{\mathcal{H}} (\phi_1 + \phi_3) d\tau = \sqrt{2}\, \beta$$

Thus

$$\begin{vmatrix} \alpha - E & \sqrt{2}\,\beta \\ \sqrt{2}\,\beta & \alpha - E \end{vmatrix} = 0 \quad \text{or} \quad \begin{vmatrix} X & \sqrt{2} \\ \sqrt{2} & X \end{vmatrix} = 0$$

$X^2 - 2 = 0$; $X = \pm\sqrt{2}$ and $E = \alpha \pm \sqrt{2}\,\beta$. For the first eigenfunction we take $X = -\sqrt{2}$:

$$-\sqrt{2}\, c_{11} + \sqrt{2}\, c_{12} = 0; \qquad \sqrt{2}\, c_{11} - \sqrt{2}\, c_{12} = 0$$

Thus, $c_{11} = c_{12}$ and $c_{11}^2(2) = 1$ so $c_{11} = 1/\sqrt{2}$. This gives

$$\psi_1 = \frac{1}{\sqrt{2}} (u_1 + u_2) = \frac{1}{2} \phi_1 + \frac{1}{\sqrt{2}} \phi_2 + \frac{1}{2} \phi_3$$

Similarly $X = +\ 2$ gives

$$\psi_2 = \frac{1}{\sqrt{2}} (u_1 - u_2) = \frac{1}{2} \phi_1 - \frac{1}{\sqrt{2}} \phi_2 + \frac{1}{2} \phi_3$$

Antisymmetric functions: The only matrix element is

$$\mathcal{H}_{33} = \frac{1}{2} \int (\phi_1 - \phi_3) \hat{\mathcal{H}} (\phi_1 - \phi_3) d\tau = \alpha$$

and the secular equation is just: $|\alpha - E| = 0$ or $E = \alpha$. There is no mixing of u_3 with other basis functions and we have

$$\psi_3 = u_3 = \frac{1}{\sqrt{2}} (\phi_1 - \phi_3)$$

9.14 The geometry of the methylenecylcopropene molecule is indicated in the figure below. This problem is concerned with the Hückel molecular orbitals for the π-electrons.

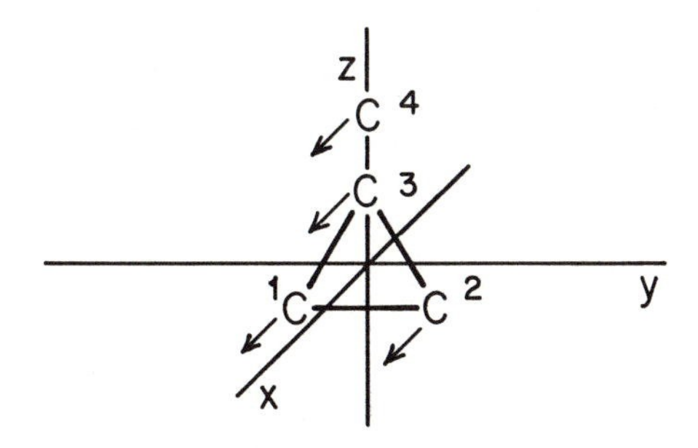

Fig. 9.14

(a) Determine the number of molecular orbitals transforming according to each of the irreducible representations of the molecular symmetry group C_{2v}.

(b) Construct normalized linear combinations of the atomic ($2p_x$) orbitals which transform according to the irreducible representations.

Solution: Ref. Appendix 8

(a) The $2p_x$ orbitals (vectors in the figure above) provide a basis for a four dimensional reducible representation Γ_π of the C_{2v} group. In order to apply the important equations of Appendix 8 we require the characters for this representation. Consider for example the operation C_2:

$$C_2 \begin{bmatrix} \phi_1 \\ \phi_2 \\ \phi_3 \\ \phi_4 \end{bmatrix} = \begin{bmatrix} 0 & -1 & 0 & 0 \\ -1 & 0 & 0 & 0 \\ 0 & 0 & -1 & 0 \\ 0 & 0 & 0 & -1 \end{bmatrix} \begin{bmatrix} \phi_1 \\ \phi_2 \\ \phi_3 \\ \phi_4 \end{bmatrix} = \begin{bmatrix} -\phi_2 \\ -\phi_1 \\ -\phi_3 \\ -\phi_4 \end{bmatrix}$$

Thus $\chi(C_2) = -2$, and it is clear that only atoms which lie on the symmetry element can contribute i.e. only atoms that are not moved by the symmetry operation. By inspection we easily discover the other characters for Γ_π which are shown below:

C_{2v}	E	$C_2(z)$	$\sigma(xz)$	$\sigma'(yz)$
Γ_π	4	-2	+2	-4

The antisymmetric behavior of the basis functions with respect to $\sigma'(yz)$ indicates that only A_2 and B_1 can be contained in Γ_π. Applying Eq. (3) of Appendix 8 we find:

$$n(A_2) = (1/4)(4 - 2 - 2 + 4) = 1$$

$$n(B_1) = (1/4)(4 + 2 + 2 + 4) = 3$$

Therefore: $\Gamma_\pi = A_2 + 3B_1$.

(b) The relevant equation for the construction of the symmetry orbitals is

$$u_j = \sum_R \chi_j^*(R) R\phi$$

First we display $R\phi$ in the table below

E	$C_2(z)$	$\sigma(xz)$	$\sigma'(yz)$
ϕ_1	$-\phi_2$	$+\phi_2$	$-\phi_1$
ϕ_3	$-\phi_3$	$+\phi_3$	$-\phi_3$
ϕ_4	$-\phi_4$	$+\phi_4$	$-\phi_4$

For $u(A_2)$ the equation then gives:

$$\phi_1 - \phi_2 - \phi_2 + \phi_1 \to \phi_1 - \phi_2$$
$$\phi_3 - \phi_3 - \phi_3 + \phi_3 \to 0$$
$$\phi_4 - \phi_4 - \phi_4 + \phi_4 \to 0$$

$$\boxed{u_1 = 1/\sqrt{2}\ (\phi_1 - \phi_2)}$$

For $u(B_1)$ the equation gives

$$\phi_1 + \phi_2 + \phi_2 + \phi_1 \to \phi_1 + \phi_2$$
$$\phi_3 + \phi_3 + \phi_3 + \phi_3 \to \phi_3$$
$$\phi_4 + \phi_4 + \phi_4 + \phi_4 \to \phi_4$$

$$\boxed{\begin{aligned} u_2 &= 1/\sqrt{2}\ (\phi_1 + \phi_2) \\ u_3 &= \phi_3 \\ u_4 &= \phi_4 \end{aligned}}$$

The normalization condition

$$\sum_i |c_{ni}|^2 = 1$$

has been used for all of the symmetry functions.

9.15 The cyclopropenyl radical has three carbon atoms situated at the corners of an equilateral triangle. Molecular orbitals for the π-electrons can be approximated by linear combinations of the $2p_z$ orbitals of the carbon atoms. Determine the eigenvalues and eigenfunctions for this radical in the Hückel approximation and find the total energy of the three π-electrons for the ground state.

Solution:

For this radical symmetry requires that: $\alpha_1 = \alpha_2 = \alpha_3 = \alpha$ and $\beta_{12} = \beta_{23} = \beta_{31} = \beta$. Using the basis functions $\phi_1 = (2p_z)_1$, $\phi_2 = (2p_z)_2$, and $\phi_3 = (2p_z)_3$ we find:

$$\begin{vmatrix} (\alpha - E) & \beta & \beta \\ \beta & (\alpha - E) & \beta \\ \beta & \beta & (\alpha - E) \end{vmatrix} = 0 \quad \text{or} \quad \begin{vmatrix} X & 1 & 1 \\ 1 & X & 1 \\ 1 & 1 & X \end{vmatrix} = 0; \quad X = (\alpha - E)/\beta$$

Thus $X^3 - 3X + 2 = 0$ and $X = -2, 1, 1$ so $E_1 = \alpha + 2\beta$, $E_2 = \alpha - \beta$, $E_3 = \alpha - \beta$. The ground state energy is: $E_{total} = 2(\alpha + 2\beta) + (\alpha - \beta) = 3(\alpha + \beta)$. Here we have used the fact that $\beta < 0$.

The first eigenfunction is obtained with $X = -2$. Thus

$$-2c_{11} + c_{12} + c_{13} = 0; \qquad c_{11} - 2c_{12} + c_{13} = 0; \qquad c_{11} + c_{12} - 2c_{13} = 0$$

These equations require that $c_{11} = c_{12}$ and $c_{11} = c_{13}$ so that the normalized eigenfunction is

$$\boxed{\psi_1 = \frac{1}{\sqrt{3}}(\phi_1 + \phi_2 + \phi_3)}$$

For the degenerate state having X = +1 we find that

$$c_{21} + c_{22} + c_{23} = 0$$

There are no unique solutions and we are free to choose any pair which are linearly independent and are orthogonal to ψ_1. For example $c_{22} + c_{23} = -c_{21} = 0$ gives one eigenfunction.

$$\boxed{\psi_2 = \frac{1}{\sqrt{2}}(\phi_2 - \phi_3)}$$

We choose ψ_3 to be orthogonal to ψ_2 and to ψ_1. Letting $\psi_3 = c_{31}\phi_1 + c_{32}\phi_2 + c_{33}\phi_3$ we have

$$\int\psi_2\psi_3 d\tau = \frac{1}{\sqrt{2}}\int(\phi_2 - \phi_3)(c_{31}\phi_1 + c_{32}\phi_2 + c_{33}\phi_3)d\tau = 0 = \frac{1}{\sqrt{2}}(c_{32} - c_{33})$$

Thus $c_{32} = c_{33}$ and ψ_3 has the form: $\psi_3 = N(c\phi_1 + \phi_2 + \phi_3)$. The second orthogonality relation gives

$$\int\psi_1\psi_3 d\tau = \frac{1}{\sqrt{3}}\int(\phi_1 + \phi_2 + \phi_3)N(c\phi_1 + \phi_2 + \phi_3)d\tau = 0 = \frac{N}{\sqrt{3}}(c + 1 + 1)$$

Therefore c = -2 and we obtain the normalized function:

$$\boxed{\psi_3 = \frac{1}{\sqrt{6}}(-2\phi_1 + \phi_2 + \phi_3)}$$

where the condition

$$\sum_{i=1}^{3} c_{3i}^2 = 1$$

has been used.

Using group theory the solution is much easier. Suppose we make use only of the C_3 axis. The $2p_z$ functions provide a basis for a three dimensional representation of the C_3 group. Equation (1) of Appendix 8 when properly normalized then gives

$$u(A) = \frac{1}{\sqrt{3}}(\phi_1 + \phi_2 + \phi_3); \qquad u(E_a) = \frac{1}{\sqrt{3}}(\phi_1 + \varepsilon^*\phi_2 + \varepsilon\phi_3)$$

$$u(E_b) = \frac{1}{\sqrt{3}}(\phi_1 + \varepsilon\phi_2 + \varepsilon^*\phi_3) \quad \text{where} \quad \varepsilon = e^{2\pi i/3}$$

These functions are orthogonal and there are no matrix elements of $\hat{\mathcal{H}}$ between them. Thus

$$E(A) = \int u^*(A)\hat{\mathcal{H}}u(A)d\tau = \alpha + 2\beta$$

$$E(E_a) = \int u^*(E_a)\hat{\mathcal{H}}u(E_a)d\tau = \alpha - \beta$$

$$E(E_b) = \int u^*(E_b)\hat{\mathcal{H}}u(E_b)d\tau = \alpha - \beta$$

Linear combinations of $u(E_a)$ and $u(E_b)$ can, of course, be constructed which give ψ_2 and ψ_3 as obtained above.

9.16 Obtain the energy eigenvalues and eigenfunctions for the π-electrons in benzene in the Hückel approximation. (Hint: Use the molecular symmetry subgroup C_6 to simplify the secular equation.)

Solution: Ref. Appendix 8

The benzene molecule has D_{6h} symmetry and this group can be used to construct symmetry orbitals. It is, however, somewhat simpler to use the subgroup C_6 which suffices for the determination of all of the linear combinations. The effects of the symmetry operations on the atomic orbital $\phi_1 = (2p_x)_1$ are shown below for C_6:

	E	C_6	C_3	C_2	C_3^2	C_6^5
$R\phi_1$	ϕ_1	ϕ_2	ϕ_3	ϕ_4	ϕ_5	ϕ_6

Thus Eq. (1) of Appendix 8 gives: (after normalization)

$$\psi_1 = 1/\sqrt{6}\,(\phi_1 + \phi_2 + \phi_3 + \phi_4 + \phi_5 + \phi_6) \qquad A$$

$$\psi_2 = 1/\sqrt{6}\,(\phi_1 - \phi_2 + \phi_3 - \phi_4 + \phi_5 - \phi_6) \qquad B$$

$$\left.\begin{aligned}\psi_3 &= 1/\sqrt{6}\,(\phi_1 + \varepsilon^*\phi_2 - \varepsilon\phi_3 - \phi_4 - \varepsilon^*\phi_5 + \varepsilon\phi_6)\\ \psi_4 &= 1/\sqrt{6}\,(\phi_1 + \varepsilon\phi_2 - \varepsilon^*\phi_3 - \phi_4 - \varepsilon\phi_5 + \varepsilon^*\phi_6)\end{aligned}\right\} E_1$$

$$\left.\begin{aligned}\psi_5 &= 1/\sqrt{6}\,(\phi_1 - \varepsilon\phi_2 - \varepsilon^*\phi_3 + \phi_4 - \varepsilon\phi_5 - \varepsilon^*\phi_6)\\ \psi_6 &= 1/\sqrt{6}\,(\phi_1 - \varepsilon^*\phi_2 - \varepsilon\phi_3 + \phi_4 - \varepsilon^*\phi_5 - \varepsilon\phi_6)\end{aligned}\right\} E_2$$

where $\varepsilon = \exp(2\pi i/6)$. The secular equations are all 1x1 and the Hamiltonian matrix is already diagonal. Thus the eigenvalues are just

$$E_i = \int \psi_i^* \hat{\mathcal{H}} \psi_i d\tau$$

These integrals can be evaluated directly, but it is perhaps easier to use linear combinations of the ψ_i which are real for this purpose.

The sum of the degenerate functions associated with E_1 is:

$$\psi(E_{1a}) \propto \psi_3 + \psi_4 = (1/\sqrt{6})[2\phi_1 + (\varepsilon + \varepsilon^*)\phi_2 - (\varepsilon + \varepsilon^*)\phi_3 - 2\phi_4 - (\varepsilon + \varepsilon^*)\phi_5 + (\varepsilon + \varepsilon^*)\phi_6]$$

$$= (1/\sqrt{6})[2\phi_1 + \phi_2 - \phi_3 - 2\phi_4 - \phi_5 + \phi_6]$$

since $\varepsilon + \varepsilon^* = 2\cos 60° = 1$. And the difference gives:

$$\psi(E_{1b}) \propto \psi_3 - \psi_4 = (1/\sqrt{6})[(\varepsilon^* - \varepsilon)\phi_2 - (\varepsilon - \varepsilon^*)\phi_3 - (\varepsilon^* - \varepsilon)\phi_5 + (\varepsilon - \varepsilon^*)\phi_6]$$

$$= -(i/\sqrt{6})[\sqrt{3}\,\phi_2 + \sqrt{3}\,\phi_3 - \sqrt{3}\,\phi_5 - \sqrt{3}\,\phi_6]$$

Similar combinations can be made for the degenerate E_2 functions so that after normalization we obtain the set:

$$\psi(A) = (1/\sqrt{6})(\phi_1 + \phi_2 + \phi_3 + \phi_4 + \phi_5 + \phi_6)$$

$$\psi(B) = (1/\sqrt{6})(\phi_1 - \phi_2 + \phi_3 - \phi_4 + \phi_5 - \phi_6)$$

$$\psi(E_{1a}) = (1/\sqrt{12})(2\phi_1 + \phi_2 - \phi_3 - 2\phi_4 - \phi_5 + \phi_6)$$

$$\psi(E_{1b}) = (1/2)(\phi_2 + \phi_3 - \phi_5 - \phi_6)$$

$$\psi(E_{2a}) = (1/\sqrt{12})(2\phi_1 - \phi_2 - \phi_3 + 2\phi_4 - \phi_5 - \phi_6)$$

$$\psi(E_{2b}) = (1/2)(\phi_2 - \phi_3 + \phi_5 - \phi_6)$$

The corresponding eigenvalues are:

$$E(A) = \int \psi^*(A)\hat{\mathcal{H}}\psi(A)d\tau = (1/6)(6\alpha + 12\beta) = \alpha + 2\beta$$

$$E(B) = \int \psi^*(B)\hat{\mathcal{H}}\psi(B)d\tau = (1/6)(6\alpha - 12\beta) = \alpha - 2\beta$$

$$E(E_{1a}) = (1/12)(12\alpha + 12\beta) = \alpha + \beta$$

$$E(E_{1b}) = \alpha + \beta$$

$$E(E_{2a}) = \alpha - \beta$$

$$E(E_{2b}) = \alpha - \beta$$

9.17 This problem is concerned with the π-electrons in the pyrazine molecule in the LCAO approximation.

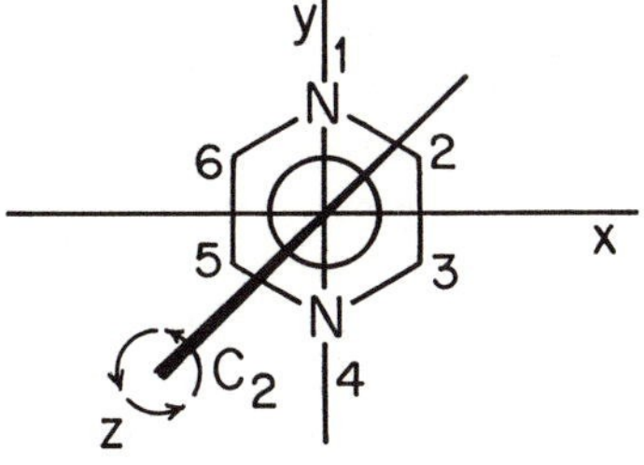

(a) Use group theory to determine symmetry orbitals for the pyrazine molecule and list the irreducible representations of the group D_{2h} to which they belong. The orbitals may be determined by using an appropriate subgroup of D_{2h}.

(b) Calculate the energies of the molecular orbitals in the Hückel approximation for $h = 0$, 0.5, and 1.0 and make a graph to show energy vs. h. Here $\alpha_N = \alpha_C + h\beta$, and you may assume that $\beta_{CN} = \beta_{CC} = \beta$.

(c) Write out the molecular orbital occupied by the unpaired electron in the mononegative radical-anion of pyrazine assuming that $h = 0.77$.

Solution: Ref. Appendix 8

(a) For convenience we use the C_{2v} symmetry group. The effects of the symmetry operations on ϕ_1 and ϕ_2 are shown below.

C_{2v}	E	C_2	$\sigma_v(xz)$	$\sigma_v'(yz)$
$R\phi_1$	ϕ_1	ϕ_4	ϕ_4	ϕ_1
$R\phi_2$	ϕ_2	ϕ_5	ϕ_3	ϕ_6
Γ_π	6	0	0	2

Using Eq. (3) of Appendix 8 we find:

$$\Gamma_\pi = 2A_1 + A_2 + B_1 + 2B_2$$

Then Eq. (1) of the same appendix yields (after normalization):

C_{2v}		D_{2h}
A_1	$u_1 = \frac{1}{\sqrt{2}}(\phi_1 + \phi_4)$, $u_2 = \frac{1}{2}(\phi_2 + \phi_3 + \phi_5 + \phi_6)$	B_{1u}
A_2	$u_1 = \frac{1}{2}(\phi_2 - \phi_3 + \phi_5 - \phi_6)$	A_u
B_1	$u_1 = \frac{1}{2}(\phi_2 + \phi_3 - \phi_5 - \phi_6)$	B_{2g}
B_2	$u_1 = \frac{1}{\sqrt{2}}(\phi_1 - \phi_4)$, $u_2 = \frac{1}{2}(\phi_2 - \phi_3 - \phi_5 + \phi_6)$	B_{3g}

The symmetry behavior of each of the functions above gives a set of characters which can be identified with one of the irreducible representations of D_{2h} taking into account the antisymmetric behavior of the $2p_z$ orbitals with respect to the plane of the molecule.

(b) We take $\beta_{ij} = \beta$ when i and j represent adjacent atoms and $\alpha_N = \alpha_1 = \alpha_4 = \alpha + h\beta$ where α is the coulomb integral for a carbon atom. The secular problems are as follows:

$$B_{1u}: \quad \mathcal{H}_{11} = \alpha_N, \quad \mathcal{H}_{22} = \alpha + \beta, \quad \mathcal{H}_{12} = \sqrt{2}\,\beta$$

Thus:

$$\begin{vmatrix} (\alpha_N - E) & \sqrt{2}\,\beta \\ \sqrt{2}\,\beta & (\alpha + \beta - E) \end{vmatrix} = 0 \quad \text{or} \quad \begin{vmatrix} (X + h) & \sqrt{2} \\ \sqrt{2} & (X + 1) \end{vmatrix} = 0$$

With $Y = X + 1$, $\lambda = h - 1$, and $X = (\alpha - E)/\beta$ we obtain

$$\begin{vmatrix} Y + \lambda & \sqrt{2} \\ \sqrt{2} & Y \end{vmatrix} = Y^2 + Y\lambda - 2 = 0$$

Therefore

$$X = \left(\frac{-\lambda \pm \sqrt{\lambda^2 + 8}}{2} - 1 \right) \tag{1}$$

$$A_u{:}\ \mathcal{H}_{11} = \alpha - \beta = E(A_u) \tag{2}$$

$$B_{2g}{:}\ \mathcal{H}_{11} = \alpha + \beta = E(B_{2g}) \tag{3}$$

$$B_{3g}{:}\ \mathcal{H}_{11} = \alpha_N,\quad \mathcal{H}_{22} = \alpha - \beta,\quad \mathcal{H}_{12} = \sqrt{2}\,\beta$$

Thus

$$\begin{vmatrix} (\alpha_N - E) & \sqrt{2}\,\beta \\ \sqrt{2}\,\beta & (\alpha - \beta - E) \end{vmatrix} = 0 \quad \text{or} \quad \begin{vmatrix} (X + h) & \sqrt{2} \\ \sqrt{2} & X - 1 \end{vmatrix} = 0$$

With $Z = X - 1$, $\eta = h + 1$, and $X = (\alpha - E)/\beta$ we obtain:

$$\begin{vmatrix} (Z + \eta) & \sqrt{2} \\ \sqrt{2} & Z \end{vmatrix} = Z^2 + \eta Z - 2 = 0$$

Therefore

$$X = \frac{-7 \pm \sqrt{\eta^2 + 8}}{2} + 1 \tag{4}$$

Equations (1) through (4) then give:

h	$X(B_{1u})$	$X(A_u)$	$X(B_{2g})$	$X(B_{3g})$
0	+1, −2	+1	−1	+2, −1
0.5	0.686, −2.186	+1	−1	1.851, −1.351
1.0	0.414, −2.414	+1	−1	1.732, −1.732

In each case $E = \alpha - X\beta$. The eigenvalues are shown in Fig. 9.17. As expected those orbitals having symmetry behavior which excludes contributions from ϕ_1 and ϕ_4 are unaffected by the perturbation $h\beta$.

(c) The unpaired (seventh) electron occupies the B_{1u} orbital in the radical anion. Taking $h = 0.77$ we find using Eq. (1) that $X = 0.534$, and the associated eigenfunction is determined by the equations:

$$(X + 0.77)c_{11} + \sqrt{2}\,c_{21} = 0; \qquad \sqrt{2}\,c_{11} + (X + 1)c_{21} = 0$$

These equations yield $c_{21} = -0.922\,c_{11}$ and after normalization we have:

$$\psi(B_{1u}) = 0.735\,u_1 - 0.678\,u_2 \cong 0.52(\phi_1 + \phi_4) - 0.34(\phi_2 + \phi_3 + \phi_5 + \phi_6)$$

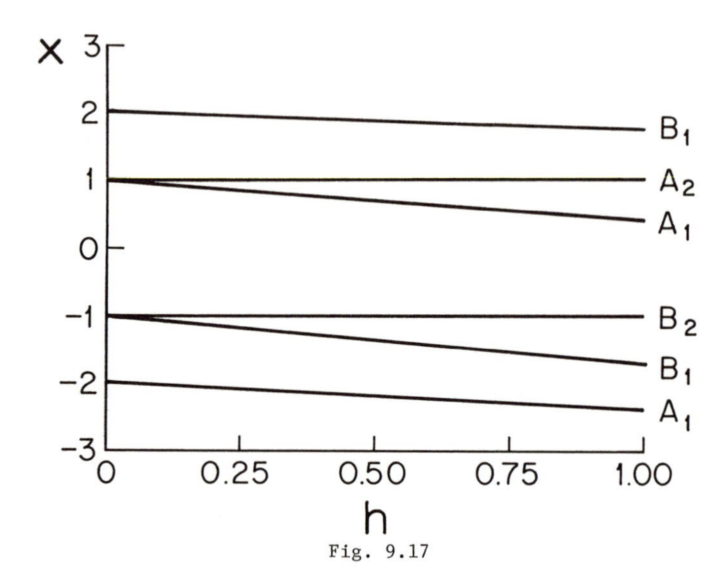

Fig. 9.17

9.18 The atomic orbitals s, p_x, p_y, p_z, etc. pertain to atoms in field free space. When atoms experience sufficiently strong interactions with their environment, the orbitals of lowest energy may correspond more closely to linear combinations of the atomic orbitals which conform to the symmetry of the environment. For the construction of such orbitals we use the angular parts of hydrogen-like orbitals which are normalized to unity. For convenience the first few of these functions are shown below:

$\Theta(\theta)\Phi(\phi)$

s: $1/2\sqrt{\pi}$

p_x: $(\sqrt{3}/2\sqrt{\pi})\sin\theta\cos\phi$

p_y: $(\sqrt{3}/2\sqrt{\pi})\sin\theta\sin\phi$

p_z: $(\sqrt{3}/2\sqrt{\pi})\cos\theta$

d_{xz}: $(\sqrt{15}/2\sqrt{\pi})\sin\theta\cos\theta\cos\phi$

d_{yz}: $(\sqrt{15}/2\sqrt{\pi})\sin\theta\cos\theta\sin\phi$

d_{xy}: $(\sqrt{15}/4\sqrt{\pi})\sin^2\theta\sin 2\phi$

d_{z^2}: $(\sqrt{5}/4\sqrt{\pi})(3\cos^2\theta - 1)$

$d_{x^2-y^2}$: $(\sqrt{15}/4\sqrt{\pi})\sin^2\theta\cos 2\phi$

(a) Show that 3 equivalent hybrid orbitals can be formed from the set 2s, $2p_x$, and $2p_y$ so that the directions of maximum extent lie in the xy-plane and are separated by 120°, i.e. construct orbitals of the form $h_i = a_i s + b_i p_x + c_i p_y$. (Hint: Make use of normalization, orthogonality, and "completeness of use" of orbitals as required in a unitary transformation.)

(b) Use the vectors $\underset{\sim}{h}$ in the figure below as a basis for the three dimensional representation Γ of D_{3h} and determine the number of times each irreducible representation is contained in Γ.

Then by associating atomic orbitals with the irreducible representations determine which sets of atomic orbitals can be used to construct the hybrid orbitals h_i.

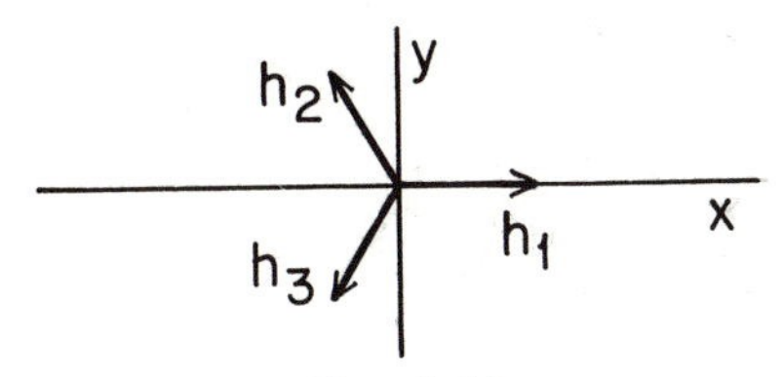

Fig. 9.18

Solution: Ref. PA3,116; C. Y. Hsu and M. Orchin, J. Chem. Ed. 50, 114 (1973).

(a) The orbital h_1 must have the form: $h_1 = a_1 s + b_1 p_x$. Normalization then gives

$$a_1^2 + b_1^2 = 1 \quad \text{or} \quad b_1 = \pm\sqrt{1 - a_1^2}$$

We choose the + sign to represent h_1. The magnitude of a_1 can be established since s is expected to make the same contribution to each hybrid orbital. Thus $a_1 = a_2 = a_3$ and "completeness of use" requires that

$$a_1^2 + a_2^2 + a_3^2 = 1 \quad \text{or} \quad a_i = 1/\sqrt{3}$$

For h_1 we have:

$$h_1 = a_1^2 s + \sqrt{1 - a_1^2}\, p_x = \frac{1}{\sqrt{3}} s + \sqrt{\frac{2}{3}}\, p_x$$

To determine the form of h_2 we make use of the orthogonality condition:

$$\int h_1 h_2 d\Omega = \int\left(\frac{1}{\sqrt{3}} s + \sqrt{\frac{2}{3}}\, p_x\right)(a_2 s + b_2 p_x + c_2 p_y) d\Omega = 0 = \frac{a_2}{\sqrt{3}} + b_2\sqrt{\frac{2}{3}}$$

But $a_2 = 1/\sqrt{3}$ so $b_2 = -1/\sqrt{6}$. This condition applies to h_3 as well and so $b_3 = -1/\sqrt{6}$. The remaining constants can be determined by the normalization condition.

$$\int |h_2|^2 d\Omega = a_2^2 + b_2^2 + c_2^2 = \frac{1}{3} + \frac{1}{6} + c_2^2 = 1$$

Thus $c_2^2 = 1/2$ or $c_2 = \pm 1/\sqrt{2}$. We choose the + sign for h_2 and the - sign for c_3 and finally obtain:

$$h_1 = \frac{1}{\sqrt{3}} s + \sqrt{\frac{2}{3}}\, p_x$$

$$h_2 = \frac{1}{\sqrt{3}} s - \frac{1}{\sqrt{6}} p_x + \frac{1}{\sqrt{2}} p_y$$

$$h_3 = \frac{1}{\sqrt{3}} s - \frac{1}{\sqrt{6}} p_x - \frac{1}{\sqrt{2}} p_y$$

(b) The transformation properties of the basis functions h_1, h_2, and h_3 can be determined by inspection. Consider first the rotation $R = C_3(z)$:

$$R\begin{pmatrix} h_1 \\ h_2 \\ h_3 \end{pmatrix} = \begin{pmatrix} h_2 \\ h_3 \\ h_1 \end{pmatrix} \quad \text{or} \quad R = \begin{pmatrix} 0 & 1 & 0 \\ 0 & 0 & 1 \\ 1 & 0 & 0 \end{pmatrix}; \quad \chi(R) = 0$$

The complete set of characters is shown below:

D_{3h}	E	$2C_3$	$3C_2$	σ_h	$2S_3$	$3\sigma_v$
Γ	3	0	1	3	0	1

Since the vectors are symmetric with respect to reflection in the xy-plane, only A_1', A_2', and E' can contribute. Application of Eq. (3) from Appendix 8 gives:

$$n(A_1') = (1/12)(3 + 3 + 3 + 3) = 1; \qquad n(E') = (1/12)(6 + 6) = 1$$

Thus

$$\Gamma = A_1' + E'$$

The transformation properties of the atomic orbitals are as follows:

A_1': s

E': p_x, p_y, $d_{x^2-y^2}$, d_{xy}

Thus sp^2 and sd^2 are possible sets.

9.19 Assume that the potential energy function for nuclear motion in a diatomic molecule is given by:

$$E(R) = \frac{A}{R^{12}} - \frac{B}{R}$$

where R is the internuclear separation. In the Born-Oppenheimer approximation the kinetic energy of nuclear motion is taken to be negligible compared to the electronic kinetic energy.

(a) Determine A and B so that at the equilibrium internuclear distance $R = R_{eq} = 1$ atomic unit and $E(R_{eq}) = -1$ Rydberg.

(b) Use the virial theorem to determine the forms of the total kinetic energy $\langle T \rangle$ and the total potential energy $\langle V \rangle$ and plot E, $\langle T \rangle$, and $\langle V \rangle$ in the range R = 0.7 to 5.0.

(c) Discuss the causes of binding in such a system.

Solution: Ref. KP,293; S5,29

(a) The constants A and B are determined by the conditions:

$$\left(\frac{dE(R)}{dR}\right)_{R=1} = -12A + B = 0; \qquad E(R_{eq}) = A - B = -1$$

Therefore, A = 1/11 and B = 12/11.

(b) For diatomic molecules the virial theorem takes the form:

$$\langle T\rangle = -E - R\,\frac{dE}{dR}\ ; \qquad \langle V\rangle = 2E + R\,\frac{dE}{dR}$$

Thus

$$\langle T\rangle = \frac{1}{R^{12}} \quad \text{and} \quad \langle V\rangle = -\,\frac{10}{11R^{12}} - \frac{12}{11R}$$

These functions are shown in Fig. 9.19.

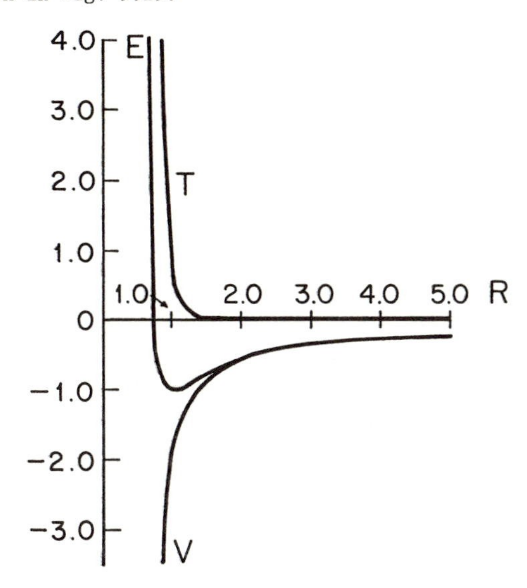

Fig. 9.19

(c) As R decreases from infinity the potential energy decreases monotonically because of the attractive term in the total energy expression. This provides the binding effect when ionic bonds are formed by bringing together ions as in Cs^+Cl^-. The kinetic energy rises as R approaches R_{eq} and is responsible for the minimum in E(R).

SUPPLEMENTARY PROBLEMS

9.20 Consider the following crude model for chemical bonding. Initially, two electrons are placed in two independent potential boxes each having infinite walls and the width a/2. These represent two hydrogen atoms. Then, the boxes are united to form a single box of width a which represents a molecule.

(a) Express the energy change for the reaction in units of h^2/ma^2.

(b) Suppose that the potential energy inside one of the boxes before the reaction is zero, and inside the other it is $V_o << h^2/ma^2$. What is the energy change when these boxes are united to form a potential like that shown in Fig. 6.5?

(c) What energy change is expected for the reaction in part (b) when $V_o >> h^2/ma^2$? What is the main cause of binding in each case?

Ans: Ref. Problem 6.5; L. Melander, J. Chem. Ed. 49, 686 (1972)

(a) $\Delta E = -\frac{3}{4}\left(\frac{h^2}{ma^2}\right)$, (b) $\Delta E = -\frac{3}{4}\left(\frac{h^2}{ma^2}\right)$, (c) $\Delta E \cong -V_o$

In parts (a) and (b) the energy decrease results from electron delocalization; while in part (c), where $V_o >> h^2/ma^2$, the energy change results from the transfer of an electron to a box of low potential energy.

9.21 According to the Born-Oppenheimer approximation (see Problem 9.1):

$$(\hat{\mathcal{H}}_T - \hat{T}_n)\psi(r;\ R) = E(R)\psi(r;\ R) \tag{1}$$

$$[\hat{T}_n + E(R)]\chi(R) = E_a\chi(R) \tag{2}$$

where the exact molecular wave function $\Psi(r,\ R) = \psi(r;\ R)\chi(R)$ and E_a is an approximation to the lowest eigenvalue E_T of the molecular Hamiltonian. Show that $E_a \lessgtr E_T$ (i.e. that E_a is a lower bound for E_T) by applying the variation principle to Eqs. (1) and (2). (Hint: First hold R fixed when dealing with Eq. (1) and r fixed when dealing with Eq. (2).)

Ref. S. T. Epstein, J. Chem. Phys. 44, 836 (1966)

9.22 Show that

$$\int \frac{1}{\pi}\ e^{-(r_A+r_B)}\ \frac{2}{r_A}\ d\tau = 2e^{-R}(1 + R)$$

by using elliptical coordinates. This is the integral $\langle u_A|2r_A^{-1}|u_B\rangle$ which appears in the H_2^+ and H_2 problems. A diagram is given in Problem 9.2.

Ref. Problem 9.3, Appendix 5

9.23 In the Heitler-London (valence bond) treatment for H_2 the ground state wave function is assumed to have the form:

$$\Psi_{VB} = N[u_A(1)u_B(2) + u_B(1)u_A(2)] \tag{1}$$

This is, of course, a singlet wave function; but the spin part, which is normalized, does not affect the calculation and can be ignored. The geometry of the problem is shown in Fig. 9.5.

(a) Determine the constant N in Eq. (1) in terms of the overlap integral S.

(b) Obtain an expression for the expectation value of $\hat{\mathcal{H}}$ for H_2 using Ψ_{VB}. Your answer should be expressed in terms of one and two electron integrals involving u_A and u_B.

(c) Compare the result of part (b) with that from simple MO theory (Problem 9.5) and in particular consider the energy as $R \to \infty$ and the dependence of E on S.

Ans: Ref. S5,49; Problem 9.5

(a) $N = 1/\sqrt{2(1 + S^2)}$

(b) $$E = 2E_{1s} + \frac{2}{(1 + S^2)}\left[-\langle u_A|\frac{2}{r_{B1}}|u_A\rangle - S\langle u_A|\frac{2}{r_{A1}}|u_B\rangle + \frac{1}{2}\langle u_A u_B|\frac{2}{r_{12}}|u_A u_B\rangle + \frac{1}{2}\langle u_A u_B|\frac{2}{r_{12}}|u_B u_A\rangle\right] + \frac{2}{R}$$

(c) In the limit that $R \to \infty$, $E \to 2E_H$. This is different from the MO theory where the integral $\langle u_A u_A|2r_{12}^{-1}|u_A u_A\rangle$ contributes for all values of R.

9.24 Consider the H_2^+ ground state function $\psi_{H_2^+} = N[u_A(1) + u_B(1)]$ where

$$u_A(1) = \sqrt{\frac{\alpha^3}{\pi}}\, e^{-\alpha r_1}$$

(a) Show that the energy resulting from the function may be written

$$E = \alpha^2 E_1(\alpha R) + \alpha E_2(\alpha R) + \frac{2}{R} \text{ (in Rydbergs)}$$

where

$$E_1(\alpha R) = (1 - S + \langle u_A|2r_A^{-1}|u_B\rangle)/(1 + S)$$

and

$$E_2(\alpha R) = (-2 - 2\langle u_A|2r_A^{-1}|u_B\rangle - \langle u_B|2r_A^{-1}|u_B\rangle)/(1 + S)$$

(b) Let $w = \alpha R$. Minimize the electronic energy with respect to α. Show that

$$\alpha = -\frac{E_2(w) + w\dfrac{dE_2(w)}{dw}}{2E_1(w) + w\dfrac{dE_1(w)}{dw}}$$

How would you use this expression to find the "best" value of α at a given R?

Ans: Ref. S5,22

(b) Evaluate $E_1(w)$, $E_2(w)$, $\dfrac{dE_1(w)}{dw}$, and $\dfrac{dE_2(w)}{dw}$. From the expression for α in terms of w one can find α for a given value of w and thus for a given value of $R = \dfrac{w}{\alpha}$.

9.25 If the wave function for an n-electron system is denoted by

$$\psi = \psi(\mathbf{r}_1, \cdots \mathbf{r}_n, m_{s_1} \cdots m_{s_n})$$

then the probability of finding an electron at $\mathbf{r}$ is given by

$$\rho(\mathbf{r}) = n\int\cdots\int|\psi(\mathbf{r}, \mathbf{r}_2, \cdots \mathbf{r}_n, m_s, \cdots m_{s_n})|^2 d\mathbf{r}_2 \cdots d\mathbf{r}_n d\sigma$$

where the integration is over the spatial coordinates of n - 1 electrons and the spin coordinates of n electrons. Show that for the ground state of H_2 the difference between the valence bond and molecular orbital electron densities midway between the two nuclei is

$$\rho_{VB} - \rho_{MO} = -\frac{2(1 - S)^2}{(1 + S)(1 + S^2)} u_A^{\ 2}$$

where u_A = 1s.
Ref. LE1,399

9.26 Obtain the molecular orbitals and their energies for the π-electrons in cis-butadiene in the Hückel approximation (neglecting overlap). The geometry of this molecule is shown below where the two-fold rotation axis is in the z-direction. Also calculate the charge densities and bond orders for the π-electrons.

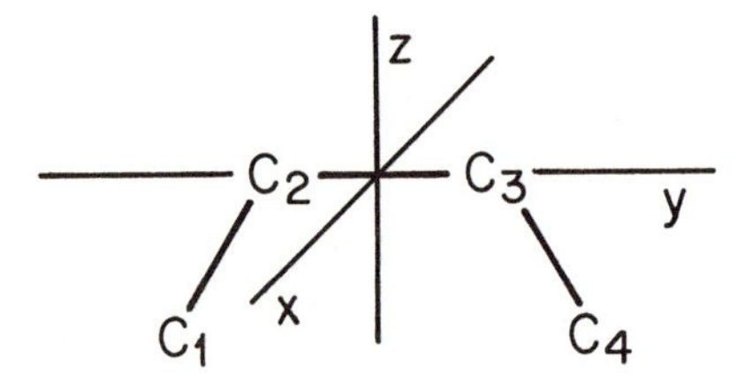

Fig. 9.26

Ans: Ref. Problem 9.28

$\psi_i = \sum_j c_{ji} u_j$; charge density $q_\alpha = 2 \sum_i^{occ} c_{\alpha i}^2$, and $q_1 = q_2 = q_3 = q_4 = 1$; bond order

$p_{\alpha\beta} = 2 \sum_i^{occ} c_{\alpha i} c_{\beta i}$ and $p_{12} = p_{34} = 0.896$, $p_{23} = 0.448$

9.27 This problem is concerned with the Hückel molecular orbital treatment of the π-electrons in naphthalene. Use the following coordinate system for answering the questions below:

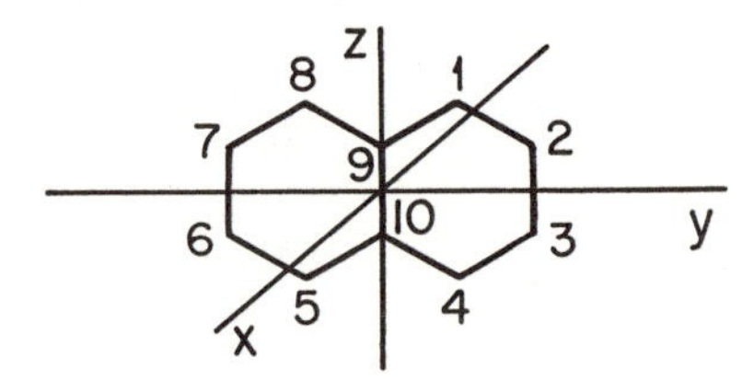

Fig. 9.27

(a) Determine the number of molecular orbitals which belong to each irreducible representation of D_{2h}, and construct normalized linear combinations of the atomic orbitals $\phi_i = (2p_x)_i$ which transform according to these irreducible representations.
(b) Write out the secular equations in terms of $X = (\alpha - E)/\beta$ which must be solved to obtain the energy eigenvalues associated with each irreducible representation.
(c) Determine all of the eigenvalues for the secular equations in part (b) and then calculate the π-electron energy for the neutral naphthalene molecule in terms of the integrals α and β.

Ans: Ref. Appendix 8; KP,382

(a) $\Gamma_\pi = 2B_{1g} + 3B_{2g} + 2A_u + 3B_{3u}$

B_{1g}: $u_1 = (1/2)(\phi_1 + \phi_4 - \phi_5 - \phi_8)$
$u_2 = (1/2)(\phi_2 + \phi_3 - \phi_6 - \phi_7)$

B_{2g}: $u_1 = (1/2)(\phi_1 - \phi_4 - \phi_5 + \phi_8)$
$u_2 = (1/2)(\phi_2 - \phi_3 - \phi_6 + \phi_7)$
$u_3 = (1/\sqrt{2})(\phi_9 - \phi_{10})$

A_u: $u_1 = (1/2)(\phi_1 - \phi_4 + \phi_5 - \phi_8)$
$u_2 = (1/2)(\phi_2 - \phi_3 + \phi_6 - \phi_7)$

B_{3u}: $u_1 = (1/2)(\phi_1 + \phi_4 + \phi_5 + \phi_8)$
$u_2 = (1/2)(\phi_2 + \phi_3 + \phi_6 + \phi_7)$
$u_3 = (1/\sqrt{2})(\phi_9 + \phi_{10})$

(b) B_{1g}: $\begin{vmatrix} X & 1 \\ 1 & (X + 1) \end{vmatrix} = 0$

B_{2g}: $\begin{vmatrix} X & 1 & \sqrt{2} \\ 1 & X - 1 & 0 \\ \sqrt{2} & 0 & X - 1 \end{vmatrix} = 0$

A_u: $\begin{vmatrix} X & 1 \\ 1 & (X - 1) \end{vmatrix} = 0$

B_{3u}: $\begin{vmatrix} X & 1 & \sqrt{2} \\ 1 & X + 1 & 0 \\ \sqrt{2} & 0 & X + 1 \end{vmatrix} = 0$

(c)

	$(E - \alpha)$		$(E - \alpha)$
B_{2g}	-2.303β	A_u	0.618β
A_u	-1.618β	B_{3u}	1.0β
B_{3u}	-1.303β	B_{2g}	1.303β
B_{2g}	-1.0β	B_{1g}	1.618β
B_{1g}	-0.618β	B_{3u}	2.303β

$E_\pi = 10\alpha + 13.684\beta$ for naphthalene.

9.28 The π-electron charge densities and bond orders are defined by the equations:

$$\text{(charge density for atom } \alpha) = q_\alpha = 2\sum_i^{occ} c_{\alpha i}^2$$

$$\text{(order of bond between atoms } \alpha \text{ and } \beta) = p_{\alpha\beta} = 2\sum_i^{occ} c_{\alpha i}c_{\beta i}$$

Here the sums run over all occupied orbitals (assumed to be doubly occupied) and the coefficients are assumed to be real. Derive an equation for the total Hückel energy of the π-electrons in a hydrocarbon in terms of the parameters q_α and $p_{\alpha\beta}$.

Ans:

$$E_\pi = \sum_\nu q_\nu \alpha_\nu + \sum_{\nu \neq \mu}\sum p_{\nu\mu}\beta_{\nu\mu}$$

9.29 Carry out a simple MO calculation of the total energy of H_3 when the hydrogen atoms are located at the vertices of an equilateral triangle. Use the framework of the Hückel method in this calculation except that overlap should be included so that the energy will be a function of the overlap integral S.

Ans:

$$E_\pi = 2E_1 + E_2 = \frac{3[\alpha + \beta(1 - 2S)]}{(1 - S)(2S + 1)}$$

9.30 A pair of equivalent sp hybrid orbitals, h_1 and h_2, can be constructed from the 2s and $2p_z$ orbitals of a carbon atom. Equivalence implies that each orbital has the same s character.

(a) Derive the necessary coefficients for the hybrid orbitals h_1 and h_2.

(b) Determine the angle between the directions of maximum extent of h_1 and h_2.

(c) Prepare a polar plot of h_1 which shows the magnitude of h_1 for each value of θ as a distance from the origin.

Ans: Ref. Problem 9.18

(a) $h_1 = \frac{1}{\sqrt{2}}(s + p_z)$, $h_2 = \frac{1}{\sqrt{2}}(s - p_z)$; (b) 180°

9.31 An approximate potential energy curve can be constructed for the H_2 molecule as a function of the overlap integral S by means of the following assumptions: (1) the energy of the two electron molecule can be calculated by the simple MO theory (HMO) including overlap with $E = 2\varepsilon_1$, where ε_1 is the lowest eigenvalue, and (2) the resonance integral H_{AB} can be written in terms of H_{AA}, H_{BB}, and S.

(a) Obtain an equation for V(R), the molecular energy referred to the energy of the separated atoms ($2H_{AA}$), as a function of S by using Cusachs' approximation for H_{AB}:

$$H_{AB} = \frac{S}{2}(2 - |S|)(H_{AA} + H_{BB})$$

Plot $-V(R)/H_{AA}$ versus S for S = 0 to 1.

(b) Determine the value of S which leads to the minimum energy.

(c) Using the results from parts (a) and (b) and Koopmans' theorem, calculate the ionization potential I.P. for H_2 in eV. You can assume that $H_{AA} = -13.6$ eV.

(d) Calculate the energy of the first electronic transition in eV again using the results of parts (a) and (b).

Ans: Ref. W. F. Cooper, G. A. Clark, and C. R. Hare, J. Chem. Ed. 48, 247 (1971)

(a) $V(R) = 2H_{AA}S(1 - S)/(1 + S)$ (b) $S = 0.414$

(c) I.P. $= -\varepsilon_1 = -1.172\ H_{AA} = 15.93$ eV (d) $\Delta E = \varepsilon_2 - \varepsilon_1 = -0.586\ H_{AA} = 7.97$ eV

9.32 A one dimensional δ-function model for chemical binding has been proposed in which the coulombic potential $V(x) = -2Z/|x|$ is replaced by $-2Z\delta(x)$.

(a) Set up the Schrödinger equation for the hydrogen atom in one dimension using the δ-function potential (in atomic units with energy in Rydbergs) and obtain the ground state eigenfunction.

(b) Derive an equation for the ground state energy in terms of Z by first showing that:

$$\lim_{\varepsilon \to 0} \left[\left(\frac{du}{dx}\right)_{+\varepsilon} - \left(\frac{du}{dx}\right)_{-\varepsilon} \right] = -2Zu(0) \tag{1}$$

Ans: Ref. A. A. Frost, J. Chem. Phys. 25, 1150 (1956)

(a) $u = (-E)^{1/4}\exp[-\sqrt{-E}\ |x|]$; (b) $E = -Z^2$

9.33 This problem concerns the application of the δ-function potential to the H_2^+ molecule ion in one dimension. Nucleus a is located at $x = -R/2$ and b is at $x = +R/2$. The exact solution can be written as an LCAO function involving $\exp(-cr_a)$ and $\exp(-cr_b)$ where $r_a = |x + R/2|$ and $r_b = |x - R/2|$.

(a) Write the energy in terms of c.

(b) Use the LCAO functions and the method of Problem 9.32 to obtain an equation for c in terms of R.

(c) Construct a plot of E versus R for the two lowest levels (Z = 1) neglecting the nuclear repulsion term. In doing this it is convenient to write both E and R in terms of the parameter $\alpha = cR$ so that the solution of a transcendental equation can be avoided.

Ans: Ref. Problem 9.32

(a) $E = -c^2$ (b) $c = Z(1 \pm e^{-cR})$

(c) $R = \dfrac{\alpha}{Z(1 \pm e^{-\alpha})}$, $E = -Z^2(1 \pm e^{-\alpha})^2$

CHAPTER 10

RADIATION AND MATTER

Radiation: A semi-classical theory for the interaction of radiation with matter can be developed in which atomic and molecular systems are treated with quantum mechanics and the radiation field is described by Maxwell's equations. This is the procedure that we have adopted. In this theory the Hamiltonian for a particle of charge e_i and mass m_i in an electromagnetic field is:

$$\hat{\mathcal{H}} = -\frac{\hbar^2}{2m_i}\nabla^2 + V(\underset{\sim}{r}_i) + \frac{i\hbar e_i}{m_i}\underset{\sim}{A}\cdot\underset{\sim}{\nabla} + \frac{e_i{}^2A^2}{2m_i} \tag{1}$$

when the gauge is chosen so that $\underset{\sim}{\nabla}\cdot\underset{\sim}{A} = 0$ and the scalar potential $\phi = 0$. The interaction of the particle with the radiation field is described by the perturbing Hamiltonian:

$$\hat{\mathcal{H}}' = \frac{i\hbar e_i}{m_i}\underset{\sim}{A}\cdot\underset{\sim}{\nabla} + \frac{e_i{}^2}{2m_i}\underset{\sim}{A}^2 \tag{2}$$

The problems in this chapter explore the development of Eq. (1) and the time evolution of atomic and molecular systems under the action of $\hat{\mathcal{H}}'$. The concepts of time independent transition rates and selection rules are also developed. For this purpose we must consider time-dependent perturbation theory.

Time-Dependent Perturbation Theory: We suppose that the Hamiltonian can be written as $\hat{\mathcal{H}}(t) = \hat{\mathcal{H}}^o + \hat{\mathcal{H}}'(t)$ where all of the time dependence is contained in $\hat{\mathcal{H}}'$ and $\hat{\mathcal{H}}' << \hat{\mathcal{H}}^o$. The state function $|\Psi(\underset{\sim}{r}, t)\rangle$ must obey the time-dependent Schrödinger equation, $\hat{\mathcal{H}}|\Psi\rangle = i\hbar\partial|\Psi\rangle/\partial t$, and following Dirac we write:

$$|\Psi(\underset{\sim}{r}, t)\rangle = \sum_{k=0}^{\infty} a_k(t)e^{-iE_k^o t/\hbar}|k\rangle \tag{3}$$

Here $|k\rangle$ is an eigenket of $\hat{\mathcal{H}}^o$ with the eigenvalue E_k^o, and $a_k(t)$ contains all of the time dependence that arises from the presence of $\hat{\mathcal{H}}'$. When Eq. (3) is substituted into the wave equation and both sides are multiplied from the left with $\langle\ell|$, it is found that:

$$\dot{a}_\ell = (i\hbar)^{-1}\sum_{k=0}^{\infty} a_k\mathcal{H}'_{\ell k}e^{i\omega_{\ell k}t} \tag{4}$$

where $\mathcal{H}'_{\ell k} = \langle\ell|\hat{\mathcal{H}}'|k\rangle$ and $\omega_{\ell k} = (E_\ell^o - E_k^o)/\hbar$. The perturbation approach is introduced by writing

$$a_k = a_k^o + a_k^{(1)} + a_k^{(2)} + \cdots \tag{5}$$

where $a_k^{(n)}$ is the nth order correction to a_k^o which results from $\hat{\mathcal{H}}'$. When Eq. (5) is substituted into Eq. (4) and the terms are collected according to order one finds:

$$\dot{a}_\ell^o = 0, \quad \dot{a}_\ell^{(1)} = (i\hbar)^{-1}\sum_k a_k^o \mathcal{H}'_{\ell k} e^{i\omega_{\ell k}t} \tag{6}$$

$$\vdots$$

$$\dot{a}_\ell^{(n)} = (i\hbar)^{-1}\sum_k a_k^{(n-1)} \mathcal{H}'_{\ell k} e^{i\omega_{\ell k}t}$$

In the first order theory we set $a_k^o = \delta_{ko}$, which specifies that the system is in the state 0 before the perturbation is applied. Then, if the perturbation is turned on sufficiently slowly (adiabatic limit), Eq. (6) can be integrated to obtain the first order equation.

$$\boxed{a_\ell^{(1)} = (i\hbar)^{-1}\int_{-\infty}^{t} \mathcal{H}'_{\ell o}(t')e^{i\omega_{\ell o}t'}dt'} \tag{7}$$

Equation (7) can then be substituted into the equation for $\dot{a}_\ell^{(2)}$ to obtain the second order equation and so on.

The population of the ℓth state ($\ell \neq 0$) is given by $|a_\ell(t)|^2$, which must be much less than unity for the theory to hold. The transition rate $W_{\ell o}$ is given by t^{-1} times an appropriate average of $|a_\ell(t)|^2$ over the distribution of transition frequencies. This point is discussed in detail in the problems that follow.

Ref. SC1,280,397; S1,144

P. W. Langhoff, S. T. Epstein, and M. Karplus, Rev. Mod. Phys. 44, 602 (1972).

A. Dalgarno in Perturbation Theory and its Applications in Quantum Mechanics, edited by C. H. Wilcox (John Wiley & Sons, Inc., New York, 1966).

PROBLEMS

10.1 Suppose that $\hat{\mathcal{H}}^o u_n = E_n u_n$ and that $\hat{\mathcal{H}} = \hat{\mathcal{H}}^o + \hat{\mathcal{H}}'(t)$ where $\hat{\mathcal{H}}' << \hat{\mathcal{H}}^o$. The solutions Ψ of the time-dependent wave equation, $\hat{\mathcal{H}}\Psi = i\hbar\partial\Psi/\partial t$, can be expanded as follows:

$$\Psi(x,t) = \sum_m a_m(t)e^{-iE_m t/\hbar}u_m(x) \tag{1}$$

Given the initial conditions, $a_n(0) = 1$, $a_k(0) = 0$ for $k \neq n$, show that

$$\boxed{a_k(t) = -\frac{i}{\hbar}\int_o^t \langle k|\hat{\mathcal{H}}'(t')|n\rangle e^{i\omega_{kn}t'}dt'} \tag{2}$$

where $\omega_{kn} = (E_k - E_n)/\hbar$. You may assume that $a_k(t) << 1$ for $k \neq n$. As stated in the Introduction to Chapter 10 these initial conditions imply that we are dealing with first order perturbation theory.

Solution: Ref. SC1,280

Substitution of Eq. (1) into the wave equation gives:

$$(\hat{\mathcal{H}}^o + \hat{\mathcal{H}}')\sum_m a_m e^{-iE_m t/\hbar} u_m = i\hbar\left[\sum_m \dot{a}_m e^{-iE_m t/\hbar} u_m + \sum_m a_m \left(-\frac{iE_m}{\hbar}\right) e^{-iE_m t/\hbar} u_m\right]$$

Therefore, using $\hat{\mathcal{H}}^o u_m = E_m u_m$ we find:

$$\sum_m a_m \mathcal{H}' u_m e^{-iE_m t/\hbar} = i\hbar \sum_m \dot{a}_m u_m e^{-iE_m t/\hbar}$$

Multiplication from the right by u_k^* and integration then gives:

$$\sum_m a_m \langle k|\hat{\mathcal{H}}'|m\rangle e^{-iE_m t/\hbar} = i\hbar \dot{a}_k e^{-iE_k t/\hbar} \tag{3}$$

Here the orthogonality condition $\langle k|m\rangle = \delta_{km}$ has been used. Rearranging Eq. (3) leads to:

$$\dot{a}_k = -\frac{i}{\hbar}\sum_m a_m \langle k|\hat{\mathcal{H}}'|m\rangle e^{i\omega_{km}t} \tag{4}$$

But $a_n(t) \sim 1$ and $a_m(t) << 1$, so that Eq. (4) can be approximated by setting $a_m = \delta_{nm}$ on the r.h.s. Finally, integrating from 0 to t:

$$a_k(t) = -\frac{i}{\hbar}\int_0^t \langle k|\hat{\mathcal{H}}'|n\rangle e^{i\omega_{kn}t} dt$$

10.2 Using the results of Problem 10.1 and assuming that

$$\hat{\mathcal{H}}'(t) = 2\hat{V}_o \cos\omega t \tag{1}$$

show that:

(a) The probability of finding a system in the state f at time t given that it was in the state i at t = 0 is:

$$P_f(t) \equiv |a_f|^2 = \frac{|V_{fi}|^2}{\hbar^2}\left(\frac{\sin^2 x}{x^2}\right)t^2 \tag{2}$$

where $V_{fi} = \langle f|\hat{V}_o|i\rangle$ and $x = (\omega_{fi} - \omega)t/2$.

(b) The transition rate from state i to state f (per system) can be written as:

$$\boxed{W_{fi} \equiv \frac{P_f(t)}{t} = \frac{2\pi}{\hbar}|V_{fi}|^2 n(h\nu_{fi})} \tag{3}$$

where $n(h\nu_{fi})$ is either the density of final states in the neighborhood of E_f, <u>i.e.</u> $n(h\nu) = dN(E)/dE$, or the density of modes in the radiation field in the neighborhood of $h\nu_{fi}$. Equation (3) is known as <u>Fermi's Golden Rule</u>. (Hint: Integrate $P_f(t)$ over the distribution of final states E_f and assume that $n(h\nu)$ is a slowly varying function of $h\nu$ for $\nu \sim \nu_{fi}$.)

Solution: Ref. S1,148; AT1,220

(a) Using $a_f(t)$ from Problem 10.1 and $\hat{\mathcal{H}}'(t)$ for Eq. (1) we have:

$$a_f(t) = -\frac{i}{\hbar}\int_0^t \langle f|\hat{V}_o|i\rangle e^{i\omega_{fi}t'} 2\cos\omega t' dt' = -\frac{i}{\hbar} V_{fi}\int_0^t e^{i\omega_{fi}t'}(e^{i\omega t'} + e^{-i\omega t'})dt'$$

$$= -\frac{i}{\hbar} V_{fi}\left[\frac{e^{i(\omega_{fi}+\omega)t} - 1}{i(\omega_{fi} + \omega)} + \frac{e^{i(\omega_{fi}-\omega)t} - 1}{i(\omega_{fi} - \omega)}\right] \qquad (4)$$

For $\omega_{fi} \sim \omega$ the second term in the brackets above can be shown to dominate by use of l'Hospital's rule. This term is associated with absorption of energy since $E_{final} - E_{initial} \cong \hbar\omega$. Then keeping only the larger term and rearranging gives

$$a_f(t) = -\frac{i}{\hbar} V_{fi} e^{i(\omega_{fi}-\omega)t/2}\left[\frac{e^{i(\omega_{fi}-\omega)t/2} - e^{-i(\omega_{fi}-\omega)t/2}}{i(\omega_{fi} - \omega)t/2}\right]\frac{t}{2}$$

$$= -i\frac{V_{fi}}{\hbar} e^{i(\omega_{fi}-\omega)t/2}\left(\frac{\sin[(\omega_{fi} - \omega)t/2]}{[(\omega_{fi} - \omega)t/2]}\right)t$$

Finally

$$P_f(t) = |a_f|^2 = \frac{|V_{fi}|^2}{\hbar^2}\left(\frac{\sin^2 x}{x^2}\right)t^2 \quad \text{where } x = (\omega_{fi} - \omega)t/2$$

(b) The transition rate to a band of states close to E_f can be written as

$$W_{fi} \equiv t^{-1}\int_0^\infty n(E)P_f(t)dE = \frac{|V_{fi}|^2}{t\hbar^2}\int n(\hbar\omega_{fi})\left(\frac{\sin^2 x}{x^2}\right)2t\hbar dx$$

$$= \frac{2|V_{fi}|^2}{\hbar} n(h\nu_{fi})\int\frac{\sin^2 x}{x^2}dx = \frac{2\pi}{\hbar}|V_{fi}|^2 n(h\nu_{fi}) \qquad (5)$$

For long times t, the function $(\sin x)/x$ is sharply peaked at $\omega = \omega_{fi}$ and the density of states function $n(h\nu_{fi})$ can safely be removed from the integral. The integral from tables is then

$$\int_{-\infty}^{+\infty}\frac{\sin^2 x}{x^2}dx = \pi$$

If $n(h\nu_{fi})$ were sharply peaked at E_f and t were short, it would be appropriate to replace $(\sin x)/x$ by unity in Eq. (3). The rate would then not be independent of time (See Problem 10.9). For a derivation of Eq. (3) in terms of the density of modes rather than the density of final states see reference S1 above.

10.3 The Lagrangian of a particle of charge e_i in an external field is

$$\mathcal{L} = \frac{m}{2}(\underset{\sim}{v}\cdot\underset{\sim}{v}) + e_i(\underset{\sim}{v}\cdot\underset{\sim}{A}) - e_i\phi \qquad (1)$$

where ϕ and $\underset{\sim}{A}$ are the scalar and vector potentials, respectively. Show that the Lagrangian function in Eq. (1) is consistent with the Lorentz force equation:

$$\underset{\sim}{F} = e_i(\underset{\sim}{\mathcal{E}} + \underset{\sim}{v} \times \underset{\sim}{B})$$

(Hint: Recall Maxwell's equations $\underset{\sim}{\mathcal{E}} = -\underset{\sim}{\nabla}\phi - \dot{\underset{\sim}{A}}$ and $\underset{\sim}{B} = \underset{\sim}{\nabla}\times\underset{\sim}{A}$.)

Solution: Ref. S3,265, Appendix 9

Lagrange's equations of motion are

$$\frac{d}{dt}\frac{\partial L}{\partial \dot{x}_i} - \frac{\partial L}{\partial x_i} = 0 \quad (i = 1, 2, 3) \tag{2}$$

For the x component we have from Eq. (1):

$$\frac{d}{dt}\frac{\partial L}{\partial \dot{x}} = \frac{d}{dt}(m\dot{x} + e_i A_x) = m\ddot{x} + e_i\frac{dA_x}{dt} = m\ddot{x} + e_i\left[\frac{\partial A_x}{\partial t} + \dot{x}\frac{\partial A_x}{\partial x} + \dot{y}\frac{\partial A_x}{\partial y} + \dot{z}\frac{\partial A_x}{\partial z}\right] \tag{3}$$

$$\frac{\partial L}{\partial x} = -e_i\frac{\partial \phi}{\partial x} + e_i\left[\dot{x}\frac{\partial A_x}{\partial x} + \dot{y}\frac{\partial A_y}{\partial x} + \dot{z}\frac{\partial A_z}{\partial x}\right] \tag{4}$$

Substitution of Eqs. (3) and (4) into Eq. (2) then gives (after rearrangement)

$$m\ddot{x} = -e_i\left[\frac{\partial \phi}{\partial x} + \frac{\partial A_x}{\partial t}\right] - e_i\left[\dot{y}\left(\frac{\partial A_x}{\partial y} - \frac{\partial A_y}{\partial x}\right) + \dot{z}\left(\frac{\partial A_x}{\partial z} - \frac{\partial A_z}{\partial x}\right)\right] \tag{5}$$

Now, the magnetic induction is given by

$$\underset{\sim}{\nabla}\times\underset{\sim}{A} = \hat{i}\left[\frac{\partial A_z}{\partial y} - \frac{\partial A_y}{\partial z}\right] + \hat{j}\left[\frac{\partial A_x}{\partial z} - \frac{\partial A_z}{\partial x}\right] + \hat{k}\left[\frac{\partial A_y}{\partial x} - \frac{\partial A_x}{\partial y}\right]$$

So that the x component of $\underset{\sim}{v}\times\underset{\sim}{\nabla}\times\underset{\sim}{A}$ is

$$\dot{y}\left[\frac{\partial A_y}{\partial x} - \frac{\partial A_x}{\partial y}\right] - \dot{z}\left[\frac{\partial A_x}{\partial z} - \frac{\partial A_z}{\partial x}\right]$$

This is just the negative of the second bracketed term on the right hand side of Eq. (5). Thus,

$$F_x = m\ddot{x} = e_i[\mathcal{E}_x + (\underset{\sim}{v}\times\underset{\sim}{\nabla}\times\underset{\sim}{A})_x] \quad \text{or} \quad \underset{\sim}{F} = e_i(\underset{\sim}{\mathcal{E}} + \underset{\sim}{v}\times\underset{\sim}{B})$$

In classical mechanics this equation defines the Lorentz force.

10.4 The classical Hamiltonian is related to the Lagrangian by the equation

$$H(p,q,t) = \sum_i p_i\dot{q}_i - \mathcal{L}(q,\dot{q},t) \tag{1}$$

where the momentum conjugate to q_i is defined by $p_i = \partial\mathcal{L}/\partial\dot{q}_i$. Use the Lagrangian of Problem 10.3 to show that the Hamiltonian for a particle of mass m and charge e_i in a radiation field is given by

$$H = \frac{1}{2m}[(p_x - e_iA_x)^2 + (p_y - e_iA_y)^2 + (p_z - e_iA_z)^2] + e_i\phi \tag{2}$$

Solution: Ref. Problem 10.3; GO,53

The Lagrangian from Problem 10.3 is

$$\mathcal{L} = \frac{m}{2}(v_x^2 + v_y^2 + v_z^2) + e_i(\underset{\sim}{v}\cdot\underset{\sim}{A}) - e_i\phi \tag{3}$$

where e_i defines the particle charge. The conjugate momenta are:

$$p_x = m\dot{x} + e_iA_x, \quad p_y = m\dot{y} + e_iA_y, \quad p_z = m\dot{z} + e_iA_z \tag{4}$$

From Eq. (1) we have

$$H = \sum_i p_i\dot{q}_i - \mathcal{L} = (m\dot{x} + e_iA_x)\dot{x} + (m\dot{y} + e_iA_y)\dot{y} + (m\dot{z} + e_iA_z)\dot{z}$$

$$- \frac{m}{2}(\dot{x}^2 + \dot{y}^2 + \dot{z}^2) - e_i(\dot{x}A_x + \dot{y}A_y + \dot{z}A_z) + e_i\phi = \frac{m}{2}(\dot{x}^2 + \dot{y}^2 + \dot{z}^2) + e_i\phi \tag{5}$$

Substitution of $\dot{x}$, $\dot{y}$, and $\dot{z}$ from Eq. (4) into Eq. (5) then gives:

$$H = \frac{1}{2m}[(p_x - e_iA_x)^2 + (p_y - e_iA_y)^2 + (p_z - e_iA_z)^2] + e_i\phi$$

10.5 Show that the $\hat{\mathcal{H}}$ operator for a particle of mass m and charge e_i in an electromagnetic field is

$$\hat{\mathcal{H}} = -\frac{\hbar^2}{2m}\nabla^2 + \frac{i\hbar e_i}{2m}(2\underset{\sim}{A}\cdot\underset{\sim}{\nabla}) + \frac{e_i^2A^2}{2m} + e_i\phi + V(\underset{\sim}{r})$$

(Hint: The gauge can be chosen so that $\underset{\sim}{\nabla}\cdot\underset{\sim}{A} = 0$ and $\phi = 0$ for a radiation field.)

Solution: Ref. Problem 10.4; EWK,108; AT2,422; HA1,10

Substitution of

$$\hat{p}_i = -i\hbar\frac{\partial}{\partial x_i}$$

into the classical Hamiltonian generated in Problem 10.4 gives the desired result. First we find the components

$$(\hat{p}_x - e_iA_x)^2 f = (-i\hbar\frac{\partial}{\partial x} - e_iA_x)^2 f = (i\hbar\frac{\partial}{\partial x} + e_iA_x)(i\hbar\frac{\partial}{\partial x} + e_iA_x)f$$

$$= (i\hbar\frac{\partial}{\partial x} + e_iA_x)(i\hbar f' + e_iA_xf)$$

$$= [-\hbar^2\frac{\partial^2}{\partial x^2}f + i\hbar e_i(A_xf' + fA_x') + i\hbar e_iA_xf' + e_i^2A_x^2f]$$

$$= (-\hbar^2\frac{\partial^2}{\partial x^2} + 2i\hbar e_iA_x\frac{\partial}{\partial x} + i\hbar e_i\frac{\partial A_x}{\partial x} + e_i^2A_x^2)f$$

Similar equations result for the y and z components so that we have

$$\hat{\mathcal{H}} = \frac{1}{2m}\{-\hbar^2\nabla^2 + i\hbar e_i[(2\underset{\sim}{A}\cdot\underset{\sim}{\nabla}) + (\underset{\sim}{\nabla}\cdot\underset{\sim}{A})] + e_i^2A^2\} + e_i\phi$$

Then choosing the gauge so that $\underset{\sim}{\nabla}\cdot\underset{\sim}{A} = 0$ and $\phi = 0$ gives

$$\hat{\mathcal{H}} = \frac{1}{2m}[-\hbar^2\nabla^2 + 2ie_i\hbar\underset{\sim}{A}\cdot\underset{\sim}{\nabla} + e_i^2A^2]$$

and for a many-particle system in a radiation field we have

$$\hat{\mathcal{H}} = \sum_i \frac{1}{2m_i} [-\hbar^2\nabla_i^2 + 2i\hbar e_i \underset{\sim}{A}_i \cdot \underset{\sim}{\nabla}_i + e_i^2 A_i^2] + V(\underset{\sim}{r}_1, \cdots \underset{\sim}{r}_n)$$

where V is the particle-particle potential energy function.

10.6 In spectroscopy the interaction between an electron (neglecting spin) and the radiation field can often be described by the perturbing Hamiltonian:

$$\hat{\mathcal{H}}'(t) = -\frac{i\hbar e}{m_e} \underset{\sim}{A} \cdot \underset{\sim}{\nabla} = \frac{e}{m_e} \underset{\sim}{A} \cdot \hat{\underset{\sim}{p}} \tag{1}$$

Equation (1) is appropriate for weak radiation fields where the term $e^2A^2/2m_e$ can be neglected or more precisely for single photon processes.

(a) Show that $\langle f|\hat{p}_x|i\rangle = im_e\omega_{fi}\langle f|x|i\rangle$ where $\omega_{fi} = (E_f - E_i)/\hbar$ and u_f and u_i are eigenfunctions of the unperturbed Hamiltonian.

(b) Assume that the vector potential for the radiation field can be expressed as $\underset{\sim}{A}(t) = 2\underset{\sim}{A}^\circ\cos\omega t$ where $\omega \sim \omega_{fi}$ and show that:

$$|\langle f| \frac{e}{m_e} \underset{\sim}{A}^\circ \cdot \hat{\underset{\sim}{p}}|i\rangle|^2 \cong |\langle f|-\underset{\sim}{\mu}\cdot\underset{\sim}{\mathcal{E}}^\circ|i\rangle|^2 \tag{2}$$

where $\underset{\sim}{\mathcal{E}}^\circ$ gives the amplitude and direction of the electric field associated with the radiation. This justifies the use of the classical energy expression for a dipole in an electric field as the perturbing Hamiltonian when the dependence of $\underset{\sim}{A}$ on $\underset{\sim}{r}$ can be neglected, i.e. when $\cos(\omega t - \underset{\sim}{k}\cdot\underset{\sim}{r}) \simeq \cos\omega t$. This is called the long wave approximation (see Problem 10.15). (Hint: Recall that $\underset{\sim}{\mathcal{E}} = -\partial\underset{\sim}{A}/\partial t$ when ϕ is zero and use the energy expression $W = \varepsilon_o\langle\mathcal{E}^2\rangle$.)

Solution: Ref. Appendix 9

(a) From Chapter 3 we have $\partial\hat{R}/\partial t = (i/\hbar)[\hat{\mathcal{H}}, \hat{R}]$ or

$$\langle f| \frac{\partial\hat{R}}{\partial t} |i\rangle = \frac{i}{\hbar} \langle f|[\hat{\mathcal{H}}, \hat{R}]|i\rangle$$

Therefore

$$\langle f|\hat{p}_x|i\rangle = m_e\langle f|\hat{\dot{x}}|i\rangle = \frac{im_e}{\hbar} \langle f|\hat{\mathcal{H}}x - x\hat{\mathcal{H}}|i\rangle \tag{3}$$

But

$$\langle f|\hat{\mathcal{H}}x|i\rangle = \sum_j \langle f|\hat{\mathcal{H}}|j\rangle\langle j|x|i\rangle = E_f\langle f|x|i\rangle \quad \text{and} \quad \langle f|x\hat{\mathcal{H}}|i\rangle = E_i\langle f|x|i\rangle$$

Equation (3) then gives:

$$\langle f|\hat{p}_x|i\rangle = \frac{im_e}{\hbar} (E_f - E_i)\langle f|x|i\rangle = im_e\omega_{fi}\langle f|x|i\rangle \tag{4}$$

(b) Using Equations (1) and (4) we obtain:

$$\langle f| \frac{e}{m_e} \underset{\sim}{A}^\circ\cdot\hat{\underset{\sim}{p}}|i\rangle = i\omega_{fi}e\underset{\sim}{A}^\circ\cdot\langle f|x\hat{i} + y\hat{j} + z\hat{k}|i\rangle = i\omega_{fi}\underset{\sim}{A}^\circ\cdot\langle f|\underset{\sim}{\mu}|i\rangle$$

We take $\underset{\sim}{A}^\circ = A^\circ\hat{k}$ for convenience and write

$$|\langle f|\frac{e}{m_e}\underset{\sim}{A}^\circ\cdot\hat{p}|i\rangle|^2 = \omega_{fi}^2(A^\circ)^2|\langle f|\mu_z|i\rangle|^2 \quad (5)$$

But

$$\underset{\sim}{\mathcal{E}} = -\frac{\partial \underset{\sim}{A}}{\partial t} = +2\omega \underset{\sim}{A}^\circ \sin\omega t = 2\underset{\sim}{\mathcal{E}}^\circ \sin\omega t \quad \text{and} \quad W = \varepsilon_o\langle\mathcal{E}^2\rangle = 4\varepsilon_o\omega^2(A^\circ)^2\langle\sin^2\omega t\rangle = 2\varepsilon_o\omega^2(A^\circ)^2$$

Equation (5) can thus be put in the form:

$$\omega_{fi}^2 \frac{W}{2\varepsilon_o\omega^2}|\langle f|\mu_z|i\rangle|^2 \simeq \frac{W}{2\varepsilon_o}|\langle f|\mu_z|i\rangle|^2 \quad (6)$$

when $\omega \sim \omega_{fi}$. Starting with $|\langle f|-\underset{\sim}{\mu}\cdot\underset{\sim}{\mathcal{E}}^\circ|i\rangle|^2$ and choosing $\underset{\sim}{\mathcal{E}}^\circ$ to lie along the z direction:

$$(\mathcal{E}^\circ)^2|\langle f|\mu_z|i\rangle|^2 = \omega^2(A^\circ)^2|\langle f|\mu_z|i\rangle|^2 = \frac{W}{2\varepsilon_o}|\langle f|\mu_z|i\rangle|^2 \quad (7)$$

Comparison of Eqs. (6) and (7) then confirms Eq. (2). For a general discussion at an advanced level see: C. J. Ballhausen and A. E. Hansen in Annual Review of Physical Chemistry 23, 15 (1972) and references cited therein.

10.7 Use Fermi's Golden Rule (Problem 10.2) with the perturbing Hamiltonian $\hat{\mathcal{H}}'(t) = -\underset{\sim}{\mu}\cdot\underset{\sim}{\mathcal{E}}(t)$ with $\underset{\sim}{\mathcal{E}}(t) = 2\underset{\sim}{\mathcal{E}}^\circ\cos\omega t$ to show that the rate of electric dipole transitions from the state i to the state f is given by:

$$\boxed{W_{fi} = \frac{1}{6\varepsilon_o\hbar^2}|\mu_{fi}|^2\rho(\nu_{fi})} \quad (1)$$

Here $\rho(\nu)$ is the energy density function which was used in Problem 1.12. (Hint: Recall that the energy density in the radiation field is $W = \varepsilon_o\langle\mathcal{E}^2(t)\rangle$ and that $\rho(\nu) = Whn(h\nu)$.)

Solution: Ref. AT1,226

Using $\hat{\mathcal{H}}'$ with Eq. (3) of Problem 10.2 gives

$$W_{fi} = \frac{2\pi}{\hbar}|\langle f|-\underset{\sim}{\mu}\cdot\underset{\sim}{\mathcal{E}}^\circ|i\rangle|^2 n(h\nu_{fi}) \quad (2)$$

We can define the direction of $\underset{\sim}{\mathcal{E}}^\circ$ as the x-direction and assume that the field is polarized without loss of generality. In the event that the radiation field is isotropic the average of the squares of the components of $\underset{\sim}{\mathcal{E}}$ will add to give the same result. Equation (2) becomes

$$W_{fi} = \frac{2\pi}{\hbar}(\mathcal{E}^\circ)^2|\langle f|\mu_x|i\rangle|^2 n(h\nu_{fi}) \quad (3)$$

For randomly oriented systems $|\mu_{fi}|^2 = 3|(\mu_x)_{fi}|^2$. The energy density can be introduced as follows:

$$W = \varepsilon_o\langle\mathcal{E}^2(t)\rangle = \varepsilon_o\int_0^{2\pi/\omega} 4(\mathcal{E}^\circ)^2\cos^2\omega t'dt' \Big/ \int_0^{2\pi/\omega} dt' = 4\varepsilon_o(\mathcal{E}^\circ)^2(1/2) = 2\varepsilon_o(\mathcal{E}^\circ)^2$$

Combining these relations Eq. (3) becomes

$$W_{fi} = \frac{2\pi}{3\hbar}\frac{W}{2\varepsilon_o}|\mu_{fi}|^2\left(\frac{n(\nu_{fi})}{h}\right) = \frac{1}{6\varepsilon_o\hbar^2}|\mu_{fi}|^2\rho(\nu_{fi})$$

In the last step above we have used the definitions: $hn(h\nu) = n(\nu)$ and $Wn(\nu) = \rho(\nu)$. Here $n(h\nu)dE$ gives the number of states in the energy range E to E + dE while $n(\nu)d\nu$ gives the number of states in the frequency range ν to $\nu + d\nu$. The energy density is obtained by multiplying the number of oscillators in the radiation field at ν by the energy of each. As mentioned in Problem 10.2, $n(h\nu)$ can be interpreted either as the density of final states in the neighborhood of E_f or the density of oscillators in the radiation field in the neighborhood of ν_{fi}.

10.8 Consider a system having only two states, i.e. $\hat{\mathcal{H}}^o u_n = E_n u_n$ with n = 1, 2. At t = 0 a perturbation $\hat{V}(t) = \hat{V}^o$, which has no diagonal matrix elements, is turned on. Here $\hat{V}^o$ is independent of time and is real. Derive equations for the populations of the two states as functions of time for the special case that the states are degenerate, i.e. $E_1 = E_2$. For the initial conditions take $P_1(0) = 1$ and $P_2(0) = 0$. (Hint: Expand $\Psi(x,t)$ in terms of the complete set u_1 and u_2 as shown in Problem 3.23, but let the coefficients be functions of time. The coefficients will then be determined by the time dependent Schrödinger equation.)

Solution: Ref. FE3,8-7

As suggested we write:

$$\Psi(x,t) = \sum_{n=1}^{2} c_n(t)e^{-iE_nt/\hbar}u_n(x)$$

Then with $\hat{\mathcal{H}} = \hat{\mathcal{H}}^o + \hat{V}^o$ the time-dependent Schrödinger equation gives:

$$(\hat{\mathcal{H}}^o + \hat{V}^o)\sum_{n=1}^{2} c_n e^{-iE_nt/\hbar}u_n = i\hbar\sum_{n=1}^{2}\left(\frac{dc_n}{dt}e^{-iE_nt/\hbar} - \frac{i}{\hbar}c_nE_ne^{-iE_nt/\hbar}\right)u_n$$

and

$$\hat{V}^o(c_1u_1e^{-iE_1t/\hbar} + c_2u_2e^{-iE_2t/\hbar}) = i\hbar\left(\frac{dc_1}{dt}u_1e^{-iE_1t/\hbar} + \frac{dc_2}{dt}u_2e^{-iE_2t/\hbar}\right) \tag{1}$$

Multiplying Eq. (1) by u_1^* and integrating and then with u_2^* and integrating gives the two equations:

$$i\hbar\frac{dc_1}{dt} = c_2\langle 1|\hat{V}^o|2\rangle e^{i(E_1-E_2)t/\hbar} = c_2V_{12}^o \tag{2a}$$

$$i\hbar\frac{dc_2}{dt} = c_1\langle 2|\hat{V}^o|1\rangle e^{i(E_2-E_1)t/\hbar} = c_1V_{12}^o \tag{2b}$$

where we have used the fact that $E_1 = E_2$. Equation (2) can be solved quite simply. First taking (2a + 2b):

$$i\hbar(\dot{c}_1 + \dot{c}_2) = V^o_{12}(c_1 + c_2) \quad \text{and} \quad (c_1 + c_2) = a\exp\left(-\frac{i}{\hbar}\, V^o_{12}t\right) \tag{3a}$$

Then (2a - 2b) gives:

$$i\hbar(\dot{c}_1 - \dot{c}_2) = -V^o_{12}(c_1 - c_2) \quad \text{and} \quad (c_1 - c_2) = b\exp\left(+\frac{i}{\hbar}\, V^o_{12}t\right) \tag{3b}$$

Combining 3a and 3b we obtain:

$$c_1(t) = \frac{a}{2}\, e^{-iV^o_{12}t/\hbar} + \frac{b}{2}\, e^{+iV^o_{12}t/\hbar} \tag{4a}$$

$$c_2(t) = \frac{a}{2}\, e^{-iV^o_{12}t/\hbar} - \frac{b}{2}\, e^{+iV^o_{12}t/\hbar} \tag{4b}$$

The populations are given by: $P_1(t) = |c_1(t)|^2$ and $P_2(t) = |c_2(t)|^2$. Therefore, the initial conditions require that $c_1(0) = 1$ and $c_2(0) = 0$ or that $a = b = 1$. We conclude that:

$$c_1(t) = \cos\left(\frac{V^o_{12}t}{\hbar}\right); \qquad c_2(t) = -i\sin\left(\frac{V^o_{12}t}{\hbar}\right)$$

Finally, the populations are: (see Fig. 10.8)

$$P_1(t) = |c_1(t)|^2 = \cos^2\left(\frac{V^o_{12}t}{\hbar}\right) \quad \text{and} \quad P_2(t) = |c_2(t)|^2 = \sin^2\left(\frac{V^o_{12}t}{\hbar}\right)$$

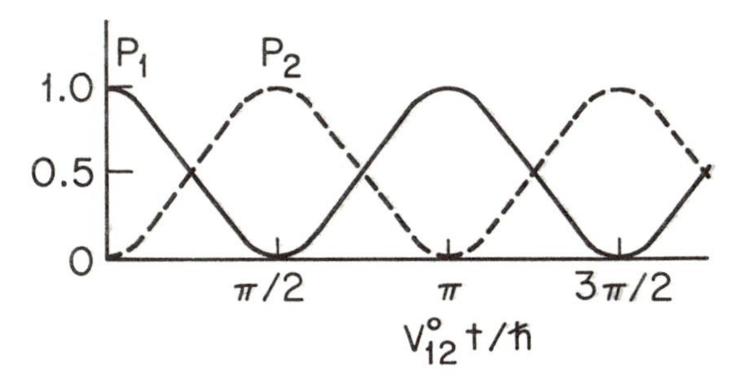

Fig. 10.8

10.9 The method of Problem 10.8 can be applied in certain cases where degeneracy is not involved. Consider for example a spin 1/2 particle with the magnetic moment $\mu = \gamma\hbar \underset{\sim}{I}$ in the magnetic field $\underset{\sim}{B} = B_1\cos\omega t\hat{i} + B_1\sin\omega t\hat{j} + B_o\hat{k}$. The Hamiltonian can be written as $\hat{\mathcal{H}} = \hat{\mathcal{H}}^o + \hat{V}$ where:

$$\hat{\mathcal{H}}^o = -\gamma\hbar B_o\hat{I}_z \quad \text{and} \quad \hat{V} = -\gamma\hbar B_1(\cos\omega t\hat{I}_x + \sin\omega t\hat{I}_y)$$

Given the states $|u_1\rangle = |m_I=+1/2\rangle$ and $|u_2\rangle = |m_I=-1/2\rangle$ and the initial conditions $P_1(0) = 1$ and $P_2(0) = 0$, derive expressions for $P_1(t)$ and $P_2(t)$ for the resonance condition

$-\omega = (E_2 - E_1)/\hbar$. (Hint: It is convenient to use the identity $2(\cos\omega t\hat{I}_x + \sin\omega t\hat{I}_y) \equiv \hat{I}_+\exp(-i\omega t) + \hat{I}_-\exp(+i\omega t)$.)

Solution: Ref. Problem 5.11

We adopt the analysis of Problem 10.8 through Eqs. (2) which in the present notation are:

$$i\hbar\dot{c}_1 = c_2\langle 1|\hat{V}|2\rangle e^{-i\omega_{21}t} \tag{1a}$$

$$i\hbar\dot{c}_2 = c_1\langle 2|\hat{V}|1\rangle e^{+i\omega_{21}t} \tag{1b}$$

where $\omega_{21} = (E_2 - E_1)/\hbar$. The required matrix elements are:

$$\langle 1|\hat{V}|2\rangle = -\frac{\gamma\hbar B_1}{2}\langle 1/2|\hat{I}_+e^{-i\omega t} + \hat{I}_-e^{+i\omega t}|-1/2\rangle = -\frac{\hbar\omega_1}{2}\langle 1/2|\hat{I}_+|-1/2\rangle e^{-i\omega t}$$

$$\langle 2|\hat{V}|1\rangle = -\frac{\gamma\hbar B_1}{2}\langle -1/2|\hat{I}_+e^{-i\omega t} + \hat{I}_-e^{+i\omega t}|1/2\rangle = -\frac{\hbar\omega_1}{2}\langle -1/2|\hat{I}_-|1/2\rangle e^{+i\omega t}$$

where $\omega_1 = \gamma B_1$. Using the resonance condition $\omega = -\omega_{21}$ Eqs. (1) then give:

$$i\hbar\dot{c}_1 = -\frac{\hbar\omega_1}{2}c_2 \quad \text{and} \quad i\hbar\dot{c}_2 = -\frac{\hbar\omega_1}{2}c_1$$

As in Problem 10.9 these equations are easily solved to obtain:

$$c_1(t) = \frac{a}{2}e^{+i\omega_1 t/2} + \frac{b}{2}e^{-i\omega_1 t/2}\ ; \qquad c_2(t) = \frac{a}{2}e^{+i\omega_1 t/2} - \frac{b}{2}e^{-i\omega_1 t/2}$$

and the initial conditions give $a = b$. Then:

$$P_1(t) = |c_1(t)|^2 = \cos^2\left(\frac{\omega_1 t}{2}\right) \quad \text{and} \quad P_2(t) = |c_2(t)|^2 = \sin^2\left(\frac{\omega_1 t}{2}\right)$$

10.10 Derive an equation for the rate of transitions from state 1 to state 2 for the system in Problem 10.9 by means of Fermi's Golden Rule. Compare the result with $P_2(t)/t$ from the exact treatment and attempt to explain the difference.

Solution: Ref. AB,27; LO,364; Problems 10.2 and 10.19

Fermi's Golden Rule as derived in Problem 10.2 states that

$$W_{21} = \frac{|a_{21}|^2}{t} = \frac{2\pi}{\hbar}|\langle 2|\hat{V}|1\rangle|^2 n(E) \tag{1}$$

Here

$$|\langle 2|\hat{V}|1\rangle| = \hbar\frac{\omega_1}{2}|\langle -1/2|\hat{I}_+e^{-i\omega t} + \hat{I}_-e^{+i\omega t}|+1/2\rangle| = \frac{\hbar\omega_1}{2}$$

Therefore,

$$W_{21} = \frac{\pi}{2}\hbar\omega_1^2\frac{n(\omega)}{\hbar} = \frac{\pi}{2}\omega_1^2 n(\omega) \tag{2}$$

This "rate" is independent of time as expected. On the other hand the exact treatment of Problem 10.9 gives $P_2(t) = \sin^2(\omega_1 t/2)$ and for short times:

$$W_{21} = \frac{P_2(t)}{t} = \frac{\omega_1^2 t}{4} \tag{3}$$

The difference in Eqs. (2) and (3) is striking and its cause is subtle. In the derivation of the perturbation equation (1) an integration has been carried out over a distribution of E_f values either explicitly or implicitly. This introduces an incoherence into the time evolution of an assembly of systems and the transition probabilities become additive. With the exact solution no distribution of E_f values was allowed and the wave functions were found to evolve in a coherent fashion to give $c_2(t) = \sin(\omega_1 t/2)$. This point has been discussed for magnetic resonance by Abragam and in general by Löwdin. Abragam shows that the perturbation theory result can be obtained from the exact expression for $P_2(t)$ as a function of ω by integrating over the appropriate distribution function.

10.11 A diatomic molecule has the permanent dipole moment $\underset{\sim}{\mu}(R)$ where R is the internuclear distance. The wave function for nuclear motion can be written (approximately) as

$$\psi_{v,J,M}(R,\theta,\phi) = R^{-1}S_v(R)Y_{JM}(\theta,\phi) \tag{1}$$

where S_v is a solution for the harmonic oscillator and the $Y_{\ell m}$'s are spherical harmonics. Electric dipole transitions between the rotational states are not allowed unless $\Delta J = \pm 1$, $\Delta M = 0, \pm 1$. Use first order perturbation theory to verify the selection rule $\Delta J = \pm 1$ in the special case that $M = 0$. (Hint: A useful recurrence relation for the Legendre function $P_\ell(z)$ can be found in Problem 10.27.)

<u>Solution</u>: Ref. BO,433; PW,270,306; Appendix 4

According to Fermi's golden rule (Problem 10.2):

$$W_{J'J''} \propto |\langle v,J',0|\underset{\sim}{\mu}\cdot\underset{\sim}{\mathcal{E}}|v,J'',0\rangle|^2$$

Taking $\underset{\sim}{\mathcal{E}}$ in the z-direction then gives

$$W_{J'J''} \propto \left|\int_0^{2\pi} d\phi \int_0^{\pi} \sin\theta d\theta Y^*_{J'0}(\theta,\phi)\cos\theta Y_{J''0}(\theta,\phi)\right|^2 \propto \left|\int_{-1}^{+1} P^*_{J'}(z)zP_{J''}(z)dz\right|^2 \tag{2}$$

From Problem 10.27 the recursion relation for $P_{J''}$ is

$$zP_{J''}(z) = \frac{J''+1}{2J''+1}P_{J''+1}(z) + \frac{J''}{2J''+1}P_{J''-1}(z)$$

and the integral in (2) becomes

$$\frac{J''+1}{2J''+1}\int_{-1}^{+1} P^*_{J'}(z)P_{J''+1}(z)dz + \frac{J''}{2J''+1}\int_{-1}^{+1} P^*_{J'}(z)P_{J''-1}(z)dz \tag{3}$$

The orthogonality condition (Appendix 4) causes the first term in (3) to vanish unless $J' = J'' + 1$ and the second term to vanish unless $J' = J'' + 1$. Thus $W_{J'J''} = 0$ unless $\Delta J = +1$ or $\Delta J = -1$.

10.12 Derive an equation for the rate of transitions between the $v = 0$ and $v = 1$ vibrational levels of a heteronuclear diatomic molecule in a radiation field having the energy density $\rho(\nu_{10})$. You may assume that the vibrations are harmonic and that the dipole moment can be expressed as:

$$\mu = \mu_{r=r_e} + \left(\frac{d\mu}{dr}\right)_{r=r_e}(r - r_e)$$

where r_e is the equilibrium internuclear distance.

Solution:

From Problem 10.7 we have the equation for the rate of electric dipole transitions:

$$W_{\ell m} = \frac{1}{6\varepsilon_o \hbar^2} |\mu_{\ell m}|^2 \rho(\nu_{\ell m}) \tag{1}$$

Then letting $q = r - r_e$, $dq = dr$ we can write:

$$\mu_{10} = \int_{-\infty}^{+\infty} \psi^{v'=1} \left[\mu_o + \left(\frac{d\mu}{dq}\right)_o q\right] \psi^{v''=0} dq$$

The approximate wave functions from Problem 4.10 are:

$$\psi^{v''=0} = \left(\frac{\alpha}{\pi}\right)^{1/4} e^{-\alpha q^2/2}, \qquad \psi^{v'=1} = \left(\frac{\alpha}{\pi}\right)^{1/4} \sqrt{2\alpha}\, q e^{-\alpha q^2/2}$$

Thus

$$\mu_{10} = \left(\frac{\alpha}{\pi}\right)^{1/2} \sqrt{2\alpha} \int_{-\infty}^{+\infty} e^{-\alpha q^2} q \left[\mu_o + \left(\frac{d\mu}{dq}\right)_o q\right] dq$$

$$= \left(\frac{\alpha}{\pi}\right)^{1/2} \sqrt{2\alpha} \int_{-\infty}^{+\infty} q^2 e^{-\alpha q^2} dq \left(\frac{d\mu}{dq}\right)_o = \frac{1}{\sqrt{2\alpha}} \left(\frac{d\mu}{dq}\right)_{q=0} \tag{2}$$

where $\alpha = 2\pi m\nu_o/\hbar$ and $h\nu_o = (E_1 - E_o)$. Substitution of Eq. (2) into Eq. (1) gives:

$$W_{\ell m} = \frac{1}{6\varepsilon_o \hbar^2} \left(\frac{1}{2\alpha}\right) \left(\frac{d\mu}{dq}\right)_{q=0}^2 \rho(\nu_{\ell m})$$

10.13 The intensities of electronic transitions are often expressed in terms of oscillator strengths f_{nm}. For an allowed transition from state n to state m, $f_{nm} \sim 1$; while for a forbidden transition, $f_{nm} << 1$. The <u>oscillator strength</u> f_{nm} is defined by the expression:

$$f_{nm} = \frac{8\pi^2 m_e \nu_{mn}}{e^2 h} |(\mu_x)_{nm}|^2 \tag{1}$$

or in terms of the matrix element $\underset{\sim}{\mu}_{nm}$ by the relation $|\underset{\sim}{\mu}_{nm}|^2 = 3|(\mu_x)_{nm}|^2$ which holds for randomly oriented systems.

(a) Show that $f_{01} = 1$ for a harmonic oscillator consisting of an electron harmonically bound to the origin, i.e. with $f = -kx$ and $\mu_x = -ex$.

(b) For an N electron system

$$(\mu_x)_{nm} = -e\int u_n^*\left(\sum_{i=1}^{N} x_i\right) u_m dv$$

Derive the Kuhn-Thomas sum rule for such a system which states that

$$\sum_m f_{nm} = N \tag{2}$$

(Hint: First evaluate the expression $\left[\sum_i \hat{p}_{x_i}, \sum_j x_j\right]$.)

Solution: Ref. Problem 10.12

(a) Starting with Eq. (2) of Problem 10.12 and using $\frac{d\mu}{dx} = -e$ we have

$$|\mu_{01}|^2 = \frac{\hbar e^2}{4\pi m_e \nu_o}$$

Therefore,

$$f_{01} = \frac{4\pi m_e \nu_o}{e^2\hbar} \cdot \frac{\hbar e^2}{4\pi m_e \nu_o} = 1$$

where we have used the fact that $\nu_{10} = (E_1 - E_o)/h = \nu_o$ for a harmonic oscillator. The dimensionless quantity f_{01} is a convenient measure of relative intensities of transitions.

(b) As suggested we write:

$$\left(\sum_i p_{x_i}\right)\sum_j x_j - \sum_j x_j\left(\sum_i p_{x_i}\right) = \sum_i [\hat{p}_{x_i}, x_i] = -i\hbar N \tag{3}$$

Here the relation

$$[\hat{p}_{x_i}, x_j] = -i\hbar\delta_{ij}$$

has been used. The diagonal matrix element of Eq. (3) is

$$\sum_m (\hat{P}_{nm}X_{mn} - X_{nm}\hat{P}_{mn}) = -i\hbar N$$

where

$$X = \sum_i x_i \qquad \text{and} \qquad P = \sum_j (\hat{p}_x)_j$$

Then to obtain the form of Eq. (2) we write

$$X_{nm} = -(\mu_x)_{nm}/e \qquad \text{and} \qquad \hat{P}_{nm} = -\frac{im_e}{\hbar e}(E_n - E_m)(\mu_x)_{nm}$$

where the last expression comes from Eq. (4) of Problem 10.6. Combining these results we obtain

$$\sum_m \frac{im_e}{\hbar e^2}[(E_n - E_m)(\mu_x)_{nm}(\mu_x)_{mn} - (\mu_x)_{nm}(\mu_x)_{mn}(E_m - E_n)] = -i\hbar N$$

$$\sum_m \frac{2im_e}{e^2}\frac{(E_n - E_m)}{\hbar}|(\mu_x)_{nm}|^2 = -i\hbar N$$

and finally

$$\sum_m \frac{4\pi m_e \nu_{mn}}{e^2\hbar}|(\mu_x)_{nm}|^2 = N \qquad \text{or} \qquad \sum_m f_{nm} = N \tag{4}$$

Notice the exchange of indices on ν_{mn} in Eq. (4) because of the minus sign.

10.14 Suppose that the probability of a transition from the ith excited state in the time interval Δt is proportional to Δt.

(a) Show that the population of the ith state depends on time as $N_i(t) = N_i(0)\exp(-k_i t)$ where $N_i(0)$ is the population at $t = 0$ and k_i is the decay constant.

(b) Show that the mean lifetime of the ith state, $\tau = \langle t\rangle$, is given by $\tau_i = k_i^{-1}$.

Solution:

(a) Let the probability of a transition for one system in the time Δt be $k_i\Delta t$. Then the change in the population of the ith state in this time interval must be $\Delta N_i = -N_i k_i \Delta t$. In the limit that ΔN_i and Δt become infinitesimally small we can write

$$\frac{dN_i}{N_i} = -k_i dt$$

which immediately gives the solution

$$N_i(t) = N_i(0)e^{-k_i t}$$

(b) The mean lifetime is given by

$$\tau_i = \langle t\rangle = \int_0^\infty tN_i(t)dt \Big/ \int_0^\infty N_i(t)dt = \int_0^\infty tN_i(0)e^{-k_i t}dt \Big/ \int_0^\infty N_i(0)e^{-k_i t}dt = k_i^{-1}$$

10.15 For linearly polarized radiation propagating in the x-direction the vector potential can be written as:

$$\underset{\sim}{A} = \hat{k}[A^\circ e^{i(\omega t - kx)} + (A^\circ)^* e^{-i(\omega t - kx)}] \tag{1}$$

where $\omega = 2\pi\nu$ and $k = 2\pi/\lambda$. In the following i and f denote states of a one-electron system.

(a) Assume that e^{-ikx} can be approximated by $1 - ikx$ and show that $(e/m_e)\langle f|\underset{\sim}{A}\cdot\underset{\sim}{p}|i\rangle$ can be written in terms of the electric dipole matrix element $\langle f|u_z|i\rangle$, the electric quadrupole matrix element $\langle f|exz|i\rangle$, and the magnetic dipole matrix element $\langle f|z\hat{p}_x - x\hat{p}_z|i\rangle$. (Hint: It is useful to write $x\hat{p}_z = (1/2)[(x\hat{p}_z + z\hat{p}_x) + (x\hat{p}_z - z\hat{p}_x)]$.)

(b) For resonance radiation with $\omega \sim \omega_{fi}$ use the electric quadrupole and magnetic dipole terms from part (a) separately to obtain transition probabilities from state i to state f.

(c) Estimate the relative intensities of electric dipole, electric quadrupole, and magnetic dipole transitions when each is allowed for $\lambda = 5{,}000$ Å. (Hint: Take $x \sim a_o$ and $|\underset{\sim}{r}\times\underset{\sim}{p}| \sim \hbar$.)

Solution: Ref. BO,435

(a) Using Eq. (1) we write:

$$\underset{\sim}{A}\cdot\hat{\underset{\sim}{p}} = A^\circ e^{i\omega t}(1 - ikx)\hat{p}_z + (A^\circ)^* e^{-i\omega t}(1 + ikx)\hat{p}_z \tag{2}$$

Now taking the second term on the r.h.s. of Eq. (2):

$$<f|(A^\circ)^* e^{-i\omega t}(1 + ikx)\hat{p}_z|i> = (A^\circ)^* e^{-i\omega t}[<f|\hat{p}_z|i> + ik<f|x\hat{p}_z|i>]$$

With the hint given above this becomes:

$$(A^\circ)^* e^{-i\omega t}[im_e\omega_{fi}<f|z|i> + \frac{ik}{2}<f|(x\hat{p}_z - z\hat{p}_x) + (x\hat{p}_z + z\hat{p}_x)|i>] \tag{3}$$

where Eq. (4) of Problem 10.6 has been used in the first term inside the brackets. But notice that

$$m_e \frac{d(xz)}{dt} = xp_z + zp_x = \frac{im_e}{\hbar}(\mathcal{H}xz - xz\mathcal{H})$$

Then

$$<f|x\hat{p}_z + z\hat{p}_x|i> = \frac{im_e}{\hbar}(E_f - E_i)<f|xz|i>$$

Also,

$$x\hat{p}_z - z\hat{p}_x = -(\underset{\sim}{r}\times\underset{\sim}{p})_y$$

Combining these relations with (3) we can write:

$$(e/m_e)<f|\underset{\sim}{A}\cdot\hat{\underset{\sim}{p}}|i> = (A^\circ)^* e^{-i\omega t}[i\omega_{fi}<f|ez|i> + \frac{ik}{2}\frac{e}{m_e}<f|(\underset{\sim}{r}\times\hat{\underset{\sim}{p}})_y|i> + \frac{ik}{2}\omega_{fi}<f|exz|i>] \tag{4}$$

We have neglected the contribution from the first term on the right side of Eq. (2), which could of course be treated the same way. This was done in anticipation of our interest in the absorption term for resonance radiation ($\omega \sim \omega_{fi}$). See Problem 10.2.

(b) The transition rate is given by Eq. (3) of Problem 10.2. For $\omega \sim \omega_{fi}$ the three terms in (4) give the following contributions:

electric dipole:
$$W_{fi} = \frac{2\pi}{\hbar}|A^\circ|^2\omega_{fi}^2|<f|\mu_z|i>|^2 n(h\nu_{fi}) = \frac{2\pi}{\hbar}\left(\frac{W}{2\varepsilon_o}\right)|<f|\mu_z|i>|^2 n(h\nu_{fi}) \tag{5}$$

magnetic dipole:
$$W_{fi} = \frac{2\pi}{\hbar}|A^\circ|^2\frac{k^2}{4}\frac{e^2}{m_e^2}|<f|(\underset{\sim}{r}\times\hat{\underset{\sim}{p}})_y|i>|^2 n(h\nu_{fi})$$

$$= \frac{2\pi}{\hbar^2}\left(\frac{W}{2\varepsilon_o}\right)\frac{k^2}{4\omega_{fi}^2}\frac{e^2}{m_e^2}|<f|(\underset{\sim}{r}\times\hat{\underset{\sim}{p}})_y|i>|^2 n(h\nu_{fi}) \tag{6}$$

electric quadrupole: $W_{fi} = \frac{2\pi}{\hbar} |A°|^2 \frac{k^2\omega_{fi}^2}{4} |\langle f|exz|i\rangle|^2 n(h\nu_{fi})$

$$= \frac{2\pi}{\hbar}\left(\frac{W}{2\varepsilon_o}\right)\frac{k^2}{4} |\langle f|exz|i\rangle|^2 n(h\nu_{fi}) \tag{7}$$

(c) From Eqs. (5), (6), and (7) we obtain:

electric dipole: $W_{fi} \simeq Ka_o^2 = K(2.8\text{x}10^{-21})$

magnetic dipole: $W_{fi} \simeq K\frac{\hbar^2}{4m_e^2c^2} = K(3.7\text{x}10^{-26})$

electric quadrupole: $W_{fi} \simeq \frac{K(2\pi)^2a_o^4}{4\lambda^2} = K(3.1\text{x}10^{-28})$

where

$$K = \frac{2\pi}{\hbar}\left(\frac{W}{2\varepsilon_o}\right)e^2 n(h\nu_{fi})$$

The relative intensities for these three types of transitions are, therefore, 1, 10^{-5}, and 10^{-7} based on the order of magnitude estimates above.

10.16 The induced electric dipole moment, $\mu = e_i\langle z\rangle$, and the polarizability α are related by the equation $\mu = \alpha\mathcal{E}$ where $\mathcal{E}$ is an electric field (see Problem 7.18). Let $\underset{\sim}{\mathcal{E}}(t) = \hat{k}2\mathcal{E}^o\cos\omega t$ and use time-dependent perturbation theory to derive an expression for α, the space fixed polarizability, in terms of ω, i.e. work out that part of $e_i\langle x\rangle$ which is linear in $\underset{\sim}{\mathcal{E}}$.

Solution: Ref. Problem 10.2; S1,154; AT2,410; EWK,118

The induced moment is given by $\langle\mu_z(t)\rangle = \langle\Psi(\underset{\sim}{r},t)|ez|\Psi(\underset{\sim}{r},t)\rangle$ where the time-dependent state function can be written as:

$$\Psi(\underset{\sim}{r},t) = a_o(t)u_o(\underset{\sim}{r})e^{-iE_ot/\hbar} + \sum_{n\neq 0}^{\infty} a_n(t)u_n(\underset{\sim}{r})e^{-iE_nt/\hbar} \tag{1}$$

Now assuming that the perturbation, $\hat{\mathcal{H}}' = -\underset{\sim}{\mu}\cdot\underset{\sim}{\mathcal{E}}(t)$, is small we have $a_o \sim 1$ and $a_n \sim 0$ for $n \neq 0$ and from Problem 10.2:

$$a_n(t) = +\frac{\mathcal{E}^o}{\hbar}\langle n|\mu_z|0\rangle\left[\frac{e^{i(\omega_{no}+\omega)t} - 1}{(\omega_{no} + \omega)} + \frac{e^{i(\omega_{no}-\omega)t} - 1}{(\omega_{no} - \omega)}\right] \tag{2}$$

Substitution of the Eq. (1) into the expression for $\langle\mu_z(t)\rangle$ gives

$$\langle\mu_z(t)\rangle = \langle 0|e_iz|0\rangle + \sum_{n\neq 0}^{\infty} a_n e^{-i\omega_{no}t}\langle 0|e_iz|n\rangle + \sum_{n\neq 0}^{\infty} a_n^* e^{i\omega_{no}t}\langle n|e_iz|0\rangle \tag{3}$$

The first term on the right side of Eq. (3) is zero if there is no permanent dipole moment. Here we have only retained terms linear in $\mathcal{E}^o$. The next step is to substitute (2) for a_n

and to extract terms linear in $\underset{\sim}{\mathcal{E}}(t)$. Thus

$$\langle\mu_z(t)\rangle = \mu_z^o + 2\mathrm{Re}\sum_{n\neq 0}^{\infty} \frac{\mathcal{E}^o}{\hbar} \langle n|e_i z|0\rangle\langle 0|e_i z|n\rangle \left\{\frac{e^{i\omega t}}{\omega_{no} + \omega} + \frac{e^{-i\omega t}}{\omega_{no} - \omega}\right\}$$

$$= \mu_z^o + \frac{2}{\hbar}(2\mathcal{E}^o\cos\omega t)\sum_{n\neq 0}^{\infty} \omega_{no} \frac{\langle 0|\mu_z|n\rangle\langle n|\mu_z|0\rangle}{\omega_{no}^2 - \omega^2}$$

This expression can also be written as

$$\langle\underset{\sim}{\mu}(t)\rangle = \underset{\sim}{\mu}^o + \frac{2}{\hbar}\sum_{n\neq 0}^{\infty} \omega_{no} \frac{(\mu_z)_{on}(\mu_z)_{no}}{\omega_{no}^2 - \omega^2} \mathcal{E}_z(t)$$

and the polarizability becomes

$$\boxed{\alpha(\omega) = \sum_{n\neq 0}^{\infty} \frac{2\omega_{no}}{\hbar} |(\mu_z)_{no}|^2 \frac{1}{\omega_{no}^2 - \omega^2}} \tag{4}$$

By writing Eq. (4) as

$$\alpha(\omega) = \frac{e_i^2}{m}\sum_{n\neq 0}^{\infty} \frac{f_{no}}{(\omega_{no}^2 - \omega^2)}$$

we can at once see the classical formula (Problem 10.23) and the oscillator strength from Problem 10.13.

Note: We have assumed that the perturbation has been turned on so slowly that there is no response to the transient, <u>i.e.</u> the application of the perturbation is an <u>adiabatic process</u> (see for example RA,424).

10.17 When an observer situated on the +x axis detects radiation emitted from an atom located near the origin, the apparent frequency is $\nu_o(1 + v_x/c)$. Here ν_o is equal to $(E_f - E_i)/h$ for the transition and v_x is the x-component of the velocity of the atom. This is, of course, the <u>Doppler Effect</u> which is encountered in elementary physics. The simple equation also holds in the theory of relativity when $v_x/c \ll 1$. Suppose that an ensemble of identical emitters in the gas phase are at equilibrium at temperature T. What line shape is expected for the detected radiation, <u>i.e.</u> what is the form of $I(\nu - \nu_o)$ versus $(\nu - \nu_o)$? Derive an equation for $(\Delta\nu)_{1/2}/\nu_o$, where $(\Delta\nu)_{1/2}$ is the half-width at half-height of the $I(\nu)$ function, as a function of T. You may assume that the emitters behave as an ideal gas.

<u>Solution</u>: Ref. MO,133; RU,78

For motion in one dimension the kinetic energy is just $E = mv_x^2/2$ and the Maxwell-Boltzmann velocity distribution function is:

$$p(v_x) = \sqrt{\frac{m}{2\pi k_B T}} \exp\left(-\frac{mv_x^2}{2k_B T}\right)$$

As indicated above the frequency shift is $(\nu - \nu_o) = \nu_o(v_x/c)$ or

$$v_x = \frac{c(\nu - \nu_o)}{\nu_o} = c\,\frac{\Delta\nu}{\nu_o}$$

The form of the intensity distribution is thus:

$$I(\nu - \nu_o) = A\exp\left[\frac{-m(c\Delta\nu)^2}{2k_B T\nu_o^2}\right] \quad (1)$$

An expression for $(\Delta\nu)_{1/2}$ is easily found by setting the exponential function in Eq. (1) equal to A/2. This yields

$$\boxed{\frac{(\Delta\nu)_{1/2}}{\nu_o} = \frac{\sqrt{2\ell n2}}{c}\sqrt{\frac{k_B T}{m}}}$$

for the width of a Doppler broadened line.

SUPPLEMENTARY PROBLEMS

10.18 Consider a two state system having the eigenvalues $E_n > E_m$. The possible transitions are (i) induced $m \to n$, (ii) induced $n \to m$, and (iii) spontaneous $n \to m$. The rates of absorption and emission are given by:

$$\frac{-dN_m}{dt} = N_m B_{nm} \rho(\nu_{nm}) \quad \text{(absorption)} \quad (1)$$

$$\frac{-dN_n}{dt} = N_n B_{mn} \rho(\nu_{nm}) + N_n A_{mn} \quad \text{(emission)} \quad (2)$$

where N_j is the number of systems in state j, $\rho(\nu_{nm})$ is the energy density in the radiation field, and B_{mn} and A_{mn} are the Einstein coefficients of induced and spontaneous emission, respectively. Use Eqs. (1) and (2) under equilibrium conditions with Planck's radiation law to show that:

(a) $B_{nm} = B_{mn}$, i.e. the Einstein coefficients of induced absorption and emission are equal.

(b) The Einstein coefficients of spontaneous and induced emission are related by the equation

$$A_{mn} = \left(\frac{8\pi h\nu_{nm}^3}{c^3}\right)B_{mn}$$

(Hint: Recall that the populations at thermal equilibrium are simply related.)

Ref. MO,753; S1,137; H. R. Lewis, Am. J. Phys. 41, 38 (1973).

10.19 In Problem 10.2 it was shown that the transition rate W_{fi} can be written as:

$$W_{fi} = \frac{2}{\hbar} |V_{fi}|^2 \int n(\hbar\omega_{fi}) \frac{\sin^2 x}{x^2} dx \tag{1}$$

where $x = (\omega_{fi} - \omega)t/2$. Determine the time dependence of W_{fi} in the special case that $n(\hbar\omega_{fi}) = U\delta[\hbar(\omega_{fi} - \omega)]$ with U = constant, i.e. when $n(\hbar\omega_{fi})$ is a much more strongly peaked function than is $\sin^2 x/x^2$.

Ans:

$$W_{fi} = |V_{fi}|^2 \frac{U}{\hbar^2} t$$

10.20 Consider a potential box of length a which has a charge +e uniformly distributed. When an electron is added to the box, the electric dipole moment of the system is given by $\mu = -ex$ where x is the displacement from the center of the box. Derive an equation for the transition moment $\langle \ell|\mu|m\rangle$ and construct a table to show the relative rates of the transitions 1 → 2, 3, 4, 5, 6. Neglect the coulombic attraction of the charges.

Ans:

$$\mu_{\ell m} = \frac{2ea}{\pi^2}\left[\frac{\sin[(\ell + m)\pi/2]}{(\ell + m)^2} + \frac{\sin[(\ell - m)\pi/2]}{(\ell - m)^2}\right]$$

Table 10.20

$\ell \leftarrow m$	$\lvert\mu_{\ell m}\rvert$	$F_{\ell m}$
2 ← 1	$\frac{16}{9}\frac{ea}{\pi^2}$	1
4 ← 1	$\frac{32}{225}\frac{ea}{\pi^2}$	6.40×10^{-3}
6 ← 1	$\frac{48}{1225}\frac{ea}{\pi^2}$	4.86×10^{-4}

10.21 According to the Franck-Condon Principle, electronic transitions occur so quickly that nuclear coordinates cannot change appreciably during the transitions. In quantitative terms this means that the transition probability depends on the "overlap" of the vibrational wave functions for the initial and final states. In general for electric dipole transitions the intensity of absorption I_{abs} is related to the transition moment $\langle\psi'|\underset{\sim}{\mu}|\psi''\rangle$ as follows:

$$I_{abs} = (\text{constant})\,|\langle\psi'|\underset{\sim}{\mu}|\psi''\rangle|^2 \tag{1}$$

Confirm the Franck-Condon Principle for a simultaneous electronic and vibrational transition by showing that:

$$I_{abs} \simeq (\text{constant})|\bar{\mu}_e|^2 |\langle\psi_n'|\psi_n''\rangle|^2 \qquad (2)$$

where $\bar{\mu}_e$ is the average value of the electronic moment $\langle\psi_e'|\mu_e|\psi_e''\rangle$. You may assume that $\mu = \mu_e + \mu_n$ and $\psi = \psi_e\psi_n$ where the subscripts e and n refer to electronic and nuclear, respectively. The functions ψ_e and ψ_n are solutions of the electronic and nuclear Hamiltonians in the Born-Oppenheimer approximation.
Ref. HZ2,199

10.22 According to Heisenberg's uncertainty principle $\Delta E \Delta t \sim \hbar$. In spectroscopy it is common to find this expression written as $\Delta E \tau \sim \hbar$ in order to relate the mean lifetime τ of a state to the spread in its energy. Actually, Δt should be written as

$$\sqrt{\langle t^2\rangle - \langle t\rangle^2}$$

where t is the lifetime of a system in the state and <> means an average over an ensemble of identical systems. Show that

$$\tau_i = \sqrt{\langle t^2\rangle - \langle t\rangle^2} \qquad (1)$$

in the event that the decay of the population of the ith state is described by a first order rate equation.
Ref. Problem 10.14; W. H. Fink, J. Chem. Ed. 48, 544 (1971)

10.23 Consider a particle of charge e_i and mass m restrained by a linear restoring force, $f = -kx$, and subjected to a uniform electric field $\mathcal{E} \propto e^{i\omega t}$. The differential equation for this system is:

$$m\frac{d^2x}{dt^2} = -kx + e_i\mathcal{E}_o e^{i\omega t} \qquad (1)$$

Derive an equation for the induced electric dipole moment, $\mu = qx = \alpha\mathcal{E}$, as a function of ω and t. This calculation is based on the Lorentz-Drude model of the electron where of course $e_i = -e$. (Hint: Assume that the displacement x is proportional to $e^{i\omega t}$ and then solve Eq. (1) for x.)
Ans: Ref. Problems 4.9 and 10.16; S1,156

$$\mu = e_i x = \left(\frac{e_i^2}{m}\right)\frac{\mathcal{E}}{(\omega_o^2 - \omega^2)}$$

10.24 A hydrogen atom in its ground state is subjected to an electric field which has the time dependence

$\mathcal{E} = 0, \quad t < 0; \qquad \mathcal{E} = \mathcal{E}_o e^{-at}, \quad t \geqslant 0 \quad (a = \text{constant})$

(a) Find the probability $|a_2(t)|^2$ in the limit that $t \to \infty$.

(b) Repeat (a) assuming that $\mathcal{E} = \mathcal{E}_o$, $t \geqslant 0$ and $\mathcal{E} = 0$ otherwise.

Ans: Ref. BO,412

$$\text{(a)} \quad |a_2|^2 = \frac{2^{15}a_o^2\mathcal{E}_o^2(e^2/4\pi\varepsilon_o)}{3^{10}\hbar^2\left(\dfrac{3^2(e^2/4\pi\varepsilon_o)^2}{2^6\hbar^2a_o^2} + a^2\right)} \qquad \text{(b)} \quad |a_2|^2 = \frac{2^{21}}{3^{12}}\left(\frac{a_o^2\mathcal{E}_o}{e}\right)^2(4\pi\varepsilon_o)$$

10.25 Consider an atomic system having the allowed energies $E_1 < E_2 < E_3$ with the populations N_1, N_2, and N_3, respectively. When radiation is introduced with the frequency $\nu_{31} = (E_3 - E_1)/h$, it is found that under steady state conditions, i.e. when $\dot{N}_1 = \dot{N}_2 = \dot{N}_3 = 0$, $N_3 > N_2$. Thus there is a population inversion, and the system can function as a laser at the frequency $\nu_{32} = (E_3 - E_2)/h$. Derive an expression for the ratio N_3/N_2 in terms of the Einstein coefficients A_{ij} and B_{ij} in the special case that no external radiation is present at any frequency other than ν_{31}. Also, determine the ratio N_2/N_1 under the same conditions.

Ans:

$$N_3/N_2 = A_{21}/A_{32}, \qquad \frac{N_2}{N_1} = \frac{B_{13}\rho_{13}A_{32}}{A_{21}[A_{32} + A_{31} + B_{31}\rho_{31}]}$$

10.26 In studying molecular motion it is very useful to express the intensity of absorption (or emission) as the Fourier transform of a correlation function. Thus

$$I(\omega) = \frac{1}{2\pi}\int_{-\infty}^{+\infty} \langle \underset{\sim}{\mu}(0)\cdot\underset{\sim}{\mu}(t)\rangle_{av} e^{-i\omega t}dt \tag{1}$$

where $\underset{\sim}{\mu}(t)$ is the electric dipole moment at time t and $\langle\rangle_{av}$ implies the average over an ensemble of identical systems. Verify Eq. (1) by the following procedure:

(1) Define the unnormalized intensity function $I(\omega)$ as:

$$I(\omega) = 3\sum_{i,f}\sum \rho_i |\langle f|\mu_x|i\rangle|^2 \delta(\omega_{fi} - \omega)$$

where ρ_i is the probability of occupation of the initial state i. This equation follows from first order perturbation theory.

(2) Insert the integral definition of the delta function:

$$\delta(\omega) = \frac{1}{2\pi}\int_{-\infty}^{+\infty} e^{i\omega\tau}d\tau$$

(3) Define

$$\mu_x(t) = e^{i\hat{\mathcal{H}}t/\hbar}\mu_x e^{-i\hat{\mathcal{H}}t/\hbar}$$

Ref. R. G. Gordon, Adv. Mag. Resonance 3, 1 (1968)

10.27 The Legendre polynomial $P_\ell(z)$ can be defined in terms of the generating function:

$$\frac{1}{\sqrt{1 - 2zs + s^2}} = \sum_{\ell=0}^{\infty} P_\ell(z)s^\ell \qquad (1)$$

Use this function to show that

$$zP_\ell(z) = \frac{\ell + 1}{2\ell + 1} P_{\ell+1}(z) + \frac{\ell}{2\ell + 1} P_{\ell-1}(z) \qquad (2)$$

(Hint: Differentiate Eq. (1) with respect to s.)

Ref. Appendix 4

10.28 The dipole moment operator for a molecule is

$$\hat{\mu} = e(\sum_i^p Z_i \underset{\sim}{R}_i - \sum_i^q \underset{\sim}{r}_i) = \hat{\underset{\sim}{\mu}}_e + \hat{\underset{\sim}{\mu}}_N$$

where p is the number of nuclei, q is the number of electrons and Z_i the nuclear charge of atom i. The permanent electric dipole moment is $\overline{\underset{\sim}{\mu}_e} \equiv \langle u_e|\hat{\underset{\sim}{\mu}}|u_e\rangle$ where u_e is the electronic wave function.

(a) Show that for $^1\Sigma$ states of diatomic molecules the rotational selection rules are $\Delta J = \pm 1$, $\Delta M = 0, \pm 1$.

(b) Expand $\overline{\underset{\sim}{\mu}_e}$ in a Taylor series about R_e and determine the selection rules for vibration. (Hint: Use the nuclear functions $u_N = R^{-1}S_N(x)Y_{J,M}$ where $x = R - R_e$ and $S_N(x)$ are harmonic oscillator solutions. The solution is greatly simplified by use of the recursion relations for $\cos\theta P_\ell^{|m|}$ and $\sin\theta P_\ell^{|m|}$ listed in Appendix 4.)

Ans: Ref. PW,306; LE2,157; Problems 10.11 and 10.27

(b) $\Delta N = 0, \pm 1$, if $\overline{\mu}_e'(R_e) \neq 0$

Weak transitions can occur at $\Delta N = \pm 2, \pm 3$, etc. if the higher derivatives of $\overline{\mu}_e$ are nonzero.

10.29 Assume that the potential function for the ground electronic state for a molecule can be written as $V_{gnd}(r) = (k/2)(r - r_e)^2$ and that the potential function for the first excited electronic state is $V_{exc}(r) = E_{el} + (k'/2)(r - r_e')^2$. Evaluate the Franck-Condon factors $|\langle u_o'|u_o''\rangle|^2$, $|\langle u_o'|u_1''\rangle|^2$, and $|\langle u_o'|u_2''\rangle|^2$ for $r_e = r_e'$ but $k \neq k'$.

Ans: Ref. Problem 10.21

$$|\langle u_o'|u_o''\rangle|^2 = 2\sqrt{\alpha'\alpha''}/(\alpha' + \alpha'')$$

$$|\langle u_o'|u_1''\rangle|^2 = 0$$

$$|\langle u_o'|u_2''\rangle|^2 = \frac{4\sqrt{\alpha'\alpha''}}{(\alpha' + \alpha'')}\left[\frac{\alpha''}{(\alpha' + \alpha'')} - \frac{1}{2}\right]^2$$

10.30 Evaluate the Franck-Condon factors $|\langle u_o'|u_o''\rangle|^2$ and $|\langle u_o'|u_1''\rangle|^2$ for the molecule in Problem 10.29 in the case that $k = k'$ but $r_e \neq r_e'$.

Ans: Ref. Problems 10.21, 10.29

$$|\langle u_o'|u_o''\rangle|^2 = e^{-\alpha\delta^2/2}$$

$$|\langle u_o'|u_1''\rangle|^2 = \frac{\delta^2\alpha}{2}\, e^{-\alpha\delta^2/2}$$

CHAPTER **11**

MOLECULAR SPECTROSCOPY

In spectroscopy the absorption or emission of radiation is used to study the microscopic details of the structure of matter. As discussed in Chapter 10 the transition frequencies depend on the eigenvalues of the system while the intensities are governed by the nature of the radiation-matter interaction and the wave functions of the states involved. Thus, quantum mechanics or an approximation to it must be used in all but the most superficial spectroscopic analyses. In a very real way, spectroscopy is a form of applied quantum mechanics.

Spectroscopy involving electric dipole transitions is usually subdivided according to the frequency of the associated radiation. Thus we find ultraviolet (UV) and visible spectroscopy, infrared (IR) spectroscopy, microwave (MW) spectroscopy, *etc.* The spectroscopies involving the interactions of electron spins or nuclear moments with radiation fields are usually given more explicit names, *e.g.* electron spin resonance (ESR), nuclear magnetic resonance (NMR), nuclear quadrupole resonance (NQR). Raman spectroscopy is unique in that inelastically scattered radiation is analyzed to determine frequency shifts.

The major fields of spectroscopy are now so well developed that their expositions fill many volumes. We cannot hope to cover even briefly all of the fields of spectroscopy in a few pages. Instead we have collected a small set of problems which are relevant to some of the fields of spectroscopy, and which illustrate the principles of quantum mechanics. Included are some more or less standard problems from vibrational (IR) and rotational (MW) spectroscopy as well as a scattering of other problems (mainly concerning magnetic resonance) which we think are instructive.

General References on Spectroscopy:

IR and Raman: HZ2, HZ3, WDC

MW: TS

NQR: DH

UV and Visible: JO

NMR and ESR: CM, SL

Mössbauer: FR

PROBLEMS

11.1 (a) Set up the Schrödinger equation for nuclear motion in a diatomic molecule when the potential energy function can be written as $V(r) = (k/2)(r - r_e)^2$ where r_e is the equilibrium internuclear distance.

(b) Obtain the eigenfunctions and eigenvalues in the special case that there is no rotational angular momentum, i.e. $J = 0$. You may assume that $\rho = r - r_e << r_e$ and thus that the wave function is very small at $r = 0$.

Solution: Ref. PW,267

(a) The radial equation for a two-body problem can be written as

$$-\frac{\hbar^2}{2\mu}\frac{1}{r^2}\frac{d}{dr}\left(r^2\frac{dR}{dr}\right) + \left[\frac{J(J+1)\hbar^2}{2\mu r^2} + V(r)\right] R = ER$$

or in terms of the function $P(r) = rR(r)$ as (see Problem 7.2)

$$-\frac{\hbar^2}{2\mu}\frac{d^2P}{dr^2} + \left[\frac{J(J+1)\hbar^2}{2\mu r^2} + V(r)\right] P = EP \qquad (1)$$

In the special case that $J = 0$ and $V(r) = (k/2)(r - r_e)^2$ this becomes

$$-\frac{\hbar^2}{2\mu}\frac{d^2P}{dr^2} + \frac{k}{2}(r - r_e)^2P = EP \qquad \text{or} \qquad -\frac{\hbar^2}{2\mu}\frac{d^2P}{d\rho^2} + \frac{k}{2}\rho^2P = EP \qquad (2)$$

(b) Here $-r_e \lessgtr \rho \lessgtr \infty$ since $r \gtrless 0$. But P is assumed to be vanishingly small at $\rho = -r_e$, and we can safely use the limits $-\infty \lessgtr \rho \lessgtr +\infty$. Equation (2) thus represents a simple harmonic oscillator in one dimension. The solutions from Problem 4.10 are

$$R(r) = \frac{P(r)}{r} = \frac{1}{r}\left[\sqrt{\frac{\alpha}{\pi}}\frac{1}{2^v v!}\right]^{1/2} e^{-\alpha\rho^2/2} H_v(\sqrt{\alpha}\,\rho), \qquad v = 0, 1, 2 \cdots$$

$$E(v) = (v + 1/2)h\nu_o; \qquad \nu_o = \frac{1}{2\pi}\sqrt{\frac{k}{\mu}}$$

where $\alpha = 2\pi\mu\nu_o/\hbar$ and the functions H_v are Hermite polynomials.

11.2 Determine the vibrational energy levels for a molecule having a potential function of the form proposed by Morse:

$$V(r) = D[1 - e^{-a(r - r_e)}]^2$$

Consider only the case for $J = 0$, and assume that $\rho = r - r_e << r_e$. (Hint: First transform the equation for $P(r) = rR(r)$ by the substitution $y = \exp[-a(r - r_e)]$ to obtain:

$$-\frac{\hbar^2}{2\mu}\left(\frac{d^2P}{dy^2} + \frac{1}{y}\frac{dP}{dy}\right) + \frac{1}{a^2}\left[\frac{(D - E)}{y^2} - \frac{2D}{y} + D\right]P = 0$$

Then make the substitution

$$P(y) = e^{-z/2}z^{b/2}F(z)$$

where

$$z = 2\sqrt{2\mu D}\ y/a\hbar \qquad \text{and} \qquad b = \frac{2}{a\hbar}\sqrt{2\mu(D - E)}$$

in order to obtain an equation for $F(z)$ which can be recognized as a standard form.)

Solution: Ref. PW,272; EWK,272; P. M. Morse, Phys. Rev. 34, 57 (1929); Problem 11.1

With $J = 0$ the equation to be solved is

$$-\frac{\hbar^2}{2\mu}\frac{d^2P}{dr^2} + \{D + D[e^{-2a(r-r_e)} - 2e^{-a(r-r_e)}]\}P = EP \tag{1}$$

With the substitution $y = e^{-a(r-r_e)}$ we find:

$$\frac{dP}{dr} = -ay\frac{dP}{dy} \qquad \text{and} \qquad \frac{d^2P}{dr^2} = ay^2\left(\frac{d^2P}{dy^2} + \frac{1}{y}\frac{d^2P}{dy^2}\right)$$

Thus Eq. (1) can be written as

$$-\frac{\hbar^2}{2\mu}\left(\frac{d^2P}{dy^2} + \frac{1}{y}\frac{dP}{dy}\right) + \frac{1}{a^2}\left[\frac{D - E}{y^2} - \frac{2D}{y} + D\right]P = 0 \tag{2}$$

The substitution $P(y) = e^{-z/2}z^{b/2}F(z)$ yields

$$y^{-1}\frac{dP}{dy} = \beta^2 e^{-z/2}z^{b/2}\left[\frac{1}{z}\frac{dF}{dz} + \frac{1}{2}\left(\frac{b}{z^2} - \frac{1}{z}\right)F\right]$$

$$\frac{d^2P}{dy^2} = \beta^2 e^{-z/2}z^{b/2}\left\{\frac{d^2F}{dz^2} + \left(\frac{b}{z} - 1\right)\frac{dF}{dz} + \left[\frac{b^2}{4z^2} - \frac{b}{2z^2} - \frac{b}{2z} + \frac{1}{4}\right]F\right\}$$

where $z = \beta y$. Combining these results with Eq. (2) gives

$$\frac{d^2F}{dz^2} + \left(\frac{b + 1}{z} - 1\right)\frac{dF}{dz} + \left[\frac{\beta}{2} - \frac{(b + 1)}{2}\right]\frac{F}{z} = 0 \tag{3}$$

Equation (3) is now in the form of the equation that is satisfied for the Associated Laguerre polynomials $L(\rho)$. It should be compared with

$$L'' + \left(\frac{2\ell + 1}{\rho} - 1\right)L' + \left[\frac{\lambda - \ell - 1}{\rho}\right]L = 0 \tag{4}$$

Well behaved solutions are known to exist when $\lambda - \ell - 1 - n = 0$ where n is a positive integer. Equations (3) and (4) are, of course, identical if we set

$$b + 1 = 2\ell + 1 \qquad \text{and} \qquad \lambda = \frac{\beta + 1}{2}$$

Solutions for Eq. (3) then exist when

$$\frac{\beta}{2} - \frac{(b + 1)}{2} = n \qquad \text{or} \qquad \frac{\sqrt{2\mu D}}{a\hbar} - \frac{1}{2}\left(\frac{2}{a\hbar}\sqrt{2\mu(D - E)} + 1\right) = n$$

This condition then gives

$$\frac{E}{hc} = (n + 1/2)\omega_e - (n + 1/2)^2\omega_e x_e \tag{5}$$

where

$$\omega_e = \frac{2a\hbar}{hc}\sqrt{\frac{D}{2\mu}} \qquad \text{and} \qquad \omega_e x_e = \frac{1}{hc}\left(\frac{a^2\hbar^2}{2\mu}\right)$$

The solution given by Eq. (5) is rigorous for Eq. (3) if $0 \lessapprox z \lessapprox \infty$ and F(z) vanishes at $z = 0$ and $z = \infty$. In fact, $r \gtrapprox 0$ so that y and z can never be identically zero. Equation (5) is a very good approximation to the exact solution for molecules. A discussion of this point can be found in: D. ter Haar, Phys. Rev. 70, 222 (1946).

11.3 Consider an anharmonic oscillator with the potential function

$$V(x) = ax^2 + bx^3 + cx^4 \tag{1}$$

where $a >> b >> c$. Use perturbation theory to derive an equation for the energy of this oscillator as a function of $(v + 1/2)$ where v is the vibrational quantum number. It is only necessary to calculate the effect of the term cx^4 to first order. The term bx^3 will contribute in second but not in first order. Show that

$$\boxed{\omega_e x_e \simeq \frac{\hbar^2}{48\mu(hc)}\left[5\,\frac{(V^{III})^2}{(V^{II})^2} - 3\,\frac{V^{IV}}{V^{II}}\right]} \tag{2}$$

where $V^N = (\partial^N V(r - r_e)/\partial r^N)_{r=r_e}$ and $x = r - r_e$.

Solution: Ref. Problems 6.6, 6.7; KI,160

With $x = r - r_e$ we can write

$$V(x) = V(0) + \left(\frac{dV}{dx}\right)_o x + \frac{1}{2}\left(\frac{d^2V}{dx^2}\right)_o x^2 + \cdots \frac{1}{n!}\left(\frac{d^nV}{dx^n}\right)_o x^n$$

and comparison with Eq. (1) shows that:

$$a = V^{II}/2, \quad b = V^{III}/6, \quad \text{and} \quad c = V^{IV}/24$$

The eigenvalues for the harmonic oscillator (Problem 4.10) and the perturbation results of Problems 6.6 and 6.7 give immediately:

$$E(v) = hcG(v) = \left(v + \frac{1}{2}\right)h\nu - \frac{30b^2}{h\nu(2\alpha)^3}\left(v^2 + v + \frac{11}{30}\right) + \frac{3c}{2\alpha^2}\left(v^2 + v + \frac{1}{2}\right) \tag{3}$$

This equation can be written in the form

$$hcG(v) = \left(v + \frac{1}{2}\right)h\nu - \left[\frac{30b^2}{h\nu(2\alpha)^3} - \frac{3c}{2\alpha^2}\right]\left(v + \frac{1}{2}\right)^2 - \frac{30b^2}{h\nu(2\alpha)^3}\left(\frac{7}{60}\right) + \frac{3c}{8\alpha^2} \tag{4}$$

to facilitate comparison with the spectroscopic equation

$$G(v) = \left(v + \frac{1}{2}\right)\omega_e - \left(v + \frac{1}{2}\right)^2\omega_e x_e + \cdots$$

We conclude that

$$\omega_e x_e \simeq \frac{1}{hc}\left[\frac{30b^2}{h\nu(2\alpha)^3} - \frac{3c}{2\alpha^2}\right] \tag{5}$$

Use of the relation $\alpha = V^{II}/h\nu$ and the equations for b and c permit Eq. (5) to be written in the form of Eq. (2) as required.

11.4 The rotational constant for acetylene ($^{12}C_2{}^1H_2$) is $B_e = 1.17692\ cm^{-1}$ and the bond lengths are $r(C{\equiv}C) = 1.207$ Å and $r(C\text{-}H) = 1.059$ Å. Here $B_e = h/8\pi^2 cI_e$.

(a) What type of pure rotational spectrum is expected for this molecule?

(b) A strong absorption centered at 3287 cm^{-1} is ascribed to the antisymmetric stretching vibration where the nuclear motion is parallel to the symmetry axis. Describe the rotational fine structure in this absorption and determine the spacings of the lines neglecting centrifugal distortions.

Solution: Ref. HZ2,66; KP,458, Problem 11.29

(a) Since no permanent electric dipole moment is present, there is no pure rotational spectrum for the gas at low pressures. At very high pressures collision induced moments may give rise to absorption.

(b) From Problem 10.11 the selection rules are $\Delta v = +1$, $\Delta J = \pm 1$. The frequencies (in cm^{-1}) for the allowed transitions are thus: (with $F(J) = J(J+1)B_e$)

$$\tilde{\nu} \cong \omega_e + F(J') - F(J'')$$

$$\Delta J = +1: \quad \tilde{\nu} \simeq \omega_e + B_e(J+1)(J+2) - B_e J(J+1) = \omega_e + 2B_e(J+1)$$

$$\Delta J = -1: \quad \tilde{\nu} \simeq \omega_e + B_e(J-1)J - B_e J(J+1) = \omega_e - 2B_e J$$

The spectrum is centered at $\omega_e = 3287\ cm^{-1}$ and the lines appear on each side with the spacing of $2B_e = 2.35\ cm^{-1}$. Since $\Delta J = 0$ is not allowed there is a gap of $4B_e$ at the center. The intensities are determined by the quantum numbers of the initial and final states (J" and J'), the population of the initial state, and the exclusion principle (see Ref. HZ2).

11.5 As shown in Problem 11.2 the vibrational energy levels in a Morse potential can be written (in units of cm^{-1}) as

$$G(v) = \left(v + \frac{1}{2}\right)\omega_e - \left(v + \frac{1}{2}\right)^2 \omega_e x_e \tag{1}$$

The separation between the v and v + 1 energy levels is defined as

$$\Delta G_{v+\frac{1}{2}} = G(v+1) - G(v)$$

As v increases, $\Delta G_{v+\frac{1}{2}}$ decreases until it reaches zero when the molecule dissociates. The dissociation energy measured from the lowest vibrational level is defined as D_o, and from the minimum in the potential curve it is D_e where

$$D_o = \sum_v \Delta G_{v+\frac{1}{2}} \qquad \text{and} \qquad D_e = G(v_{max})$$

Here v_{max} is the vibrational quantum number for the last discrete energy level before dissociation. Assuming the validity of Eq. (1) show that for $\omega_e >> \omega_e x_e$:

$$D_e \simeq \frac{\omega_e^2}{4\omega_e x_e}$$

Solution: Ref. KI,160; HZ2,98

First we rewrite $\Delta G_{v+\frac{1}{2}}$ in terms of v. Thus

$$\Delta G_{v+\frac{1}{2}} = G(v+1) - G(v) = \omega_e - \left(v+\frac{3}{2}\right)^2 \omega_e x_e + \left(v+\frac{1}{2}\right)^2 \omega_e x_e = \omega_e - 2(v+1)\omega_e x_e$$

For $v = v_{max}$ then $\Delta G_{v+\frac{1}{2}} = 0$ and $(v_{max} + 1) = \omega_e/2\omega_e x_e$. Then using Eq. (1) we have

$$D_e = G(v_{max}) = \left(\frac{\omega_e}{2\omega_e x_e} - \frac{1}{2}\right)\omega_e - \left(\frac{\omega_e}{2\omega_e x_e} - \frac{1}{2}\right)^2 \omega_e x_e = \frac{\omega_e^2}{4\omega_e x_e} - \frac{\omega_e x_e}{4} \simeq \frac{\omega_e^2}{4\omega_e x_e}$$

11.6 Consider a particle of mass m and charge q which moves under the influence of the potential:

$$V(x) = ax^2 - bx^3$$

where $b << a$. When radiation is present the most intense transitions are expected to correspond to $\Delta v = +1$; however the $\Delta v = +2$ transitions are also allowed for $b \neq 0$. Calculate the ratio of the intensity of the $v'' = 0 \rightarrow v' = 2$ transition to that of the $v'' = 0 \rightarrow v' = 1$ transition in terms of the constants a and b. First order perturbation theory should be used in this calculation.

Solution: Ref. Problems 10.2, 10.20, and 6.7

From the "Golden Rule" (Problem 10.2) and the definition of intensity we have

$$\frac{\text{Intensity }(0\rightarrow 2)}{\text{Intensity }(0\rightarrow 1)} = \frac{|\langle 0'|x|2'\rangle|^2 \nu_{20}}{|\langle 0'|x|1'\rangle|^2 \nu_{10}} \tag{1}$$

where $|0'\rangle$, $|1'\rangle$, and $|2'\rangle$ refer to the eigenfunctions of the Hamiltonian

$$\hat{\mathcal{H}} = \frac{\hat{p}^2}{2m} + ax^2 - bx^3$$

and $\nu_{ij} = (E_i - E_j)/h$. The transition $0 \rightarrow 1$ is allowed for the harmonic oscillator and we can use the relation $\langle 0|x|1\rangle \simeq \langle 0'|x|1'\rangle$ where $|n\rangle$ specifies the nth eigenket of a harmonic oscillator with the force constant $k = 2a$. For the $0 \rightarrow 2$ transition, which is not allowed for the harmonic oscillator, we can first order perturbation theory to express the kets $|0'\rangle$ and $|2'\rangle$ in terms of the set $|n\rangle$. Thus, using the results of Problem 6.7 we find:

$$|0'\rangle = |0\rangle + \frac{b}{(2\alpha)^{3/2} h\nu}\left[3|1\rangle + \frac{\sqrt{6}}{3}|3\rangle + \cdots\right]$$

$$|2'\rangle = |2\rangle + \frac{b}{(2\alpha)^{3/2} h\nu}\left[-6\sqrt{2}\,|1\rangle + 9\sqrt{3}\,|3\rangle + \frac{2\sqrt{15}}{3}|5\rangle + \cdots\right]$$

where

$$\nu = (1/2\pi)\sqrt{2a/m} \qquad \text{and} \qquad \alpha = \sqrt{2am}/\hbar$$

The matrix element required in the numerator of Eq. (1) thus becomes

$$\langle 0'|x|2'\rangle = \frac{b}{(2\alpha)^{3/2}h\nu}\left[-6\sqrt{2}\langle 0|x|1\rangle + 3\langle 1|x|2\rangle + \frac{\sqrt{6}}{3}\langle 3|x|2\rangle + \cdots\right]$$

where we have used the fact that $\langle n|x|m\rangle = 0$ if $n \neq m \pm 1$. Then inserting the matrix elements from Problem 4.12, i.e.

$$\langle n|x|n+1\rangle = \sqrt{(n+1)/2\alpha}$$

we obtain

$$\langle 0'|x|2'\rangle = -\frac{b\sqrt{2}}{2h\nu\alpha^2}, \qquad \text{also} \qquad \langle 0'|x|1'\rangle \simeq 1/\sqrt{2\alpha}$$

Combining these results and noting that $\nu_{20}/\nu_{10} \simeq 2$ for the anharmonic oscillator, Eq. (1) gives

$$\frac{I(0\to 2)}{I(0\to 1)} = \frac{2b^2}{(h\nu)^2\alpha^3} = \frac{b^2}{a^{5/2}}\left(\frac{h}{4\pi\sqrt{2m}}\right)$$

11.7 Ammonia is a symmetric top molecule, i.e. two of the three principal moments of inertia are equal, $I_A \neq I_B = I_C$. Show that the moment of inertia for rotation about the symmetry axis is given by:

$$I_A = 2m_p\ell^2(1 - \cos\theta)$$

where m_p is the mass of the proton, ℓ is the N-H bond length, and θ is the HNH angle. The ammonia molecule (NH_3) has C_{3v} symmetry.

Solution:

By definition the moment of inertia about the symmetry axis is given by

$$I_A = \sum_i m_i r_{iA}^2$$

where m_i is the mass of the ith particle and r_{iA} is the perpendicular distance from the symmetry axis to the ith particle.

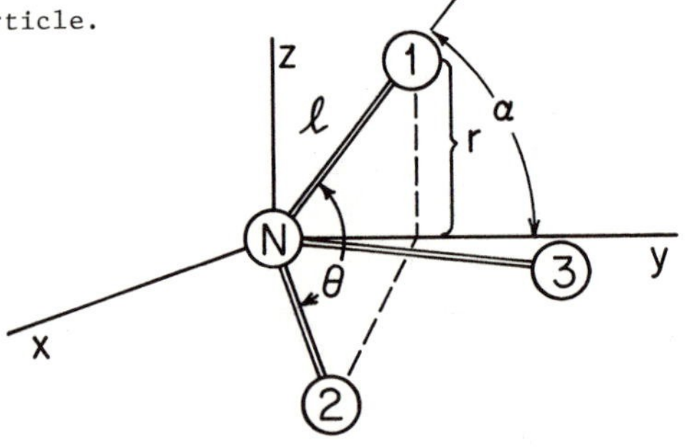

Also, the moment of inertia can be written as $I_A = 3m_p(\ell\sin\alpha)^2$ where α is the angle between the symmetry axis and the NH bonds (see Fig. 11.7). In order to express α in terms of θ it is convenient to introduce the vectors $\underset{\sim}{v}_1$ and $\underset{\sim}{v}_2$ directed from N to protons 1 and 2, respectively. Then

$$\underset{\sim}{v}_1 = \ell\cos\alpha\hat{j} + \ell\sin\alpha\hat{k}, \qquad \underset{\sim}{v}_2 = (\ell\sin\alpha)\sin 60°\hat{i} + \ell\cos\alpha\hat{j} - (\ell\sin\alpha)\cos 60°\hat{k}$$

and

$$\underset{\sim}{v}_1 \cdot \underset{\sim}{v}_2 = \ell^2\cos\theta = \ell^2\cos^2\alpha - \ell^2\sin^2\alpha\cos 60°$$

Then

$$\cos\theta = 1 - \frac{3}{2}\sin^2\alpha \quad \text{and} \quad I_A = 3m_p\ell^2\,\frac{2}{3}\,(1 - \cos\theta) = 2m_p\ell^2(1 - \cos\theta)$$

11.8 The moment of inertia matrix for a molecule is written as

$$\begin{pmatrix} I_{xx} & -I_{xy} & -I_{xz} \\ -I_{xy} & I_{yy} & -I_{yz} \\ -I_{xz} & -I_{yz} & I_{zz} \end{pmatrix}$$

where

$$I_{xx} = \sum_i m_i(y_i^2 + z_i^2) \qquad I_{xy} = I_{yx} = \sum_i m_i x_i y_i$$

$$I_{yy} = \sum_i m_i(x_i^2 + z_i^2) \qquad I_{xz} = I_{zx} = \sum_i m_i x_i z_i$$

$$I_{zz} = \sum_i m_i(x_i^2 + y_i^2) \qquad I_{yz} = I_{zy} = \sum_i m_i y_i z_i$$

Here m_i is the mass of the ith atom and x_i, y_i, z_i are the coordinates of the ith atom with respect to the center of mass. The principal moments of inertia are given by the eigenvalues of the moment of inertia matrix.

Use the definitions given above to determine the principal moments of inertia for FNO where d(FN) = 1.52 Å, d(NO) = 1.13 Å, and <FNO = 110°.

Solution: Ref. GO,149; HO,123

Placing N at the origin, O on the +x axis, and F in the xy plane we obtain the coordinates (in Å units):

$$x'_F = -0.520; \quad x'_O = 1.13; \quad x'_N = 0$$

$$y'_F = 1.428; \quad y'_O = 0; \quad y'_N = 0$$

The center of mass has the coordinates:

$$x_{CM} = \sum_i m_i x'_i / \sum_i m_i = \frac{16(1.13) - 19(0.520)}{49.01} = 0.167$$

$$y_{CM} = \sum_i m_i y'_i / \sum_i m_i = \frac{19(1.428)}{49.01} = 0.554$$

The atomic coordinates with respect to the CM are thus:

$$x_F = x'_F - x_{CM} = -0.687; \qquad x_O = x'_O - x_{CM} = 0.963; \qquad x_N = -x_{CM} = -0.167$$

$$y_F = y'_F - y_{CM} = 0.875; \qquad y_O = y'_O - y_{CM} = -0.554; \qquad y_N = -y_{CM} = -0.554$$

The elements of the moment of inertia matrix are given by:

$$I_{xx} = \sum_i m_i y_i^2 = 16(0.554)^2 + 14.01(0.554)^2 + 19(0.875)^2 = 23.76$$

$$I_{yy} = \sum_i m_i x_i^2 = 16(0.963)^2 + 14.01(0.167)^2 + 19(0.687)^2 = 24.19$$

$$I_{zz} = \sum_i m_i (x_i^2 + y_i^2) = 16[(0.963)^2 + (0.554)^2] + 14.01[(0.167)^2 + (0.554)^2] + 19[(0.687)^2 + (0.875)^2] = 47.95$$

$$I_{xy} = \sum_i m_i x_i y_i = -16(0.963)(0.554) + 14.01(0.167)(0.554) - 19(0.687)(0.875) = -18.66$$

All of these numbers are in units of $(\text{Å})^2/N_A$. The eigenvalues of the moment of inertia matrix are computed as described in Appendix 6. Thus we have the secular equation:

$$\begin{vmatrix} 23.76 - \lambda & -18.66 & 0 \\ -18.66 & 24.19 - \lambda & 0 \\ 0 & 0 & 47.95 - \lambda \end{vmatrix} = 0$$

which gives three roots:

$$\lambda_{1,2} = \frac{47.95 \pm \sqrt{2299.20 - 906.24}}{2} = 42.63,\ 5.31$$

$$\lambda_3 = 47.95$$

In cgs units the principal moments of inertia are

$$I_x = 42.63\text{x}10^{-16}/N_A = 7.077\text{x}10^{-39}\ \text{g}\cdot\text{cm}^2$$

$$I_y = 5.31\text{x}10^{-16}/N_A = 8.816\text{x}10^{-40}\ \text{g}\cdot\text{cm}^2$$

$$I_z = 47.95\text{x}10^{-16}/N_A = 7.961\text{x}10^{-39}\ \text{g}\cdot\text{cm}^2$$

To convert to units of $\text{kg}\cdot\text{m}^2$ these numbers should be multiplied by $10^{-7}\ \text{kg}\cdot\text{m}^2/\text{g}\cdot\text{cm}^2$.

11.9 The Hamiltonian for a rotating molecule can be written as

$$\hat{\mathcal{H}} = \frac{\hbar^2}{2}\left(\frac{\hat{J}_x^2}{I_x} + \frac{\hat{J}_y^2}{I_y} + \frac{\hat{J}_z^2}{I_z}\right) \tag{1}$$

where I_x, I_y, and I_z are the principal moments of inertia for rotation about the molecule fixed axes (x,y,z). When referred to the molecule fixed "gyrating" axis, the angular momenta obey the modified commutation relations:

$$\hat{J}_x\hat{J}_y - \hat{J}_y\hat{J}_x = -i\hat{J}_z \tag{2}$$

where x, y, z are cyclic and only the sign on the r.h.s. differs from the usual relation for space fixed axes (X,Y,Z). Here $\hat{J}^{+} = \hat{J}_x + i\hat{J}_y$ is a <u>lowering</u> operator in contrast to $\hat{J}_{+} = \hat{J}_X + i\hat{J}_Y$ which is a raising operator. If the kets $|JKM\rangle$ are introduced so that:

$$\hat{J}^2|JKM\rangle = J(J+1)|JKM\rangle, \quad \hat{J}_z|JKM\rangle = K|JKM\rangle, \quad \hat{J}_Z|JKM\rangle = M|JKM\rangle$$

then the basic matrix elements for the gyrating frame are:

$$\langle JKM|\hat{J}_z|JKM\rangle = K \tag{3a}$$

$$\langle J,K\mp 1,M|\hat{J}^{\pm}|JKM\rangle = \sqrt{(J \pm K)(J \mp K + 1)} \tag{3b}$$

For a symmetric top molecule $I_z = I_A$ and $I_x = I_y = I_B$. Use Eqs. (1) and (3) to show that the energy levels are given by

$$\frac{E(J,K)}{hc} = J(J+1)B + K^2(A - B) \tag{4}$$

where $A = h/8\pi^2 cI_A$ and $B = h/8\pi^2 cI_B$.

Solution: Ref. TI,251; J. H. Van Vleck, Rev. Mod. Phys. <u>23</u>, 213 (1951)

For a symmetric top molecule $I_x = I_y = I_B$ and $I_z = I_A$, thus Eq. (1) gives

$$\hat{\mathcal{H}} = \frac{\hbar^2\hat{J}_z^2}{2I_A} + \frac{\hbar^2}{2I_B}(\hat{J}_x^2 + \hat{J}_y^2) \tag{5}$$

To construct the Hamiltonian matrix we need the following matrix elements:

$$\langle JKM|\hat{J}_z|J'K'M'\rangle = \delta_{JJ'}\delta_{KK'}\delta_{MM'}K$$

$$\langle JKM|\hat{J}_x^2 + \hat{J}_y^2|J'K'M'\rangle = \langle JKM|\hat{J}^2 - \hat{J}_z^2|J'K'M'\rangle = \delta_{JJ'}\delta_{MM'}\delta_{KK'}[J(J+1) - K^2]$$

The matrix turns out to be diagonal and the eigenvalues are thus

$$E(J,K) = \langle JKM|\hat{\mathcal{H}}|JKM\rangle = \frac{\hbar^2K^2}{2I_A} + \frac{\hbar^2}{2I_B}[J(J+1) - K^2]$$

or

$$\frac{E(J,K)}{hc} = J(J+1)B + K^2(A - B)$$

11.10 The N-H bond distance in NH_3 is 1.012 Å and each N-H bond makes an angle of 67.85° with the three-fold axis. Calculate the principal moments of inertia I_A and I_B and plot the rotational energy levels for $0 \lessgtr J \lessgtr 3$ and $0 \lessgtr |K| \lessgtr J$.

Solution:

A diagram of the plane containing N, one proton, and the symmetry axis is shown below along with a view along the symmetry axis.

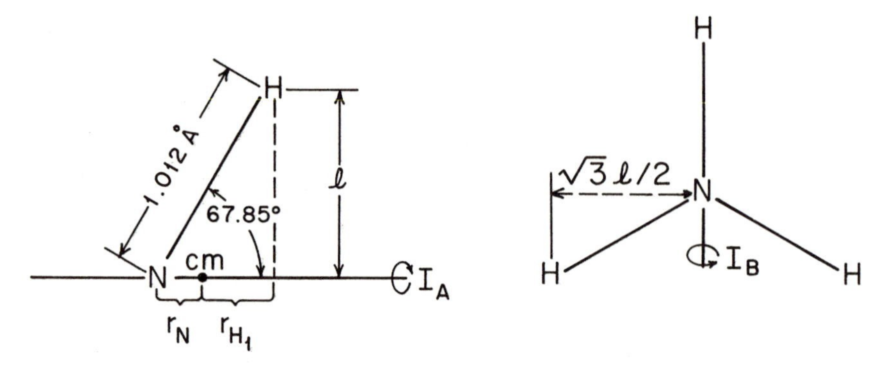

It is evident that $\ell = 1.012\sin 67.85° = (1.012)(0.926) = 0.937$ Å. Then

$$I_A = \sum_{i=1}^{3} m_{H_i}\ell^2 = \frac{3(1.008)(0.937\text{x}10^{-8})^2}{6.023\text{x}10^{23}} = 4.41\text{x}10^{-40}\ \text{gcm}^2$$

In order to locate the center of mass we need the relations:

$$d = r_N + r_H = 1.012\cos 67.85° = 0.381\ \text{Å}$$

$$3m_p r_H - m_N r_N = 0$$

$$3m_p r_H - m_N(d - r_H) = 0, \qquad r_H = \frac{m_N d}{(3m_p + m_N)}$$

Therefore

$$r_H \simeq \frac{(14.007)(0.381)}{3(1.008) + 14.007} = 0.314\ \text{Å}, \qquad r_N = d - r_H \cong 0.067\ \text{Å}$$

The moment of inertia I_B can be calculated using any axis through cm which is perpendicular to the symmetry axis. Choosing an axis in the plane with one N–H bond we have:

$$I_B = m_N r_N^2 + m_p(r_{H1}^2 + r_{H2}^2 + r_{H3}^2)$$

with

$$r_{H2}^2 = r_{H3}^2 = r_{H1}^2 + \left(\frac{\sqrt{3}}{2}\ell\right)^2$$

Therefore

$$I_B = \{(14.007)(0.067)^2 + (1.008[3(0.314)^2 + \frac{3}{2}(0.937)^2]\}\frac{10^{-16}}{6.023\text{x}10^{23}} = 2.78\text{x}10^{-40}\ \text{gcm}^2$$

Since $I_A > I_B$, NH_3 is an oblate symmetric top.

The rotational energy levels of a symmetric top molecule are given by

$$F(J,K) = J(J + 1)B + K^2(A - B)\ \text{in cm}^{-1}$$

where

$$A = \frac{h}{8\pi^2 c I_A} = 6.35 \text{ cm}^{-1}, \qquad B = \frac{h}{8\pi^2 c I_B} = 10.07 \text{ cm}^{-1}$$

The energy levels are shown in Figure 11.10.

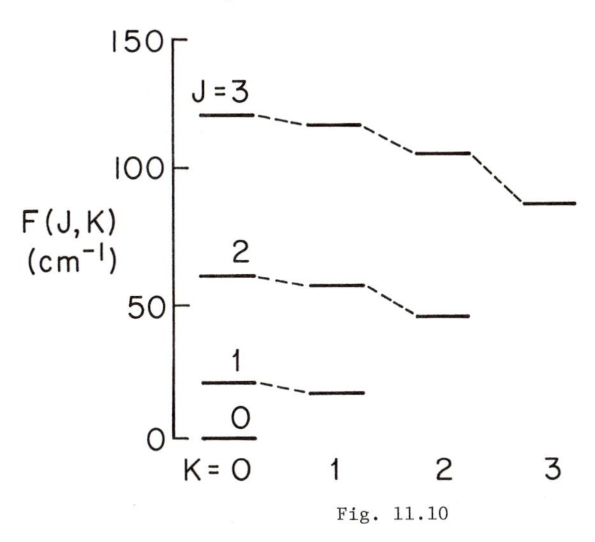

Fig. 11.10

11.11 Rotators which are hindered by barriers having three-fold symmetry are often described with the Hamiltonian:

$$\hat{\mathcal{H}} = -\frac{\hbar^2}{2I}\frac{\partial^2}{\partial\phi^2} + \frac{V_o}{2}(1 - \cos 3\phi) \tag{1}$$

(a) Obtain approximate eigenvalues E_n for Eq. (1) in terms of the barrier height V_o for the case where $E_n \ll V_o$. Use first order perturbation theory taking the zero order Hamiltonian to be that for a harmonic oscillator.

(b) Obtain approximate eigenvalues E_n for the case where $E_n \gg V_o$. Use perturbation theory through second order and take the zero order Hamiltonian to be that for a plane rotor.

Solution:

(a) For $E_n \ll V_o$ the motion will be restricted to the vicinity of the minima in the potential curve. These occur when $3\phi = m2\pi$ or $\phi = m2\pi/3$ with $m = 0, \pm 1, \cdots$. It is, therefore, reasonable to expand $\hat{\mathcal{H}}$ about a minimum to obtain:

$$\hat{\mathcal{H}} = -\frac{\hbar^2}{2I}\frac{\partial^2}{\partial\phi^2} + \frac{V_o}{2}\{1 - [1 - \frac{(3\phi)^2}{2!} + \frac{(3\phi)^4}{4!} - \cdots]\} \cong \left(-\frac{\hbar^2}{2I}\frac{\partial^2}{\partial\phi^2} + \frac{9}{4}V_o\phi^2\right) - \frac{27}{16}V_o\phi^4 \tag{2}$$

Equation (2) is of the form $\hat{\mathcal{H}}^o + \hat{\mathcal{H}}'$ where $\hat{\mathcal{H}}^o$ is the Hamiltonian for a harmonic oscillator having $k = (9V_o/2)$ and $\hat{\mathcal{H}}' = -27V_o\phi^4/16$. The approximate eigenvalues are thus

$$E_n = \left(n + \frac{1}{2}\right)h\nu_o - \frac{27}{16} V_o \langle n|\phi^4|n\rangle$$

where $\nu_o = (1/2\pi)\sqrt{k/I}$ and $|n\rangle$ is the nth eigenket for the harmonic oscillator. Using the result of Problem 6.6 we find:

$$E_n = 3\hbar\sqrt{\frac{V_o}{2I}}\left(n + \frac{1}{2}\right) - \frac{9}{16}\frac{\hbar^2}{I}\left(n^2 + n + \frac{1}{2}\right)$$

Each level has 3-fold degeneracy.

(b) With $V_o = 0$ Eq. (1) yields the eigenvalues $E_n^o = n^2\hbar^2/2I$ and plane rotor eigenfunctions. For $E_n >> V_o$ perturbation theory gives

$$E_n = \frac{n^2\hbar^2}{2I} + \langle n|\hat{\mathcal{H}}'|n\rangle + \sum_{\ell \neq n} \frac{|\langle n|\hat{\mathcal{H}}'|\ell\rangle|^2}{E_n^o - E_\ell^o} \tag{3}$$

with $\hat{\mathcal{H}}' = (V_o/2)(1 - \cos 3\phi)$. The required matrix elements are

$$\langle n|\frac{V_o}{2}(1 - \cos 3\phi)|n'\rangle = \frac{V_o}{2}\delta_{nn'} - \frac{V_o}{4}\left(\frac{1}{2\pi}\right)\int_o^{2\pi} e^{-in\phi}(e^{i3\phi} + e^{-i3\phi})e^{in'\phi}d\phi$$

$$= \frac{V_o}{2}\delta_{nn'} - \frac{V_o}{4}\delta_{n,n'\pm 3}$$

and Eq. (3) gives

$$E_n = \frac{n^2\hbar^2}{2I} + \frac{V_o}{2} - \frac{V_o^2}{16}\left(\frac{2I}{\hbar^2}\right)\left[\frac{1}{n^2 - (n+3)^2} + \frac{1}{n^2 - (n-3)^2}\right]$$

$$E_n = \frac{n^2\hbar^2}{2I} + \frac{V_o}{2} - \frac{V_o^2 I}{4\hbar^2(4n^2 - 9)}$$

11.12 Consider the rotator in Problem 11.11 in the limit that $V_o >> E_n$. In this limit the torsional eigenfunctions are strongly localized at $\phi = 0$, $2\pi/3$, and $4\pi/3$. The wave function for the lowest torsional level can be approximated by

$$\Psi = \sum_{k=1}^{3} c_k\psi_k$$

where $\psi_k = \psi(\phi - \phi_k)$ is a harmonic oscillator wave function centered at ϕ_k with $n = 0$. Use degenerate perturbation theory to determine the eigenvalues and eigenfunctions associated with the lowest torsional level. The eigenvalues should be given in terms of the integrals: $\langle\psi_k|\hat{\mathcal{H}}|\psi_k\rangle = W$, $\langle\psi_k|\hat{\mathcal{H}}|\psi_\ell\rangle = \Delta$, $(k \neq \ell)$. The overlap integrals can be neglected, i.e. $\langle\psi_k|\psi_\ell\rangle = \delta_{k\ell}$. (Hint: It is convenient to use linear combinations of the functions ψ_k which transform according to the irreducible representations of C_3 as basis functions.)

Solution: Ref. Problem 9.15

The group C_3 contains the operations E, C_3, C_3^2 which have the following effects on ψ_1: $E\psi_1 = \psi_1$, $C_3\psi_1 = \psi_2$, $C_3^2\psi_1 = \psi_3$. Equation (1) of Appendix 8 thus gives (when normalized)

$$u(A) = \frac{1}{\sqrt{3}}(\psi_1 + \psi_2 + \psi_3) \qquad u(E_a) = \frac{1}{\sqrt{3}}(\psi_1 + \varepsilon^*\psi_2 + \varepsilon\psi_3)$$

$$u(E_b) = \frac{1}{\sqrt{3}}(\psi_1 + \varepsilon\psi_2 + \varepsilon^*\psi_3) \quad \text{with } \varepsilon = \exp(2\pi i/3)$$

The matrix elements of $\hat{\mathcal{H}}$ with respect to these functions are:

$$\langle u(A)|\hat{\mathcal{H}}|u(A)\rangle = W + 2\Delta, \qquad \langle u(E_a)|\hat{\mathcal{H}}|u(E_a)\rangle = W - \Delta, \qquad \langle u(E_b)|\hat{\mathcal{H}}|u(E_b)\rangle = W - \Delta$$

All other elements are zero and the symmetry functions are the approximate energy eigenfunctions. Thus $E(A) = W + 2\Delta$, $E(E_a) = E(E_b) = W - \Delta$. The quantity $E(E_a) - E(A) = -3\Delta$ is called the tunneling splitting. For the set of levels with $n = 0$, $\Delta < 0$, i.e. in the torsional ground state the degenerate set of states (E_a, E_b) have higher energies than the nondegenerate A state. The ordering is the reverse of this for odd values of n.

11.13 At $t = 0$ the rotator in Problem 11.12 is localized in the vicinity of $\phi = 0$, i.e. $\Psi(0) = \psi_1$. Obtain an expression for $\Psi(t)$ in the form:

$$\Psi(t) = \sum_i c_i(t)\psi_i \tag{1}$$

and plot $|c_i(t)|^2$ versus time for $i = 1, 2, 3$.

Solution: Problem 11.12

As shown in Problem 11.12 the time independent eigenfunctions are $u(A)$, $u(E_a)$, and $u(E_b)$. The time-dependent state function can be written as

$$\Psi(t) = \sum_j a_j u_j e^{-iE_j t/\hbar} = e^{-i(w+2\Delta)t/\hbar}\{a_1 u(A) + e^{+i3\Delta t/\hbar}[a_2 u(E_a) + a_3 u(E_b)]\}$$

At $t = 0$ we have $\Psi(0) = \psi_1$ which permits the a's to be determined. Thus

$$a_1 = \langle u(A)|\psi_1\rangle = \frac{1}{\sqrt{3}}\langle\psi_1 + \psi_2 + \psi_3|\psi_1\rangle = \frac{1}{\sqrt{3}}$$

$$a_2 = \langle u(E_a)|\psi_1\rangle = \frac{1}{\sqrt{3}}, \qquad a_3 = \langle u(E_b)|\psi_1\rangle = \frac{1}{\sqrt{3}}$$

Equation (1) then gives

$$\Psi(t) = \frac{1}{\sqrt{3}} e^{-i(w+2\Delta)t/\hbar}\{u(A) + e^{+i3\Delta t/\hbar}[u(E_a) + u(E_b)]\}$$

$$= \frac{1}{3} e^{-i(w+2\Delta)t/\hbar}\{(1 + 2e^{+i3\Delta t/\hbar})\psi_1 + (1 - e^{+i3\Delta t/\hbar})(\psi_2 + \psi_3)\}$$

The probabilities are:

$$|c_1(t)|^2 = \frac{1}{9}\,[5 + 4\cos\Omega t] = \frac{1}{9}\left[1 + 8\cos^2\left(\frac{\Omega t}{2}\right)\right]$$

$$|c_2(t)|^2 = |c_3(t)|^2 = \frac{2}{9}\,[1 - \cos\Omega t] = \frac{4}{9}\,\sin^2\left(\frac{\Omega t}{2}\right)$$

where $\Omega = (3\Delta/\hbar)$. The variation of probability with time is shown in Fig. 11.13.

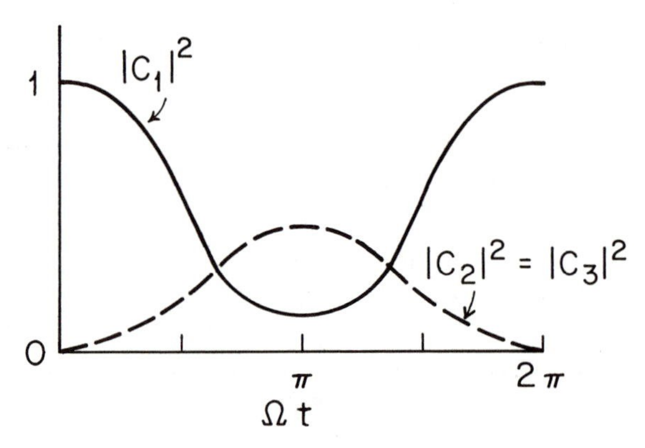

Fig. 11.13.

11.14 The NH_3 molecule has C_{3v} symmetry. The Cartesian displacements of the four atoms provide a basis for the reducible 12 dimensional representation Γ_{12} of C_{3v}.

(a) Determine the characters of Γ_{12}.

(b) How many times are each of the irreducible representations of C_{3v} contained in Γ_{12}?

(c) What are the symmetries of the normal modes of vibration of NH_3?

(d) How many vibrations are infrared active?

(e) How many vibrations are Raman active?

Solution: Ref. Appendix 8

(a) As explained in Example 3 of Appendix 8 only atoms which lie on a symmetry element can contribute to the character for the associated operation. Every atom contributes +3 to the character for E. For an atom on a mirror plane the contribution is +1. The situation is slightly more complicated for an atom on a C_3 axis. Consider Fig. 11.14. Before rotation the displacement $\underset{\sim}{r}$ has the components x_1 and y_1. After rotation of $\underset{\sim}{r}$ by 120° we have $\underset{\sim}{r} = x_2\hat{i} + y_2\hat{j} + z_2\hat{k}$ where:

$$x_1 = r\cos\theta, \quad y_1 = r\sin\theta, \quad x_2 = r\cos(\theta + 120°) = x_1\cos(120°) - y_1\sin(120°)$$

$$y_2 = r\sin(\theta + 120°) = x_1\sin(120°) + y_1\cos(120°)$$

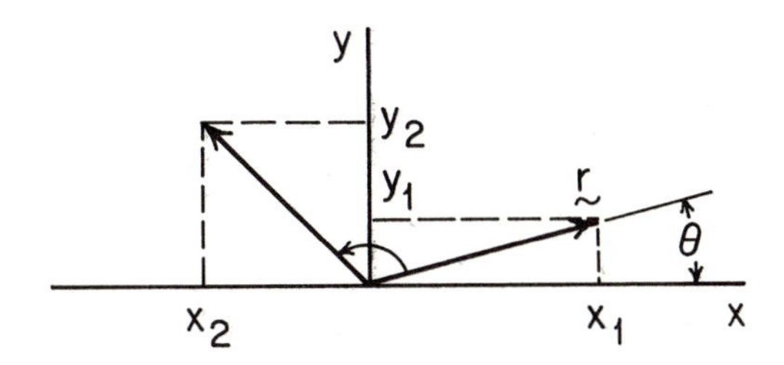

Fig. 11.14

The matrix representing the rotation is thus

$$\begin{pmatrix} -1/2 & -\sqrt{3}/2 & 0 \\ +\sqrt{3}/2 & -1/2 & 0 \\ 0 & 0 & 1 \end{pmatrix}$$

and the contribution to the character for this atom is zero. These considerations give the following characters for Γ_{12}:

	E	$2C_3$	$3\sigma_v$
Γ_{12}	12	0	2

(b) Using

$$n_i = \frac{1}{h}\sum_i \chi(R)\chi_i^*(R)$$

we obtain

$$n(A_1) = \frac{1}{6}\,[12 + (2)(3)] = 3, \qquad n(A_2) = \frac{1}{6}\,(12 - 6) = 1, \qquad n(E) = \frac{1}{6}\,(24) = 4$$

Thus

$$\Gamma_{12} = 3A_1 + A_2 + 4E$$

(c) From the character table for C_{3v} we find:

$\Gamma(\text{translation}) = A_1 + E, \qquad \Gamma(\text{rotation}) = A_2 + E$

$\Gamma(\text{vibration}) = \Gamma_{12} - \Gamma(\text{trans}) - \Gamma(\text{rot}) = 2A_1 + 2E$

There are $3N - 6 = 6$ normal modes of vibration: 2 have A_1 symmetry and 4 are associated with E symmetry (E is doubly degenerate, i.e. the representations have dimension 2.)

(d) All of the fundamentals are infrared active.

(e) All of the fundamentals are Raman active.

The answers for parts d and e are obtained by inspection of the character table for the point group C_{3v} as explained in Appendix 8. Those normal modes which transform according to the same irreducible representation as a translation (x,y,z) are infrared active, while those that belong to the same representation as one of the products x^2, xy, xz, etc. are Raman active.

11.15 The HCN molecule has $C_{\infty v}$ symmetry.

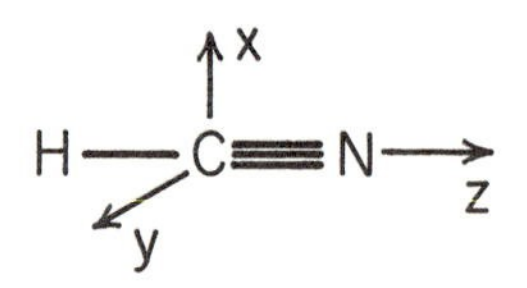

(a) Determine the symmetries of the normal modes of vibration of HCN. (Hint: Since Eq. (3) of Appendix 8 fails for infinite groups, it is convenient to use a subgroup of $C_{\infty v}$ in resolving Γ(vib). The irreducible representations of the subgroup can then be related to the irreducible representations of $C_{\infty v}$ by considering the behavior of the basis vectors.)

(b) How many vibrations are infrared active?

(c) How many vibrations are Raman active?

Solution: Ref. D. P. Strommen and E. R. Lippincott, J. Chem. Ed. 49, 341 (1972) and references contained therein.

(a) As suggested above we choose the subgroup C_{2v}. Proceeding as in Example 3 of Appendix 8 with the 3N dimensional representation Γ_{3N} we find the characters:

C_{2v}	E	C_2	$\sigma_v(xz)$	$\sigma_v'(yz)$
Γ_{3N}	9	-3	3	3

Equation (3) of Appendix 8 then gives:

$$\Gamma_{3N} = 3A_1 + 3B_1 + 3B_2, \qquad \Gamma(\text{translation}) = A_1 + B_1 + B_2, \qquad \Gamma(\text{rotation}) = B_1 + B_2$$

and

$$\Gamma(\text{vibration}) = \Gamma_{3N} - \Gamma(\text{trans}) - \Gamma(\text{rot}) = 2A_1 + B_1 + B_2$$

Notice that A_2 does not appear in Γ(rot) because the nuclei are unaffected by rotation about the z-direction. Since the z vector transforms according to the A_1 representation of C_{2v} and the Σ^+ representation of $C_{\infty v}$, we make the association $A_1 \rightarrow \Sigma^+$. Similarly x and y transform according to B_1 and B_2 of C_{2v} and Π of $C_{\infty v}$, thus $B_1, B_2 \rightarrow \Pi$ and

$$\Gamma(\text{vibration}) = 2\Sigma^+ + \Pi$$

(b) All four modes are infrared active.

(c) All four modes are also Raman active since $x^2 + y^2$ and z^2 transform like Σ^+ and xz and yz transform like Π (see Appendix 8).

11.16 Use the coordinate system shown below in answering questions about the electronic structure and transitions in ethylene.

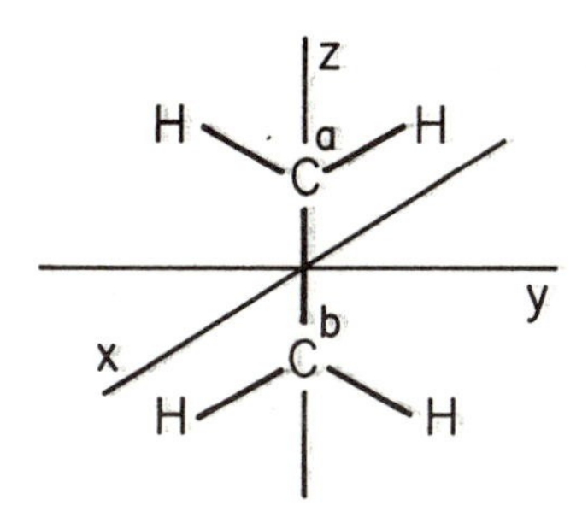

(a) Write out the Hückel molecular orbitals for the π-electrons ($2p_x$) and determine the representations of D_{2h} to which they belong.
(b) Determine the symmetry species of all of the possible electronic states (ground and excited in the context of the Hückel theory) for the two π-electrons in ethylene.
(c) Construct a table to show the transitions from the ground state and qualitatively indicate their expected intensities.

Solution: Ref. Appendix 8; JO,96

(a) The Hückel MO's are of the form

$$\psi_{\pm} = \frac{1}{\sqrt{2}}(u_A \pm u_B)$$

where $u_i = (2p_x)_i$ in strict analogy to the MO's for H_2. The functions ψ_+ and ψ_- can each be used as the basis for one-dimensional representations of D_{2h}. Consider ψ_+ first:

$$C_2(z)\psi_+ = \frac{1}{\sqrt{2}}[C_2(z)u_A + C_2(z)u_B] = \frac{1}{\sqrt{2}}(-u_A - u_B) = -\psi_+$$

$$C_2(y)\psi_+ = \frac{1}{\sqrt{2}}[C_2(y)u_A + C_2(y)u_B] = \frac{1}{\sqrt{2}}(-u_B - u_A) = -\psi_+$$

Continuing this procedure for ψ_+ and ψ_- we obtain the following characters:

D_{2h}	E	$C_2(z)$	$C_2(y)$	$C_2(x)$	i	$\sigma(xy)$	$\sigma(xz)$	$\sigma(yz)$
$\Gamma(\psi_+)$	1	-1	-1	1	-1	1	1	-1
$\Gamma(\psi_-)$	1	-1	1	-1	1	-1	1	-1

Inspection of the character table shows immediately that:

$$\Gamma(\psi_+) = B_{3u}, \quad \Gamma(\psi_-) = B_{2g}$$

(b) The Hückel energies are:

$$E(B_{3u}) = \int \psi_+^* \hat{\mathcal{H}} \psi_+ d\tau = \alpha + \beta \qquad E(B_{2g}) = \int \psi_-^* \hat{\mathcal{H}} \psi_- d\tau = \alpha - \beta$$

Since $\beta < 0$, we have the following states:

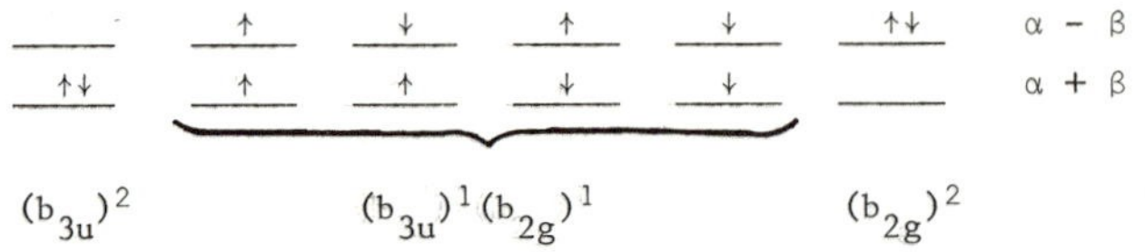

___	↑	↓	↑	↓	↑↓	$\alpha - \beta$
↑↓	↑	↑	↓	↓	___	$\alpha + \beta$
$(b_{3u})^2$	$(b_{3u})^1(b_{2g})^1$				$(b_{2g})^2$	

All of the doubly occupied states are totally symmetric (A_g). For the others we use the rule: $\chi(A \times B) = \chi(A)\chi(B)$. Thus:

$(b_{3u})^2$ belongs to ^1A_g; $(b_{3u})^1(b_{2g})^1$ gives $^1B_{1u}$, $^3B_{1u}$; $(b_{2g})^2$ belongs to ^1A_g

As in atomic term symbols the superscript indicates the multiplicity: $2S + 1$.

(c) The transitions are:

	Symmetry Allowed	Spin Allowed	Intensity
$b_{3u}b_{2g}(^1B_{1u}) \leftarrow b_{3u}^2(^1A_g)$	yes, z	yes	strong
$b_{3u}b_{2g}(^3B_{1u}) \leftarrow b_{3u}^2(^1A_g)$	yes, z	no	weak
$b_{2g}^2(^1A_g) \leftarrow b_{3u}^2(^1A_g)$	no	yes	weak

The assignments are based on the following arguments. The transition moment has the form

$$M_{if} = \int u_i^*(\sum_i e\underset{\sim}{r}_i)u_f dv \int \chi_i^* \chi_f d\sigma$$

where u_i is the spatial wave function for the ith state and χ_i is the corresponding spin function. In this problem u_i, the ground state, is always totally symmetric. The spatial integral will vanish unless u_f has the same symmetry behavior as one of the Cartesian coordinates. Orthogonality will cause the integral over spin functions to vanish unless $\chi_i = \chi_f$, <u>i.e.</u> unless $\Delta S = 0$. The selection rules are weakened by electronic-vibrational interactions and spin-orbit coupling.

11.17 The ground state configuration for the π-electrons in benzene is $(a_{2u})^2(e_{1g})^4$. Show that only one state is possible for this configuration and that its symmetry is given by $^1A_{1g}$. (Hint: Be sure to take into account the exclusion principle.)

<u>Solution</u>: Ref. KA,470; Appendix 8, Problem 9.16

The Slater determinant for the $(a_{2u})^2(e_{1g})^4$ configuration is

$$\Psi = |\psi_1\bar{\psi}_1\psi_2\bar{\psi}_2\psi_3\bar{\psi}_3|$$

where from Problem 9.16 we have

$$a_{2u}: \quad \psi_1 = \frac{1}{\sqrt{6}}(\phi_1 + \phi_2 + \phi_3 + \phi_4 + \phi_5 + \phi_6)$$

$$e_{1g}: \quad \psi_2 = \frac{1}{\sqrt{12}}(2\phi_1 + \phi_2 - \phi_3 - 2\phi_4 - \phi_5 + \phi_6)$$

$$\psi_3 = \frac{1}{2}(\phi_2 + \phi_3 - \phi_5 - \phi_6)$$

In dealing with the transformation properties of ψ_2 and ψ_3 it is convenient to introduce the functions χ_1, χ_2, and χ_3 so that

$$\psi_2 = \frac{1}{\sqrt{3}} (\chi_1 - \chi_3) \qquad\qquad \psi_3 = \chi_2$$

where

$$\chi_1 = \frac{1}{2} (\phi_1 + \phi_2 - \phi_4 - \phi_5), \qquad \chi_2 = \frac{1}{2} (\phi_2 + \phi_3 - \phi_5 - \phi_6), \qquad \chi_3 = \frac{1}{2} (\phi_3 + \phi_4 - \phi_6 - \phi_1)$$

The effects of the operations of the group D_{6h} on the sets $\{\phi\}$ and $\{\chi\}$ are shown below:

D_{6h}	E	C_6	C_6^5	C_3	C_3^2	C_2	$_1C_2'$	$_2C_2'$	$_3C_2'$	$_1C_2''$	$_2C_2''$	$_3C_2''$	i
ϕ_1	1	2	6	3	5	4	1	3	5	2	4	6	4
ϕ_2	2	3	1	4	6	5	6	2	4	1	3	5	5
ϕ_3	3	4	2	5	1	6	5	1	3	6	2	4	6
ϕ_4	4	5	3	6	2	1	4	6	2	5	1	3	1
ϕ_5	5	6	4	1	3	2	3	5	1	4	6	2	2
ϕ_6	6	1	5	2	4	3	2	4	6	3	5	1	3
χ_1	χ_1	χ_2	$-\chi_3$	χ_3	$-\chi_2$	$-\chi_1$	$-\chi_3$	χ_2	$-\chi_1$	χ_1	χ_3	$-\chi_2$	$-\chi_1$
χ_2	χ_2	χ_3	χ_1	$-\chi_1$	$-\chi_3$	$-\chi_2$	$-\chi_2$	χ_1	χ_3	$-\chi_3$	χ_2	$-\chi_1$	$-\chi_2$
χ_3	χ_3	$-\chi_1$	χ_2	$-\chi_2$	χ_1	$-\chi_3$	$-\chi_1$	$-\chi_3$	χ_2	$-\chi_2$	χ_1	χ_3	$-\chi_3$

The improper rotations add nothing new and have been omitted. We now examine the effects of the operations on Ψ.

$$E\Psi = E|\psi_1\bar{\psi}_1\psi_2\bar{\psi}_2\psi_3\bar{\psi}_3| = |\psi_1\bar{\psi}_1\psi_2\bar{\psi}_2\psi_3\bar{\psi}_3| = \Psi$$

$$C_6\Psi = C_6|\psi_1\bar{\psi}_1\psi_2\bar{\psi}_2\psi_3\bar{\psi}_3| = \Big|\psi_1\bar{\psi}_1 \left(\frac{\sqrt{3}}{2}\psi_3 + \frac{1}{2}\psi_2\right)\left(\frac{\sqrt{3}}{2}\bar{\psi}_3 + \frac{1}{2}\bar{\psi}_2\right) \left(\frac{1}{2}\psi_3 - \frac{\sqrt{3}}{2}\psi_2\right)\left(\frac{1}{2}\bar{\psi}_3 - \frac{\sqrt{3}}{2}\bar{\psi}_2\right)\Big|$$

Here we have used the relation $\chi_1 + \chi_3 = \chi_2$. Multiplying out the functions in the determinant and using the rule for adding determinants we find:

$$C_6\Psi = \frac{9}{16}|\psi_1\bar{\psi}_1\psi_3\bar{\psi}_3\psi_2\bar{\psi}_2| - \frac{3}{16}|\psi_1\bar{\psi}_1\psi_3\bar{\psi}_2\psi_2\bar{\psi}_3| - \frac{3}{16}|\psi_1\bar{\psi}_1\psi_2\bar{\psi}_3\psi_3\bar{\psi}_2| + \frac{1}{16}|\psi_1\bar{\psi}_1\psi_2\bar{\psi}_2\psi_3\bar{\psi}_3| = \Psi$$

The last step follows from the fact that one exchange of a pair of rows or columns changes the sign of a determinant. Following the same procedure we can show that $\hat{R}\Psi = \Psi$ for all of the operations of D_{6h}. Thus Ψ is a basis for the A_{1g} irreducible representation. This state is a singlet since all of the electrons are paired, i.e. $S = 0$. The complete state symbol is thus ${}^1A_{1g}$. The rule is general. All closed shell configurations of electrons belong to the totally symmetric representation of the molecular symmetry group.

11.18 The benzene anion radical in tetrahydrofuran solution shows a strong absorption at 4,200 Å and additional absorptions at wavelengths less than 4,000 Å which cannot be resolved. Base your answers to this problem on Hückel MO theory assuming D_{6h} symmetry.

(a) To which transition would you assign the band at 4,200 Å? Give the symmetry representations of the ground and excited configurations involved and include the multiplicities.

(b) Which transition would be next lowest in energy? Is this transition allowed?

(c) What other spin states might arise in the excited configurations?

Solution: Ref. Problem 9.16 and Appendix 8

By comparing the eigenfunctions from Problem 9.16 with the character table for D_{6h} we obtain the following states:

b_{2g} ——— $\alpha - 2\beta$

e_{2u} ↑ ——— $\alpha - \beta$

e_{1g} ↑↓ ↑↓ $\alpha + \beta$

a_{2u} ↑↓ $\alpha + 2\beta$

where lower case letters are used for the symmetry types of individual orbitals. The ground state and the excited states of lowest energy can be described by the configurations:

ground state : $(a_{2u})^2(e_{1g})^4(e_{2u})^1 = {}^2E_{2u}$

excited state ($\Delta E = \beta$) : $(a_{2u})^2(e_{1g})^4(b_{2g})^1 = {}^2B_{2g}$

excited state ($\Delta E = 2\beta$): $(a_{2u})^2(e_{1g})^3(e_{2u})^2 = {}^2E_{1g}, {}^4E_{1g}$

Here the capital letters denote many-electron states and the superscripts give the multiplicities, i.e. 2S + 1. We have used two rules in deriving these representations. First, a k-fold degenerate state containing 2k electrons transforms like the totally symmetric singlet term ${}^1A_{1g}$. And, second, the states arising when 2k-n electrons are present are the same as those arising from n electrons.

(a) We assign the absorption at 4,200 Å to the transition ${}^2B_{2g} \leftarrow {}^2E_{2u}$ which has the energy 2β. The characters for the representation $\Gamma_1 = B_{2g} x E_{2u}$ are shown in the table below.

D_{6h}	E	$2C_6$	$2C_3$	C_2	$3C_2'$	$3C_2''$	i	$2S_3$	$2S_6$	σ_h	$3\sigma_d$	$3\sigma_v$
Γ_1	2	1	-1	-2	0	0	-2	-1	1	2	0	0
Γ_2	4	-1	1	-4	0	0	-4	1	-1	4	0	0

Then applying Eq. (3) of Appendix 8 we find that $\Gamma_1 = E_{1u}$. Therefore, this transition is allowed and is x or y polarized.

(b) The next lowest transition is ${}^2E_{1g} \leftarrow {}^2E_{2u}$. The representation $\Gamma_2 = E_{1g} x E_{2u}$ is also shown in the table. We find that

$$\Gamma_2 = B_{1u} + B_{2u} + E_{1u}$$

Because E_{1u} is contained in Γ_2 this transition is also allowed.

(c) When three separate orbitals are singly occupied, the resultant spin can be $S = 1/2$ or $3/2$. Thus, the multiplicities are $2S + 1 = 2$ or 4. An example occurs in the transition $E_{1g} \leftarrow E_{2u}$.

11.19 Two protons separated by a distance R are placed in a magnetic field $\underset{\sim}{B}$ as shown in Fig. 11.19. The effective Hamiltonian for this system is

$$\hat{\mathcal{H}} = -\gamma\hbar(\hat{I}_{1z} + \hat{I}_{2z})B + \frac{1}{2}\left(\frac{\mu_o}{4\pi}\right)\frac{\gamma^2\hbar^2}{R^3}(1 - 3\cos^2\theta)(3\hat{I}_{1z}\hat{I}_{2z} - \hat{\underset{\sim}{I}}_1\cdot\hat{\underset{\sim}{I}}_2) \qquad (1)$$

where γ is the magnetogyric ratio for a proton. Equation (1) is sufficient only when the second term on the r.h.s. is much smaller than the first term. Use the basis functions $\alpha_1\alpha_2$, $\alpha_1\beta_2$, $\beta_1\alpha_2$, and $\beta_1\beta_2$, which form a complete set, to set up the secular equation for this problem. Determine the eigenvalues and eigenfunctions for this two spin problem. Here $\hat{I}_{1z}\alpha_1 = +(1/2)\alpha_1$, $\hat{I}_{1z}\beta_1 = -(1/2)\beta_1$, etc.

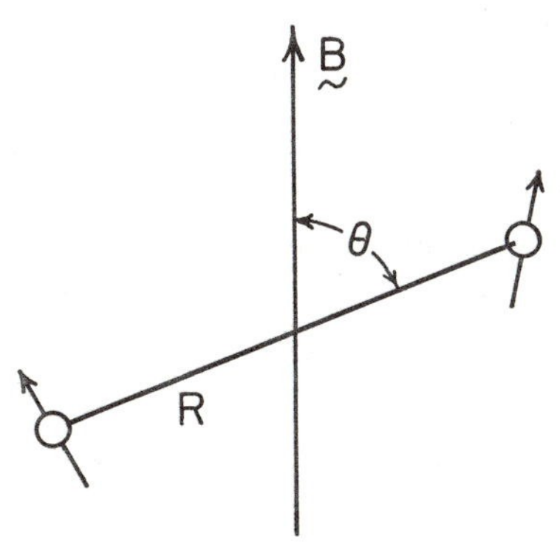

Fig. 11.19

Solution: Ref. SL,46; CM,29

It is convenient to write Eq. (1) so that

$$\hat{\mathcal{H}} = -\gamma\hbar B(\hat{I}_{1z} + \hat{I}_{2z}) + \left(\frac{\mu_o}{4\pi}\right)\frac{\gamma^2\hbar^2}{R^3}(1 - 3\cos^2\theta)[\hat{I}_{1z}\hat{I}_{2z} - \frac{1}{4}(\hat{I}_{1+}\hat{I}_{2-} + \hat{I}_{1-}\hat{I}_{2+})]$$

The required matrix elements are:

$$\langle\alpha\alpha|\hat{\mathcal{H}}|\alpha\alpha\rangle = -\hbar\omega_o(1/2 + 1/2) + \hbar A(1/4) = -\hbar\omega_o + \hbar A/4$$

$$\langle\alpha\beta|\hat{\mathcal{H}}|\alpha\beta\rangle = \langle\beta\alpha|\hat{\mathcal{H}}|\beta\alpha\rangle = -\hbar A/4, \qquad \langle\beta\beta|\hat{\mathcal{H}}|\beta\beta\rangle = +\hbar\omega_o + \hbar A/4$$

$$\langle\alpha\beta|\hat{\mathcal{H}}|\beta\alpha\rangle = \langle\beta\alpha|\hat{\mathcal{H}}|\alpha\beta\rangle = -\hbar A/4$$

where $\omega_o = \gamma B$ and $A = (\gamma^2\hbar/R^3)(1 - 3\cos^2\theta)$. The secular equation is thus

$$\begin{array}{r} \alpha\alpha: \\ \alpha\beta: \\ \beta\alpha: \\ \beta\beta: \end{array} \begin{vmatrix} (-\hbar\omega_o + \hbar A/4) - E & 0 & 0 & 0 \\ 0 & -\hbar A/4 - E & -\hbar A/4 & 0 \\ 0 & -\hbar A/4 & -\hbar A/4 - E & 0 \\ 0 & 0 & 0 & (\hbar\omega_o + \hbar A/4) - E \end{vmatrix} = 0$$

As shown in Appendix 6 the eigenvalues turn out to be:

$$E_1 = -\hbar\omega_o + \hbar A/4, \qquad E_2 = 0, \qquad E_3 = -\hbar A/2, \qquad E_4 = +\hbar\omega_o + \hbar A/4$$

The corresponding eigenfunctions are:

$$u_1 = \alpha_1\alpha_2, \qquad u_2 = \frac{1}{\sqrt{2}}(\alpha_1\beta_2 + \beta_1\alpha_2), \qquad u_3 = \frac{1}{\sqrt{2}}(\alpha_1\beta_2 - \beta_1\alpha_2), \qquad u_4 = \beta_1\beta_2$$

11.20 The spin system in Problem 11.19 is subjected to an oscillating magnetic field of the form $\underset{\sim}{B}_1 = \hat{i}2B_1\cos\omega t$ in addition to the static field $\underset{\sim}{B} = B_o\hat{k}$. The resulting perturbation is

$$\hat{\mathcal{H}}' = -\gamma\hbar\hat{I}_x B_1 2\cos\omega t \tag{1}$$

Determine the frequencies and relative intensities of the transitions which are induced when $\omega_o >> A$.

Solution: Ref. Problems 10.7 and 5.11

The derivation of Problem 10.7 with $\underset{\sim}{\mu}$ replaced by $\gamma\hbar\hat{\underset{\sim}{I}}$ and $\underset{\sim}{\mathcal{E}}$ by $\underset{\sim}{B}_1$ shows that:

$$W_{fi} \propto |\langle f|\hat{I}_x|i\rangle|^2 \delta(E_f - E_i - \hbar\omega) \tag{2}$$

where ω can in principle be swept through the region of transition frequencies. Since $E_f - E_i \simeq \hbar\omega_o$ for all nonvanishing matrix elements, the relative intensities are determined by the squares of the matrix elements.

From Problem 11.19 we find the eigenfunctions:

$$|u_1\rangle = |I=1, m_I=1\rangle, \qquad |u_2\rangle = |1,0\rangle, \qquad |u_4\rangle = |1,-1\rangle, \qquad \text{and} \qquad |u_3\rangle = |0,0\rangle$$

There are no nonvanishing matrix elements of the operator $\hat{I}_x = (\hat{I}_+ + \hat{I}_-)/2$ between states of different I and we are left with the transitions in the triplet (I = 1) manifold.

$$\langle 1,0|\hat{I}_x|1,-1\rangle = \frac{1}{2}\langle 1,0|\hat{I}_+|1,-1\rangle = \frac{\sqrt{2}}{2}, \qquad \langle 1,1|\hat{I}_x|1,0\rangle = \frac{1}{2}\langle 1,1|\hat{I}_+|1,0\rangle = \frac{\sqrt{2}}{2}$$

There are two allowed transitions each having the same intensity. The frequencies are:

$$\nu_{43} = \frac{1}{h}(E_4 - E_3) = \nu_o + \frac{3}{4}\left(\frac{\gamma^2\hbar}{2\pi R^3}\right)(1 - 3\cos^2\theta)$$

$$\nu_{31} = \frac{1}{h}(E_3 - E_1) = \nu_o - \frac{3}{4}\left(\frac{\gamma^2\hbar}{2\pi R^3}\right)(1 - 3\cos^2\theta)$$

11.21 The contact interaction between an electron and a nucleus is described by the Hamiltonian

$$\hat{\mathcal{H}}_{IS} = a\hat{\underset{\sim}{I}}\cdot\hat{\underset{\sim}{S}} \tag{1}$$

where $\hat{\underset{\sim}{I}}$ and $\hat{\underset{\sim}{S}}$ are the angular momentum operators for the nucleus and the electron, respectively. The hyperfine splitting constant is the expectation value of the operator

$$\hat{a} = \left(\frac{\mu_o}{4\pi}\right)\frac{8\pi}{3}\, g\mu_B\gamma\hbar\delta(\underset{\sim}{r}) \tag{2}$$

with respect to the spatial wave function of the electron. Here $\underset{\sim}{r}$ gives the position of the electron with respect to the nucleus, <u>i.e.</u> $\underset{\sim}{r} = \underset{\sim}{r}_e - \underset{\sim}{r}_N$. Evaluate $a = \langle n|\hat{a}|n\rangle$ for the 1s and 2p states of the hydrogen atom.

<u>Solution</u>: Ref. SL,84; CM,14,81; BS,151 (δ-function)

First, we consider s type wave functions, which can be written as $\psi(r)$. For these functions Eq. (2) gives

$$a = \left(\frac{\mu_o}{4\pi}\right)\frac{8\pi}{3}\, g\mu_B\gamma\hbar\langle\psi(r)|\delta(r)|\psi(r)\rangle = \left(\frac{\mu_o}{4\pi}\right)\frac{8\pi}{3}\, g\mu_B\gamma\hbar|\psi(0)|^2 \tag{3}$$

The hydrogen 1s wave function is:

$$\psi_{1s} = \frac{1}{\sqrt{\pi a_o^3}}\, e^{-r/a_o}$$

and with this function Eq. (3) gives

$$a = \frac{2(4\pi\text{x}10^{-7})(2.0023)(9.274\text{x}10^{-24})(2.675\text{x}10^{8})(1.0546\text{x}10^{-34})}{3\pi(5.292\text{x}10^{-11})^3} = 9.429\text{x}10^{-25} \text{ joules}$$

$$a/h = 1{,}422 \text{ MHz}$$

For $\psi_{2p}(r,\theta,\phi)$ Eq. (3) is somewhat ambiguous since there are three variables. When dealing with functions having spherical symmetry so that $f(\underset{\sim}{r}) = f(r)$, the three-dimensional δ function takes the form $\delta(\underset{\sim}{r}) = (2\pi r^2)^{-1}\delta(r)$. Otherwise, $\delta(\underset{\sim}{r})$ must be expanded as a function of r, cosθ, and ϕ. It will, however, still have $\delta(r)/r^2$ as a factor so that non-s states will give a vanishing value of a. For example with 2p we find

$$a \propto \int_0^\infty r^2 e^{-r/2a_o}\, \frac{\delta(r)}{r^2}\, r^2 dr = 0$$

Thus only electrons in s-type orbitals or hybrid orbitals having some s character can have a <u>direct</u> scalar interaction with a nucleus.

11.22 For a hydrogen atom in a magnetic field B the spin Hamiltonian is

$$\hat{\mathcal{H}} = g\mu_B B\hat{S}_z - \gamma_p\hbar B\hat{I}_z + a\hat{\underset{\sim}{I}}\cdot\hat{\underset{\sim}{S}} \tag{1}$$

where $a/h = 1{,}420$ MHz is the hyperfine splitting constant.

(a) Obtain the eigenvalues and eigenvectors for $\hat{\mathcal{H}}$ in the limit that B = 0.

(b) Determine the frequencies and intensities of the transitions that are induced by the oscillating field $\underset{\sim}{B}_1 = \hat{i}2B_1\cos\omega t$.

Solution: Ref. CM,72

(a) The Hamiltonian shown above is of the same form as that treated in Problem 11.44. If we make the associations $-\hbar\omega_1 \to g\mu_B B$, $\hbar\omega_2 \to \gamma_p \hbar B$, and $\hbar J \to a$, the results of that problem can be taken over directly.

Starting anew, however, it is desirable to treat the limit $B \to 0$ with a new representation. In particular notice that:

$$\hat{\mathcal{H}} = a\hat{\underset{\sim}{I}}\cdot\hat{\underset{\sim}{S}} = \frac{a}{2}(\hat{F}^2 - \hat{I}^2 - \hat{S}^2) \quad \text{where} \quad \hat{\underset{\sim}{F}} = \hat{\underset{\sim}{I}} + \hat{\underset{\sim}{S}}$$

We choose the basis functions $|ISFM_F\rangle$ and immediately obtain

$$E(I,S,F) = \frac{a}{2}[F(F+1) - I(I+1) - S(S+1)]$$

$$E(1/2,1/2,F) = \frac{a}{2}\left[F(F+1) - \frac{3}{2}\right]$$

The vector model of addition shows that $F = 0,1$ and we have

$$E(1/2,1/2,1) = \frac{a}{4}, \qquad E(1/2,1/2,0) = -\frac{3}{4}a$$

The energy splitting is thus

$$\Delta E = E(1/2,1/2,1) - E(1/2,1/2,0) = a$$

The eigenvectors can be written in terms of the product functions $|m_I m_S\rangle = |m_I\rangle|m_S\rangle$ by means of the relation:

$$|F,m_F\rangle = \sum_{m_I m_S} |m_I m_S\rangle\langle m_I m_S|F,m_F\rangle$$

According to Problem 5.22 the proper combinations are

$$|T_1\rangle = |1,1\rangle = |\alpha_e\alpha_p\rangle$$

$$|T_o\rangle = |1,0\rangle = (|\alpha_e\beta_p\rangle + |\beta_e\alpha_p\rangle)/\sqrt{2}; \qquad |S\rangle = |0,0\rangle = (|\alpha_e\beta_p\rangle - |\beta_e\alpha_p\rangle)/\sqrt{2}$$

$$|T_{-1}\rangle = |1,-1\rangle = |\beta_e\beta_p\rangle$$

(b) The perturbation resulting from the oscillating field is

$$\hat{\mathcal{H}}' = (g\mu_B\hat{S}_x - \gamma_p\hbar\hat{I}_x)B_1 2\cos\omega t$$

Following the method of Problem 11.20 we determine the matrix elements of $(g\mu_B\hat{S}_x - \gamma_p\hbar\hat{I}_x) = \hat{\mu}$.

$$\langle 1,1|\hat{\mu}|1,0\rangle = \frac{1}{\sqrt{2}}\langle\alpha_e\alpha_p|\hat{\mu}(|\alpha_e\beta_p\rangle + |\beta_e\alpha_p\rangle) = \frac{1}{2\sqrt{2}}(-\gamma_p\hbar + g\mu_B)$$

$$\langle 1,1|\hat{\mu}|0,0\rangle = \frac{1}{\sqrt{2}}\langle\alpha_e\alpha_p|\hat{\mu}(|\alpha_e\beta_p\rangle - |\beta_e\alpha_p\rangle) = \frac{1}{2\sqrt{2}}(-\gamma_p\hbar - g\mu_B)$$

$$\langle 1,0|\hat{\mu}|1,-1\rangle = \frac{1}{\sqrt{2}}(\langle\alpha_e\beta_p| + \langle\beta_e\alpha_p|)\hat{\mu}|\beta_e\beta_p\rangle = \frac{1}{2\sqrt{2}}(g\mu_B - \gamma_p\hbar)$$

$$\langle 0,0|\hat{\mu}|1,-1\rangle = \frac{1}{\sqrt{2}}(\langle\alpha_e\beta_p| - \langle\beta_e\alpha_p|)\hat{\mu}|\beta_e\beta_p\rangle = \frac{1}{2\sqrt{2}}(g\mu_B + \gamma_p\hbar)$$

$<1,0|\hat{\mu}|0,0> = 0, \qquad <1,1|\hat{\mu}|1,-1> = 0$

Since $g\mu_B >> \gamma_p \hbar$, all of the allowed transitions have approximately the same intensity (assuming that the population differences between the initial and final states are equal). In summary we find the following allowed transitions:

$S \to T_1, \quad S \to T_{-1} \quad$ with $\quad \nu = \Delta E/h = a$

$T_{-1} \to T_o, \quad T_o \to T_{+1} \quad$ with $\quad \nu = 0$

11.23 The unpaired electron in an aromatic free radical is delocalized over the π-system of orbitals. A hyperfine splitting constant $a_H^{(i)}$ can be assigned to each proton, and the relationship below has been found to hold reasonably well:

$$a_H^{(i)} = Q\rho_\pi^{(i)} \qquad (1)$$

Here Q is a proportionality constant and $\rho_\pi^{(i)}$ is the electron spin-density on the adjacent carbon atom, e.g. in the CH_3 radical $\rho_\pi = 1$. The constant $a_H^{(i)}$ is usually reported in units of gauss where 1 gauss = 10^{-4} T.

(a) The ESR spectrum of the benzene anion radical ($C_6H_6^-$) consists of seven lines having a uniform spacing of 3.75 gauss and the relative intensities 1:6:15:20:15:6:1. Calculate the magnitude of Q in Eq. (1).

(b) Assuming the Q is transferable from the benzene anion to the naphthalene anion ($C_{10}H_8^-$), predict the magnitudes of the hyperfine splitting constants for naphthalene and determine the line positions and intensities in the associated ESR spectrum. (Hint: Use Hückel MO's for your calculation.)

Solution: Ref. Problem 9.27; CM,72

(a) The intensity pattern given for $C_6H_6^-$ indicates that the six protons are equally coupled to the unpaired electron. Note that the relative intensities are binomial coefficients, i.e. coefficients in the expansion of $(a + b)^6$. According to Eq. (1) we have 3.75 gauss = $|Q|(1/6)$ or $|Q|$ = 22.5 gauss. Actually, Q = −22.5 gauss. According to simple MO theory Q = 0 since the p-orbitals vanish at the hydrogen nuclei.

(b) From Problem 9.27 it is clear that the unpaired electron should occupy the B_{1g} orbital. The associated eigenvalue is $E(B_{1g}) = \alpha - 0.618\beta$ and substitution of this quantity into the equation of coefficients gives:

$$\psi(B_{1g}) = c_1(u_1 - 0.618\, u_2), \qquad c_1 = (1.382)^{-1/2} = 0.851$$

and

$$\psi(B_{1g}) = 0.426(\phi_1 - \phi_4 + \phi_5 - \phi_8) - 0.263(\phi_2 - \phi_3 + \phi_6 - \phi_7)$$

The spin-density on atom 1 is $\rho_\pi^{(1)} = |c_1|^2 = 0.18$ and on atom 2 is $\rho_\pi^{(2)} = |c_2|^2 = 0.069$.

Thus using Eq. (1)

$$|a_H^{(1)}| = 4.07 \text{ gauss}, \qquad |a_H^{(2)}| = 1.55 \text{ gauss}$$

There are four equivalent protons of type 1. The splitting pattern for the energy levels and for the spectrum (since $\Delta m_I = 0$) can be constructed as shown below.

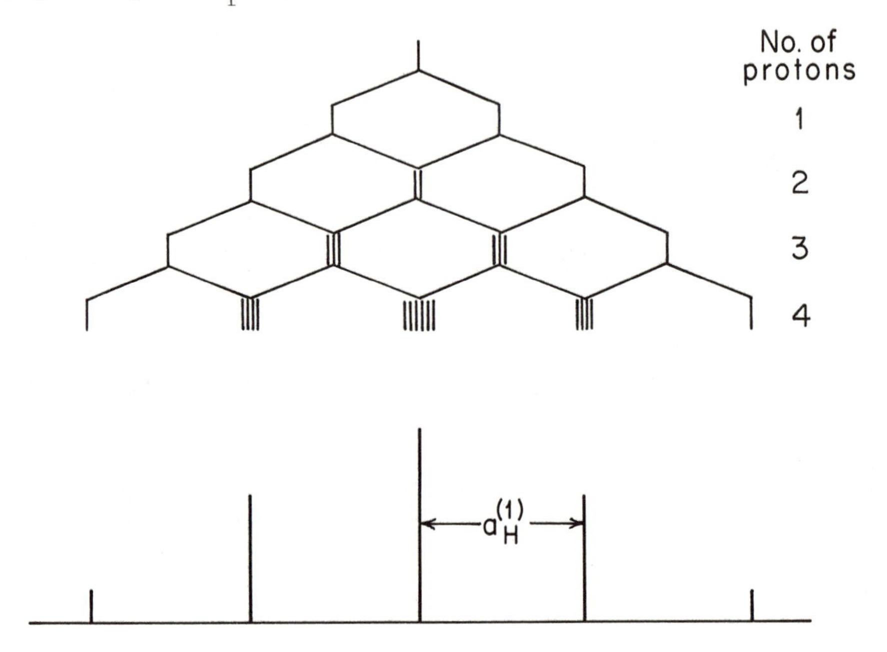

When the four equivalent protons of type 2 are added, each line in the spectrum will split again into 5 lines having the spacing $a_H^{(2)}$ and the intensity distribution 1:4:6:4:1. The resulting spectrum thus has 25 lines as shown below.

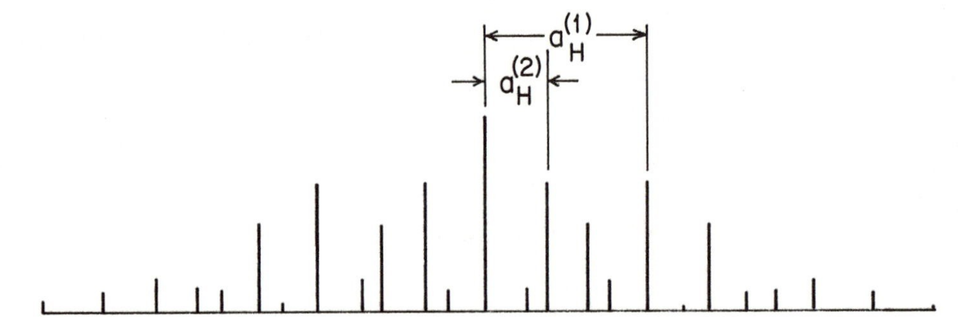

11.24 A calculation has been reported which shows that the ^{13}C hyperfine splitting constant should be about 1,200 gauss for a carbon atom having an unpaired 2s electron. The ESR spectrum for methyl radicals containing ^{13}C, i.e. $^{13}CH_3$, shows that the ^{13}C hyperfine splitting constant is 41 gauss.

(a) What conclusion can you draw about the structure of the methyl radical?

(b) What splitting constant would you predict for the protons?

(c) Construct the ESR spectrum for a methyl group containing ^{13}C giving the line positions and relative intensities. (Carbon-13 has a spin of 1/2.)

Solution: Ref. Problems 11.21 and 11.23; CM,79

(a) The very small hyperfine splitting constant for ^{13}C indicates that the unpaired electron resides in an orbital have little or no s-character. A planar structure using sp^2 for the hybridization of the bonding orbitals on carbon, and a $2p_\pi$ orbital $\perp$ to the plane for the unpaired electron would be consistent with this result. The small residual coupling must result from a spin-polarization of the electrons in orbitals having s-character. The exchange interactions of the unpaired electron with electrons in σ-orbitals provide a mechanism for a polarization.

(b) The coupling of the unpaired electron in a $2p_\pi$ orbital with the protons in the plane must be similar to that found in aromatic radicals, e.g. the benzene anion. This suggests that $a_H = -22.5\ \rho_\pi = -22.5$ gauss. The experimental value is -23.04 gauss.

(c) As illustrated in Problem 11.23 a single spin 1/2 particle (^{13}C) causes a splitting of the electron resonance line into two lines of equal intensity. The three protons then cause each of these lines to split into four components having the relative intensities 1:3:3:1.

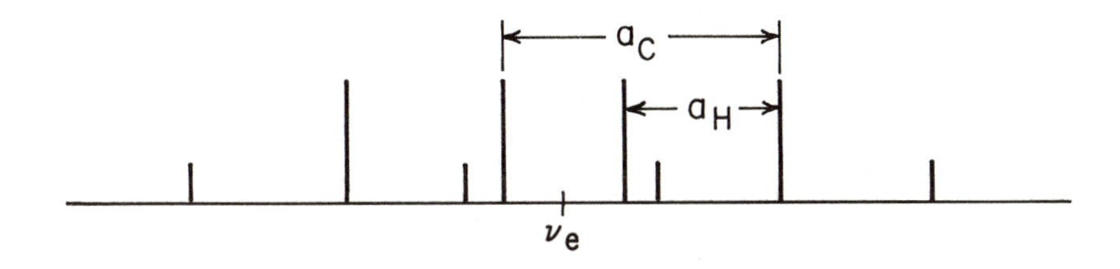

11.25 For a free radical in solution in which the unpaired electron interacts with a nucleus having spin I, the Hamiltonian is:

$$\hat{\mathcal{H}} = g\mu_B B_o \hat{S}_z - \gamma\hbar B_o \hat{I}_z + a\hat{\underset{\sim}{I}}\cdot\hat{\underset{\sim}{S}} \qquad (1)$$

(a) Use second order perturbation theory to obtain expressions for the eigenvalues of (1) in the case that $g\mu_B B_o >> \gamma\hbar B_o$, a.

(b) In the high field limit the electron spin resonance (ESR) transitions obey the selection rules: $\Delta m_S = \pm 1$, $\Delta m_I = 0$. (See Problem 11.47) Show that the frequencies of the allowed transitions are given by

$$\nu \cong \nu_e + \frac{a}{h} m_I + \frac{a^2}{2h^2\nu_e} (I^2 + I - m_I^2)$$

Solution: Ref. R. W. Fessenden, J. Chem. Phys. 37, 747 (1962)

(a) The Hamiltonian can be written as

$$\hat{\mathcal{H}} = g\mu_B B_o \hat{S}_z - \gamma\hbar B_o \hat{I}_z + a\hat{I}_z\hat{S}_z + \frac{a}{2}(\hat{I}_+\hat{S}_- + \hat{I}_-\hat{S}_+)$$

which can be separated to give the form $\hat{\mathcal{H}} = \hat{\mathcal{H}}^o + \hat{\mathcal{H}}'$. We choose the first two terms on the r.h.s. as $\hat{\mathcal{H}}^o$, but the first three terms could be choosen if desired. The second order correction will be slightly affected by the choice. According to perturbation theory

$$E_n = E_n^o + E_n^{(1)} + E_n^{(2)} + \cdots$$

Using the basis functions $|S,m_S>|I,m_I> = |m_S,I,m_I>$ where $S = 1/2$ we obtain:

$$E^o(m_S,I,m_I) = g\mu_B B_o m_S - \gamma\hbar B_o m_I = h\nu_e m_S - h\nu_I m_I \tag{2}$$

$$E^{(1)}(m_S,I,m_I) = <m_S,I,m_I|\hat{\mathcal{H}}'|m_S,I,m_I> = am_I m_S \tag{3}$$

The equation for the second order energy is:

$$E^{(2)}(m_S,I,m_I) = \sum_{\substack{m_I',m_S' \\ \neq m_I,m_S}} \frac{|<m_S,I,m_I|\hat{\mathcal{H}}'|m_S',I,m_I'>|^2}{E^o(m_S,I,m_I) - E^o(m_S',I,m_I')}$$

and the only nonvanishing off-diagonal matrix elements of $\hat{\mathcal{H}}'$ are:

$$<\frac{1}{2},I,m_I-1|\ \frac{a}{2}(\hat{I}_+\hat{S}_- + \hat{I}_-\hat{S}_+)|-\frac{1}{2},I,m_I> = \frac{a}{2}\sqrt{(I + m_I)(I - m_I + 1)}$$

$$<-\frac{1}{2},I,m_I+1|\ \frac{a}{2}(\hat{I}_+\hat{S}_- + \hat{I}_-\hat{S}_+)|\ \frac{1}{2},I,m_I> = \frac{a}{2}\sqrt{(I - m_I)(I + m_I + 1)}$$

Thus the second order corrections for $m_S = \pm 1/2$ are

$$E^{(2)}\left(\frac{1}{2},I,m_I\right) = \frac{a^2}{4h(\nu_e + \nu_I)}(I - m_I)(I + m_I + 1) \tag{4}$$

$$E^{(2)}\left(-\frac{1}{2},I,m_I\right) = -\frac{a^2}{4h(\nu_e + \nu_I)}(I + m_I)(I - m_S + 1) \tag{5}$$

Of course, $\nu_e >> \nu_I$. If the term $a\hat{I}_z\hat{S}_z$ had been included in $\hat{\mathcal{H}}^o$, the denominators would become $h(\nu_e + \nu_I) + am_I$.

(b) According to the selection rules for absorption, ($\Delta m_S = +1$, $\Delta m_I = 0$), the energies of the allowed transitions are given by

$$\Delta E(I,m_I) = h\nu = E(1/2,I,m_I) - E(-1/2,I,m_I) \tag{6}$$

Then combining Eqs. (2) through (6) we obtain:

$$\nu = \nu_e + \frac{a}{h} m_I + \frac{a^2}{4h^2\nu_o} [(I - m_I)(I + m_I + 1) + (I + m_I)(I - m_I + 1)]$$

$$= \nu_e + \frac{a}{h} m_I + \frac{a^2}{2h^2\nu_o} (I^2 + I - m_I^2)$$

Here we replace $h(\nu_e + \nu_I)$ or $h(\nu_e + \nu_I) + am_I$ by ν_o in the denominator, which can safely be set equal to the frequency at the center of the ESR spectrum. Of course, this approximation leads to little error for $a/h\nu_e \ll 1$.

11.26 A symmetric top molecule having the electric dipole moment $\underset{\sim}{\mu}$ is placed in the electric field $\underset{\sim}{\mathcal{E}}$ so that the perturbation is

$$\hat{\mathcal{H}}' = -\mu\mathcal{E}\cos\theta$$

where θ is the angle between the dipole moment and the electric field.

(a) Use first order perturbation theory to obtain the energies for a symmetric top molecule in the field $\underset{\sim}{\mathcal{E}}$ for $J \neq 0$.

(b) Obtain an equation for the frequencies in the rotational absorption spectrum in terms of J, K, and M_J for the ground state when the microwave field is in the z-direction. The selection rules are: $\Delta J = +1$, $\Delta K = 0$, $\Delta M_J = 0$. (Hint: An integral of the type $\mu\langle J,K,M_J|\cos\theta|J,K,M_J\rangle$ can be evaluated by geometric arguments based on the fact that only the time independent component of $\underset{\sim}{\mu}$ in the direction of $\underset{\sim}{J}$ can be observed.)

Solution: Ref. TS,248; Problem 11.9

(a) The Hamiltonian can be written as $\hat{\mathcal{H}}^o + \hat{\mathcal{H}}'$ where $\hat{\mathcal{H}}' = -\mu\mathcal{E}\cos\theta$. From Problem 11.9 we have

$$E^o(J,K,M_J) = \langle J,K,M_J|\hat{\mathcal{H}}^o|J,K,M_J\rangle = [J(J + 1)B + K^2(A - B)]hc$$

According to first order perturbation theory

$$E^{(1)}(J,K,M_J) = -\mathcal{E}\langle J,K,M_J|\mu\cos\theta|J,K,M_J\rangle$$

The geometrical situation is shown below.

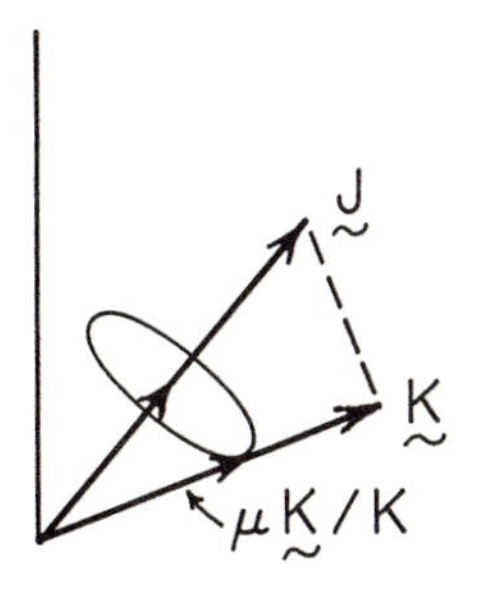

The component of $\underset{\sim}{\mu}$ along $\underset{\sim}{J}$ is given by:

$$\mu \frac{\underset{\sim}{K}}{K} \cdot \frac{\underset{\sim}{J}}{\sqrt{J(J+1)}} = \frac{\mu K}{\sqrt{J(J+1)}}$$

Then

$$\underset{\sim}{\mu}(\text{eff}) = \frac{\mu K}{\sqrt{J(J+1)}} \frac{\underset{\sim}{J}}{\sqrt{J(J+1)}} \qquad \text{and} \qquad \underset{\sim}{\mu}(\text{eff}) \cdot \hat{k} = \frac{\mu K M_J}{J(J+1)}$$

where we have used the relations:

$$\text{unit vector in direction of } \underset{\sim}{J} = \underset{\sim}{J} / \sqrt{J(J+1)} = \hat{n}_J$$

$$\text{unit vector in direction of } \underset{\sim}{K} = \underset{\sim}{K}/K = \hat{n}_K$$

and

$$\underset{\sim}{J} \cdot \hat{n}_K = K, \qquad \underset{\sim}{J} \cdot \hat{k} = M_J$$

we conclude that

$$E^{(1)}(J,K,M_J) = \langle J,K,M_J | \mu \mathcal{E} \cos\theta | J,K,M_J \rangle = -\frac{\mu \mathcal{E} K M_J}{J(J+1)}$$

The general analytical method for deriving matrix elements for symmetric top molecules makes use of the fact that the eigenfunctions are proportional to the Wigner rotation matrices, $D^J_{-K,-M}(\alpha\beta\gamma)$. See for example the discussions in RO3,55.

(b) $\nu(\text{cm}^{-1}) = (hc)^{-1}[E(J',K',M_J') - E(J'',K'',M_J'')]$

With $\Delta K = \Delta M_J = 0$ and $\Delta J = J' - J'' = +1$ we obtain

$$\nu(\text{cm}^{-1}) = [(J+1)(J+2) - J(J+1)]B - \frac{\mu \mathcal{E} K M_J}{hc} \left[\frac{1}{(J+1)(J+2)} - \frac{1}{J(J+1)}\right]$$

$$= 2(J+1)B + \frac{\mu \mathcal{E} K M_J}{hc} \left[\frac{2}{J(J+1)(J+2)}\right]$$

where $J = J''$ = rotational quantum number for the ground state.

11.27 When a nucleus having a spin I and a <u>quadrupole moment</u> Q is placed in an electric field having axial symmetry the Hamiltonian is

$$\hat{\mathcal{H}}_Q = \frac{e^2 qQ}{4I(2I-1)} (3\hat{I}_z^2 - \hat{I}^2) \tag{1}$$

Here q is the field gradient defined so that $eq = (\partial^2 V/\partial z^2)_{r=0}$ where V is the potential resulting from external charges.

(a) Obtain an equation for the eigenvalues of $\hat{\mathcal{H}}_Q$.

(b) Determine the transition frequencies and relative probability when the perturbation $\hat{\mathcal{H}}' = -\gamma\hbar \hat{I}_x 2B_1 \cos\omega t$ is applied and $I = 3/2$.

Solution: Ref. DH,3

(a) We choose the kets $|I,m\rangle$ for which $\hat{\mathcal{H}}_Q$ is diagonal. Thus

$$E_{I,m} = \langle I,m|\hat{\mathcal{H}}_Q|I,m\rangle = \frac{e^2qQ}{4I(2I-1)}[3m^2 - I(I+1)] \tag{2}$$

This is, of course, an exact result.

(b) Equation (2) gives the energies (for $I = 3/2$)

$$E_{3/2,\pm 3/2} = A\left[3\left(\frac{3}{2}\right)^2 - \frac{3}{2}\left(\frac{5}{2}\right)\right] = 3A, \qquad A = \frac{e^2qQ}{12}$$

$$E_{3/2,\pm 1/2} = A\left[3\left(\frac{1}{2}\right)^2 - \frac{15}{4}\right] = -3A$$

The transition rate is proportional to $|\langle I,m+1|\hat{I}_x|I,m\rangle|^2$ thus we require the integrals:

$$\langle 3/2,3/2|\frac{\hat{I}_+}{2}|3/2,1/2\rangle = \frac{1}{2}\sqrt{(I-m)(I+m+1)} = \frac{1}{2}\sqrt{(3/2-1/2)(3/2+1/2+1)} = \frac{\sqrt{3}}{2}$$

$$\langle 3/2,1/2|\frac{\hat{I}_+}{2}|3/2,-1/2\rangle = \frac{1}{2}\sqrt{(3/2+1/2)(3/2-1/2+1)} = \frac{\sqrt{4}}{2} = 1$$

$$\langle 3/2,-1/2|\frac{\hat{I}_+}{2}|3/2,-3/2\rangle = \frac{1}{2}\sqrt{(3/2+3/2)(3/2-3/2+1)} = \frac{\sqrt{3}}{2}$$

Transition	Frequency	Probability
$3/2 \leftrightarrow 1/2$	$6A/h$	3/4
$1/2 \leftrightarrow -1/2$	0	1
$-1/2 \leftrightarrow -3/2$	$6A/h$	3/4

11.28 Consider an emitter of radiation, either an atom or nucleus, which is constrained to move along the x axis and is bound to the origin by a linear restoring force. The vector potential of the emitted radiation can be written as

$$\underset{\sim}{A}(t) = \underset{\sim}{A}_o \exp[i\int_o^t \omega(t')dt'] \qquad \text{where} \qquad \omega(t') \cong \omega_o[1 + v(t')/c]$$

(a) Assume that the motion of the center of mass of the emitter can be treated classically with $x(t) = x_o \sin\Omega t$ and resolve $\underset{\sim}{A}(t)$ into the frequency components $\exp(-in\Omega t)$, i.e. expand $\underset{\sim}{A}(t)$ in a complex Fourier series.

(b) Determine the relative intensities of emission at the frequencies $\omega_o \pm n\Omega$ for $n = 0, 1, 2, 3$ in the special case that $x_o\omega_o/c = 2\pi x_o/\lambda = 1$. This is a model for the Mössbauer effect, and the radiation at ω_o corresponds to recoilless emission.

Solution: Ref. FR,26

(a) The vector potential is given by:

$$\underset{\sim}{A}(t) = \underset{\sim}{A}_o \exp\{i\int_o^t \omega_o[1 + v(t')/c]dt\} = \underset{\sim}{A}_o e^{i\omega_o t} e^{i\omega_o x(t)/c} = \underset{\sim}{A}_o e^{i\omega_o t} e^{iy\sin\theta} \tag{1}$$

where $y = 2\pi x_o/\lambda = x_o/\lambdabar$ and $\theta = \Omega t$. Equation (1) can be resolved into frequency components by writing

$$e^{iy\sin\theta} = \sum_{n=-\infty}^{+\infty} c_n e^{-in\theta} \tag{2}$$

with

$$c_n = \frac{1}{2\pi}\int_0^{2\pi} e^{in\theta} e^{iy\sin\theta} d\theta = J_n(y)$$

For the last step see for example: G. N. Watson, Theory of Bessel Functions (Cambridge University Press, 1945), p. 20.

Combining Eqs. (1) and (2) we obtain

$$\underset{\sim}{A} = \underset{\sim o}{A} e^{i\omega_o t} \sum_{n=-\infty}^{+\infty} J_n(x_o/\lambdabar) e^{in\Omega t} = \underset{\sim o}{A} \sum_{n=-\infty}^{+\infty} J_n(x_o/\lambdabar) e^{i(\omega_o + n\Omega)t}$$

(b) The intensity is proportional to $\mathcal{E}^2$ where $\underset{\sim}{\mathcal{E}} = -\partial\underset{\sim}{A}/\partial t$. Thus, for the nth frequency the intensity is proportional to $J_n^2(x_o/\lambdabar)$. The necessary numbers are: $J_o^2(1) = 0.59$; $J_1^2(1) = 0.19$; $J_2^2(1) = 0.013$; and $J_3^2(1) = 3.8\times10^{-4}$. The fraction of recoilless emission is thus ~0.59.

SUPPLEMENTARY PROBLEMS

11.29 Repeat Problem 11.1 for a molecule having $J \neq 0$ and show that the eigenvalues can be written as

$$E(v,J) = (v + 1/2)h\nu_o + J(J+1)\frac{\hbar^2}{2I_e} - \frac{J^2(J+1)^2}{(1/2)kr_e^2}\left(\frac{\hbar}{2I_e}\right)^2 + \cdots$$

where $I_e = \mu r_e^2$ and $J(J+1)\hbar^2/2I_e \ll kr_e^2$. In this derivation again assume that $\rho = (r - r_e) \ll r_e$ and retain only terms through second order in ρ/r_e. (Hint: In order to remove terms linear in ρ from the wave equation try the substitution $\rho = \xi + a$ where a is an appropriate constant. Then show that

$$a = \frac{(J+1)r_e(\hbar^2/2I_e)}{3(\hbar^2/2I_e)J(J+1) + kr_e^2/2}\ .\ )$$

Ref. PW,267; EWK,268

11.30 The potential function describing the vibration of the H^1Cl^{35} molecule can be approximated by the Morse function:

$$V(r - r_e) = D_e[e^{-2a(r-r_e)} - 2e^{-a(r-r_e)}] \tag{1}$$

where r_e is the equilibrium internuclear distance (1.27 Å), D_e is approximately 4.4 eV, and $a = 1.85 \times 10^8$ cm^{-1}. The rotational constant B_e is approximately 10.6 cm^{-1}, and the reduced mass μ is 1.64×10^{-24} grams.

(a) What are the values in eV of the average potential energy $\langle V \rangle$, the average kinetic energy $\langle T \rangle$, and the total energy E when the internuclear distance is 1.27 Å? The kinetic energy of nuclear motion can be neglected in this problem.

(b) Determine the fundamental vibrational frequency ω_e.

(c) How will the vibration-rotation spectrum for $^1H^{37}Cl$ differ from that of $^1H^{35}Cl$?

Ans: Ref. Problem 9.19

(a) $\langle V \rangle = -8.8$ eV, $\langle T \rangle = +4.4$ eV, $E = -4.4$ eV

(b) $\omega_e \simeq 2,880$ cm^{-1}

(c) $\omega_e^{37} - \omega_e^{35} \simeq -2.16$ cm^{-1}; $B_e^{37} - B_e^{35} \simeq -0.01594$ cm^{-1}

11.31 The potential function $V(x) = -Ax^2 + Bx^4$ with $A,B > 0$ exhibits double minima as shown in Fig. 11.31.

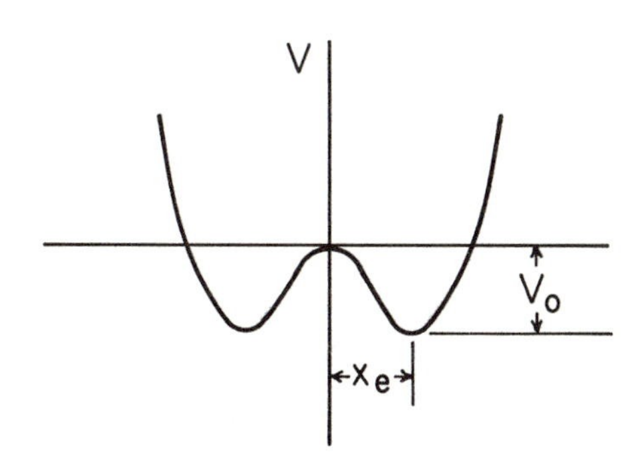

Fig. 11.31

(a) Give the values of x at which V(x) reaches minima.

(b) Obtain an expression for the barrier height V_o in terms of A and B.

(c) Expand V(x) about $x = x_e$ and thus obtain the vibrational frequency in the high barrier limit, i.e. $V_o \gg \omega_e hc$.

Ans:

(a) $x = \pm\sqrt{A/2B}$; (b) $V_o = A^2/4B$; (c) $\omega_e = \frac{1}{\pi c}\sqrt{\frac{A}{\mu}}$

11.32 The Rydberg potential function for vibration can be written as

$$V(x) = -D_e(1 + bx)e^{-bx}$$

where $x = r - r_e$. For low amplitude vibrations the energies can be expressed as

$$hcG(v) = \left(v + \frac{1}{2}\right)\omega_e - \left(v + \frac{1}{2}\right)^2 \omega_e x_e$$

Use the results of perturbation theory to obtain expressions for ω_e and $\omega_e x_e$ in terms of D_e, b, and μ.

Ans: Ref. Problem 11.3; KI,165

$$\omega_e = \frac{b}{2\pi c}\sqrt{\frac{D_e}{\mu}} \quad ; \qquad \omega_e x_e = \frac{\hbar^2(11b^2)}{48\mu hc}$$

11.33 In CH_3Cl (methyl chloride) the HCH bond angle is 110°20' and the bond distances are: d(CH) = 1.103 Å and d(CCl) = 1.782 Å. Two of the principal moments of inertia equal 57×10^{-40} g cm^2. Calculate the rotational constants A and B (in cm^{-1}) for this symmetric top and plot the first few energy levels.

Ans: Ref. Problem 11.9

$A = 5.10$ cm^{-1}; $B = 0.49$ cm^{-1}; $F(J,K) = 0.49\ J(J + 1) + 4.61\ K^2$

11.34 The inertial dyadic for a molecule is defined as

$$\begin{aligned} \underset{\sim}{I} = &\ I_{xx}\hat{i}\hat{i} - I_{xy}\hat{i}\hat{j} - I_{xz}\hat{i}\hat{k} \\ &- I_{yx}\hat{j}\hat{i} + I_{yy}\hat{j}\hat{j} - I_{yz}\hat{j}\hat{k} \\ &- I_{zx}\hat{k}\hat{i} - I_{zy}\hat{k}\hat{j} + I_{zz}\hat{k}\hat{k} \end{aligned} \qquad (1)$$

so that the moment of inertia about the direction of the unit vector $\hat{n}$ is given by

$$I = \hat{n}\cdot\underset{\sim}{I}\cdot\hat{n} \qquad (2)$$

Here $\hat{n} = \alpha\hat{i} + \beta\hat{j} + \gamma\hat{k}$ where α, β, and γ are direction cosines. Let $\underset{\sim}{R} = \hat{n}/\sqrt{I}$ and show that the $\underset{\sim}{R}$ vector generates an ellipsoid such that each point on the surface of the ellipsoid is the distance $1/\sqrt{I}$ away from the origin.

Ref. GO,149; HO,123

11.35 The molecule CH_2Br_2 has roughly tetrahedral geometry with two equivalent CH bonds and two equivalent CBr bonds. The molecular symmetry group is C_{2v}. Determine the symmetry types of the normal modes of vibration and indicate which modes are infrared and Raman active.

Ans: Ref. Appendix 8

$\Gamma(\text{vib}) = 4A_1 + A_2 + 2B_1 + 2B_2$

All nine modes are Raman active and the following eight are active in the infrared: $4A_1$, $2B_1$, and $2B_2$.

11.36 Two NO_2 molecules having C_{2v} symmetry can form an N-N bond to yield N_2O_4 which has four equivalent oxygen atoms. Assume that N_2O_4 is planar and determine:

(a) the number of infrared active fundamentals.

(b) the number of Raman active fundamentals.

(c) the number of infrared lines expected to coincide in frequency with Raman lines.

Ans: Ref. Appendix 8

(a) 5 infrared active modes; B_{1u}, $2B_{2u}$, and $2B_{3u}$

(b) 6 Raman active modes; $3A_g$, $2B_{1g}$, and B_{2g}

(c) No Raman and infrared fundamentals coincide in frequency.

11.37 The H_2O_2 molecule, which is nonplanar, has two equivalent OH bonds and two equal OOH angles. The planes defined by $O^{(1)}O^{(2)}H^{(2)}$ and $O^{(2)}O^{(1)}H^{(1)}$ are rotated with respect to each other by the angle $\alpha < 90°$.

(a) Determine the number of infrared and Raman active vibrational bands.

(b) How many infrared bands coincide with Raman bands in frequency?

Ans: Ref. Appendix 8

(a) $\Gamma(vib) = 4A + 2B$

There are six Raman active and six infrared active fundamentals.

(b) 6

11.38 In the infrared spectrum of CH_3Cl the absorption at 3042 cm^{-1} has been assigned to a doubly degenerate mode ($\nu_4^{CH}(e)$). Determine the symmetries of the states which arise when ν_4 is doubly excited.

Ans: Ref. HZ3,313; Appendix 8

$(E \times E)^+ = A_1 + E$

11.39 The electron configuration for doubly excited benzene is $(a_{2u})^2(e_{1g})^2(e_{2u})^2$. Determine the symmetries (including multiplicities) for all of the states associated with this configuration.

Ans: Ref. Appendix 8; R. L. Ellis and H. H. Jaffe, J. Chem. Ed. 48, 93 (1971)

$3\,{}^1A_{1g} + {}^3A_{1g} + {}^5A_{1g} + {}^1A_{2g} + 2\,{}^3A_{2g} + 3\,{}^1E_{2g} + 2\,{}^3E_{2g}$

11.40 The rotational constant for HCN is $B = 1.48\ cm^{-1}$. Which rotational level will have the greatest population at (a) 25°C, (b) 300°C?

Ans: Ref. KP,471; HZ2,124

(a) $J_{max} = 8$ (b) $J_{max} = 11$

11.41 The rotational wave function ψ_r for a linear molecule can be written as $\psi_r = \Theta_{JM}(\theta)\Phi_M(\phi)$ where

$$\Theta_{JM} = \left[\frac{(2J+1)(J-|M|)!}{2(J+|M|)!}\right]^{1/2} P_J^{|M|}(\cos\theta), \qquad \Phi_M = \frac{1}{\sqrt{2\pi}} e^{iM\phi}$$

Show that for a diatomic molecule in any rotational state J, the expectation value of every component of the dipole moment in a space fixed coordinate system must vanish. This statement applies to nondegenerate electronic states.

11.42 The cyanine analogs can be represented by the formula:

$$(CH_3)_2\overset{+}{N} = CH - (CH = CH)_k - N(CH_3)_2$$

This structure is equivalent to the one obtained by interchanging the single and double bonds and shifting the + charge to the nitrogen atom on the right. The molecule is a "resonance hybrid" of these two forms.

(a) Use a particle-in-box or free electron model (FEM) for the π-electrons in such molecules to derive an expression for the wave length of the transition having lowest energy. In the derivation let d be the average length of both the CC and CN bonds, and assume that the conjugated system extends one bond length beyond each terminal nitrogen atom. (Hint: First show that the number of occupied levels is $n = k + 2$.)

(b) Calculate the wavelengths of the lowest energy transitions for k = 1, 2, 3 assuming that $d = 1.4$ Å.

Ans: Ref. JO,220; KP,536; KA,673

(a) $\lambda = \dfrac{hc}{\Delta E} = \dfrac{8m_e c}{h}\,\dfrac{4d^2(k+2)^2}{(2k+5)}$

(b) $k = 1$, $\lambda = 332$ nm; $k = 2$, $\lambda = 459$ nm; $k = 3$, $\lambda = 582$ nm

11.43 Repeat Problem 11.19 using the basis functions $|I,M_I\rangle$ where $\hat{\underset{\sim}{I}} = \hat{\underset{\sim}{I}}_1 + \hat{\underset{\sim}{I}}_2$, $\hat{I}_z|I,M_I\rangle = M_I|I,M_I\rangle$, and $\hat{I}^2|I,M_I\rangle = I(I+1)|I,M_I\rangle$.
Ref. Problem 5.22; BS,59

11.44 Suppose that the interaction of two nuclei in a magnetic field can be described by the Hamiltonian:

$$\hat{\mathcal{H}} = -\hbar\omega_1\hat{I}_{1z} - \hbar\omega_2\hat{I}_{2z} + \hbar J\hat{\underset{\sim}{I}}_1\cdot\hat{\underset{\sim}{I}}_2 \qquad (1)$$

Here ω_1 and ω_2 are the Larmor frequencies of nuclei 1 and 2, respectively, and J is the spin coupling constant. Obtain the exact eigenvalues and eigenvectors for this system in the special case that $I_1 = I_2 = 1/2$. (Hint: Proceed as in Problem 11.19 using basis functions of the type $|m_1\rangle|m_2\rangle = |m_1m_2\rangle$.)

Ans: Ref. Problem 11.19; Appendix 6

$$E_1 = -\frac{\hbar\omega_1}{2} - \frac{\hbar\omega_2}{2} + \frac{\hbar J}{4}, \qquad \psi_1 = |\alpha\alpha\rangle$$

$$E_2 = \hbar C - \frac{\hbar J}{4}, \qquad \psi_2 = \cos\theta|\alpha\beta\rangle + \sin\theta|\beta\alpha\rangle$$

$$E_3 = -\hbar C - \frac{\hbar J}{4}, \qquad \psi_3 = -\sin\theta|\alpha\beta\rangle + \cos\theta|\beta\alpha\rangle$$

$$E_4 = \frac{\hbar\omega_1}{2} + \frac{\hbar\omega_2}{2} + \frac{\hbar J}{4}, \qquad \psi_4 = |\beta\beta\rangle$$

where $C = (1/2)\sqrt{\delta^2 + J^2}$, $\delta = \omega_2 - \omega_1$, $\cos\theta = (1 + Q^2)^{-1/2}$, $\sin\theta = Q(1 + Q^2)^{-1/2}$, and $Q = J/(2C + \delta)$.

11.45 Determine the frequencies and relative intensities of the NMR transitions which are induced in the spin system of Problem 11.44 by the perturbation $\hat{\mathcal{H}}' = -\gamma\hbar\hat{I}_x B_1 2\cos\omega t$ where $\hat{I}_x = \hat{I}_{1x} + \hat{I}_{2x}$.

Ans: Ref. Problems 11.20, 11.44

Transition (i→j)	$(E_i - E_j)/h$	Relative Intensity
(4→3)	$\nu_o + \frac{1}{2}\left(\frac{J}{2\pi}\right) + \frac{C}{2\pi}$	$(1 - \sin2\theta)$
(4→2)	$\nu_o + \frac{1}{2}\left(\frac{J}{2\pi}\right) - \frac{C}{2\pi}$	$(1 + \sin2\theta)$
(3→1)	$\nu_o - \frac{1}{2}\left(\frac{J}{2\pi}\right) - \frac{C}{2\pi}$	$(1 - \sin2\theta)$
(2→1)	$\nu_o - \frac{1}{2}\left(\frac{J}{2\pi}\right) + \frac{C}{2\pi}$	$(1 + \sin2\theta)$

11.46 A rigid rotator with a quadrupole moment Θ experiences an external field gradient such as that midway between charged wires in the z-direction.

$$\underset{\sim}{\mathcal{E}'} = \begin{pmatrix} \mathcal{E}' & 0 & 0 \\ 0 & -\mathcal{E}' & 0 \\ 0 & 0 & 0 \end{pmatrix}$$

The perturbation is

$$\hat{\mathcal{H}}' = -\frac{1}{2}\Theta\mathcal{E}'(\ell_x^2 - \ell_y^2)$$

where $\hat{\ell}$ is the unit vector along the axis of the rotator. Show that to first order the J = 0 state is unaffected, J = 1 is split into states of energy 2B, $2B \pm \frac{1}{5}\Theta\mathcal{E}'$, and J = 2 into states of energy 6B, $6B \pm \frac{1}{7}\Theta\mathcal{E}'$, and $6B \pm \frac{2\sqrt{3}}{21}\Theta\mathcal{E}'$. The necessary matrix elements are:

$$\langle JM|\ell_x^2 - \ell_y^2|JM\rangle = 0, \quad \langle JM|\ell_x^2 - \ell_y^2|J,M\pm1\rangle = 0$$

$$\langle JM|\ell_x^2 - \ell_y^2|J,M\pm 2\rangle = \frac{[(J-1\mp M)(J\mp M)(J+1\pm M)(J+2\pm M)]^{1/2}}{(2J-1)(2J+3)}$$

Ref. A. D. Buckingham and M. Pariseau, Trans. Faraday Soc. 62, 1 (1966)

11.47 Investigate the validity of the selection rules given for the spin system in Problem 11.25 by considering the transition rates resulting from the external field $2B_1\cos\omega t\hat{i}$. Again assume that $g\mu_B B_o \gg \gamma\hbar B_o$, a. Wavefunctions correct to first order are sufficient for this calculation.

11.48 The spin Hamiltonian for the unpaired electron in a methyl radical in fluid solution is given by

$$\hat{\mathcal{H}} = g\mu_B B_o \hat{S}_z + a\hat{S}\cdot(\hat{I}_1 + \hat{I}_2 + \hat{I}_3) \quad (1)$$

where the nuclear Zeeman interaction has been neglected.

(a) Treat the contact interaction $\hat{\mathcal{H}}' = a\hat{S}\cdot(\hat{I}_1 + \hat{I}_2 + \hat{I}_3)$ as a perturbation and use first order perturbation theory to determine the frequencies and relative intensities of the transitions in the ESR spectrum, i.e. when $\Delta m_s = +1$.

(b) Apply second order perturbation theory to determine the splittings of the spectral lines and the relative intensities of the components when $(a/h) \simeq 64.5$ MHz and $B_o = 0.35$ T. The g-factor for a free electron can be used in this calculation.

Ans: Ref. R. W. Fessenden, J. Chem. Phys. 37, 747 (1962); Problem 5.25

(a)

Frequency	Relative Intensity
$\nu_e \pm 3a/2h$	1
$\nu_e \pm a/2h$	3

(b)

Nuclear Quantum Numbers	Frequencies	Relative Intensities
$I = 3/2,\ M_I = \pm 3/2$:	$\nu = \nu_e \pm (3a/2h) + 3\Delta$	1
$= \pm 1/2$:	$\nu = \nu_e \pm (a/2h) + 7\Delta$	1
$I = 1/2,\ M_I = \pm 1/2$:	$\nu = \nu_e \pm (a/2h) + \Delta$	2

where $\Delta = \frac{1}{4\nu_o}\left(\frac{a}{h}\right)^2 = 106$ kHz

11.49 Suppose that the vibration of a diatomic molecule having the reduced mass μ is determined by the Fues potential given below:

$$V(R) = D_e\left[\frac{-2}{(R/R_e)} + \frac{1}{(R/R_e)^2}\right] \quad (1)$$

(a) Show that $R = R_e$ at the minimum, and that $V(R_e) = -D_e$.

(b) Obtain expressions for ω_e and $\omega_e x_e$.

(c) Use the virial theorem to show that the average kinetic energy (of the electrons) is proportional to R^{-2} while the average potential energy (of electrons and nuclei) is proportional to R^{-1}.

<u>Ans</u>: Ref. S5,34; Problems 11.3 and 9.19

(b) $\omega_e = \frac{1}{2\pi c R_e}\sqrt{\frac{2D_e}{\mu}}$, $\omega_e x_e = \frac{3}{2}\left[\frac{\hbar^2}{\mu(hc)R_e^2}\right]$

(c) $(K.E.)_{av} = D_e/(R/R_e)^2$; $(P.E.)_{av} = -2D_e/(R/R_e)$

CHAPTER 12
SCATTERING THEORY

All of chemistry is ultimately describable in terms of the interaction potentials between molecules. Thus the study of these potentials is one of the most fundamental scientific pursuits. In experimental investigations collisions between molecules are arranged, and the resulting scattering is examined as a function of initial energies and geometry. Mathematical models of the interaction potentials are then constructed which are consistent with the experimental observations.

In this chapter we consider some of the simpler ideas of classical and quantum mechanical scattering. In general we assume that only two-body interactions are involved, and that the flux densities are sufficiently low that interactions between particles in an incident or scattered beam can be neglected.

The scattering event as viewed in classical mechanics is shown below where only the

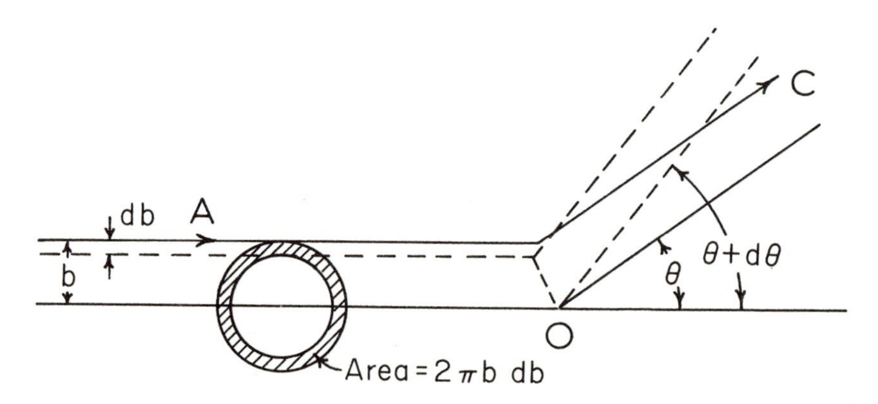

asymptotes are indicated. A beam of particles (A) having the intensity I approaches a target at 0. The impact parameter is b, the angle of deflection is θ, and the <u>differential scattering cross-section</u> is defined as

$$\boxed{\sigma_d(\theta,\phi) = \frac{I(\theta,\phi)}{I} \equiv \frac{d\sigma}{d\Omega}} \tag{1}$$

In Eq. (1) $I(\theta,\phi)$ is the number of particles per sec scattered into unit solid angle and σ is the total cross-section for scattering into all angles. Thus

$$d\sigma = \sigma_d(\theta,\phi)d\Omega \tag{2}$$

For central potentials, i.e. potentials depending only on the separation distance r, σ_d is independent of ϕ and Eq. (2) gives

$$\boxed{\sigma = 2\pi\int_0^\pi \sigma_d(\theta)\sin\theta d\theta} \tag{3}$$

A view into the beam shows that the number of particles scattered into the solid angle $2\pi\sin\theta d\theta$ is equal to the number which approached in the annulus $2\pi bdb$, i.e.

$$d\sigma = 2\pi b|db| \tag{4}$$

Then combining Eqs. (2) and (4) and maintaining $\sigma_d \gtrless 0$ gives

$$\sigma_d = \frac{b}{\sin\theta}\left|\frac{db}{d\theta}\right| \tag{5}$$

In the time-independent quantum mechanical treatment of elastic scattering, both the incident and scattered particles are described by plane waves for $r \gg b$. The wave at A is assumed to have the form $\psi_{in}(\underset{\sim}{r}) \propto e^{i\underset{\sim}{k}\cdot\underset{\sim}{r}}$ and that at C is taken to be

$$\psi_{sc}(\underset{\sim}{r}) \propto e^{i\underset{\sim}{k}\cdot\underset{\sim}{r}} + f(\Omega)\frac{e^{ikr}}{r}, \quad kr \to \infty \tag{6}$$

The fluxes of these waves can be evaluated with the current density formula

$$\underset{\sim}{j} = \frac{i\hbar}{2\mu}(\psi\underset{\sim}{\nabla}\psi^* - \psi^*\underset{\sim}{\nabla}\psi) \tag{7}$$

Connection can then be made with σ_d as defined in Eq. (1) through the definitions:

$$I = |\underset{\sim}{j}_{in}|$$

$$I(\theta,\phi) = \lim_{r\to\infty}(|\underset{\sim}{j}_{sc}|r^2)$$

Equations (1), (6), and (7) can then be combined to give the differential cross-section equation:

$$\sigma_d = |f(\Omega)|^2 \tag{8}$$

In this chapter we will be concerned with some elementary methods for solving the Schrödinger equation to obtain scattering amplitudes and cross-sections. For incident particles having low kinetic energy the partial wave method is useful. At high energies we will resort to the exact integral equation approach and the associated Born approximation.

Mathematical Preliminaries

In the following problems it is useful to recognize that the Schrödinger equation for two-body scattering can often be placed in the standard form:

$$\boxed{\left(\frac{d^2}{dx^2} - \frac{\ell(\ell+1)}{x^2} + 1\right)y_\ell = 0} \tag{9}$$

where x = kr. The general solution to this equation is

$$y_\ell = Ax^{1/2}J_{\ell+\frac{1}{2}}(x) + Bx^{1/2}J_{-\ell-\frac{1}{2}}(x) \tag{10}$$

where A and B are constants, and the J_ℓ's are standard cylindrical Bessel functions of order ℓ as shown below

$$J_\ell(x) = \sum_{m=0}^{\infty} \frac{(-1)^m (x/2)^{\ell+2m}}{m!\Gamma(m + \ell + 1)} \tag{11}$$

where $\Gamma(m + \ell + 1)$ is the gamma function with argument $(m + \ell + 1)$.

The spherical Bessel functions, which are more useful for scattering calculations, are defined as

$$j_\ell(x) = \sqrt{\frac{\pi}{2x}}\, J_{\ell+\frac{1}{2}}(x) \tag{12}$$

Similarly, the spherical Neumann functions are defined as

$$n_\ell(x) = (-1)^{\ell+1} \sqrt{\frac{\pi}{2x}}\, J_{-\ell-\frac{1}{2}}(x) \tag{13}$$

The j_ℓ function is regular at the origin, i.e. it remains finite, while the n_ℓ function is irregular and blows up at the origin. These functions have the following important limits:

$$j_\ell(x) \xrightarrow[x\to 0]{} \frac{x^\ell}{1\cdot 3\cdot 5\cdots(2\ell + 1)}, \qquad j_\ell(x) \xrightarrow[x\to\infty]{} \frac{1}{x}\cos[x - \frac{1}{2}(\ell + 1)\pi]$$

$$n_\ell(x) \xrightarrow[x\to 0]{} \frac{-1\cdot 3\cdot 5\cdots(2\ell - 1)}{x^\ell}, \qquad n_\ell(x) \xrightarrow[x\to\infty]{} \frac{1}{x}\sin[x - \frac{1}{2}(\ell + 1)\pi]$$

For $\ell = 0$ Eq. (9) describes simple harmonic motion and we have:

$$j_o(x) = \frac{\sin x}{x}, \qquad n_o(x) = -\frac{\cos x}{x}$$

General References: PR,128; CHB,43; BR1,chapter 1; MM2,chapters 2,3; FA

PROBLEMS

12.1 Consider a structureless particle of mass μ interacting with a fixed scattering center through the potential V(r). Figure 12.1 describes an event in which the particle is located initially at $x = -\infty$ with the velocity v_R.

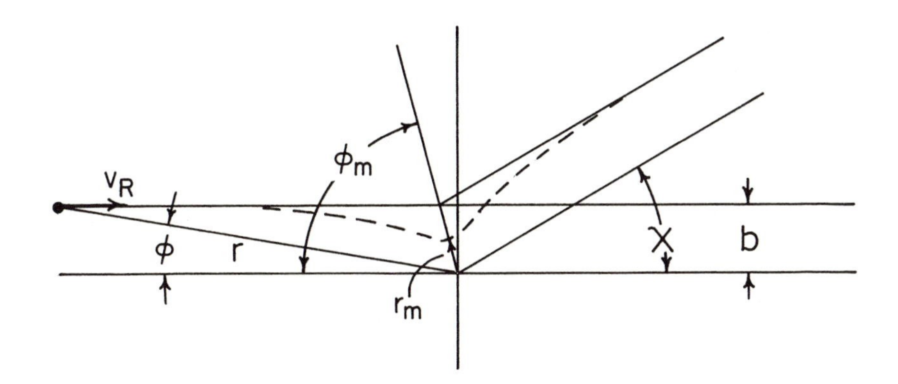

Fig. 12.1

(a) Use the conservation of energy and angular momentum to verify the following equations for the relative velocity.

$$|v_R| = \sqrt{(\dot{r}^2 + r^2\dot{\phi}^2) + 2V(r)/\mu} \tag{1}$$

$$v_R = r^2\dot{\phi}/b \tag{2}$$

(b) At the point of closest approach $r = r_m$ and

$$\left(\frac{r^2}{b}\right)\left[1 - \left(\frac{b}{r}\right)^2 + \frac{2V(r)}{\mu v_R^2}\right]^{1/2} = 0 \tag{3}$$

Derive Eq. (3) by evaluating $dr/d\phi$. This equation can in principle be solved to obtain r_m.

(c) Show that

$$\chi(b,v_R) = \pi - 2b\int_{r_m}^{\infty} \frac{r^{-2}dr}{\sqrt{1 - \frac{2V(r)}{\mu r^2} - \left(\frac{b}{r}\right)^2}} \tag{4}$$

where χ is the angle of deflection. (Hint: Notice that $\chi = \pi - 2\phi_m$.)

Solution: Ref. HCB,49; PR,128

(a) As shown in Problem 1.3 the classical Hamiltonian for the internal motion of the reduced two-particle system is

$$H = \frac{\mu}{2}(\dot{r}^2 + r^2\dot{\phi}^2) + V(r) \tag{5}$$

At $r = \infty$, $V(r) = 0$ and the total energy is equal to the kinetic energy which is determined by the initial conditions. Thus

$$\frac{\mu v_R^2}{2} = \frac{\mu}{2}(\dot{r}^2 + r^2\dot{\phi}^2) + V(r) \tag{6}$$

and rearrangement gives Eq. (1).

The angular momentum is initially

$$|\underset{\sim}{\mathcal{L}}| = |\underset{\sim}{r}\times\underset{\sim}{p}| = b\mu v_R$$

and by definition $\mathcal{L} = \partial L/\partial\dot{\phi} = \mu r^2\dot{\phi}$ where L is the Lagrangian. This is, of course, equivalent to the definition $\mathcal{L} = I\omega$ where I is the moment of inertia and ω is the angular frequency. Using the conservation of angular momentum we find

$$b\mu v_R = \mu r^2\dot{\phi} \quad \text{or} \quad v_R = r^2\dot{\phi}/b \tag{7}$$

(b) First note that

$$\frac{dr}{d\phi} = \frac{dr/dt}{d\phi/dt}$$

Then from Eq. (6) we have

$$\dot{r} = \frac{dr}{dt} = \pm\left[v_R^2 - \frac{b^2v_R^2}{r^2} - \frac{2V(r)}{\mu}\right]^{1/2}$$

and from Eq. (7) $\dot{\phi} = bv_R/r^2$. Thus

$$\frac{dr}{d\phi} = \pm\left(\frac{r^2}{b}\right)\left[1 - \left(\frac{b}{r}\right)^2 - \frac{2V(r)}{\mu v_R^2}\right]^{1/2} = \pm f(r)$$

At the point of closest approach $dr/d\phi = 0$ and we obtain Eq. (3).

(c) Since $\chi = \pi - 2\phi_m$ we must evaluate ϕ_m. From part (b) we obtain

$$\frac{dr}{d\phi} = -f(r)$$

where the minus sign is appropriate before the point of closest approach is reached. Then $d\phi = -dr/f(r)$ and

$$\phi_m = \int_0^{\phi_m} d\phi = -\int_\infty^{r_m} f(r)^{-1}dr$$

The deflection angle χ is thus given by

$$\chi = \pi - 2b\int_{r_m}^{\infty}\left[1 - \left(\frac{b}{r}\right)^2 - \frac{2V(r)}{\mu v_R^2}\right]^{-1/2} r^{-2}dr$$

12.2 Consider the collision

$$m_1 + m_2(\text{at rest}) \rightarrow m_1 + m_2(\text{in motion})$$

In the laboratory frame the collision is described by Fig. 12.2a and in the center-of-mass frame (where the center of mass is always at rest) it is described by Fig. 12.2b.

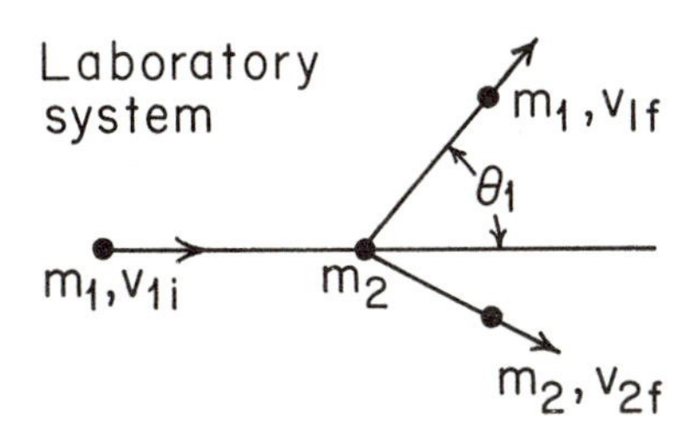

Fig. 12.2a

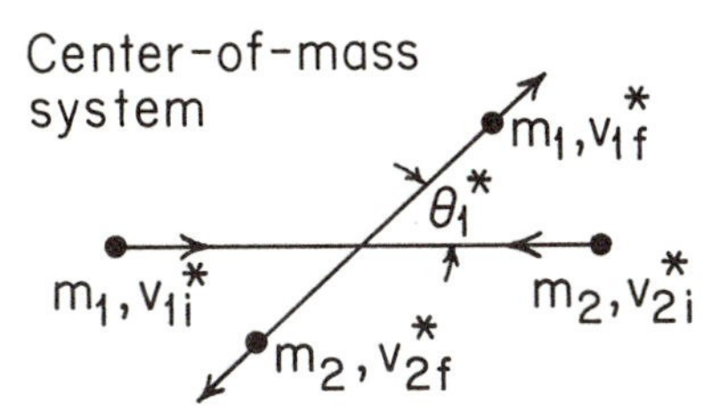

Fig. 12.2b

Show that the velocity of particle 1 in the laboratory system is related to its velocity in the center-of-mass system by the equation

$$\mathbf{v}_1 = \mathbf{v}_1^* + \mathbf{V}_c$$

where

$$\mathbf{v}_1^* = \frac{m_2}{(m_1 + m_2)}\mathbf{v}_R \qquad \text{and} \qquad \mathbf{V}_c = \frac{m_1}{(m_1 + m_2)}\mathbf{v}_{1i}$$

Here $\mathbf{v}_R = \mathbf{v}_1 - \mathbf{v}_2$ is the relative velocity and the labels i and f denote initial and final, respectively.

Solution: Ref. SC1,111; BR1,50

The center of mass is given by

$$\mathbf{R} = \frac{m_1\mathbf{r}_1 + m_2\mathbf{r}_2}{M} \qquad \text{where} \quad M = m_1 + m_2$$

and the relative position vector is $\mathbf{r} = \mathbf{r}_1 - \mathbf{r}_2$. Thus, taking derivatives and rearranging we have

$$\dot{\mathbf{r}}_1 = \mathbf{v}_1 = \mathbf{V}_c + \frac{m_2}{M}\mathbf{v}_R = \mathbf{V}_c + \mathbf{v}_1 \tag{2a}$$

$$\dot{\mathbf{r}}_2 = \mathbf{v}_2 = \mathbf{V}_c - \frac{m_1}{M}\mathbf{v}_R = \mathbf{V}_c - \mathbf{v}_2 \tag{2b}$$

where $\mathbf{V}_c = \dot{\mathbf{R}}$ and $\mathbf{v}_R = \dot{\mathbf{r}}$. The velocity of the center of mass in the laboratory system is obtained by considering the situation in which particle 2 is initially motionless. Then from Eq. (1)

$$\mathbf{V}_c = \dot{\mathbf{R}} = \frac{m_1}{M}\mathbf{v}_{1i} = \frac{m_1}{M}\mathbf{v}_{Ri} \tag{3}$$

12.3 If χ is the deflection angle for the reduced one-body problem (see Problem 12.1), show that

$$\tan\theta_1 = \frac{\sin\chi}{\cos\chi + (m_1 v_{Ri}/m_2 v_{Rf})}$$

for either elastic or inelastic collisions and

$$\tan\theta_1 = \frac{\sin\chi}{\cos\chi + (m_1/m_2)}$$

for elastic collisions only. The notation follows that in Problem 12.2. (Hint: First show that θ_1^* and χ are identical.)

Solution: Ref. Problem 12.2

From Problem 12.2 we have

$$\underset{\sim}{v}_1^* = \frac{m_2}{M}\underset{\sim}{v}_R$$

Thus, the velocity vector in the center-of-mass system is always parallel to the relative velocity vector and θ_1^* in Fig. 12.2b is equal to the deflection angle χ as specified in Problem 12.1. The pertinent vector relations are shown in Fig. 12.3.

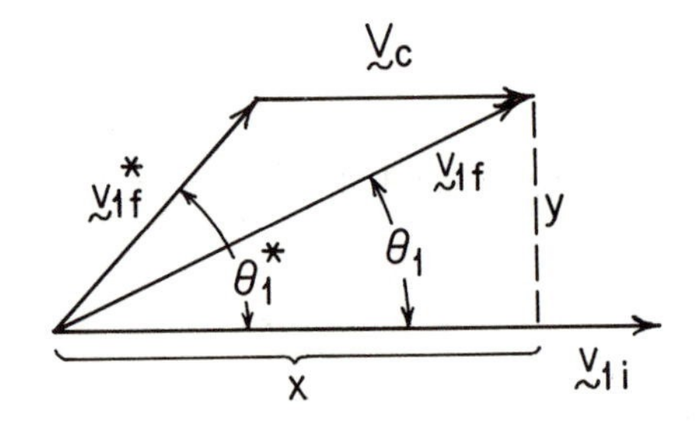

Fig. 12.3

Since particle 2 is initially at rest $\underset{\sim}{v}_{Ri} = \underset{\sim}{v}_{1i}$, which implies that $\underset{\sim}{v}_1^*$ is parallel to $\underset{\sim}{v}_1$. Thus the angle between $\underset{\sim}{v}_{1i}$ and $\underset{\sim}{v}_{1f}^*$ is θ_1^*. From the figure we can immediately write

$$\tan\theta_1 = \frac{y}{x} = \frac{v_{1f}^*\sin\chi}{v_{1f}^*\cos\chi + V_c} = \frac{\sin\chi}{\cos\chi + (V_c/v_{1f}^*)} \tag{1}$$

Since

$$|\underset{\sim}{V}_c| = V_c = \frac{m_1}{M}|\underset{\sim}{v}_{1i}| = \frac{m_1}{M}v_{1i} = \frac{m_1}{M}v_{Ri}$$

and

$$|\underset{\sim}{v}_{1f}^*| = v_{1f}^* = \frac{m_2}{M}|\underset{\sim}{v}_{Rf}| = \frac{m_2}{M}v_{Rf}$$

Eq. (1) can be written as

$$\tan\theta_1 = \frac{\sin\chi}{\cos\chi + (m_1 v_{Ri}/m_2 v_{Rf})} \tag{2}$$

For elastic collisions $v_{Ri} = v_{Rf}$, thus

$$\tan\theta_1 = \frac{\sin\chi}{\cos\chi + (m_1/m_2)}$$

12.4 The differential cross-section in the laboratory coordinate system is denoted by $\sigma_d(\theta,\phi)$ and that in the center-of-mass system is $\sigma_d^*(\theta^*,\phi^*)$. From the discussion in the Introduction we have $d\sigma = \sigma_d(\theta,\phi)d\Omega = \sigma_d^*(\theta^*,\phi^*)d\Omega^*$ where $d\Omega = \sin\theta d\theta d\phi$, $d\Omega^* = \sin\theta^* d\theta^* d\phi^*$, and the asterisks refer to the center-of-mass system.

(a) Show that

$$\boxed{\sigma_d(\theta,\phi) = \frac{(1 + \gamma^2 + 2\gamma\cos\theta^*)^{3/2}}{|1 + \gamma\cos\theta|} \sigma_d^*(\theta^*,\phi^*)} \tag{1}$$

where $\gamma = (m_1 v_{Ri}/m_2 v_{Rf})$. (Hint: First obtain an equation for $\cos\theta$ using the results of Problem 12.3 and then derive $\sin\theta d\theta$. Note that θ and θ^* here are equal to θ_1 and χ of Problem 12.3, respectively.)

(b) In the special case that $\gamma = 1$, <u>i.e.</u> elastic scattering for equal masses, show that

$$\sigma_d(\theta,\phi) = 4\cos\theta\sigma_d^*(2\theta,\phi) \tag{2}$$

What is the range of scattering angles in the laboratory system?

<u>Solution</u>: Ref. SC1,111; BR1,50; Problem 12.3

(a) In the present notation Eq. (2) of Problem 12.3 becomes

$$\tan\theta = \frac{\sin\theta^*}{\cos\theta^* + \gamma}$$

The construction below suggests the following expression for $\cos\theta$:

$$\cos\theta = \frac{\cos\theta^* + \gamma}{(1 + 2\gamma\cos\theta^* + \gamma^2)^{1/2}} \tag{3}$$

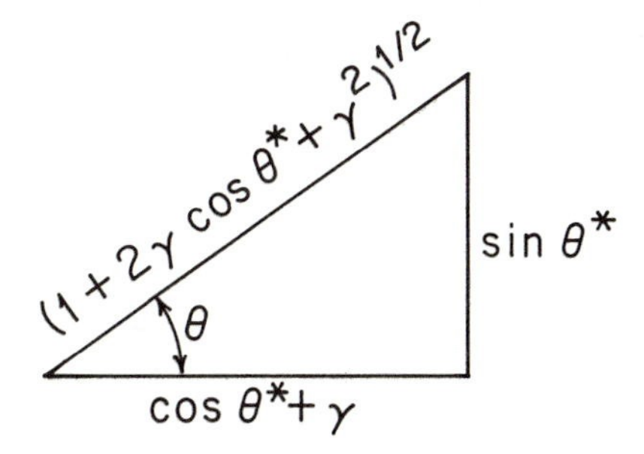

Differentiation of Eq. (3) then gives

$$\sin\theta d\theta = \frac{\sin\theta^* d\theta^* (1 + \gamma\cos\theta)}{(1 + 2\gamma\cos\theta^* + \gamma^2)^{3/2}} \tag{4}$$

From the differential cross-section equation we have

$$\sigma_d(\theta,\phi)\sin\theta d\theta d\phi = \sigma_d^*(\theta^*,\phi^*)\sin\theta^* d\theta^* d\phi^* \tag{5}$$

where $\phi = \phi^*$ and $d\phi = d\phi^*$. The substitution of Eq. (4) into Eq. (5) then gives

$$\sigma_d(\theta,\phi) = \frac{(1 + 2\gamma\cos\theta^* + \gamma^2)^{3/2}}{|1 + \gamma\cos\theta^*|} \sigma_d^*(\theta^*,\phi^*) \tag{1}$$

The absolute magnitude of the denominator is required since θ^* ranges from 0 to π.

(b) If $\gamma = 1$, one can show that

$$\tan\theta = \frac{\sin\theta^*}{\cos\theta^* + 1} = \frac{\sin 2\theta}{\cos 2\theta + 1}$$

or $\theta^* = 2\theta$. Equation (1) then becomes

$$\sigma_d(\theta,\phi) = \frac{(1 + 2\cos 2\theta + 1)^{3/2}}{|1 + \cos 2\theta|} \sigma_d^*(2\theta,\phi) \tag{2}$$

But $\cos 2\theta = 2\cos^2\theta - 1$, so Eq. (2) can be written as

$$\sigma_d(\theta,\phi) = \frac{(4\cos^2\theta)^{3/2}}{2\cos^2\theta} \sigma_d^*(2\theta,\phi) = 4\cos\theta\sigma_d^*(2\theta,\phi) \tag{3}$$

The full range of values for θ^* is 0 to π. For $\gamma = 1$ it is clear that θ cannot exceed $\pi/2$. Thus, there is no back-scattering.

12.5 If $V(r)$ is a potential energy function which decreases in magnitude faster than r^{-1} as $r \to \infty$, the solution of the Schrödinger equation

$$\left[-\frac{\hbar^2}{2\mu}\nabla^2 + V(r) - E\right]\psi(\underset{\sim}{r}) = 0$$

for a scattered wave in the limit that $kr \to \infty$ can be written as

$$\psi_{sc}(\underset{\sim}{r}) \propto e^{i\underset{\sim}{k}\cdot\underset{\sim}{r}} + f(\theta,\phi)e^{ikr}/r \tag{1}$$

Here it is assumed that the incident wave has the form $\psi_{in}(\underset{\sim}{r}) \propto \exp(i\underset{\sim}{k}\cdot\underset{\sim}{r})$. In the event that forward scattering is not considered and the incident beam is collimated, Eq. (1) can be replaced with

$$\psi_{sc}(r) \propto f(\theta,\phi)e^{ikr}/r \tag{2}$$

Show that under these conditions the differential cross-section is given by

$$\boxed{\sigma_d = |f(\theta)|^2}$$

Solution: Ref. Problems 5.6 and 4.2; BR1,4

From the Introduction we have

$$\sigma_d = \frac{I(\theta,\phi)}{I} = \lim_{r\to\infty} (|\underset{\sim}{j}_{sc}|r^2)/|\underset{\sim}{j}_{in}| \tag{3}$$

where

$$\underset{\sim}{j} = \frac{i\hbar}{2\mu}(\psi\underset{\sim}{\nabla}\psi^* - \psi^*\underset{\sim}{\nabla}\psi) \tag{4}$$

and (Problem 5.6)

$$\nabla = \hat{e}_r \frac{\partial}{\partial r} + \hat{e}_\theta \frac{1}{r}\frac{\partial}{\partial \theta} + \hat{e}_\phi \frac{1}{r\sin\theta}\frac{\partial}{\partial \phi}$$

Substituting $\psi_{sc}(r)$ from Eq. (2) into Eq. (4) gives

$$j_{sc} \propto \frac{\hbar k}{\mu}\frac{|f(\theta)|^2}{r^2}\hat{e}_r + \frac{i\hbar}{2\mu r^3}\left(f\frac{df^*}{d\theta} - f^*\frac{df}{d\theta}\right)\hat{e}_\theta \tag{5}$$

where we have used the fact that there is no ϕ dependence for a central potential. Using the plane wave expression for $\psi_{in}(r)$ with Eq. (4) we find:

$$j_{in} \propto \frac{\hbar k}{\mu}(\cos\theta\hat{e}_r - \sin\theta\hat{e}_\theta)$$

which for $\theta = 0$ is equivalent to the result of Problem 4.2. Then Eq. (3) gives

$$\sigma_d = |f(\theta)|^2$$

The second term in $j_{sc}r^2$ makes no contribution since it vanishes as $r \to \infty$.

12.6 In Problem 12.5 it was shown that $\sigma_d = |f(\theta,\phi)|^2$ under the assumption that there is no interference between the incident and the scattered waves. Show that this equation is valid for σ_d even when interference is included by using

$$\psi_{sc}(r) \propto e^{ik\cdot r} + f(\theta,\phi)\frac{e^{ikr}}{r}$$

(Hint: Assume that the detector intercepts a finite solid angle $\Delta\Omega$ where the scattering angle varies from θ to $\theta + \Delta\theta$ and that $\theta \neq 0$.)

Solution: Ref. FL1,208

Following the procedure of Problem 12.5 we obtain

$$j_{sc} \propto j_r\hat{e}_r + j_\theta\hat{e}_\theta$$

where

$$j_r = \frac{\hbar k}{\mu}\left(\cos\theta + \frac{|f|^2}{r^2}\right) - \frac{i\hbar}{2\mu}\left\{f\frac{e^{ik(r-z)}}{r}\left[ik(\cos\theta + 1) - \frac{1}{r}\right] + f^*\frac{e^{ik(z-r)}}{r}\left[ik(\cos\theta + 1) + \frac{1}{r}\right]\right\} \tag{1}$$

and

$$j_\theta = -\frac{\hbar k}{\mu}\sin\theta + \frac{i\hbar}{2\mu}\left\{\frac{e^{ik(r-z)}}{r}\left(ikf\sin\theta - \frac{f'}{r}\right) + \frac{e^{ik(z-r)}}{r}\left[\frac{(f^*)'}{r} + ikf^*\sin\theta\right] + \frac{1}{r^3}[f(f^*)' - (f^*)f']\right\} \tag{2}$$

The new terms appearing here which were not present in Eq. (5) of Problem 12.5 are of the form

$$F_k^{\pm}(\Omega,r)e^{\pm ik(r-z)} \tag{3}$$

where $F_k^{\pm}(\Omega,r)$ represents the slowly varying functions of Ω from Eqs. (1) and (2) which constitute the coefficients of the exponential terms. For a detector of finite size and a finite scattering angle the contribution of such terms will vanish because of the oscillatory nature of the complex exponential terms, i.e.

$$\lim_{r\to\infty} F^{\pm}(\Omega,r)\,\frac{\int e^{\pm ik(r-z)}d\Omega}{\int d\Omega} = 0; \quad \text{if } \theta \neq 0$$

12.7 For a two-body problem involving only central forces, i.e. a spherically symmetric potential, the Schrödinger equation has the same form as that for a hydrogen-like atom. Thus

$$\left[-\frac{\hbar^2}{2\mu}\nabla^2 + V(r) - E\right]\psi_{\underset{\sim}{k}}(\underset{\sim}{r}) = 0 \tag{1}$$

In scattering problems it is convenient to expand $\psi_{\underset{\sim}{k}}(\underset{\sim}{r})$ in terms of Legendre functions as follows

$$\psi_{\underset{\sim}{k}}(\underset{\sim}{r}) = \sum_{\ell=0} r^{-1}g_\ell(r)P_\ell(\cos\theta) \tag{2}$$

Use this partial wave expansion to show that the coefficients g_ℓ satisfy the equation

$$\left[\frac{d^2}{dr^2} - \frac{\ell(\ell+1)}{r^2} - U(r) + k^2\right]g_\ell(r) = 0 \tag{3}$$

where $U(r) = 2\mu V(r)/\hbar^2$ and $k^2 = 2\mu E/\hbar^2$.

Solution: Ref. BR1,6; SC1,117; Problems 7.2,7.4

From Problem 5.5 we have

$$\nabla^2 = \frac{1}{r^2}\left[\frac{\partial}{\partial r}r^2\frac{\partial}{\partial r} - \hat{L}^2\right]$$

Using this expression and substituting Eq. (2) into Eq. (1) gives

$$\sum_{\ell=0}^{\infty}\left[\frac{1}{r^2}\left(\frac{\partial}{\partial r}r^2\frac{\partial}{\partial r} - \hat{L}^2\right) - U(r) + k^2\right]g_\ell P_\ell(\cos\theta)r^{-1} = 0$$

Then using the relation $\hat{L}^2P_\ell(\cos\theta) = \ell(\ell+1)P_\ell(\cos\theta)$ and following the steps given in part (a) of Problem 7.4 with $P(r)$ replaced by $g_\ell(r)$ yields

$$\sum_\ell \frac{P_\ell(\cos\theta)}{r}\left[\frac{d^2}{dr^2} - \frac{\ell(\ell+1)}{r^2} - U(r) + k^2\right]g_\ell(r) = 0 \tag{4}$$

Multiplication by $P_{\ell'}(\cos\theta)$ and the use of the orthogonality condition

$$\int_{-1}^{+1} P_{\ell'}(x)P_{\ell}(x)dx = \frac{2}{2\ell + 1}\delta_{\ell\ell'}$$

reduces Eq. (4) to an infinite set of differential equations for $g_{\ell}(r)$:

$$\left[\frac{d^2}{dr^2} - \frac{\ell(\ell + 1)}{r^2} - U(r) + k^2\right] g_{\ell}(r) = 0$$

12.8 For elastic scattering from a spherically symmetric potential (see the Introduction) the scattered wave can be written as

$$\psi_{sc}(\underset{\sim}{r}) = N[\psi_{in}(\underset{\sim}{r}) + f(\theta)e^{ikr}/r] \tag{1}$$

in the limit that $kr \to \infty$ with N equal to a constant. Expand $\psi_{sc}(\underset{\sim}{r})$ in terms of Legendre functions and make use of the asymptotic expansion of $j_{\ell}(kr)$

$$j_{\ell}(kr) \xrightarrow[r\to\infty]{} \frac{1}{kr}\cos\left[kr - (\ell + 1)\frac{\pi}{2}\right] = \frac{1}{kr}\sin\left(kr - \frac{\ell\pi}{2}\right) \tag{2}$$

to show that the scattering amplitude $f(\theta)$ can be written in a partial wave expansion as

$$\boxed{f(\theta) = \frac{1}{2ik}\sum_{\ell=0}^{\infty}(2\ell + 1)(e^{i2\delta_{\ell}} - 1)P_{\ell}(\cos\theta)} \tag{3}$$

where δ_{ℓ} is the phase shift of the ℓth partial wave. (Hint: Expand $\psi_{in}(\underset{\sim}{r})$ as shown in Problem 12.24, and expand $\psi_{sc}(\underset{\sim}{r})$ allowing for a phase shift δ_{ℓ} for the ℓth wave. Then use Eq. (1) to solve for $f(\theta)$.)

Solution: Ref. FA,18; BR1,9; SC1,118; ME,231

As suggested we write

$$\begin{aligned}\psi_{sc}(\underset{\sim}{r}) &= N\sum_{\ell=0}^{\infty} A_{\ell}(2\ell + 1)\frac{i^{\ell}}{kr}\sin(kr - \frac{\ell\pi}{2} + \delta_{\ell})P_{\ell}(\cos\theta) \\ &= N\left\{\sum_{\ell=0}^{\infty}(2\ell + 1)\frac{i^{\ell}}{kr}\sin\left(kr - \frac{\ell\pi}{2}\right)P_{\ell}(\cos\theta) + f(\theta)\frac{e^{ikr}}{r}\right\}\end{aligned}$$

Then

$$f(\theta) = re^{-ikr}\sum_{\ell=0}^{\infty}(2\ell + 1)\frac{i^{\ell}}{kr}\left[A_{\ell}\sin(kr - \frac{\ell\pi}{2} + \delta_{\ell}) - \sin\left(kr - \frac{\ell\pi}{2}\right)\right]P_{\ell}(\cos\theta) \tag{4}$$

and writing the sine functions in exponential form permits the terms depending on r to be separated. Thus

$$f(\theta) = \sum_{\ell=0}^{\infty} \frac{(2\ell + 1)}{2ik} i^{\ell} \left\{ A_{\ell} \left[e^{i(-\frac{\ell\pi}{2} + \delta_{\ell})} - e^{i(-2kr + \frac{\ell\pi}{2} - \delta_{\ell})} \right] - \left[e^{-i\frac{\ell\pi}{2}} - e^{i(-2kr + \frac{\ell\pi}{2})} \right] \right\} P_{\ell}(\cos\theta) \tag{5}$$

Since $f(\theta)$ is independent of r, the terms on the r.h.s. which depend on r must vanish. This gives

$$\sum_{\ell=0}^{\infty} \frac{(2\ell + 1)}{2ik} i^{\ell} e^{i(\frac{\ell\pi}{2} - 2kr)} (A_{\ell} e^{-i\delta_{\ell}} - 1) P_{\ell}(\cos\theta) = 0 \tag{6}$$

and since Eq. (6) must hold for all values of θ we conclude that

$$A_{\ell} e^{-i\delta_{\ell}} - 1 = 0 \qquad \text{or} \qquad A_{\ell} = e^{+i\delta_{\ell}}$$

Inserting this expression for A_{ℓ} into Eq. (5) gives

$$f(\theta) = \sum_{\ell=0}^{\infty} \frac{(2\ell + 1)}{2ik} i^{\ell} e^{-i\frac{\ell\pi}{2}} (e^{2i\delta_{\ell}} - 1) P_{\ell}(\cos\theta)$$

The derivation of Eq. (3) is completed by using the identity

$$\exp\left(-\frac{i\ell\pi}{2}\right) = (-i)^{\ell}$$

Equation (3) is particularly valuable for low energy particles where convergence is rapid. To understand this recall that $\mathcal{L} = \sqrt{\ell(\ell + 1)}\,\hbar$, $p = mv = \hbar(2\pi/\lambda) = \hbar k$, and classically $\mathcal{L} = bp$ where b is the impact parameter. For a scattering potential which vanishes beyond a radius r_o it is clear that particles having $b > r_o$ will not be scattered, thus scattering will be small for

$$b = \frac{\mathcal{L}}{p} > r_o \qquad \text{or} \qquad \ell > kr_o$$

12.9 Show that the total elastic scattering cross-section σ is given by

$$\boxed{\sigma = \frac{4\pi}{k^2} \sum_{\ell=0}^{\infty} (2\ell + 1) \sin^2\delta_{\ell}} \tag{1}$$

Solution: Ref. Problem 12.8 and the Introduction

From the definitions we write

$$\sigma = 2\pi \int_o^{\pi} \sigma_d(\theta) \sin\theta d\theta = 2\pi \int_o^{\pi} |f(\theta)|^2 \sin\theta d\theta$$

and using Eq. (3) of Problem 12.8 for $f(\theta)$ this becomes

$$\sigma = \frac{\pi}{2k^2} \sum_{\ell=0,\ell'=0} \sum (2\ell + 1)(2\ell' + 1)(e^{-2i\delta_\ell} - 1)(e^{+2i\delta_{\ell'}} - 1) \int_0^\pi P_\ell(\cos\theta)P_{\ell'}(\cos\theta)\sin\theta d\theta$$

$$= \frac{\pi}{k^2} \sum_\ell (2\ell + 1)[2 - (e^{2i\delta_\ell} + e^{-2i\delta_\ell})] = \frac{2\pi}{k^2} \sum_\ell (2\ell + 1)(1 - \cos 2\delta_\ell) \tag{2}$$

Equation (2) is, of course, equivalent to Eq. (1). We have used the orthogonality condition from Appendix 4 in this derivation.

12.10 When Schrödinger's equation for two-body scattering can be written in the form of Eq. (9) of the Introduction, the general solution has the form

$$g_{\ell,k}(r) = krB_\ell[j_\ell(kr) - K_\ell n_\ell(kr)] \tag{1}$$

Here we assume that $U(r)$ is chosen so that $g_{\ell,k}(r)$ from Eq. (1) satisfies the equation

$$\left[\frac{d^2}{dr^2} - \frac{\ell(\ell + 1)}{r^2} - U(r) + k^2\right] g_{\ell,k}(r) = 0 \tag{2}$$

in the limits that $kr \to 0$ and $kr \to \infty$. When kr approaches zero only the regular solution $krj_\ell(kr)$ is acceptable since $\psi_{\tilde{k}}(\tilde{r})$ must remain finite; however, for large values of kr both terms in Eq. (1) may be necessary, i.e. the exact solution to Eq. (2) approaches the form of Eq. (1) with $K_\ell = 0$ for small values of kr and with $K_\ell \neq 0$ for large values of kr. Thus the phase shift δ_ℓ determines the value of K_ℓ. Obtain expressions for B_ℓ in terms of K_ℓ and K_ℓ in terms of δ_ℓ by comparing the asymptotic form of Eq. (1) with the expansions for $\psi_{sc}(\tilde{r})$ and $f(\theta)$ obtained in Problem 12.8.

Solution: Ref. BR1,6; Problem 12.8; Introduction; SC1,85

First write

$$\psi_{sc}(\tilde{r}) = e^{ikr\cos\theta} + f(\theta)e^{ikr}/r \tag{3}$$

with

$$e^{ikr\cos\theta} = \sum_{\ell=0}^{\infty} i^\ell (2\ell + 1)j_\ell(kr)P_\ell(\cos\theta)$$

and

$$f(\theta) = \frac{1}{2ik} \sum_{\ell=0}^{\infty} (2\ell + 1)(e^{i2\delta_\ell} - 1)P_\ell(\cos\theta)$$

Then equate the ℓth term in the expansion of Eq. (3) with Eq. (1) and make use of the asymptotic expressions for $j_\ell(kr)$ and $n_\ell(kr)$ to obtain

$$i^\ell \frac{(2\ell + 1)}{kr} \sin\left(kr - \frac{\ell\pi}{2}\right) + \frac{e^{ikr}}{r} \frac{(2\ell + 1)}{2ik} (e^{i2\delta_\ell} - 1) =$$

$$\frac{B_\ell}{r}\left[\sin\left(kr - \frac{\ell\pi}{2}\right) + K_\ell \cos\left(kr - \frac{\ell\pi}{2}\right)\right] \tag{4}$$

where $P_\ell(\cos\theta)$ has been divided out. B_ℓ can be obtained most readily by expressing the trigonometric functions in exponential form. When this is done, the coefficients can be collected as follows.

<u>coefficient of e^{ikr}:</u>

$$(2\ell + 1)e^{i2\delta_\ell} - B_\ell k e^{-i\ell\pi/2}(1 + iK_\ell) = 0 \tag{5}$$

Here we have used the identity $i^\ell e^{-i\ell\pi/2} = 1$.

<u>coefficient of e^{-ikr}:</u>

$$-i^\ell(2\ell + 1) + B_\ell k(1 - iK_\ell) = 0 \tag{6}$$

Equation (6) then gives

$$B_\ell = \frac{i^\ell(2\ell + 1)}{k(1 - iK_\ell)}$$

and the substitution of this expression into Eq. (5) gives

$$(2\ell + 1)\left\{e^{i2\delta_\ell} - \left(\frac{1 + iK_\ell}{1 - iK_\ell}\right)\right\} = 0$$

or

$$(1 - iK_\ell)e^{i2\delta_\ell} = 1 + iK_\ell$$

Solving for K_ℓ we find

$$K_\ell = \frac{(e^{i\delta_\ell} - e^{-i\delta_\ell})}{i(e^{i\delta_\ell} + e^{-i\delta_\ell})} = \tan\delta_\ell$$

12.11 A low energy particle is scattered elastically from the potential

$$V(r) = -V_o, \qquad 0 \lessgtr r \lessgtr a$$
$$= 0, \qquad r > a$$

(a) Assume that $\ell = 0$ (s-wave scattering) and show that

$$\tan\delta_o = \frac{k_2\tan(k_1a) - k_1\tan k_2a}{k_1 + k_2\tan(k_2a)\tan(k_1a)} \tag{1}$$

where

$$k_1^2 = \frac{2\mu}{\hbar^2}(E + V_o) \qquad \text{and} \qquad k_2^2 = \frac{2\mu}{\hbar^2}E$$

(Hint: Match the partial wave solution $r^{-1}g_o(r)$ and its derivative for the two regions at $r = a$.)

(b) For $k_2a \ll 1$ show that

$$\sigma_o = 4\pi a^2\left[1 - \frac{\tan(k_1a)}{k_1a}\right]^2 \tag{2}$$

where σ_o is the total cross-section for scattering for $\ell = 0$.

Solution: Ref. Problems 12.7 and 12.10; DA2,389; Introduction

(a) From Problem 12.10 we have for region 1 ($0 \lesssim r \lesssim a$)

$$\left(\frac{d^2}{dr^2} + k_1^2\right) g_o^{(1)} = 0; \qquad g_o^{(1)} = A_o k_1 r j_o(kr) = A_o \sin(k_1 r)$$

and for region 2 ($r > a$)

$$\left(\frac{d^2}{dr^2} + k_2^2\right) g_o^{(2)} = 0; \quad g_o^{(2)} = k_2 r j_o(k_2 r) - (\tan\delta_o) k_2 r n_o(k_2 r) = \sin(k_2 r) + \tan\delta_o \cos(k_2 r)$$

The continuity conditions for $\psi(r)$ and $\psi'(r)$ require that

$$g_o^{(1)}(r)/r = g_o^{(2)}(r)/r$$

and

$$\frac{d}{dr}\left(\frac{g_o^{(1)}(r)}{r}\right) = \frac{d}{dr}\left(\frac{g_o^{(2)}(r)}{r}\right)$$

at $r = a$. These conditions can be combined to give

$$\frac{1}{g_o^{(1)}(a)} \left[\frac{d}{dr}\left(\frac{g_o^{(1)}(r)}{r}\right)\right]_{r=a} = \frac{1}{g_o^{(2)}(a)} \left[\frac{d}{dr}\left(\frac{g_o^{(2)}(r)}{r}\right)\right]_{r=a}$$

Then substituting the expressions for $g_o^{(1)}(r)$ and $g_o^{(2)}(r)$ and differentiating gives:

$$-1 + k_1 a \frac{\cos(k_1 a)}{\sin(k_1 a)} = -1 + k_2 a \left[\frac{\cos(k_2 a) - \tan\delta_o \sin(k_2 a)}{\sin(k_2 a) + \tan\delta_o \cos(k_2 a)}\right]$$

Solving for $\tan\delta_o$ we obtain:

$$\tan\delta_o = \left[\frac{k_2 \tan(k_1 a) - k_1 \tan(k_2 a)}{k_1 + k_2 \tan(k_1 a) \tan(k_2 a)}\right]$$

(b) For $k_2 a \ll 1$ Eq. (1) becomes

$$\tan\delta_o \simeq k_2 a \left[\frac{\tan(k_1 a)}{k_1 a} - 1\right] \tag{3}$$

and with

$$\frac{k_2}{k_1} \tan(k_1 a) \ll 1$$

Eq. (3) becomes $\tan\delta_o \simeq \sin\delta_o \simeq \delta_o$. From Problem 12.9

$$\sigma_o = \frac{4\pi}{k_2^2} \sin^2\delta_o \simeq 4\pi a^2 \left[\frac{\tan(k_1 a)}{k_1 a} - 1\right]^2$$

Thus σ_o has the form $4\pi a_s^2$ where a_s is called the scattering length. Notice that $\sigma_o = 0$ for $\tan(k_1 a) = k_1 a$.

12.12 Consider the elastic scattering of a low energy particle from the "hard sphere" potential:

$$V(r) = \infty, \quad r \lesssim a$$
$$= 0, \quad r > a$$

(a) Derive an expression for $\tan\delta_\ell$ and show that for $\ell = 0$,

$$\tan\delta_o = -\tan ka$$

where $k^2 = 2\mu E/\hbar^2$.

(b) Show that:

$$\lim_{k\to 0} \sigma_o = 4\pi a^2$$

(c) Obtain an expression for the s-wave contribution to the scattering amplitude f(0) and verify the optical theorem (Problem 12.25) for $ka \ll 1$.

Solution: Ref. Problem 12.11

(a) We proceed as in Problem 12.11. In region 1 ($r \lesssim a$) the wave function vanishes, i.e. $g_o^{(1)} = 0$, while in region 2 ($r > a$)

$$g_o^{(2)} = \sin(kr) + \tan\delta_o \cos(kr)$$

Thus continuity requires that $g_o^{(2)}(a) = 0$ and

$$\tan\delta_o = \frac{\sin(ka)}{\cos(ka)} = -\tan ka$$

(b) Low energy implies that $ka \ll 1$. Thus $\sin\delta_o \simeq \tan\delta_o \simeq ka$ and from Problem 12.11

$$\sigma_o = \frac{4\pi}{k^2}\sin^2\delta_o \simeq \frac{4\pi}{k^2}(ka)^2 = 4\pi a^2$$

For very low energies only the s-wave ($\ell = 0$) is scattered. For this situation the total cross-section $\sigma = \sigma_o$.

(c) From Problem 12.10 with $\ell = 0$ we find using $\delta_o \simeq ka$:

$$f(0) = \frac{1}{2ik}(e^{i2\delta_o} - 1) = \frac{1}{2ik}\left[i2\delta_o + \frac{(i2\delta_o)^2}{2} + \cdots\right] = a + ika^2 + \cdots$$

According to the optical theorem

$$\sigma_o = \frac{4\pi}{k}\,\mathrm{Im} f(0) = \frac{4\pi}{k}(ka^2) = 4\pi a^2$$

This, of course, agrees with part (b).

12.13 Scattering of low energy particles from spherical barriers of the form

$$V(r) = \pm V_o \qquad 0 \lesssim r \lesssim a$$
$$= 0 \qquad r > a$$

was considered in Problems 12.11 and 12.26. In the limit that $E \to 0$ the cross-section for scattering from the attractive potential ($-V_o$) was found to be

$$\sigma_o = 4\pi a^2\left[1 - \frac{\tan(ka)}{ka}\right]^2 \tag{1}$$

while for the repulsive potential ($+V_o$)

$$\sigma_o = 4\pi a^2\left[1 - \frac{\tanh(ka)}{ka}\right]^2 \tag{2}$$

where $k = \sqrt{2\mu V_o}/\hbar$. Plot $\sigma_o/(\pi a^2)$ in the range $0 \lessapprox ka \lessapprox 10$ for each case.

Solution: Ref. MM2,30

See Figure 12.13.

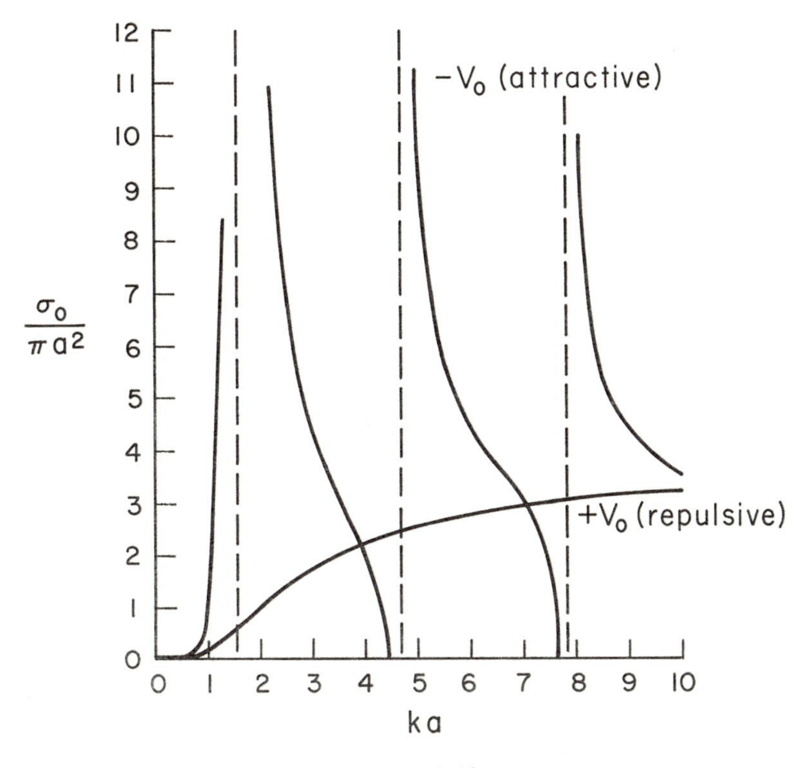

Fig. 12.13

12.14 The Schrödinger equation for scattering by the central potential V(r) can be written as

$$(\nabla^2 + k^2)\psi(\underset{\sim}{r}) = U(r)\psi(\underset{\sim}{r}) \tag{1}$$

where $k^2 = 2\mu E/\hbar^2$ and $U(r) = 2\mu V(r)/\hbar^2$. It is convenient to consider the related equation

$$(\nabla^2 + k^2)G(\underset{\sim}{r},\underset{\sim}{r}_o) = \delta(\underset{\sim}{r} - \underset{\sim}{r}_o) \tag{2}$$

Verify that

$$G(\underset{\sim}{r},\underset{\sim}{r}_o) = -\frac{\exp(ik|\underset{\sim}{r} - \underset{\sim}{r}_o|)}{4\pi|\underset{\sim}{r} - \underset{\sim}{r}_o|} \tag{3}$$

is a solution of Eq. (2). The function $G(\underset{\sim}{r},\underset{\sim}{r}_o)$ is called the free particle Green's function. (Hint: It is convenient to use the relation $\nabla^2(1/r) = -4\pi\delta(\underset{\sim}{r})$.)

Solution: Ref. FA,30; DA2,374; Appendix 5

Setting $\underset{\sim}{r}_o = 0$ for convenience and substituting Eq. (3) into Eq. (2) gives

$$(\nabla^2 + k^2)G(\underset{\sim}{r},0) = -\frac{1}{4\pi}(\nabla^2 + k^2)\frac{e^{ikr}}{r} \tag{4}$$

Then taking derivatives

$$\nabla^2 \frac{e^{ikr}}{r} = \frac{1}{r}\nabla^2 e^{ikr} + 2\left(\underset{\sim}{\nabla}\frac{1}{r}\right)\cdot(\underset{\sim}{\nabla}e^{ikr}) + e^{ikr}\nabla^2\left(\frac{1}{r}\right) = -k^2\frac{e^{ikr}}{r} + e^{ikr}\nabla^2\left(\frac{1}{r}\right)$$

where we have used

$$\nabla^2 = \frac{1}{r^2}\frac{\partial}{\partial r}r^2\frac{\partial}{\partial r} - \frac{1}{r^2}\hat{L}^2$$

and

$$\underset{\sim}{\nabla} = \hat{e}_r\frac{\partial}{\partial r} + \hat{e}_\theta\frac{1}{r}\frac{\partial}{\partial\theta} + \hat{e}_\phi\frac{1}{r\sin\theta}\frac{\partial}{\partial\phi}$$

from Appendix 5. Since $\nabla^2(1/r) = -4\pi\delta(\underset{\sim}{r})$, Eq. (4) gives

$$(\nabla^2 + k^2)G(\underset{\sim}{r},0) = -\frac{1}{4\pi}e^{ikr}[-4\pi\delta(\underset{\sim}{r})] = \delta(\underset{\sim}{r})$$

as required.

12.15 Show that

$$\psi(\underset{\sim}{r}) = e^{ikz} + \int G(\underset{\sim}{r},\underset{\sim}{r}_o)U(\underset{\sim}{r}_o)\psi(\underset{\sim}{r}_o)d\underset{\sim}{r}_o \tag{1}$$

satisfies the scattering equation of Problem 12.14:

$$(\nabla^2 + k^2)\psi(\underset{\sim}{r}) = U(\underset{\sim}{r})\psi(\underset{\sim}{r}) \tag{2}$$

when $G(\underset{\sim}{r},\underset{\sim}{r}_o)$ is defined as in Eq. (2) of Problem 12.14.

Solution: Ref. FA,32

Substituting Eq. (1) into Eq. (2) gives

$$(\nabla^2 + k^2)\psi(\underset{\sim}{r}) = (-k^2 + k^2) + \int(\nabla^2 + k^2)G(\underset{\sim}{r},\underset{\sim}{r}_o)U(\underset{\sim}{r}_o)\psi(\underset{\sim}{r}_o)d\underset{\sim}{r}_o$$

$$= \int\delta(\underset{\sim}{r} - \underset{\sim}{r}_o)U(\underset{\sim}{r}_o)\psi(\underset{\sim}{r}_o)d\underset{\sim}{r}_o = U(\underset{\sim}{r})\psi(\underset{\sim}{r})$$

12.16 The general solution of the scattering equation can be written as

$$\psi(\mathbf{r}) = e^{ikz} + \int G(\mathbf{r},\mathbf{r}_o)U(\mathbf{r}_o)\psi(\mathbf{r}_o)d\mathbf{r}_o \tag{1}$$

while the boundary conditions require that:

$$\psi(\mathbf{r}) \xrightarrow[r\to\infty]{} e^{ikz} + f(\theta)\frac{e^{ikr}}{r} \tag{2}$$

Compare Eqs. (1) and (2) in the limit of large r to show that

$$f(\theta) = -\frac{1}{4\pi}\int e^{-i\mathbf{k}'\cdot\mathbf{r}'}U(\mathbf{r}')\psi(\mathbf{r}')d\mathbf{r}' \tag{3}$$

(Hint: Express $G(\mathbf{r},\mathbf{r}_o)$ from Problem 12.14 in terms of $\mathbf{r}\cdot\mathbf{r}_o$ in the limit $r \to \infty$.)

Solution: Ref. FA,31; PC,268

From Problem 12.14 $G(\mathbf{r},\mathbf{r}_o) = -\exp(ik|\mathbf{r} - \mathbf{r}_o|)/4\pi|\mathbf{r} - \mathbf{r}_o|$. As suggested we write

$$|\mathbf{r} - \mathbf{r}_o| = \sqrt{r^2 - 2\mathbf{r}\cdot\mathbf{r}_o + r_o^2} = r\sqrt{1 - \frac{2\mathbf{r}\cdot\mathbf{r}_o}{r^2} + \frac{r_o^2}{r^2}}$$

$$\lim_{r\to\infty} |\mathbf{r} - \mathbf{r}_o| = r\left(1 - \frac{\hat{r}\cdot\mathbf{r}_o}{r}\right), \qquad \hat{r} = \mathbf{r}/r$$

Thus

$$G(\mathbf{r},\mathbf{r}_o) \xrightarrow[r\to\infty]{} -\frac{\exp(ikr - ik\hat{r}\cdot\mathbf{r}_o)}{4\pi r}$$

and Eq. (1) gives the asymptotic form

$$\psi(\mathbf{r}) = e^{ikz} - \frac{e^{ikr}}{r}\int \frac{e^{-ik\hat{r}\cdot\mathbf{r}_o}}{4\pi}U(\mathbf{r}_o)\psi(\mathbf{r}_o)d\mathbf{r}_o \tag{4}$$

Comparison of Eqs. (2) and (4) indicates that $f(\theta)$ is given by Eq. (3) if $\mathbf{k}' = k\hat{r}$ and the integration variable $\mathbf{r}_o$ is replaced with $\mathbf{r}'$.

12.17 In the first Born approximation

$$f(\theta) = -\frac{\mu}{2\pi\hbar^2}\int e^{i\mathbf{r}'\cdot(\mathbf{k}-\mathbf{k}')}V(\mathbf{r}')d\mathbf{r}' \tag{1}$$

What terms are being neglected in this approximation? (Hint: Attempt an iterated solution of the basic equation for $f(\theta)$.)

Solution: Ref. PC,272

From Problem 12.16

$$f(\theta) = -\frac{\mu}{2\pi\hbar^2}\int e^{-i\mathbf{k}'\cdot\mathbf{r}'}\psi(\mathbf{r}')V(\mathbf{r}')d\mathbf{r}' \tag{2}$$

A comparison with Eq. (1) shows that $\psi(\mathbf{r}')$ has been replaced with $\exp(i\mathbf{k}\cdot\mathbf{r}')$ in the first

Born approximation. The exact equation for $\psi(\underset{\sim}{r}')$ in the asymptotic limit is

$$\psi(\underset{\sim}{r}') = e^{i\underset{\sim}{k}\cdot\underset{\sim}{r}'} - \frac{e^{ikr}}{4\pi r}\int e^{-i\underset{\sim}{k}'\cdot\underset{\sim}{r}''}\frac{2\mu}{\hbar^2}V(\underset{\sim}{r}'')\psi(\underset{\sim}{r}'')d\underset{\sim}{r}'' \tag{3}$$

The first step in an interated solution of Eq. (2) is to substitute Eq. (3) for $\psi(\underset{\sim}{r}')$. Thus

$$f(\theta) = -\frac{\mu}{2\pi\hbar^2}\int e^{i\underset{\sim}{r}'\cdot(\underset{\sim}{k}-\underset{\sim}{k}')}V(\underset{\sim}{r}')d\underset{\sim}{r}'$$
$$+\left(\frac{\mu}{2\pi\hbar^2}\right)^2\frac{e^{ikr}}{r}\iint e^{-i\underset{\sim}{k}'\cdot(\underset{\sim}{r}'+\underset{\sim}{r}'')}V(\underset{\sim}{r}')V(\underset{\sim}{r}'')\psi(\underset{\sim}{r}'')d\underset{\sim}{r}'d\underset{\sim}{r}'' \tag{4}$$

In the next iteration $\psi(\underset{\sim}{r}'')$ is replaced by the r.h.s. of Eq. (3) with r" and r"' in place of r' and r", respectively. It is clear that the second term on the r.h.s. of Eq. (4) has been neglected in the first Born approximation. It is, in fact, assumed that the perturbation is small and that the final state can be approximated by a plane wave. Energy conservation requires that $k = k'$, and for large values of $k = |\underset{\sim}{k}|$ the complex factor in the neglected integral oscillates at much higher frequency than the other factors. This integral, therefore, makes a vanishingly small contribution for high energy collisions. The Born approximation is thus valid in the high energy region. It complements the method of partial waves which converges nicely for low energy collisions.

12.18 Show that the scattering amplitude $f(\theta)$ for a central potential $V(r)$ can be written in the Born approximation as

$$\boxed{f(\theta) = -\frac{2\mu}{\hbar^2}\int_0^\infty \frac{\sin(Kr')}{Kr'}V(r')r'^2dr'} \tag{1}$$

where $\underset{\sim}{K} = \underset{\sim}{k} - \underset{\sim}{k}'$ and $K = |\underset{\sim}{K}| = 2k\sin(\theta/2)$. Here $\hbar\underset{\sim}{K}$ is the <u>momentum transfer</u>.

<u>Solution</u>: Ref. AN3,402

From Problem 12.17

$$f(\theta) = -\frac{\mu}{2\pi\hbar^2}\int e^{i\underset{\sim}{r}'\cdot(\underset{\sim}{k}-\underset{\sim}{k}')}V(\underset{\sim}{r}')d\underset{\sim}{r}' \tag{2}$$

where $\underset{\sim}{k}$ is the initial wave vector and $\underset{\sim}{k}' = k\underset{\sim}{r}/r$, <u>i.e.</u> the lengths of $\underset{\sim}{k}$ and $\underset{\sim}{k}'$ are the same but their directions differ by the scattering angle θ. The length of $\underset{\sim}{K}$ is given by

$$K = \sqrt{\underset{\sim}{K}\cdot\underset{\sim}{K}} = (k^2 + k'^2 - 2\underset{\sim}{k}\cdot\underset{\sim}{k}')^{1/2} = [2k^2(1-\cos\theta)]^{1/2} = 2k\sin(\theta/2)$$

When written out in full Eq. (2) becomes:

$$f(\theta) = -\frac{\mu}{2\pi\hbar^2}\int_0^{2\pi}d\phi\int_0^\infty\int_0^\pi e^{i\underset{\sim}{r}'\cdot\underset{\sim}{K}}V(r')\sin\theta'd\theta'(r')^2dr' \tag{3}$$

Since V(r) is spherically symmetrical, any convenient orientation can be chosen for the coordinate system. We choose $\underset{\sim}{K} = K\hat{k}$ as the polar axis with θ' and ϕ' as the polar angles of $\underset{\sim}{r}'$. Thus

$$f(\theta) = -\frac{\mu}{2\pi\hbar^2}\,2\pi\int_0^\infty\int_0^\pi e^{iKr'\cos\theta'}\sin\theta' d\theta' V(r')r'^2 dr' = -\frac{2\mu}{\hbar^2 K}\int_0^\infty \sin(Kr')V(r')r'dr'$$

12.19 Determine the scattering amplitude $f(\theta)$ in the first Born approximation for the potential

$$V(r) = -Ae^{-br^2}$$

where A, b > 0. (Hint: Try integration by parts.)

Solution:

From Problem 12.18 we have

$$f(\theta) = -\frac{2\mu}{\hbar^2}\int_0^\infty \frac{\sin(Kr)}{K}V(r)rdr = \frac{2\mu A}{\hbar^2 K}\int_0^\infty \sin(Kr)e^{-br^2}rdr$$

Integration by parts with $u = \sin Kr$ and $dv = e^{-br^2}rdr$ gives

$$f(\theta) = \frac{2\mu A}{\hbar^2 K}\left[-\sin(Kr)\frac{e^{-br^2}}{2b}\Big|_0^\infty + \int_0^\infty \frac{e^{-br^2}}{2b}K\cos(Kr)dr\right]$$

and using integral tables we find

$$f(\theta) = \frac{2\mu A}{\hbar^2 K}\left[0 + \frac{K}{2b}\frac{\sqrt{\pi}}{2\sqrt{b}}e^{-K^2/4b}\right] = \frac{A\sqrt{\pi}\,\mu}{2\hbar^2 b^{3/2}}e^{-K^2/4b}$$

12.20 In Problem 12.8 the partial wave expansion for the scattered wave was shown to be:

$$\psi(\underset{\sim}{r}) = \sum_{\ell=0}^{\infty}(2\ell+1)\frac{i^\ell}{kr}e^{i\delta_\ell}\sin(kr - \frac{\ell\pi}{2} + \delta_\ell)P_\ell(\cos\theta) \quad (1)$$

Equation (1) represents a standing wave in which the total outgoing radial flux is equal to the incoming flux. For inelastic scattering the outgoing flux must be less than the incoming flux and it is appropriate to replace

$$e^{i\delta_\ell}\sin(kr - \frac{\ell\pi}{2} + \delta_\ell)$$

with

$$[\eta_\ell e^{i2\delta_\ell}e^{i(kr-\ell\pi/2)} - e^{-i(kr-\ell\pi/2)}](2i)^{-1} \quad (2)$$

where η_ℓ is a real number and $0 < \eta < 1$. Obtain a partial wave expansion for the scattering amplitude $f(\theta)$ in terms of η_ℓ.

Solution: Ref. Problem 12.8; BR1,26

If the incoming wave is substracted from Eq. (1), the result must be $f(\theta)\exp(ikr)/r$. Thus substituting Eq. (2) into Eq. (1) and substracting the asymptotic form of $\exp(ikz)$ gives

$$\psi(\underset{\sim}{r}) - e^{ikz} = \sum_{\ell=0}^{\infty} \frac{(2\ell + 1)}{kr} \frac{i^{\ell}}{2i} [\eta_{\ell} e^{i2\delta_{\ell}} e^{i(kr-\ell\pi/2)} - e^{-i(kr-\ell\pi/2)}$$

$$- e^{i(kr-\ell\pi/2)} + e^{-i(kr-\ell\pi/2)}] P_{\ell}(\cos\theta)$$

$$= \sum_{\ell=0}^{\infty} \frac{(2\ell + 1)}{2ik} i^{\ell} e^{-i\ell\pi/2} (\eta_{\ell} e^{i2\delta_{\ell}} - 1) P_{\ell}(\cos\theta) \frac{e^{ikr}}{r}$$

Thus

$$\boxed{f(\theta) = \frac{1}{2ik} \sum_{\ell=0}^{\infty} (2\ell + 1)(\eta_{\ell} e^{i2\delta_{\ell}} - 1) P_{\ell}(\cos\theta)} \tag{3}$$

The magnitudes of the wave vectors for the incident particles and the elastically scattered particles are equal to k. Any particle which is absorbed or scattered with a change in kinetic energy will be lost from the set of elastically scattered particles. Thus $f(\theta)$ in Eq. (3) is the scattering amplitude for elastically scattered particles only.

12.21 The wave function

$$\psi(\underset{\sim}{r}) = \sum_{\ell=0}^{\infty} \frac{(2\ell + 1)}{kr} \frac{i^{\ell}}{2i} [S_{\ell} e^{i(kr-\ell\pi/2)} - e^{-i(kr-\ell\pi/2)}] P_{\ell}(\cos\theta) \tag{1}$$

includes both incoming and outgoing waves. Here $S_{\ell} = \eta_{\ell} e^{i2\delta_{\ell}}$. The differential cross-section for inelastic scattering is defined by

$$(\sigma_d)_{inel} = \frac{d\sigma_{inel}}{d\Omega} = \lim_{r\to\infty} \frac{(|\underset{\sim}{j}_{in}| - |\underset{\sim}{j}_{out}|) r^2}{|\underset{\sim}{j}_{in}|} \tag{2}$$

where the j's are calculated using the appropriate components of Eq. (1). Use Eqs. (1) and (2) to show that

$$\boxed{\sigma_{inel} = \frac{\pi}{k^2} \sum_{\ell=0}^{\infty} (2\ell + 1)(1 - \eta_{\ell}^2)} \tag{3}$$

Solution: Ref. BR1,26; Problem 12.5

From Eq. (1) we have

$$\psi_{\ell}^{in} = - \frac{(2\ell + 1)}{kr} \frac{i^{\ell}}{2i} e^{-i(kr-\ell\pi/2)} P_{\ell}(\cos\theta)$$

$$\psi_{\ell}^{out} = \frac{(2\ell + 1)}{kr} \frac{i^{\ell}}{2i} S_{\ell} e^{i(kr-\ell\pi/2)} P_{\ell}(\cos\theta)$$

The calculation in Problem 12.5 shows that for large values of r only the radial component of the flux contributes. Thus we can write the current as:

$$j_r^{(\ell)} = \frac{i\hbar}{2\mu}\left(\psi_\ell \frac{\partial\psi_\ell^*}{\partial r} - \psi_\ell^* \frac{\partial\psi_\ell}{\partial r}\right)$$

and

$$|(j_r^\ell)_{in}| = \left| -\frac{i\hbar}{2\mu}\frac{(2\ell + 1)^2}{2k^2}(-ik)\frac{P_\ell^2(\cos\theta)}{r^2}\right| = +\frac{\hbar k}{4\mu}\frac{(2\ell + 1)^2}{k^2}\frac{P_\ell^2(\cos\theta)}{r^2}$$

$$|(j_r^\ell)_{out}| = +\frac{\hbar k}{4\mu}\frac{(2\ell + 1)^2}{k^2}|S_\ell|^2\frac{P_\ell^2(\cos\theta)}{r^2}$$

Then using the definition we find

$$(\sigma_d)_{inel} = \sum_{\ell=0}^{\infty}\frac{(2\ell + 1)^2}{4k^2}(1 - |S_\ell|^2)P_\ell^2(\cos\theta)$$

since $|j_{in}| = \hbar k/\mu$. The total inelastic cross-section is given by

$$\sigma_{inel} = 2\pi\int_0^\pi (\sigma_d)_{inel}\sin\theta d\theta = \frac{\pi}{k^2}\sum_{\ell=0}^{\infty}(2\ell + 1)(1 - \eta_\ell^2)$$

12.22 Show that

$$\sigma_{Total} = \frac{4\pi}{k}\,\text{Im}f(0) \tag{1}$$

regardless of the value of η_ℓ, and thus that the optical theorem is independent of the degree of inelasticity.

Solution: Ref. Problems 12.20, 12.21, and 12.29

Combining the results of Problems 12.21 and 12.29 gives

$$\sigma_{Total} = \sigma_{el} + \sigma_{inel} = \frac{\pi}{k^2}\sum_{\ell=0}^{\infty}(2\ell + 1)(1 + \eta_\ell^2 - 2\eta_\ell\cos 2\delta_\ell + 1 - \eta_\ell^2)$$

$$= \frac{2\pi}{k^2}\sum_{\ell=0}(2\ell + 1)(1 - \eta_\ell\cos 2\delta_\ell) \tag{2}$$

Then from Problem 12.20 with $\theta = 0$

$$f(0) = \frac{1}{2ik}\sum_{\ell=0}^{\infty}(2\ell + 1)(\eta_\ell e^{i2\delta_\ell} - 1) = -\frac{i}{2k}\sum_{\ell=0}^{\infty}(2\ell + 1)(\eta_\ell\cos 2\delta_\ell + i\eta_\ell\sin 2\delta_\ell - 1)$$

and

$$\text{Im}f(0) = \frac{1}{2k}\sum_{\ell=0}(2\ell + 1)(1 - \eta_\ell\cos 2\delta_\ell) \tag{3}$$

The comparison of Eqs. (2) and (3) immediately gives Eq. (1).

12.23 A particle (m_1, p_1), i.e. having mass m_1 and momentum $\underset{\sim}{p}_1$, collides with a particle $(m_2, \underset{\sim}{p}_2 = 0)$. The collision results in the formation of the particles $(m_3, \underset{\sim}{p}_3, \theta_3)$ and (m_4, p_4, θ_4) as shown in Fig. 12.23.

(a) Show that a measurement of θ_3 and $\underset{\sim}{p}_3$ (with explicit knowledge of all masses and the initial velocity) will permit the determination of the exothermicity of the reaction. The exothermicity Q is defined as the difference between the total kinetic energy of the final state and the initial state, i.e. $T_3 + T_4 - T_1$.

(b) Show that the maximum kinetic energy which can be transferred elastically, i.e. with $Q = 0$, in the collision is

$$\Delta E(\max) = \frac{2m_2}{(m_1 + m_2)^2} \; p_1^2$$

Here $m_3 = m_1$ and $m_4 = m_2$.

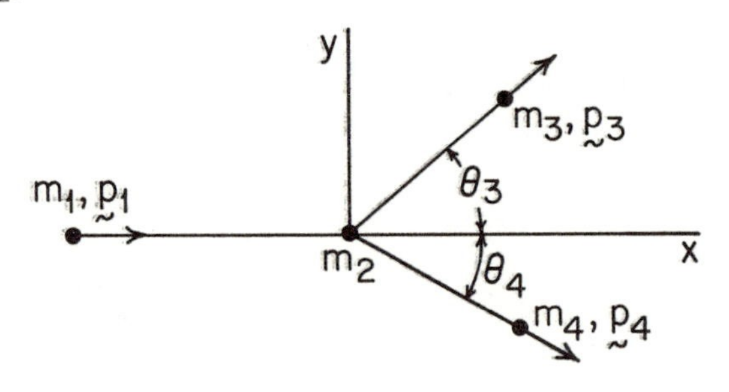

Fig. 12.23

Ans: Ref. SY,175

(a)
$$Q = -T_1\left(1 - \frac{m_1}{m_4}\right) + T_3\left(1 + \frac{m_3}{m_4}\right) - \frac{2}{m_4}\,(m_1 m_3 T_1 T_3)^{1/2}\cos\theta_3$$

A measurement of T_3 and θ_3 thus determines Q.

12.24 The following expansion is important in scattering theory

$$e^{ikz} = e^{ikr\cos\theta} = \sum_{\ell=0}^{\infty} a_\ell(kr) P_\ell(\cos\theta) \tag{1}$$

Show that

$$a_\ell(kr) = i^\ell (2\ell + 1) j_\ell(kr)$$

where

$$j_\ell(kr) = \sqrt{\frac{\pi}{2kr}}\, J_{\ell+1/2}(kr)$$

and

$$J_{\ell+1/2}(kr) = (-i)^\ell \sqrt{\frac{kr}{2\pi}} \int_{-1}^{+1} e^{ikr\cos\theta} P_\ell(\cos\theta)\, d(\cos\theta)$$

(Hint: Make use of the orthogonality relation for Legendre functions.)

Ref. WA,50 (for Bessel functions); FA,17

12.25 Use the partial wave expansion for the scattering amplitude $f(\theta)$ to show that

$$\boxed{\sigma = \frac{4\pi}{k}\, \mathrm{Im} f(0)} \qquad (1)$$

where $\mathrm{Im} f(0)$ means the imaginary part of the forward scattering amplitude. This important equation is known as the optical theorem.

Ref. SC,120,137; Problems 12.8 and 12.9

12.26 Consider the elastic scattering of a low energy particle from a repulsive barrier having the form:

$$V(r) = V_o, \qquad r \lessgtr a$$
$$= 0, \qquad r > a$$

Determine the total scattering cross-section and show that $\sigma_o = 4\pi a^2$ in the limit that $V_o \to \infty$.

Ans: Ref. Problem 12.11

$$\sigma_o \cong \frac{4\pi}{k_2^2} \sin^2\delta_o \doteq 4\pi a^2 \quad \text{for } V_o \to \infty$$

12.27 A potential function of considerable utility in both nuclear physics and chemistry is the Yukawa or "screened" coulomb potential:

$$V(r) = -\frac{ZZ'e^2}{(4\pi\varepsilon_o)} \frac{e^{-r/r_o}}{r}, \qquad r_o > 0$$

(a) Determine the scattering amplitude $f(\theta)$ in the first Born approximation for this potential.

(b) Obtain the differential scattering cross-section σ_d in the limit $r_o \to \infty$ and thus derive the Rutherford scattering formula:

$$\sigma_d = \left[\frac{ZZ'e^2}{2(4\pi\varepsilon_o)\mu v^2}\right]^2 \frac{1}{\sin^4(\theta/2)}$$

Ans: Ref. FA,37; DA2,377

(a)
$$f(\theta) = \frac{2\mu ZZ'e^2}{\hbar^2(4\pi\varepsilon_o)}\left(\frac{r_o^2}{1 + K^2 r_o^2}\right), \qquad K = 2k\sin(\theta/2) \text{ and } \hbar k = \mu v$$

12.28 Determine the scattering amplitude $f(\theta)$ in the first Born approximation for the spherical square well:

$$V(r) = -V_o \qquad 0 \lesseqgtr r \lesseqgtr a$$
$$= 0 \qquad r > a$$

Ans:

$$f(\theta) = \frac{2\mu V_o}{\hbar^2 K^3}(\sin Ka - Ka\cos Ka)$$

12.29 Show that when absorption is present, the elastic cross-section can be written as

$$\boxed{\sigma_{el} = \frac{\pi}{k^2}\sum_{\ell=0}^{\infty}(2\ell + 1)|1 - \eta_\ell e^{i2\delta_\ell}|^2}$$

Ref. Problems 12.9 and 12.20; BR1,26

12.30 Show that when the inelasticity is a maximum:

$$\sigma_{inel} = \sigma_{el} = \sigma_{Total}/2$$

Ref. Problems 12.29 and 12.21

12.31 Consider a totally absorbing sphere of radius r_o, i.e. a "black" sphere. Determine the elastic scattering amplitude $f(\theta)$ and the cross-sections σ_{el}, σ_{inel}, and σ_{Total} for the special case where $kr_o >> 1$.

Ans: Ref. CGT,174

$$f(\theta) = \frac{i}{2k}\sum_{\ell=0}^{kr_o}(2\ell + 1)P_\ell(\cos\theta), \qquad \sigma_{inel} = \sigma_{el} \cong \pi r_o^2, \qquad \sigma_{Total} = \sigma_{el} + \sigma_{inel}$$

APPENDIX 1

UNITS AND FUNDAMENTAL CONSTANTS

Ref. J. Chem. Ed. 48, 569 (1971)*

The International System of Units (SI) is used in this book. Base units in this system are shown in Table I.

Table I

Physical Quantity	*Unit*	*Symbol*
length	meter	m
mass	kilogram	kg
time	second	s
electric current	ampere	A
temperature	kelvin	K
amount of substance	mole	mol

Derived SI units of special interest are shown in Table II.

Table II

Physical Quantity	*Unit*	*Symbol*	*Definition*
force	newton	N	$kg \cdot m \cdot s^{-2}$
energy	joule	J	$kg \cdot m^2 \cdot s^{-2}$
power	watt	W	$kg \cdot m^2 \cdot s^{-3} = J \cdot s^{-1}$
electric charge	coulomb	C	$A \cdot s$
electric potential difference	volt	V	$kg \cdot m^2 \cdot s^{-3} \cdot A^{-1} = JA^{-1}s^{-1}$
magnetic flux	weber	Wb	$kg \cdot m^2 \cdot s^{-2} \cdot A^{-1} = V \cdot s$
magnetic flux density	tesla	T	$kg \cdot s^{-2} \cdot A^{-1} = V \cdot s \cdot m^{-2}$
frequency	hertz	Hz	s^{-1}
electric capacitance	farad	F	$A^2 \cdot s^4 \cdot kg^{-1} \cdot m^{-2} = A \cdot s \cdot V^{-1}$

The Angstrom unit (Å) which is equal to 10^{-10}m is often used in atomic calculations. Also the force unit dyne (dyn) equal to 10^{-5}N and energy unit erg equal to 10^{-7}J are in common use. Some useful units in terms of physical constants are given in Table III.

*This article is a reprint from the Nat. Bur. Stand. Tech. News Bull., January, 1971.

Table III

Physical Quantity	*Unit*	*Symbol*
energy	electron volt	eV $\simeq 1.6021\text{x}10^{-19}$J
mass	atomic mass unit	u $\simeq 1.6604\text{x}10^{-27}$kg
electric dipole moment	debye	D $\simeq 3.3356\text{x}10^{-30}$A·m·s

A selected list of physical constants are shown in Table IV.

Table IV

Constant	*Symbol*	*Value*
speed of light in vacuum	c	$2.997925\text{x}10^{8}$m/s
elementary charge	e	$1.602191\text{x}10^{-19}$C
Avogadro constant	N_A	$6.022169\text{x}10^{23}$mol^{-1}
atomic mass unit	u	$1.660531\text{x}10^{-27}$kg
electron rest mass	m_e	$9.109558\text{x}10^{-31}$kg
proton rest mass	m_p	$1.672614\text{x}10^{-27}$kg
Planck constant	h	$6.626196\text{x}10^{-34}$J·s
	$\hbar$	$1.054592\text{x}10^{-34}$J·s
Rydberg constant	R_∞	$1.097373\text{x}10^{7}$m^{-1}
gyromagnetic ratio of proton	γ_p	$2.675196\text{x}10^{8}$rad·s^{-1}T^{-1}
Bohr magneton	μ_B	$9.274096\text{x}10^{-24}$J/T
g-factor for free electron	g	2.0023
gas constant	R	8.31434 J·K^{-1}mol^{-1}
Boltzmann constant	k_B	$1.38062\text{x}10^{-23}$JK^{-1}
fine structure constant	α	$7.297351\text{x}10^{-3}$
Bohr radius	a_o	$5.29167\text{x}10^{-11}$m
permittivity of free space	ε_o	$8.854\text{x}10^{-12}$Fm^{-1}
	$4\pi\varepsilon_o$	$1.113\text{x}10^{-10}$Fm^{-1}
permeability of free space	μ_o	$4\pi\text{x}10^{-7}$NA^{-2}

APPENDIX 2

VECTOR FORMULAS

Ref. HO,3; SP1,116

We represent the vector $\underset{\sim}{V}$ as follows

$$\underset{\sim}{V} = V_1\hat{i} + V_2\hat{j} + V_3\hat{k}$$

where $\hat{i}$, $\hat{j}$, and $\hat{k}$ are unit vectors in the directions of the positive x, y, and z axes as shown below

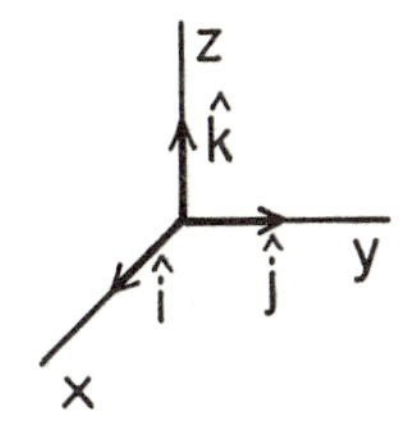

Dot or Scalar Product:

$$\underset{\sim}{A}\cdot\underset{\sim}{B} = A_1B_1 + A_2B_2 + A_3B_3 = AB\cos\theta, \quad 0 \lesseqgtr \theta \lesseqgtr \pi; \qquad \underset{\sim}{A}\cdot\underset{\sim}{B} = \underset{\sim}{B}\cdot\underset{\sim}{A}$$

Cross or Vector Product:

$$\underset{\sim}{A}\mathrm{x}\underset{\sim}{B} = \begin{vmatrix} \hat{i} & \hat{j} & \hat{k} \\ A_1 & A_2 & A_3 \\ B_1 & B_2 & B_3 \end{vmatrix} = (AB\sin\theta)\hat{n}, \quad 0 \lesseqgtr \theta \lesseqgtr \pi$$

Here θ is the angle between $\underset{\sim}{A}$ and $\underset{\sim}{B}$ and $\hat{n}$ is a unit vector that is perpendicular to the plane containing $\underset{\sim}{A}$ and $\underset{\sim}{B}$. If a right handed screw is rotated in the direction from $\underset{\sim}{A}$ to $\underset{\sim}{B}$, the screw will advance in the direction of $\hat{n}$.

$$\underset{\sim}{A}\mathrm{x}\underset{\sim}{B} = -\underset{\sim}{B}\mathrm{x}\underset{\sim}{A}$$

Scalar Triple Products:

$$\underset{\sim}{A}\cdot(\underset{\sim}{B}\mathrm{x}\underset{\sim}{C}) = (\underset{\sim}{A}\mathrm{x}\underset{\sim}{B})\cdot\underset{\sim}{C} = \begin{vmatrix} A_1 & A_2 & A_3 \\ B_1 & B_2 & B_3 \\ C_1 & C_2 & C_3 \end{vmatrix}$$

$|\underset{\sim}{A}\cdot(\underset{\sim}{B}\mathrm{x}\underset{\sim}{C})|$ = volume of parallelepiped with sides $\underset{\sim}{A}$, $\underset{\sim}{B}$, and $\underset{\sim}{C}$

Vector Triple Products:

$$\underset{\sim}{A}x(\underset{\sim}{B}x\underset{\sim}{C}) = (\underset{\sim}{A}\cdot\underset{\sim}{C})\underset{\sim}{B} - (\underset{\sim}{A}\cdot\underset{\sim}{B})\underset{\sim}{C}, \qquad (\underset{\sim}{A}x\underset{\sim}{B})x\underset{\sim}{C} = (\underset{\sim}{A}\cdot\underset{\sim}{C})\underset{\sim}{B} - (\underset{\sim}{B}\cdot\underset{\sim}{C})\underset{\sim}{A}$$

Differentiation of Vectors:

$$\frac{d}{dt}(\underset{\sim}{A} + \underset{\sim}{B} + \underset{\sim}{C} \cdots) = \frac{d\underset{\sim}{A}}{dt} + \frac{d\underset{\sim}{B}}{dt} + \frac{d\underset{\sim}{C}}{dt} + \cdots$$

$$\frac{d\underset{\sim}{A}}{dt} = \frac{dA_1}{dt}\hat{i} + \frac{dA_2}{dt}\hat{j} + \frac{dA_3}{dt}\hat{k}$$

$$\frac{d}{dt}(\underset{\sim}{A}\cdot\underset{\sim}{B}) = \frac{d\underset{\sim}{A}}{dt}\cdot\underset{\sim}{B} + \underset{\sim}{A}\cdot\frac{d\underset{\sim}{B}}{dt}$$

$$\frac{d}{dt}(\underset{\sim}{A}x\underset{\sim}{B}) = \frac{d\underset{\sim}{A}}{dt}\,x\,\underset{\sim}{B} + \underset{\sim}{A}\,x\,\frac{d\underset{\sim}{B}}{dt}$$

Differential Operators - Rectangular Coordinates:

$$\underset{\sim}{\nabla} \equiv \text{del} \equiv \hat{i}\frac{\partial}{\partial x} + \hat{j}\frac{\partial}{\partial y} + \hat{k}\frac{\partial}{\partial z}, \qquad \underset{\sim}{\nabla}\phi \equiv \text{grad}\phi \equiv \frac{\partial\phi}{\partial x}\hat{i} + \frac{\partial\phi}{\partial y}\hat{j} + \frac{\partial\phi}{\partial z}\hat{k}$$

$$\underset{\sim\sim}{\nabla A} \equiv \frac{\partial\underset{\sim}{A}}{\partial x}\hat{i} + \frac{\partial\underset{\sim}{A}}{\partial y}\hat{j} + \frac{\partial\underset{\sim}{A}}{\partial z}\hat{k}$$

where ϕ is a scalar function.

$$\underset{\sim}{\nabla}\cdot\underset{\sim}{V} \equiv \text{div}\underset{\sim}{V} \equiv \left(\hat{i}\frac{\partial}{\partial x} + \hat{j}\frac{\partial}{\partial y} + \hat{k}\frac{\partial}{\partial z}\right)\cdot(V_1\hat{i} + V_2\hat{j} + V_3\hat{k}) = \frac{\partial V_1}{\partial x} + \frac{\partial V_2}{\partial y} + \frac{\partial V_3}{\partial z}$$

$$\underset{\sim}{\nabla}x\underset{\sim}{V} \equiv \text{curl}\underset{\sim}{V} \equiv \left(\hat{i}\frac{\partial}{\partial x} + \hat{j}\frac{\partial}{\partial y} + \hat{k}\frac{\partial}{\partial z}\right)x\,(V_1\hat{i} + V_2\hat{j} + V_3\hat{k})$$

$$= \begin{vmatrix} \hat{i} & \hat{j} & \hat{k} \\ \partial/\partial x & \partial/\partial y & \partial/\partial z \\ V_1 & V_2 & V_3 \end{vmatrix} = \hat{i}\left(\frac{\partial V_3}{\partial y} - \frac{\partial V_2}{\partial z}\right) + \hat{j}\left(\frac{\partial V_1}{\partial z} - \frac{\partial V_3}{\partial x}\right) + \hat{k}\left(\frac{\partial V_2}{\partial x} - \frac{\partial V_1}{\partial y}\right)$$

$$\nabla^2\phi \equiv \text{Laplacian of } \phi \equiv \frac{\partial^2\phi}{\partial x^2} + \frac{\partial^2\phi}{\partial y^2} + \frac{\partial^2\phi}{\partial z^2}$$

$$\nabla^2\underset{\sim}{V} \equiv \text{Laplacian of } \underset{\sim}{V} \equiv \frac{\partial^2\underset{\sim}{V}}{\partial x^2} + \frac{\partial^2\underset{\sim}{V}}{\partial y^2} + \frac{\partial^2\underset{\sim}{V}}{\partial z^2}$$

APPENDIX 3

HERMITE DIFFERENTIAL EQUATION

Ref. PW,67; EWK,60; EI,256

We wish to solve the differential equation

$$H'' - 2\eta H' + (\beta/\alpha - 1)H = 0 \tag{1}$$

The solution is assumed to be a power series*

$$H(\eta) = \sum_{i=0}^{\infty} c_i \eta^i = c_o + c_1\eta + c_2\eta^2 + c_3\eta^3 + c_4\eta^4 + \cdots$$

Thus

$$H'(\eta) = \sum_{i=1}^{\infty} i c_i \eta^{i-1} = (1)c_1 + (1\cdot 2)c_2\eta + (1\cdot 3)c_3\eta^2 + (1\cdot 4)c_4\eta^3 \cdots$$

and

$$H''(\eta) = \sum_{i=2}^{\infty} (i-1) i c_i \eta^{i-2} = (1\cdot 2)c_2 + (1\cdot 2\cdot 3)c_3\eta + (1\cdot 3\cdot 4)c_4\eta^2 + \cdots$$

Substituting these quantities into (1) yields

$$\begin{aligned}(1\cdot 2)c_2 &+ (2\cdot 3)c_3\eta + (3\cdot 4)c_4\eta^2 + \cdots - (2\cdot 1)c_1\eta - (2\cdot 2)c_2\eta^2 - \cdots \\ &+ (\beta/\alpha-1)c_o + (\beta/\alpha-1)c_1\eta + (\beta/\alpha-1)c_2\eta^2 + \cdots = 0\end{aligned} \tag{2}$$

Now (2) holds for all values of η. Thus the coefficients of each power of η must vanish to satisfy the equality. The coefficients of powers of η are

$$\begin{aligned}
\eta^o&: \quad 2c_2 + (\beta/\alpha-1)c_o = 0 \\
\eta^1&: \quad 2\cdot 3c_3 + (\beta/\alpha-1-2)c_1 = 0 \\
\eta^2&: \quad 3\cdot 4c_4 + (\beta/\alpha-1-2\cdot 2)c_2 = 0 \\
\eta^3&: \quad 4\cdot 5c_5 + (\beta/\alpha-1-2\cdot 3)c_3 = 0 \\
\\
\eta^i&: \quad (i+1)(i+2)c_{i+2} + (\beta/\alpha-1-2i)c_i = 0
\end{aligned}$$

and the associated recursion relationship is

$$\boxed{c_{i+2} = -\frac{(\beta/\alpha-1-2i)}{(i+1)(i+2)}\, c_i} \tag{3}$$

*For an elementary discussion of series solutions see: L. F. Beste, J. Chem. Ed. 46, 151 (1969).

Thus we can obtain c_2, c_4, c_6, ... *etc.* successively from c_o and c_3, c_5, c_7, ... *etc.* from c_1. We may write the solution as

$$H(\eta) = \sum_{i=0}^{\infty} c_i \eta^i = \sum_{\text{even}} c_i \eta^i + \sum_{\text{odd}} c_i \eta^i \tag{4}$$

$$= c_o\left[1 + \frac{c_2}{c_o}\eta^2 + \frac{c_4}{c_2}\frac{c_2}{c_o}\eta^4 + \cdots\right] + c_1\left[\eta + \frac{c_3}{c_1}\eta^3 + \frac{c_5}{c_3}\frac{c_3}{c_o}\eta^5 + \cdots\right]$$

For very large values of η

$$\frac{c_{i+2}}{c_i} = -\frac{(\beta/\alpha-1-2i)}{(i+1)(i+2)} \simeq \frac{2i}{i^2} = \frac{2}{i}$$

This is the same ratio as one would obtain from the successive coefficients of e^{η^2}, *i.e.*

$$e^{\eta^2} = \sum_{i=0}^{\infty} \frac{(\eta^2)^i}{i!} = 1 + \eta^2 + \eta^4/2! + \eta^6/3! + \cdots + \frac{\eta^m}{(m/2)!} + \frac{\eta^{m+2}}{(m/2+1)!} + \cdots$$

thus

$$\frac{[(m/2 + 1)!]^{-1}}{[m/2!]^{-1}} = \frac{m/2!}{(m/2 + 1)!} = \frac{1}{m/2 + 1} = \frac{2}{m}$$

This means that for $|\eta| \to \infty$

$$H(\eta) = \underbrace{c_o(\text{constant})_1 e^{\eta^2}}_{\text{even}} + \underbrace{c_1(\text{constant})_2 \eta e^{\eta^2}}_{\text{odd}}$$

The Schrödinger equation for the harmonic oscillator has the solution

$$u(\eta) = e^{-\eta^2/2}H(\eta) \to a_o e^{\eta^2/2} + a_1 \eta e^{\eta^2/2} \quad \text{as} \quad |\eta| \to \infty$$

where a_o and a_1 are constants. This function is, however, not square integrable since $u(\eta) \to \infty$ as $|\eta| \to \infty$. The key step now is to realize that for certain values of β/α we find solutions which do not diverge. If we set $\beta/\alpha = 2n + 1$, then

$$c_{n+2} = -\frac{(2n + 1 - 1 - 2n)}{(n + 1)(n + 2)} c_n = 0$$

Thus for n even, the first term on the r.h.s. of Eq. (4) will terminate with $i = n$. Similarly, for n odd the second term will terminate with $i = n$. Well behaved solutions exist then for n even if $c_1 = 0$ and for n odd if $c_o = 0$. The resulting functions $H_n(\eta)$ are called Hermite polynomials of order n. The product $u^*(\eta)u(\eta)$ will remain finite as $|\eta| \to \infty$ because the factor $e^{-\eta^2/2}$ will dominate the polynomial $H_n(\eta)$.

APPENDIX 4

LEGENDRE POLYNOMIALS AND SPHERICAL HARMONICS

Ref. PW,118; SP2,242; BS,18; RO3,235

Legendre Functions:

The Associated Legendre differential equation is:

$$(1 - z^2)\frac{d^2P(z)}{dz^2} - 2z\frac{dP(z)}{dz} + \left[\ell(\ell + 1) - \frac{m^2}{1 - z^2}\right]P(z) = 0$$

This equation may be solved by the power series method to give as the particular solution

$$P(z) \rightarrow P_\ell^m(z) = (1 - z^2)^{m/2}\frac{d^m}{dz^m}P_\ell(z) \tag{1}$$

where $P_\ell(z)$ is a particular solution of the somewhat simpler Legendre differential equation

$$(1 - z^2)\frac{d^2P_\ell(z)}{dz^2} - 2z\frac{dP_\ell(z)}{dz} + \ell(\ell + 1)P_\ell(z) = 0$$

The Legendre polynomials $P_\ell(z)$ can be generated by Rodrique's formula

$$P_\ell(z) = \frac{1}{2^\ell \ell!}\frac{d^\ell}{dz^\ell}(z^2 - 1)^\ell \tag{2}$$

or by the "generating function"

$$\frac{1}{\sqrt{1 - 2zs + s^2}} = \sum_{\ell=0}^{\infty} P_\ell(z)s^\ell$$

The first few Legendre polynomials obtained with Eq. (2) are:

$$P_o(z) = 1 \qquad P_1(z) = z \qquad P_2(z) = \frac{1}{2}(3z^2 - 1)$$

$$P_3(z) = \frac{1}{2}(5z^3 - 3z) \qquad P_4(z) = \frac{1}{8}(35z^4 - 30z^2 + 3)$$

Eqs. (1) and (2) define the Associated Legendre polynomials, $P_\ell^m(z)$:

$$P_1^1(z) = (1 - z^2)^{1/2} \qquad P_2^1(z) = 3z(1 - z^2)^{1/2}$$

$$P_2^2(z) = 3(1 - z^2) \qquad P_3^1(z) = \frac{3}{2}(1 - z^2)^{1/2}(5z^2 - 1)$$

$$P_3^2(z) = 15z(1 - z^2) \qquad P_3^3(z) = 15(1 - z^2)^{3/2}$$

The functions $P_\ell(z)$ and $P_\ell^m(z)$ are orthonormal on the interval $-1 < z < 1$, thus

$$\int_{-1}^{+1} P_{\ell'}^*(z)P_\ell(z)dz = \delta_{\ell\ell'}\frac{2}{(2\ell + 1)}$$

and

$$\int_{-1}^{+1} (P_{\ell'}^{m'})^*(z)P_\ell^m(z)dz = \delta_{\ell\ell'}\delta_{mm'}\frac{2(\ell + m)!}{(2\ell + 1)(\ell - m)!}$$

Application:

The differential equation

$$\left[\frac{1}{\sin\theta}\frac{d}{d\theta}\sin\theta\frac{d}{d\theta} + \ell(\ell + 1) - \frac{m^2}{\sin^2\theta}\right] y(\theta) = 0$$

appears in many problems having fundamental spherical symmetry, e.g. rigid rotor, hydrogen atom, "flooded planet." The substitution $z = \cos\theta$ and the definition $P(z) \equiv y(\theta)$ give

$$\frac{dy(\theta)}{d\theta} = \frac{dP(z)}{dz}\frac{dz}{d\theta} = -\sin\theta\frac{dP(z)}{dz}$$

and

$$\frac{d}{dz}\left[(1 - z^2)\frac{dP(z)}{dz}\right] + \left[\ell(\ell + 1) - \frac{m^2}{1 - z^2}\right] P(z) = 0$$

which is identical to the Associated Legendre differential equation.

There exist, of course, other solutions to the two basic differential equations considered in this appendix; these functions are singular at $z = \pm 1$ and are thus excluded in many physical problems.

Spherical Harmonics:

The spherical harmonics represent the angular solution to the central potential problem. They are defined in terms of the associated Legendre functions:

$$Y_{\ell m}(\theta,\phi) = \varepsilon_\ell\left[\frac{(2\ell + 1)(\ell - |m|)!}{2(\ell + |m|)!}\right]^{1/2} P_\ell^m(\cos\theta)\frac{1}{\sqrt{2\pi}}e^{im\phi}$$

Occasionally $Y_{\ell m}$ is written as

$$Y_{\ell m}(\theta,\phi) = \Theta_{\ell m}(\theta)\Phi_m(\phi)$$

where

$$\Theta_{\ell m}(\theta) = \varepsilon_\ell\left[\frac{(2\ell + 1)(\ell - |m|)!}{2(\ell + |m|)!}\right]^{1/2} P_\ell^m(\cos\theta)$$

and

$$\Phi_m(\phi) = \frac{1}{\sqrt{2\pi}}e^{im\phi}$$

The factor ε_ℓ is chosen to be +1 by Rose while some other authors choose $(-1)^{(m+|m|)/2}$.

Useful Recursion Relations:

Legendre:

$$zP_\ell(z) = \frac{\ell + 1}{2\ell + 1} P_{\ell+1}(z) + \frac{\ell}{2\ell + 1} P_{\ell-1}(z)$$

$$(\ell + 1)P_\ell(z) = \frac{d}{dz} [P_{\ell+1}(z) - zP_\ell(z)]$$

$$\ell P_\ell(z) = z \frac{dP_\ell(z)}{dz} - \frac{dP_{\ell-1}(z)}{dz}$$

Associated Legendre:

$$zP_\ell^{|m|} = \frac{1}{2\ell + 1} \{(\ell - |m| + 1)P_{\ell+1}^{|m|}(z) + (\ell + |m|)P_{\ell-1}^{|m|}(z)\}$$

$$(1 - z^2)^{1/2}P_\ell^{|m|} = \frac{1}{2\ell + 1} \{P_{\ell+1}^{|m|+1}(z) - P_{\ell-1}^{|m|+1}(z)\}$$

$$= \frac{1}{2\ell + 1} \{(\ell + |m|)(\ell + |m| - 1)P_{\ell-1}^{|m|-1}(z)$$

$$- (\ell - |m| + 1)(\ell - |m| + 2)P_{\ell+1}^{|m|-1}(z)\}$$

APPENDIX 5

CURVILINEAR COORDINATES

Ref. SP2,140; SP3,135; NE,83

Let the following equations define a well-behaved coordinate transformation from rectangular to some other system

$$x = x(u_1,u_2,u_3) \qquad u_1 = u_1(x,y,z)$$
$$y = y(u_1,u_2,u_3) \quad \text{and} \quad u_2 = u_2(x,y,z) \tag{1}$$
$$z = z(u_1,u_2,u_3) \qquad u_3 = u_3(x,y,z)$$

A position vector to the point $P(x,y,z)$ is given by

$$\underset{\sim}{r} = x\hat{i} + y\hat{j} + z\hat{k} \tag{2}$$

and unit tangent vectors to the curves $u_1(u_2,u_3 = \text{constant})$, $u_2(u_1,u_3 = \text{constant})$, and $u_3(u_1,u_2 = \text{constant})$ are given as

$$\hat{e}_i = \frac{\partial \underset{\sim}{r}}{\partial u_i} \Bigg/ \left| \frac{\partial \underset{\sim}{r}}{\partial u_i} \right|, \quad i = 1,\ 2,\ 3 \tag{3}$$

Thus, a vector $\underset{\sim}{A}$ $(= A_x\hat{i} + A_y\hat{j} + A_z\hat{k})$ may be represented by these tangent unit vectors

$$\underset{\sim}{A} = \sum_{i=1}^{3} A_i^{(e)}\hat{e}_i \tag{4}$$

$$= \sum_{i=1}^{3} (\underset{\sim}{A}\cdot\underset{\sim}{\nabla} u_i)\frac{\partial \underset{\sim}{r}}{\partial u_i} = \sum_{i=1}^{3} (\underset{\sim}{A} \cdot \frac{\partial \underset{\sim}{r}}{\partial u_i})\underset{\sim}{\nabla} u_i \tag{5}$$

where $A_i^{(e)}$ are components of $\underset{\sim}{A}$ in the transformed system. We may then write dr as

$$d\underset{\sim}{r} = \sum_{i=1}^{3} \frac{\partial \underset{\sim}{r}}{\partial u_i} du_i \tag{6}$$

The differential element of volume, dV, is comprised of a parallelepiped of adjacent sides $\frac{\partial \underset{\sim}{r}}{\partial u_i} du_i$, and is given by

$$dV = \left| \frac{\partial \underset{\sim}{r}}{\partial u_1} \cdot \frac{\partial \underset{\sim}{r}}{\partial u_2} \times \frac{\partial \underset{\sim}{r}}{\partial u_3} \right| du_1du_2du_3 \tag{7}$$

$$= \begin{vmatrix} g_{11} & g_{12} & g_{13} \\ g_{21} & g_{22} & g_{23} \\ g_{31} & g_{32} & g_{33} \end{vmatrix}^{1/2} du_1du_2du_3 \tag{8}$$

$$= \sqrt{g}\ du_1du_2du_3 \tag{9}$$

Here we have taken advantage of the fact that the volume of a parallelepiped defined by three vectors $\underset{\sim}{A}$, $\underset{\sim}{B}$ and $\underset{\sim}{C}$ is $|\underset{\sim}{A}\cdot\underset{\sim}{B}\times\underset{\sim}{C}|$.

The components of the transformation tensor g are given by

$$g_{ij} = \left(\frac{\partial \underset{\sim}{r}}{\partial u_i} \cdot \frac{\partial \underset{\sim}{r}}{\partial u_j}\right) \qquad i, j = 1, 2, 3 \tag{10}$$

For orthogonal systems

$$\sqrt{g} = \sqrt{g_{11}g_{22}g_{33}}$$

and

$$\underset{\sim}{\nabla} = \sum_{i=1}^{3} \frac{1}{\sqrt{g_{ii}}} \frac{\partial}{\partial u_i} \hat{e}_i \tag{11}$$

$$\nabla^2 = \frac{1}{\sqrt{g}} \sum_{i=1}^{3} \frac{\partial}{\partial u_i} \frac{\sqrt{g}}{g_{ii}} \frac{\partial}{\partial u_i} \tag{12}$$

$$\underset{\sim}{\nabla}\cdot\underset{\sim}{A} = \frac{1}{\sqrt{g}} \sum_{i=1}^{3} \frac{\partial}{\partial u_i} \left[\sqrt{\frac{g}{g_{ii}}}\right] A_i^{(e)} \tag{13}$$

and

$$\underset{\sim}{\nabla}\times\underset{\sim}{A} = \frac{1}{\sqrt{g}} \begin{vmatrix} \sqrt{g_{11}}\,\hat{e}_1 & \sqrt{g_{22}}\,\hat{e}_2 & \sqrt{g_{33}}\,\hat{e}_3 \\ \dfrac{\partial}{\partial u_1} & \dfrac{\partial}{\partial u_2} & \dfrac{\partial}{\partial u_3} \\ \sqrt{g_{11}}\,A_1^{(e)} & \sqrt{g_{22}}\,A_2^{(e)} & \sqrt{g_{33}}\,A_3^{(e)} \end{vmatrix} \tag{14}$$

For non-orthogonal coordinate systems, the reader should consult an advanced vector calculus text.

∇^2 and dV for Selected Orthogonal Curvilinear Coordinate Systems

(a) Cylindrical

$$x = r\cos\theta, \quad y = r\sin\theta, \quad z = z$$

$$dV = r\,dr\,d\theta\,dz$$

$$\nabla^2 = \frac{1}{r}\frac{\partial}{\partial r}\left(r\frac{\partial}{\partial r}\right) + \frac{1}{r^2}\frac{\partial^2}{\partial\theta^2} + \frac{\partial^2}{\partial z^2}$$

(b) Parabolic

$$x = uv\cos\theta, \quad y = uv\sin\theta, \quad z = \frac{1}{2}(u^2 - v^2)$$

$$dV = uv(u^2 + v^2)\,du\,dv\,d\theta$$

$$\nabla^2 = \frac{1}{(u^2+v^2)}\left\{\frac{1}{u}\frac{\partial}{\partial u}\left(u\frac{\partial}{\partial u}\right) + \frac{1}{v}\frac{\partial}{\partial v}\left(v\frac{\partial}{\partial v}\right)\right\} + \frac{1}{(u^2v^2)}\frac{\partial^2}{\partial\theta^2}$$

(c) Elliptical (ϕ measured from xz plane)

$\mu = (r_a + r_b)/R \qquad (1 \lesssim \mu \lesssim \infty)$

$\nu = (r_a - r_b)/R \qquad (-1 \lesssim \nu \lesssim 1)$

$\phi \qquad (0 \leq \phi \leq 2\pi)$

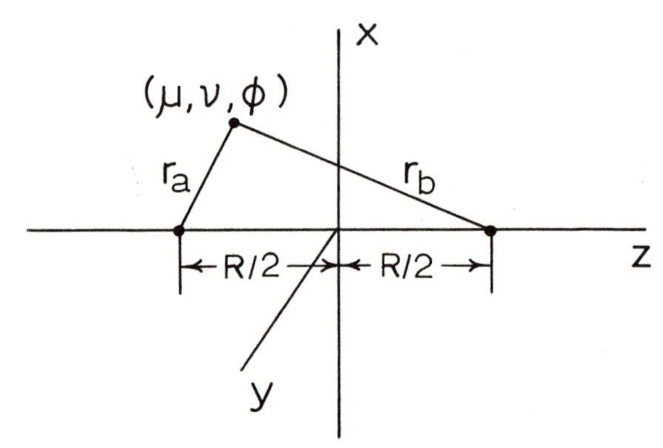

$$x = \frac{R}{2}\,[(\mu^2 - 1)(1 - \nu^2)]^{1/2}\cos\phi$$

$$y = \frac{R}{2}\,[(\mu^2 - 1)(1 - \nu^2)]^{1/2}\sin\phi$$

$$z = \frac{R}{2}\,\mu\nu$$

$$dV = \frac{R^3}{8}\,(\mu^2 - \nu^2)d\mu d\nu d\phi$$

$$\nabla^2 = \frac{4}{R(\mu^2 - \nu^2)}\left[\frac{\partial}{\partial\mu}\left\{(\mu^2 - 1)\frac{\partial}{\partial\mu}\right\} + \frac{\partial}{\partial\nu}\left\{(1 - \nu^2)\frac{\partial}{\partial\nu}\right\} + \frac{(\mu^2 - \nu^2)}{(\mu^2 - 1)(1 - \nu^2)}\frac{\partial^2}{\partial\phi^2}\right]$$

(d) Spherical Polar

$$x = r\sin\theta\cos\phi, \quad y = r\sin\theta\sin\phi, \quad z = r\cos\theta$$

$$dV = r^2\sin\theta dr d\theta d\phi$$

$$\nabla^2 = \frac{1}{r^2}\frac{\partial}{\partial r}\left(r^2\frac{\partial}{\partial r}\right) + \frac{1}{r^2\sin\theta}\frac{\partial}{\partial\theta}\left(\sin\theta\frac{\partial}{\partial\theta}\right) + \frac{1}{r^2\sin^2\theta}\frac{\partial^2}{\partial\phi^2}$$

$$\underset{\sim}{\nabla} = \frac{\partial}{\partial r}\hat{e}_r + \frac{1}{r}\frac{\partial}{\partial\theta}\hat{e}_\theta + \frac{1}{r\sin\theta}\frac{\partial}{\partial\phi}\hat{e}_\phi$$

APPENDIX 6

TRANSFORMATIONS AND DIAGONALIZATION

Ref. HO,94; AN3,235; AN2,94

The matrix $\tilde{A}$ can be diagonalized by a similarity transformation of the form

$$\tilde{S}^{-1}\tilde{A}\tilde{S} = \tilde{D} \tag{1}$$

where $\tilde{S}$ is a unitary matrix, i.e. $\tilde{S}^+ = \tilde{S}^{-1}$ where $S^+_{ij} = S^*_{ji}$. Multiplication from the left by $\tilde{S}$ gives

$$\tilde{A}\tilde{S} = \tilde{S}\tilde{D} \tag{2}$$

This is equivalent to the set of linear equations

$$\sum_{j=1}^{n} (a_{ij} - \lambda_k \delta_{ij}) S_{jk} = 0 \tag{3}$$

Nontrivial solutions exist for Eq. (3) only when the determinant

$$\det|a_{ij} - \lambda\delta_{ij}| = 0 \tag{4}$$

The roots $\lambda_1, \lambda_2, \cdots \lambda_n$ of Eq. (4) are called the eigenvalues of $\tilde{A}$ and the matrix $\tilde{D}$ has the form

$$\begin{pmatrix} \lambda_1 & 0 & \cdots & 0 \\ 0 & \lambda_2 & & 0 \\ \vdots & & & 0 \\ 0 & \cdot & \cdot\ \cdot & \lambda_n \end{pmatrix}$$

For each eigenvalue λ_k Eq. (3) can be solved for the set of elements of the transformation matrix $(S_{1k}, S_{2k}, \cdots, S_{nk})$.

Equation (2) can also be written in the form

$$\tilde{A}\tilde{u}_k = \lambda_k \tilde{u}_k \tag{5}$$

where the column vector $\tilde{u}_k = (S_{1k}, S_{2k}, \cdots, S_{nk})$ is known as the kth eigenvector of $\tilde{A}$ and λ_k is the corresponding eigenvalue.

Example: Let

$$\tilde{M} = \begin{pmatrix} A & D \\ D & B \end{pmatrix}$$

where all elements are real and solve the equation $\tilde{M}\tilde{u} = \lambda\tilde{u}$ for the eigenvalues and eigenvectors. We require that

$$\det\begin{vmatrix}(A - \lambda) & D \\ D & (B - \lambda)\end{vmatrix} = 0$$

Therefore, $\lambda^2 - \lambda(A + B) + AB - D^2 = 0$ and

$$\lambda_{\pm} = \frac{(A + B)}{2} \pm \frac{1}{2}\sqrt{(A + B)^2 - 4(AB - D^2)} = \frac{(A + B)}{2} \pm \frac{1}{2}\sqrt{(A - B)^2 + 4D^2}$$

For convenience in the following we let

$$2C = \sqrt{(A - B)^2 + 4D^2}$$

The eigenvector associated with λ_+ is determined from Eq. (3) as follows:

$$(A - \lambda_+)S_{11} + DS_{21} = 0$$

$$\left(\frac{A - B - 2C}{-2D}\right)S_{11} = QS_{11} = S_{21}$$

Now introducing the normalization condition

$$\sum_i |S_{ij}|^2 = 1$$

gives $Q^2S_{11}^2 = 1 - S_{11}^2$ and $S_{11}^2 = 1/(1 + Q^2)$. Thus

$$S_{11} = 1/\sqrt{1 + Q^2} \quad \text{and} \quad S_{21} = Q/\sqrt{1 + Q^2} \quad \text{where} \quad Q = \frac{2C - (A - B)}{2D} = \frac{2D}{2C + (A - B)}$$

Similar manipulations with λ_- give

$$S_{12} = -Q/\sqrt{1 + Q^2} \quad \text{and} \quad S_{22} = 1/\sqrt{1 + Q^2}$$

These terms can be written in a compact form by defining the angle θ so that

$$\cos\theta = 1/\sqrt{1 + Q^2} \quad \text{and} \quad \sin\theta = Q/\sqrt{1 + Q^2}$$

The transformation matrix thus becomes

$$\underset{\sim}{S} = \begin{pmatrix}\cos\theta & -\sin\theta \\ \sin\theta & \cos\theta\end{pmatrix}$$

Exercises for this example:

(a) Verify that $\underset{\sim}{S}^+\underset{\sim}{S} = 1$ or $\underset{\sim}{S}^+ = \underset{\sim}{S}^{-1}$.

(b) Verify that $\underset{\sim}{S}^{-1}\underset{\sim}{M}\underset{\sim}{S} = \begin{pmatrix}\lambda_+ & 0 \\ 0 & \lambda_-\end{pmatrix}$

APPENDIX 7

HYDROGENIC WAVE FUNCTIONS AND ENERGIES

Ref. PW,132; EWK,80

The hydrogenic wave functions are given by $u_{n\ell m_\ell}(r,\theta,\phi) = R_{n\ell}(r)\Theta_{\ell m_\ell}(\theta)\Phi_{m_\ell}(\phi)$ where (using $m = m_\ell$)

$$R_{n\ell}(r) = -\left\{\left(\frac{2Z}{na_o}\right)^3 \frac{(n-\ell-1)!}{[(n+\ell)!]^3 2n}\right\}^{1/2} e^{-\rho/2}\rho^{\ell}L_{n+\ell}^{2\ell+1}(\rho) \tag{1}$$

$$\Theta_{\ell m}(\theta) = (-1)^{(m+|m|)/2}\left[\frac{(2\ell+1)(\ell-|m|)!}{2(\ell+|m|)!}\right]^{1/2} P_\ell^m(\cos\theta) \tag{2}$$

$$\Phi_m(\phi) = \frac{1}{\sqrt{2\pi}}\, e^{im\phi} \tag{3}$$

where $\rho = 2Zr/na_o$. In Eq. (1) $L_{n+\ell}^{2\ell+1}(\rho)$ are the associated Laguerre functions and in Eq. (2) the $P_\ell^m(\cos\theta)$ are the associated Legendre polynomials. These functions are normalized so that

$$\int_o^\infty |R_{n\ell}(r)|^2 r^2 dr = 1, \qquad \int_o^\pi |\Theta_{\ell m}(\theta)|^2 \sin\theta d\theta = 1, \qquad \int_o^{2\pi} |\Phi_m(\phi)|^2 d\phi = 1$$

and thus

$$\int |u_{n\ell m}(r,\theta,\phi)|^2 d\tau = \int_o^{2\pi}\int_o^{\pi}\int_o^{\infty} |u_{n\ell m}(r,\theta,\phi)|^2 r^2 dr \sin\theta d\theta d\phi = 1$$

In the following table wave functions for n = 1 through 3 and ℓ = 0 through 2 are presented in the real form, i.e. normalized linear combinations have been taken for functions having the same value of m_ℓ^2.

$n = 1,\ \ell = 0,\ m_\ell = 0:\quad u_{1s} = \frac{\alpha}{\sqrt{\pi}}\, e^{-\sigma}$

$n = 2,\ \ell = 0,\ m_\ell = 0:\quad u_{2s} = \frac{\alpha}{4\sqrt{2\pi}}\,(2-\sigma)e^{-\sigma/2}$

$\ell = 1,\ m_\ell = 0:\quad u_{2p_z} = \frac{\alpha}{4\sqrt{2\pi}}\,\sigma e^{-\sigma/2}\cos\theta$

$m_\ell = \pm 1:\quad u_{2p_x} = \frac{\alpha}{4\sqrt{2\pi}}\,\sigma e^{-\sigma/2}\sin\theta\cos\phi$

$u_{2p_y} = \frac{\alpha}{4\sqrt{2\pi}}\,\sigma e^{-\sigma/2}\sin\theta\sin\phi$

$$n = 3,\ \ell = 0,\ m_\ell = 0: \quad u_{3s} = \frac{\alpha}{81\sqrt{3\pi}}(27 - 18\sigma + 2\sigma^2)e^{-\sigma/3}$$

$$\ell = 1,\ m_\ell = 0: \quad u_{3p_z} = \frac{\sqrt{2}\,\alpha}{81\sqrt{\pi}}(6 - \sigma)\sigma e^{-\sigma/3}\cos\theta$$

$$m_\ell = \pm 1: \quad u_{3p_x} = \frac{\sqrt{2}}{81\sqrt{\pi}}\alpha(6 - \sigma)\sigma e^{-\sigma/3}\sin\theta\cos\phi$$

$$u_{3p_y} = \frac{\sqrt{2}}{81\sqrt{\pi}}\alpha(6 - \sigma)\sigma e^{-\sigma/3}\sin\theta\sin\phi$$

$$\ell = 2,\ m_\ell = 0: \quad u_{3d_{z^2}} = \frac{\alpha}{81\sqrt{6\pi}}\sigma^2 e^{-\sigma/3}(3\cos^2\theta - 1)$$

$$m_\ell = \pm 1: \quad u_{3d_{zx}} = \frac{\sqrt{2}\,\alpha}{81\sqrt{\pi}}\sigma^2 e^{-\sigma/3}\sin\theta\cos\theta\cos\phi$$

$$u_{3d_{yz}} = \frac{\sqrt{2}\,\alpha}{81\sqrt{\pi}}\sigma^2 e^{-\sigma/3}\sin\theta\cos\theta\sin\phi$$

$$m_\ell = \pm 2: \quad u_{3d_{x^2-y^2}} = \frac{\alpha}{81\sqrt{2\pi}}\sigma^2 e^{-\sigma/3}\sin^2\theta\cos 2\phi$$

$$u_{3d_{xy}} = \frac{\alpha}{81\sqrt{2\pi}}\sigma^2 e^{-\sigma/3}\sin^2\theta\sin 2\phi$$

Note: $\alpha = (Z/a_o)^{3/2}$, $\sigma = Zr/a_o$.

The associated energy eigenvalues are:

$$E_{n\ell} = -\frac{Z^2}{n^2} R_\infty hc \left(1 + \frac{m_e}{M}\right)^{-1}$$

where

$$R_\infty = \frac{2\pi^2 m_e e^4}{(4\pi\varepsilon_o)^2 h^2}$$

$$R_\infty(\text{cm}^{-1}) = \frac{2\pi^2 m_e e^4}{(4\pi\varepsilon_o)^2 h^3 c} = 109{,}737\ \text{cm}^{-1}$$

$$R_\infty(\text{eV}) = \frac{2\pi^2 m_e e^3}{(4\pi\varepsilon_o)^2 h^2} = 13.605\ \text{eV}$$

and

$$a_o = \frac{h^2(4\pi\varepsilon_o)}{4\pi^2 m_e e^2} = 0.5292\text{x}10^{-10}\text{m}$$

M = nuclear mass

APPENDIX 8

USE OF CHARACTER TABLES

Ref. CO1; EWK,172; AT1,121; TI; LE2,420

Many excellent discussions of the applications of group theory to problems in molecular quantum mechanics are available in the literature. Here we only define the notation and list a few equations which are useful for the application of character tables to problems in this book. Our purpose is to provide material for review and easy reference.

The point groups of interest consist of symmetry operations which exist because of molecular symmetry elements. The operations are:

Symbol	Effect
E	identity operation
σ	reflection through a plane
i	inversion through the center of symmetry
C_n	rotation by $2\pi/n$
S_n	rotation by $2\pi/n$ followed by reflection in the plane $\perp$ to the rotation axis

The operation R can be represented by the matrix $\{\Gamma(R)_{nm}\}$ which produces the desired transformation of a set of basis vectors u_n. Thus

$$Ru_m = \sum_n \Gamma(R)_{nm} u_n$$

The matrix elements provide sufficient representations of the operators because they have the same multiplication table as the operators. Accordingly, $R_1R_2 = R_3$ corresponds to the matrix product

$$\sum_j \Gamma(R_1)_{ij}\Gamma(R_2)_{jk} = \Gamma(R_3)_{ik}$$

The trace of such a matrix is called the character of the representation. For example, if R is represented by $\underset{\sim}{\Gamma}(R)$, the character is

$$\chi(R) = \sum_n \Gamma(R)_{nn}$$

We are concerned with two applications of group theory; namely (a) the simplification of the solution of the Schrödinger equation for molecules having symmetry and (b) the determination of selection rules for transitions that are induced by radiation. The former is possible because the eigenfunctions of the Hamiltonian which have the same eigenvalue are bases for the irreducible representations of the symmetry group to which the molecule belongs. In general, eigenfunctions which transform according to different irreducible representations of the molecular symmetry group will have different eigenvalues, except in cases of accidental degeneracy which are not related to symmetry. The eigenvalue problem can thus be simplified and sometimes solved completely by choosing basis functions at the outset which transform according to the irreducible representations of the symmetry group. The construction of such functions is simplified by making use of the unnormalized projection operator,

$$P_j = \sum_R \chi_j^*(R)R$$

which operates on an arbitrary function ϕ to generate a linear combination of functions transforming according to the jth irreducible representation. Thus

$$\boxed{u_j = P_j\phi = \sum_R \chi_j^*(R)R\phi} \tag{1}$$

The complete basis is obtained by operating on ℓ distinct functions where ℓ is the dimension of the jth irreducible representation. The resulting functions may not be orthogonal. In some problems there are n-functions ϕ which are related by symmetry. When Eq. (1) is applied to such a set with $n > \ell$, the resulting n linear combinations are not necessarily linearly independent and the solutions may not be unique. It is always possible, however, to construct a set of ℓ linearly independent functions from such a set.

Simplifications occur in the eigenvalue problem because matrix elements of the type

$$\int u_i^* \hat{\mathcal{H}} u_j \, d\tau$$

vanish when u_i and u_j transform according to different irreducible representations of the symmetry group. Such integrals can be non-zero only when the representation of the integrand is totally symmetric or contains the totally symmetric representation. Since $\hat{\mathcal{H}}$ transforms according to the totally symmetric representation of the molecular symmetry group, the integral will vanish unless the direct product of the representations of u_i and u_j contains the totally symmetric representation. It can be shown that this only occurs when the representations of u_i and u_j are identical, *i.e.* $\Gamma_i = \Gamma_j$.

In order to determine selection rules in molecular spectroscopy we need to consider integrals of the type

$$\int u_i^* \hat{\mathcal{H}}' u_f \, d\tau$$

where $\hat{\mathcal{H}}'$ is a perturbation and u_i and u_f are eigenfunctions of the unperturbed Hamiltonian. The integral will vanish unless the direct product of the representations of u_i and u_j is equal to or contains the representation of $\hat{\mathcal{H}}'$. In specific problems the symmetry of $\hat{\mathcal{H}}'$ is known and the transformation properties of u_i and u_f must be determined. Often we are concerned only with the number of allowed transitions, e.g. in infrared and Raman spectroscopy. In such cases the apparatus of group theory can be used to obtain directly the number of eigenfunctions n_i which transform according to the ith irreducible representation of the symmetry group. The procedure is to select a set of basis functions (vectors) which are sufficient for the description of the symmetry behavior of the eigenfunctions. The resulting representation (set of matrices) $\tilde{\Gamma}(R)$ is usually reducible, but it contains the irreducible representations associated with each of the eigenfunctions; i.e.

$$\Gamma(R) = \sum_i n_i \Gamma_i(R) \tag{2}$$

The number of times the ith irreducible representation is contained in the reducible representation $\Gamma(R)$ is given by

$$\boxed{n_i = \frac{1}{h}\sum_R \chi(R)\chi_i^*(R)} \tag{3}$$

where h is the order of the group (no. of symmetry elements) and $\chi(R)$ is the character of R in the reducible representation.

If the representations of both u_i and u_f are known for the integral above, then the representation based on their direct product is required. In practice only the characters of this representation are needed. Consider the general case where the sets {u} and {v} are bases for the representations Γ_i and Γ_f, respectively. The direct product of the two sets, i.e. the set of products $u_i v_f$, is the basis for the direct product representation $\Gamma_{if} = \Gamma_i x \Gamma_f$. It can be shown that the character for the operation R in the direct product representation is given by

$$\boxed{\chi_{if}(R) = \chi_i(R)\chi_f(R)} \tag{4}$$

where $\chi_i(R)$ and $\chi_f(R)$ are the characters of R in the representations Γ_i and Γ_f, respectively.

It is possible that {u} and {v} are bases for the same irreducible representation Γ_i. Then Eq. (4) gives $\chi_{ii}(R) = \chi_i^2(R)$. This relation is sufficient for the one-dimensional representations; however, when Γ_i is degenerate new complications arise. We consider two cases of special interest in spectroscopy.

(1) Excited vibrational states: A vibrational frequency ν having d-fold degeneracy is associated with a set of d normal modes. These modes (normal coordinates) form the basis for the d-dimensional representation Γ of the molecular point group. If the ith mode of

this set has the vibrational quantum n_i, then the total excitation of these modes can be indicated by the constant

$$n = \sum_{i=1}^{d} n_i$$

It can be shown that the total vibrational wave function for the degenerate set transforms as $[\Gamma^n]^+$ which is the symmetric direct product of Γ with itself n times (LE2,422). This situation differs from the one considered above because there is only one set of basis functions. Consider for example the case where n = 2. If two different d-fold degenerate sets, x_i (i = 1, ···, d) and y_i (i = 1, ···, d), were involved then the direct product would contain d^2 elements $x_i y_i$. When $\{x\} \equiv \{y\}$, only d(d + 1)/2 products can be formed. All of the basis functions in the direct product set are symmetric functions of the types x_i^2 or $x_i y_j + x_j y_i = x_i x_j + x_j x_i$. The d(d - 1)/2 antisymmetric combinations $x_i y_j - x_j y_i$ clearly vanish in this case. It can be shown that the character for R in the symmetric direct product representation for a doubly excited state (n = 2) is given by

$$\boxed{\chi_{\Gamma\Gamma}(R)^+ = \frac{1}{2}\,[\chi_\Gamma(R)^2 + \chi_\Gamma(R^2)]} \qquad (5)$$

This equation holds for symmetric direct products $(\Gamma x \Gamma)^+$ for representations Γ of any dimension. The generalizations of Eq. (5) to triply excited (n = 3) and higher states do, however, depend on the degeneracies involved (HE2,258; WDC,152).

(2) Excited electronic states: As discussed in Chapter 9 the many-electron wave functions for molecules must obey the Pauli exclusion principle. Consider for example the ground state of the benzene molecule. According to Problem 9.16 the configuration for the π-electrons can be written as $(a_{2u})^2(e_{1g})^4$. The repeated application of Eq. (4) neglecting spin restrictions gives the following resolution of this configuration into states:

$$3A_{1g} + 3A_{2g} + 5E_{2g}$$

However, the exclusion principle severely limits the number of possibilities and, in fact, permits only the A_{1g} state with S = 0 for this or any other closed shell system (see Problem 11.17).

For a degenerate state (transforming like Γ) containing two electrons, symmetric and antisymmetric spin functions corresponding to S = 1 and S = 0, respectively, are found. The allowed states, i.e. those combinations of spatial and spin functions which satisfy the exclusion principle, are obtained by associating the singlet (S = 0) spin function with spatial functions which transform like representations resulting from the symmetric direct product of Γ with itself; and the triplet (S = 1) spin functions with the spatial functions which transform like representations resulting from the antisymmetric direct product $(\Gamma x \Gamma)^-$

of Γ with itself. The characters for $(\Gamma x \Gamma)^+$ are obtained with Eq. (5) while the characters for $(\Gamma x \Gamma)^-$ are obtained from the equation:

$$\boxed{\chi_{\Gamma\Gamma}(R)^- = \frac{1}{2}[\chi_\Gamma(R)^2 - \chi_\Gamma(R^2)]} \qquad (6)$$

When more than two equivalent electrons are present, the calculation becomes much more involved. The reader should consult the article by Ford (listed below) and the references contained therein for discussions of the general case. The symmetric group algebra which is required for the construction of properly antisymmetrized many-electron wave functions is explained in MA,158.

Ref: D. I. Ford, J. Chem. Ed. 49, 336 (1972).

Example 1: Construction of symmetry orbitals for the π-electrons in the allyl radical using $2p_x$ basis functions.

The allyl radical which is treated in Problems 9.12 and 9.13 has C_{2v} symmetry, *i.e.* it has the following symmetry elements: E, C_2, $\sigma_v(xz)$ and $\sigma_v'(yz)$.

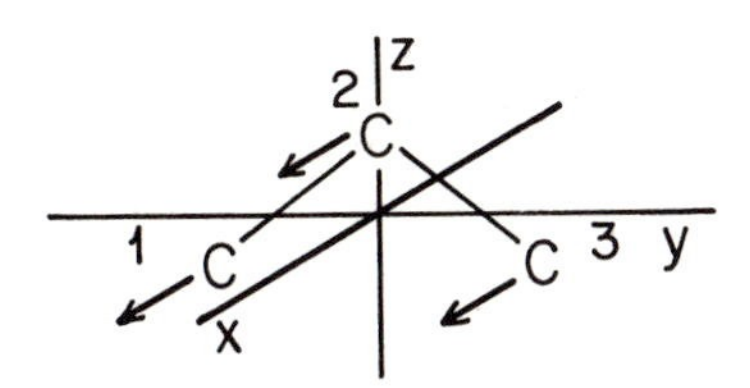

We denote the $2p_x$ orbitals by ϕ_1, ϕ_2, and ϕ_3. Then R operating on ϕ_1 gives:

$E\phi_1 = \phi_1$; $C_2\phi_1 = -\phi_3$; $\sigma_v(xz)\phi_1 = \phi_3$; $\sigma_v'(yz)\phi_1 = -\phi_1$

and using Eq. (1):

$$P(A_1)\phi_1 = \phi_1 - \phi_3 + \phi_3 - \phi_1 = 0$$
$$P(A_2)\phi_1 = \phi_1 - \phi_3 - \phi_3 + \phi_1 = 2(\phi_1 - \phi_3)$$
$$P(B_1)\phi_1 = \phi_1 + \phi_3 + \phi_3 + \phi_1 = 2(\phi_1 + \phi_3)$$
$$P(B_2)\phi_1 = \phi_1 + \phi_3 - \phi_3 - \phi_1 = 0$$

The function ϕ_3 gives nothing new, but for ϕ_2 we find:

$E\phi_2 = \phi_2$; $C_2\phi_2 = -\phi_2$; $\sigma_v(xz)\phi_2 = \phi_2$; $\sigma_v(yz)\phi_2 = -\phi_2$

and

$P(A_1) = \phi_2 - \phi_2 + \phi_2 - \phi_2 = 0$; $P(A_2) = 0$; $P(B_1) = 4\phi_2$; $P(B_2) = 0$

Thus, there are two combinations which transform according to B_1 and one which transforms according to A_2. When normalized these become

$$B_1: \frac{1}{\sqrt{2}}(\phi_1 + \phi_3), \phi_2; \qquad A_2: \frac{1}{\sqrt{2}}(\phi_1 - \phi_3)$$

These are, of course, the same functions which were derived intuitively in Problem 9.13.

Example 2: A representation for C_{2v} based on the $2p_x$ orbitals of the allyl radical.

The MO's for the π-electrons are linear combinations of the $2p_x$ orbitals which (1) diagonalize the Hamiltonian and (2) transform according to the irreducible representations of the molecular symmetry group. Suppose that we choose the functions ϕ_1, ϕ_2, and ϕ_3 as bases for a three-dimensional representation of C_{2v}. The matrices for this reducible representation can be written out quite easily:

$$E\begin{pmatrix}\phi_1\\ \phi_2\\ \phi_3\end{pmatrix} = \begin{pmatrix}1 & 0 & 0\\ 0 & 1 & 0\\ 0 & 0 & 1\end{pmatrix}\begin{pmatrix}\phi_1\\ \phi_2\\ \phi_3\end{pmatrix} = \begin{pmatrix}\phi_1\\ \phi_2\\ \phi_3\end{pmatrix}$$

$$C_2\begin{pmatrix}\phi_1\\ \phi_2\\ \phi_3\end{pmatrix} = \begin{pmatrix}0 & 0 & -1\\ 0 & -1 & 0\\ -1 & 0 & 0\end{pmatrix}\begin{pmatrix}\phi_1\\ \phi_2\\ \phi_3\end{pmatrix} = \begin{pmatrix}-\phi_3\\ -\phi_2\\ -\phi_1\end{pmatrix}$$

$$\sigma_v(xz)\begin{pmatrix}\phi_1\\ \phi_2\\ \phi_3\end{pmatrix} = \begin{pmatrix}0 & 0 & 1\\ 0 & 1 & 0\\ 1 & 0 & 0\end{pmatrix}\begin{pmatrix}\phi_1\\ \phi_2\\ \phi_3\end{pmatrix} = \begin{pmatrix}\phi_3\\ \phi_2\\ \phi_1\end{pmatrix}$$

$$\sigma_v'(yz)\begin{pmatrix}\phi_1\\ \phi_2\\ \phi_3\end{pmatrix} = \begin{pmatrix}-1 & 0 & 0\\ 0 & -1 & 0\\ 0 & 0 & -1\end{pmatrix}\begin{pmatrix}\phi_1\\ \phi_2\\ \phi_3\end{pmatrix} = \begin{pmatrix}-\phi_1\\ -\phi_2\\ -\phi_3\end{pmatrix}$$

The characters for this representation are thus

	E	C_2	$\sigma_v(xz)$	$\sigma_v(yz)$
$\Gamma(2p_x)$	3	-1	1	-3

The application of Eq. (3) then immediately gives the number of times each of the irreducible representations are contained in $\Gamma(2p_x)$ and thus the number of eigenfunctions which transform according to each of the irreducible representations. The results are (using the character table for C_{2v}):

$$n(A_1) = (1/4)(3 - 1 + 1 - 3) = 0$$
$$n(A_2) = (1/4)(3 - 1 - 1 + 3) = 1$$
$$n(B_1) = (1/4)(3 + 1 + 1 + 3) = 2$$
$$n(B_2) = (1/4)(3 + 1 - 1 - 3) = 0$$

Example 3: Determination of the symmetries of the normal modes of vibration for H_2O and the identification of infrared and Raman active modes. The structure of the H_2O molecule is shown below:

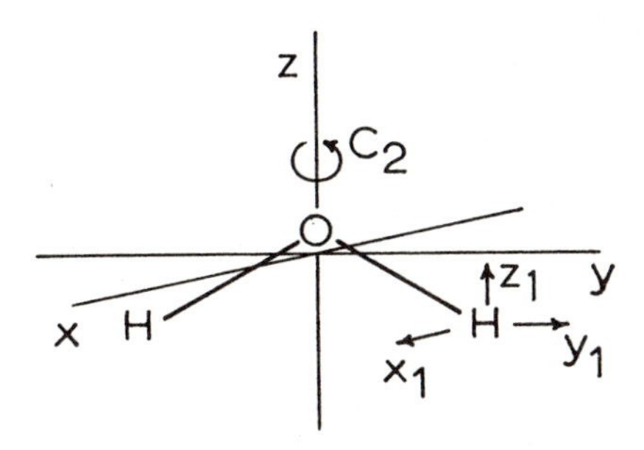

In order to represent the motion of the atoms in a molecule we use a set of 3N cartesian vectors, three of which are shown above. These vectors provide a basis for a 9 dimensional representation of C_{2v}. Consider for example the operation C_2:

$$\begin{pmatrix} 0 & 0 & 0 & -1 & 0 & 0 & 0 & 0 & 0 \\ 0 & 0 & 0 & 0 & -1 & 0 & 0 & 0 & 0 \\ 0 & 0 & 0 & 0 & 0 & +1 & 0 & 0 & 0 \\ -1 & 0 & 0 & 0 & 0 & 0 & 0 & 0 & 0 \\ 0 & -1 & 0 & 0 & 0 & 0 & 0 & 0 & 0 \\ 0 & 0 & +1 & 0 & 0 & 0 & 0 & 0 & 0 \\ 0 & 0 & 0 & 0 & 0 & 0 & -1 & 0 & 0 \\ 0 & 0 & 0 & 0 & 0 & 0 & 0 & -1 & 0 \\ 0 & 0 & 0 & 0 & 0 & 0 & 0 & 0 & +1 \end{pmatrix} \begin{pmatrix} x_{H1} \\ y_{H1} \\ z_{H1} \\ x_{H2} \\ y_{H2} \\ z_{H2} \\ x_0 \\ y_0 \\ z_0 \end{pmatrix} = \begin{pmatrix} -x_{H2} \\ -y_{H2} \\ z_{H2} \\ -x_{H2} \\ -y_{H1} \\ z_{H1} \\ -x_0 \\ -y_0 \\ z_0 \end{pmatrix}$$

The column vector after rotation was constructed by inspection and the matrix was then set up to give the desired effect. As explained previously we only need the character (trace) of this matrix which is -1. It should be clear that only atoms which are "unmoved" by the symmetry operation can contribute to the character, <u>i.e.</u> only atoms which lie on the corresponding symmetry element can contribute. For the operation $\sigma_v(xz)$ only the oxygen atom is unmoved and for it $x_0, y_0, z_0 \rightarrow x_0, -y_0, z_0$. Therefore $\chi(\sigma_{xz}) = +1$. Similarly, 0, H1, and H2 all lie on σ_{yz} and each contributes +1 to give $\chi(\sigma_{yz}) = +3$. The characters for Γ_9 are shown below:

C_{2v}	E	C_2	$\sigma_v(xz)$	$\sigma_v'(yz)$
Γ_9	9	-1	+1	+3

This representation can be reduced using Eq. (3).

$$n(A_1) = \frac{1}{4}(9 - 1 + 1 + 3) = 3$$

$$n(A_2) = \frac{1}{4}(9 - 1 - 1 - 3) = 1$$

$$n(B_1) = \frac{1}{4}(9 + 1 + 1 - 3) = 2$$

$$n(B_2) = \frac{1}{4}(9 + 1 - 1 + 3) = 3$$

Thus $\Gamma_9 = 3A_1 + A_2 + 2B_1 + 3B_2$. The character table for C_{2v} in Appendix 8 indicates that the three translations x, y, z and the three rotations R_x, R_y, R_z provide a basis for the representation

$$\Gamma(\text{trans, rot}) = A_1 + A_2 + 2B_1 + 2B_2$$

This leaves for the genuine vibrations:

$$\Gamma(\text{vib}) = 2A_1 + B_2$$

The normal coordinates for atomic motion are three in number with two transforming like A_1 and one like B_2.

The transition moment for a harmonic oscillator was considered in Problem 10.12. For the fundamental transition ($v'' = 0 \rightarrow v' = 1$) the x-component of the transition moment is proportional to

$$\int \psi^{v'=1} x \psi^{v''=0} d\tau \qquad (1)$$

But since $\psi^{v''=0}$ is totally symmetric and $\psi^{v'=1}$ transforms like the normal coordinate Q, this integral will vanish unless x transforms according to the same irreducible representation as Q, *i.e.* for the integrand to be totally symmetric the product xQ must be totally symmetric. The general rule is:

I. The fundamental vibration for a given normal coordinate will be active in the infrared spectrum only if the normal coordinate transforms according to the same irreducible representation of the molecular symmetry group as one or more of the Cartesian coordinates.

Raman scattering results from an induced electric dipole moment. In place of the Cartesian coordinate in (1) a component of the polarizability tensor must be substituted. *e.g.*

$$\int \psi^{v'=1} \alpha_{xy} \psi^{v''=0} d\tau$$

The component α_{xy} transforms as xy, α_{xz} as xz, and so on. The general rule is:

II. The fundamental vibration for a given normal coordinate will be active in the Raman spectrum only if the normal coordinate transforms according to the same irreducible representation of the molecular symmetry group as one or more of the components of the polarizability tensor.

Applying rules I and II to the H_2O molecule we find that there are two A_1 modes, both of which are infrared and Raman active. There is one B_2 mode which is also infrared and Raman active. Therefore, three frequencies will be found in the infrared spectrum and the same three frequencies will be found in the Raman spectrum. Information is also available about the polarization of the scattered light (Raman lines). It turns out that only those fundamentals associated with totally symmetric modes will be polarized. In the case of H_2O the B_2 mode will give rise to a depolarized Raman line.

Example 4: Determination of the symmetries and multiplicities of the excited states arising from the configuration $(a_1)^1(e)^2$ of the cyclopropenyl radical.

This C_{3v} system is discussed in Problem 9.15. The $(a_1)^1$ electron causes no problems. The representation is A and the spin is 1/2, i.e. the symbol is $^2A_1$. For the pair $(e)^2$, however, we must be careful with the exclusion principle. For S = 0 and S = 1 we require symmetric and antisymmetric spatial functions, respectively. The necessary characters are given in the table below for the E representation of C_{3v}:

R	E	C_3	σ_v
R^2	E	C_3^2	E
$\chi(R)$	2	-1	0
$\chi(R)^2$	4	1	0
$\chi(R^2)$	2	-1	2
$\chi(R)^+$	3	0	2
$\chi(R)^-$	1	1	-1

The characters $\chi(R)^+$ and $\chi(R)^-$ have been obtained using Eqs. (5) and (6), respectively. Equation (3) then permits the following reductions

$$S = 0: \quad (E\text{x}E)^+ = A_1 + E, \qquad S = 1: \quad (E\text{x}E)^- = A_2$$

Combining these with the appropriate multiplicities then gives

$$(e)^2 \rightarrow {}^1A_1 + {}^1E + {}^3A_2$$

The complete set of states is obtained from the product

$${}^2A_1\text{x}({}^1A_1 + {}^1E + {}^3A_2) = {}^2A_1 + ({}^2A_1\text{x}{}^1E) + ({}^2A_1\text{x}{}^3A_2) = {}^2A_1 + {}^2E + {}^4A_2 + {}^2A_2$$

This last resolution has been carried out with Eqs. (3) and (4) and the C_{3v} character table. The multiplicities are obtained from simple vector addition, i.e. $\underset{\sim}{S}_1 + \underset{\sim}{S}_2 \rightarrow S_1 + S_2$, $S_1 + S_2 - 1, \cdots, |S_1 - S_2|$.

In this appendix some of the most widely used character tables are reproduced. The irreducible representations Γ_i are listed at the left of each table in the conventional notation: A, B (one-dimensional representations), E (two-dimensional representation), T (three-dimensional representations). Across the top, the classes of symmetry operations are given. In applying Eq. (3) the coefficients of these classes must be taken into account.

Character Tables

1. The C_n Groups (n = 2,3,4,5,6)

C_2	E	C_2		
A	1	1	z, R_z	x^2, y^2, z^2, xy
B	1	-1	x, y, R_x, R_y	yz, xz

C_3	E	C_3	C_3^2		$\varepsilon = \exp(2\pi i/3)$
A	1	1	1	z, R_z	$x^2 + y^2, z^2$
E	1	ε	ε^*	$(x,y)(R_x,R_y)$	$(x^2 - y^2, xy)(yz, xz)$
	1	ε^*	ε		

C_4	E	C_4	C_2	C_4^3		
A	1	1	1	1	z, R_z	$x^2 + y^2, z^2$
B	1	-1	1	-1		$x^2 - y^2, xy$
E	1	i	-1	-i	$(x,y)(R_x,R_y)$	(yz, xz)
	1	-i	-1	i		

C_5	E	C_5	C_5^2	C_5^3	C_5^4		$\varepsilon = \exp(2\pi i/5)$
A	1	1	1	1	1	z, R_z	$x^2 + y^2, z^2$
E_1	1	ε	ε^2	ε^{2*}	ε^*	$(x,y)(R_x,R_y)$	(yz, xz)
	1	ε^*	ε^{2*}	ε^2	ε		
E_2	1	ε^2	ε^*	ε	ε^{2*}		$(x^2 - y^2, xy)$
	1	ε^{2*}	ε	ε^*	ε^2		

C_6	E	C_6	C_3	C_2	C_3^2	C_6^5		$\varepsilon = \exp(2\pi i/6)$
A	1	1	1	1	1	1	z, R_z	$x^2 + y^2, z^2$
B	1	-1	1	-1	1	-1		
E_1	1	ε	$-\varepsilon^*$	-1	$-\varepsilon$	ε^*	(x,y)	(xz, yz)
	1	ε^*	$-\varepsilon$	-1	$-\varepsilon^*$	ε	(R_x,R_y)	
E_2	1	$-\varepsilon^*$	$-\varepsilon$	1	$-\varepsilon^*$	$-\varepsilon$		$(x^2 - y^2, xy)$
	1	$-\varepsilon$	$-\varepsilon^*$	1	$-\varepsilon$	$-\varepsilon^*$		

2. The C_{nv} Groups (n = 2,3,4,5,6)

C_{2v}	E	C_2	$\sigma_v(xz)$	$\sigma_v'(yz)$		
A_1	1	1	1	1	z	x^2,y^2,z^2
A_2	1	1	-1	-1	R_z	xy
B_1	1	-1	1	-1	x,R_y	xz
B_2	1	-1	-1	1	y,R_x	yz

C_{3v}	E	$2C_3$	$3\sigma_v$		
A_1	1	1	1	z	$x^2 + y^2,z^2$
A_2	1	1	-1	R_z	
E	2	-1	0	$(x,y)(R_x,R_y)$	$(x^2 - y^2,xy)(xz,yz)$

C_{4v}	E	$2C_4$	C_2	$2\sigma_v$	$2\sigma_d$		
A_1	1	1	1	1	1	z	$x^2 + y^2,z^2$
A_2	1	1	1	-1	-1	R_z	
B_1	1	-1	1	1	-1		$x^2 - y^2$
B_2	1	-1	1	-1	1		xy
E	2	0	-2	0	0	$(x,y)(R_x,R_y)$	(xz,yz)

C_{5v}	E	$2C_5$	$2C_5^2$	$5\sigma_v$		
A_1	1	1	1	1	z	$x^2 + y^2,z^2$
A_2	1	1	1	-1	R_z	
E_1	2	2cos72°	2cos144°	0	$(x,y)(R_x,R_y)$	(xz,yz)
E_2	2	2cos144°	2cos72°	0		$(x^2 - y^2,xy)$

C_{6v}	E	$2C_6$	$2C_3$	C_2	$3\sigma_v$	$3\sigma_d$		
A_1	1	1	1	1	1	1	z	$x^2 + y^2,z^2$
A_2	1	1	1	1	-1	-1	R_z	
B_1	1	-1	1	-1	1	-1		
B_2	1	-1	1	-1	-1	1		
E_1	2	1	-1	-2	0	0	$(x,y)(R_x,R_y)$	(xz,yz)
E_2	2	-1	-1	2	0	0		$(x^2 - y^2,xy)$

3. The C_{nh} Groups (n = 2,3,4,5,6)

C_{2h}	E	C_2	i	σ_h		
A_g	1	1	1	1	R_z	x^2,y^2,z^2,xy
B_g	1	-1	1	-1	R_x,R_y	xz,yz
A_u	1	1	-1	-1	z	
B_u	1	-1	-1	1	x,y	

C_{3h}	E	C_3	C_3^2	σ_h	S_3	S_3^5		$\varepsilon = \exp(2\pi i/3)$
A'	1	1	1	1	1	1	R_z	$x^2 + y^2,z^2$
E'	1	ε	ε^*	1	ε	ε^*	(x,y)	$(x^2 - y^2,xy)$
	1	ε^*	ε	1	ε^*	ε		
A"	1	1	1	-1	-1	-1	z	
E"	1	ε	ε^*	-1	$-\varepsilon$	$-\varepsilon^*$	(R_x,R_y)	(xz,yz)
	1	ε^*	ε	-1	$-\varepsilon^*$	$-\varepsilon$		

C_{4h}	E	C_4	C_2	C_4^3	i	S_4^3	σ_h	S_4		
A_g	1	1	1	1	1	1	1	1	R_z	$x^2 + y^2,z^2$
B_g	1	-1	1	-1	1	-1	1	-1		$x^2 - y^2,xy$
E_g	1	i	-1	-i	1	i	-1	-i	(R_x,R_y)	(xz,yz)
	1	-i	-1	i	1	-i	-1	i		
A_u	1	1	1	1	-1	-1	-1	-1	z	
B_u	1	-1	1	-1	-1	1	-1	1		
E_u	1	i	-1	-i	-1	-i	1	i	(x,y)	
	1	-i	-1	i	-1	i	1	-i		

C_{5h}	E	C_5	C_5^2	C_5^3	C_5^4	σ_h	S_5	S_5^7	S_5^3	S_5^9		$\varepsilon = \exp(2\pi i/5)$
A'	1	1	1	1	1	1	1	1	1	1	R_z	$x^2 + y^2,z^2$
E_1'	1	ε	ε^2	ε^{2*}	ε^*	1	ε	ε^2	ε^{2*}	ε^*	(x,y)	
	1	ε^*	ε^{2*}	ε^2	ε	1	ε^*	ε^{2*}	ε^2	ε		
E_2'	1	ε^2	ε^*	ε	ε^{2*}	1	ε^2	ε^*	ε	ε^{2*}		$(x^2 - y^2,xy)$
	1	ε^{2*}	ε	ε^*	ε^2	1	ε^{2*}	ε	ε^*	ε^2		
A"	1	1	1	1	1	-1	-1	-1	-1	-1	z	
E_1"	1	ε	ε^2	ε^{2*}	ε^*	-1	$-\varepsilon$	$-\varepsilon^2$	$-\varepsilon^{2*}$	$-\varepsilon^*$	(R_x,R_y)	(xz,yz)
	1	ε^*	ε^{2*}	ε^2	ε	-1	$-\varepsilon^*$	$-\varepsilon^{2*}$	$-\varepsilon^2$	$-\varepsilon$		
E_2"	1	ε^2	ε^*	ε	ε^{2*}	-1	$-\varepsilon^2$	$-\varepsilon^*$	$-\varepsilon$	$-\varepsilon^{2*}$		
	1	ε^{2*}	ε	ε^*	ε^2	-1	$-\varepsilon^{2*}$	$-\varepsilon$	$-\varepsilon^*$	$-\varepsilon^2$		

C_{6h}	E	C_6	C_3	C_2	C_3^2	C_6^5	i	S_3^5	S_6^5	σ_h	S_6	S_3		$\varepsilon = \exp(2\pi i/6)$
A_g	1	1	1	1	1	1	1	1	1	1	1	1	R_z	$x^2 + y^2, z^2$
B_g	1	-1	1	-1	1	-1	1	-1	1	-1	1	-1		
E_{1g}	$\begin{cases}1 \\ 1\end{cases}$	$\begin{matrix}\varepsilon \\ \varepsilon^*\end{matrix}$	$\begin{matrix}-\varepsilon^* \\ -\varepsilon\end{matrix}$	$\begin{matrix}-1 \\ -1\end{matrix}$	$\begin{matrix}-\varepsilon \\ -\varepsilon^*\end{matrix}$	$\begin{matrix}\varepsilon^* \\ \varepsilon\end{matrix}$	$\begin{matrix}1 \\ 1\end{matrix}$	$\begin{matrix}\varepsilon \\ \varepsilon^*\end{matrix}$	$\begin{matrix}-\varepsilon^* \\ -\varepsilon\end{matrix}$	$\begin{matrix}-1 \\ -1\end{matrix}$	$\begin{matrix}-\varepsilon \\ -\varepsilon^*\end{matrix}$	$\begin{matrix}\varepsilon^* \\ \varepsilon\end{matrix}$	(R_x, R_y)	(xz, yz)
E_{2g}	$\begin{cases}1 \\ 1\end{cases}$	$\begin{matrix}-\varepsilon^* \\ -\varepsilon\end{matrix}$	$\begin{matrix}-\varepsilon \\ -\varepsilon^*\end{matrix}$	$\begin{matrix}1 \\ 1\end{matrix}$	$\begin{matrix}-\varepsilon^* \\ -\varepsilon\end{matrix}$	$\begin{matrix}-\varepsilon \\ -\varepsilon^*\end{matrix}$	$\begin{matrix}1 \\ 1\end{matrix}$	$\begin{matrix}-\varepsilon^* \\ -\varepsilon\end{matrix}$	$\begin{matrix}-\varepsilon \\ -\varepsilon^*\end{matrix}$	$\begin{matrix}1 \\ 1\end{matrix}$	$\begin{matrix}-\varepsilon^* \\ -\varepsilon\end{matrix}$	$\begin{matrix}-\varepsilon \\ -\varepsilon^*\end{matrix}$		$(x^2 - y^2, xy)$
A_u	1	1	1	1	1	1	-1	-1	-1	-1	-1	-1	z	
B_u	1	-1	1	-1	1	-1	-1	1	-1	1	-1	1		
E_{1u}	$\begin{cases}1 \\ 1\end{cases}$	$\begin{matrix}\varepsilon \\ \varepsilon^*\end{matrix}$	$\begin{matrix}-\varepsilon^* \\ -\varepsilon\end{matrix}$	$\begin{matrix}-1 \\ -1\end{matrix}$	$\begin{matrix}-\varepsilon \\ -\varepsilon^*\end{matrix}$	$\begin{matrix}\varepsilon^* \\ \varepsilon\end{matrix}$	$\begin{matrix}-1 \\ -1\end{matrix}$	$\begin{matrix}-\varepsilon \\ -\varepsilon^*\end{matrix}$	$\begin{matrix}\varepsilon^* \\ \varepsilon\end{matrix}$	$\begin{matrix}1 \\ 1\end{matrix}$	$\begin{matrix}\varepsilon \\ \varepsilon^*\end{matrix}$	$\begin{matrix}-\varepsilon^* \\ -\varepsilon\end{matrix}$	(x, y)	
E_{2u}	$\begin{cases}1 \\ 1\end{cases}$	$\begin{matrix}-\varepsilon^* \\ -\varepsilon\end{matrix}$	$\begin{matrix}-\varepsilon \\ -\varepsilon^*\end{matrix}$	$\begin{matrix}1 \\ 1\end{matrix}$	$\begin{matrix}-\varepsilon^* \\ -\varepsilon\end{matrix}$	$\begin{matrix}-\varepsilon \\ -\varepsilon^*\end{matrix}$	$\begin{matrix}-1 \\ -1\end{matrix}$	$\begin{matrix}\varepsilon^* \\ \varepsilon\end{matrix}$	$\begin{matrix}\varepsilon \\ \varepsilon^*\end{matrix}$	$\begin{matrix}-1 \\ -1\end{matrix}$	$\begin{matrix}\varepsilon^* \\ \varepsilon\end{matrix}$	$\begin{matrix}\varepsilon \\ \varepsilon^*\end{matrix}$		

4. The D_{nh} Groups ($n = 2,3,4,5,6$)

D_{2h}	E	$C_2(z)$	$C_2(y)$	$C_2(x)$	i	$\sigma(xy)$	$\sigma(xz)$	$\sigma(yz)$		
A_g	1	1	1	1	1	1	1	1		x^2, y^2, z^2
B_{1g}	1	1	-1	-1	1	1	-1	-1	R_z	xy
B_{2g}	1	-1	1	-1	1	-1	1	-1	R_y	xz
B_{3g}	1	-1	-1	1	1	-1	-1	1	R_x	yz
A_u	1	1	1	1	-1	-1	-1	-1		
B_{1u}	1	1	-1	-1	-1	-1	1	1	z	
B_{2u}	1	-1	1	-1	-1	1	-1	1	y	
B_{3u}	1	-1	-1	1	-1	1	1	-1	x	

D_{3h}	E	$2C_3$	$3C_2$	σ_h	$2S_3$	$3\sigma_v$		
A_1'	1	1	1	1	1	1		$x^2 + y^2, z^2$
A_2'	1	1	-1	1	1	-1	R_z	
E'	2	-1	0	2	-1	0	(x, y)	$(x^2 - y^2, xy)$
A_1''	1	1	1	-1	-1	-1		
A_2''	1	1	-1	-1	-1	1	z	
E''	2	-1	0	-2	1	0	(R_x, R_y)	(xz, yz)

D_{4h}	E	$2C_4$	C_2	$2C_2'$	$2C_2''$	i	$2S_4$	σ_h	$2\sigma_v$	$2\sigma_d$		
A_{1g}	1	1	1	1	1	1	1	1	1	1		$x^2 + y^2, z^2$
A_{2g}	1	1	1	-1	-1	1	1	1	-1	-1	R_z	
B_{1g}	1	-1	1	1	-1	1	-1	1	1	-1		$x^2 - y^2$
B_{2g}	1	-1	1	-1	1	1	-1	1	-1	1		xy
E_g	2	0	-2	0	0	2	0	-2	0	0	(R_x, R_y)	(xz, yz)
A_{1u}	1	1	1	1	1	-1	-1	-1	-1	-1		
A_{2u}	1	1	1	-1	-1	-1	-1	-1	1	1	z	
B_{1u}	1	-1	1	1	-1	-1	1	-1	-1	1		
B_{2u}	1	-1	1	-1	1	-1	1	-1	1	-1		
E_u	2	0	-2	0	0	-2	0	2	0	0	(x, y)	

D_{5h}	E	$2C_5$	$2C_5^2$	$5C_2$	σ_h	$2S_5$	$2S_5^3$	$5\sigma_v$		
A_1'	1	1	1	1	1	1	1	1		$x^2 + y^2, z^2$
A_2'	1	1	1	-1	1	1	1	-1	R_z	
E_1'	2	2cos72°	2cos144°	0	2	2cos72°	2cos144°	0	(x, y)	
E_2'	2	2cos144°	2cos72°	0	2	2cos144°	2cos72°	0		$(x^2 - y^2, xy)$
A_1''	1	1	1	1	-1	-1	-1	-1		
A_2''	1	1	1	-1	-1	-1	-1	1	z	
E_1''	2	2cos72°	2cos144°	0	-2	-2cos72°	-2cos144°	0	(R_x, R_y)	(xz, yz)
E_2''	2	2cos144°	2cos72°	0	-2	-2cos144°	-2cos72°	0		

D_{6h}	E	$2C_6$	$2C_3$	C_2	$3C_2'$	$3C_2''$	i	$2S_3$	$2S_6$	σ_h	$3\sigma_d$	$3\sigma_v$		
A_{1g}	1	1	1	1	1	1	1	1	1	1	1	1		$x^2 + y^2, z^2$
A_{2g}	1	1	1	1	-1	-1	1	1	1	1	-1	-1	R_z	
B_{1g}	1	-1	1	-1	1	-1	1	-1	1	-1	1	-1		
B_{2g}	1	-1	1	-1	-1	1	1	-1	1	-1	-1	1		
E_{1g}	2	1	-1	-2	0	0	2	1	-1	-2	0	0	(R_x, R_y)	(xz, yz)
E_{2g}	2	-1	-1	2	0	0	2	-1	-1	2	0	0		$(x^2 - y^2, xy)$
A_{1u}	1	1	1	1	1	1	-1	-1	-1	-1	-1	-1		
A_{2u}	1	1	1	1	-1	-1	-1	-1	-1	-1	1	1	z	
B_{1u}	1	-1	1	-1	1	-1	-1	1	-1	1	-1	1		
B_{2u}	1	-1	1	-1	-1	1	-1	1	-1	1	1	-1		
E_{1u}	2	1	-1	-2	0	0	-2	-1	1	2	0	0	(x, y)	
E_{2u}	2	-1	-1	2	0	0	-2	1	1	-2	0	0		

5. The S_4 Group

S_4	E	S_4	C_2	$S_4{}^3$		
A	1	1	1	1	R_z	$x^2 + y^2, z^2$
B	1	-1	1	-1	z	$x^2 - y^2, xy$
E	$\left\{\begin{matrix}1 \\ 1\end{matrix}\right.$	$\begin{matrix}i \\ -i\end{matrix}$	$\begin{matrix}-1 \\ -1\end{matrix}$	$\left.\begin{matrix}-i \\ i\end{matrix}\right\}$	(x,y) (R_x, R_y)	(xz,yz)

6. The Cubic Groups

T_d	E	$8C_3$	$3C_2$	$6S_4$	$6\sigma_d$		
A_1	1	1	1	1	1		$x^2 + y^2 + z^2$
A_2	1	1	1	-1	-1		
E	2	-1	2	0	0		$(2z^2 - x^2 - y^2, x^2 - y^2)$
T_1	3	0	-1	1	-1	(R_x, R_y, R_z)	
T_2	3	0	-1	-1	1	(x,y,z)	(xy,xz,yz)

O_h	E	$8C_3$	$6C_2$	$6C_4$	$3C_2(=C_4{}^2)$	i	$6S_4$	$8S_6$	$3\sigma_h$	$6\sigma_d$		
A_{1g}	1	1	1	1	1	1	1	1	1	1		$x^2 + y^2 + z^2$
A_{2g}	1	1	-1	-1	1	1	-1	1	1	-1		
E_g	2	-1	0	0	2	2	0	-1	2	0		$(2z^2 - x^2 - y^2, x^2 - y^2)$
T_{1g}	3	0	-1	1	-1	3	1	0	-1	-1	(R_x, R_y, R_z)	
T_{2g}	3	0	1	-1	-1	3	-1	0	-1	1		(xz,yz,xy)
A_{1u}	1	1	1	1	1	-1	-1	-1	-1	-1		
A_{2u}	1	1	-1	-1	1	-1	1	-1	-1	1		
E_u	2	-1	0	0	2	-2	0	1	-2	0		
T_{1u}	3	0	-1	1	-1	-3	-1	0	1	1	(x,y,z)	
T_{2u}	3	0	1	-1	-1	-3	1	0	1	-1		

7. The $C_{\infty v}$ and $D_{\infty h}$ Groups

$C_{\infty v}$	E	$2C_\infty{}^\Phi$	...	$\infty\sigma_v$		
$A_1 \equiv \Sigma^+$	1	1	...	1	z	$x^2 + y^2, z^2$
$A_2 \equiv \Sigma^-$	1	1	...	-1	R_z	
$E_1 \equiv \Pi$	2	$2\cos\Phi$	...	0	(x,y) (R_x, R_y)	(xz,yz)
$E_2 \equiv \Delta$	2	$2\cos 2\Phi$	...	0		$(x^2 - y^2, xy)$
$E_3 \equiv \Phi$	2	$2\cos 3\Phi$	...	0		
...	...	...	...	...		

$D_{\infty h}$	E	$2C_{\infty}^{\Phi}$	$\cdots$	$\infty\sigma_v$	i	$2S_{\infty}^{\Phi}$	$\cdots$	∞C_2		
Σ_g^+	1	1	$\cdots$	1	1	1	$\cdots$	1		$x^2 + y^2, z^2$
Σ_g^-	1	1	$\cdots$	-1	1	1	$\cdots$	-1	R_z	
Π_g	2	$2\cos\Phi$	$\cdots$	0	2	$-2\cos\Phi$	$\cdots$	0	(R_x, R_y)	(xz, yz)
Δ_g	2	$2\cos 2\Phi$	$\cdots$	0	2	$2\cos 2\Phi$	$\cdots$	0		$(x^2 - y^2, xy)$
$\cdots$	$\cdots$	$\cdots$	$\cdots$	$\cdots$	$\cdots$	$\cdots$	$\cdots$	$\cdots$		
Σ_u^+	1	1	$\cdots$	1	-1	-1	$\cdots$	-1	z	
Σ_u^-	1	1	$\cdots$	-1	-1	-1	$\cdots$	1		
Π_u	2	$2\cos\Phi$	$\cdots$	0	-2	$2\cos\Phi$	$\cdots$	0	(x, y)	
Δ_u	2	$2\cos 2\Phi$	$\cdots$	0	-2	$-2\cos 2\Phi$	$\cdots$	0		
$\cdots$	$\cdots$	$\cdots$	$\cdots$	$\cdots$	$\cdots$	$\cdots$	$\cdots$	$\cdots$		

APPENDIX 9

MAXWELL'S EQUATIONS

Ref. FE2,Chap. 18; PP,142; SF,85

We give only a brief summary in order to establish the notation. The reader should consult the references listed above for detailed explanations and discussion. Again we use SI (rationalized MKS) units:

Definition of terms:

$\mathcal{E}$ = electric field intensity ($V \cdot m^{-1}$)

D = electric displacement ($C \cdot m^{-2}$)

H = magnetic field intensity ($A \cdot m^{-1}$)

B = magnetic induction (T or $Wb \cdot m^{-2}$)

P = polarization ($C \cdot m^{-2}$)

M = magnetization ($A \cdot m^{-1}$)

j = current density ($A \cdot m^{-2}$)

σ = conductivity ($A \cdot V^{-1} \cdot m^{-1}$)

ρ = charge density ($C \cdot m^{-3}$)

A = vector potential

ϕ = scalar potential

ε_o = permittivity of free space ($F \cdot m^{-1}$)

μ_o = permeability of free space ($N \cdot A^{-2}$)

k_e = dielectric constant

k_m = relative permeability

ε = permittivity

μ = permeability

The equations:

In free space

$$\nabla \cdot \mathcal{E} = 0 \qquad (1)$$

$$\nabla \cdot H = 0 \qquad (2)$$

$$\nabla \times \mathcal{E} = -\mu_o \frac{\partial H}{\partial t} \qquad (3)$$

$$\nabla \times H = \varepsilon_o \frac{\partial \mathcal{E}}{\partial t} \qquad (4)$$

In stationary matter

$$\nabla \cdot D = \rho \qquad (5)$$

$$\nabla \cdot B = 0 \qquad (6)$$

$$\nabla \times \mathcal{E} = -\frac{\partial B}{\partial t} \qquad (7)$$

$$\nabla \times H = j + \frac{\partial D}{\partial t} \qquad (8)$$

where $D = \varepsilon_o \mathcal{E} + P$ and $B = \mu_o(H + M)$.

The constitutive relations:

The solutions of the field equations in matter require additional relations between $\mathcal{E}$ and D, B and H, and j and $\mathcal{E}$. The standard set of such relations is:

$$D = \varepsilon \mathcal{E} = k_e \varepsilon_o \mathcal{E}; \qquad B = \mu H = k_m \mu_o H; \qquad j = \sigma \mathcal{E}$$

The potentials:

The magnetic induction $\underset{\sim}{B}$ can be written in terms of the vector potential $\underset{\sim}{A}$ as follows: $\underset{\sim}{B} = \underset{\sim}{\nabla} \times \underset{\sim}{A}$. The substitution of this relation into Eq. (7) gives:

$$\underset{\sim}{\nabla} \times \left(\underset{\sim}{\mathcal{E}} + \frac{\partial \underset{\sim}{A}}{\partial t} \right) = 0$$

Since $\underset{\sim}{\nabla} \times (\underset{\sim}{\nabla} \phi) = 0$ for any function ϕ we can define the scalar potential by the equation: $\underset{\sim}{\mathcal{E}} + \partial \underset{\sim}{A}/\partial t = -\underset{\sim}{\nabla}\phi$. Thus the magnetic induction and the electric field intensity can be defined by the equations:

$$\boxed{\begin{aligned} \underset{\sim}{B} &= \underset{\sim}{\nabla} \times \underset{\sim}{A} \\ \underset{\sim}{\mathcal{E}} &= -\frac{\partial \underset{\sim}{A}}{\partial t} - \underset{\sim}{\nabla}\phi \end{aligned}} \qquad \begin{aligned} (9) \\ (10) \end{aligned}$$

The gauge: Ref. HA1,10

The vector fields $\underset{\sim}{B}$ and $\underset{\sim}{\mathcal{E}}$ are single valued functions of position; however $\underset{\sim}{A}$ and ϕ are not uniquely defined by Eqs. (9) and (10). In fact, $\underset{\sim}{A}$ and ϕ can be replaced by $\underset{\sim}{A}'$ and ϕ' where

$$\underset{\sim}{A}' = \underset{\sim}{A} + \underset{\sim}{\nabla}\chi; \qquad \phi' = \phi - \frac{\partial \chi}{\partial t}$$

without changing $\underset{\sim}{B}$ or $\underset{\sim}{\mathcal{E}}$. This property is called the gauge invariance of $\underset{\sim}{B}$ and $\underset{\sim}{\mathcal{E}}$. The choice of χ defines the <u>gauge</u> of $\underset{\sim}{A}$. Two gauges commonly chosen for convenience are:

Coulomb: Here it is required that $\underset{\sim}{\nabla} \cdot \underset{\sim}{A} = 0$. With this choice it can be shown that the scalar potential is determined only by the charge density ρ.

Lorentz: Here $\underset{\sim}{\nabla} \cdot \underset{\sim}{A} + \frac{\partial \phi}{\partial t} = 0$. The advantage is that the Lorentz gauge can be written in relativistically invariant form. It is, therefore, often chosen for use in radiation theory. For non-relativistic problems dealing with radiation the Coulomb gauge, which is simpler, can be used. It should be realized that the Coulomb and Lorentz choices restrict the form of χ but do not uniquely determine it.

APPENDIX 10
USEFUL INTEGRALS

1. $$\int x^n e^{ax} dx = e^{ax} \sum_{r=0}^{n} (-1)^r \frac{n!x^{n-r}}{(n-r)!a^{r+1}}$$

2. $$\int x^n \sin(ax) dx = -\frac{x^n \cos(ax)}{a} + \frac{n}{a} \int x^{n-1} \cos(ax) dx$$

3. $$\int x^n \cos(ax) dx = \frac{x^n \sin(ax)}{a} - \frac{n}{a} \int x^{n-1} \sin(ax) dx$$

4. $$\int \sin^n(ax) dx = -\frac{\sin^{n-1}(ax)\cos(ax)}{na} + \frac{(n-1)}{n} \int \sin^{n-2}(ax) dx$$

5. $$\int \cos^n(ax) dx = \frac{\cos^{n-1}(ax)\sin(ax)}{na} + \frac{(n-1)}{n} \int \cos^{n-2}(ax) dx$$

6. $$\int_b^\infty x^n e^{-ax} dx = \frac{n!e^{-ab}}{a^{n+1}} \left[1 + ab + \frac{(ab)^2}{2!} + \cdots + \frac{(ab)^n}{n!}\right]$$

7. $$\int_o^\infty x^n e^{-ax} dx = \frac{n!}{a^{n+1}} \quad (a > 0, \quad n = 1, 2, \cdots)$$

8. $$\int_o^\infty x^{2n+1} e^{-ax^2} dx = \frac{n!}{2a^{n+1}} \quad (a > 0)$$

9. $$\int_o^\infty x^{2n} e^{-ax^2} dx = \frac{1\cdot3\cdot5\cdots(2n-1)}{2^{n+1}a^n} \sqrt{\frac{\pi}{a}}$$

10. $$\int_{-\varepsilon}^{+\varepsilon} \delta(x) dx = 1; \qquad \int_o^a \delta(x) dx = \begin{cases} +1/2, & a > 0 \\ -1/2, & a < 0 \end{cases}$$

11. $$\int_a^b f(x)\delta(x) dx = f(0) \quad (a < 0 < b)$$

12. $\int f(x)\delta(x - a)dx = f(a)$

13. $\int f(x)\delta'(x)dx = -f'(0)$

14. $\int f(\underset{\sim}{r})\delta(\underset{\sim}{r})d^3r = f(0)$ where $\delta(\underset{\sim}{r})d^3r = \delta(x)\delta(y)\delta(z)dxdydz$

$$= \frac{\delta(r)}{2\pi r^2} r^2dr = \frac{\delta(r)}{2\pi} dr$$

(for spherical symmetry)

15. $\int \delta(x - a)\delta(x - b)dx = \delta(a - b)$

REFERENCES

AB A. Abragam, The Principles of Magnetic Resonance (Oxford University Press, London, 1961).

AN1 J. M. Anderson, Intro. to Quantum Chemistry (W. A. Benjamin, New York, 1969).

AN2 J. M. Anderson, Mathematics for Quantum Chemistry (W. A. Benjamin, New York, 1966).

AN3 E. E. Anderson, Modern Physics and Quantum Mechanics (W. B. Saunders Co., 1971).

AT1 P. W. Atkins, Molecular Quantum Mechanics, Parts I and II (Clarendon Press, Oxford, 1970).

AT2 P. W. Atkins, Molecular Quantum Mechanics, Part III (Clarendon Press, Oxford, 1970).

BA1 G. M. Barrow, Intro. to Molecular Spectroscopy (McGraw-Hill Book Co., New York, 1962).

BA2 D. R. Bates, Quantum Theory, I. Elements (Academic Press, New York, 1961).

BC B. Chu, Molecular Forces (John Wiley and Sons, New York, 1967).

BL S. M. Blinder, Advanced Physical Chemistry (The Macmillan Co., New York, 1969).

BO D. Bohm, Quantum Theory (Prentice Hall, Inc., Englewood Cliffs, 1951).

BR R. Bracewell, The Fourier Transform and Its Applications (McGraw-Hill Book Co., New York, 1965).

BR1 B. H. Bransden, Atomic Collision Theory (W. A. Benjamin, Inc., New York, 1970).

BS D. M. Brink and G. R. Satchler, Angular Momentum, 2nd Ed. (Clarendon Press, Oxford, 1968).

BS2 H. A. Bethe and E. E. Salpeter, Quantum Mechanics of One- and Two-Electron Atoms (Academic Press, New York, 1957).

BW M. Born and E. Wolf, Principles of Optics, 3rd Ed. (Pergamon Press, Oxford, 1964).

CGT J. A. Cronin, D. F. Greenberg, and V. L. Telegdi, University of Chicago Graduate Problems in Physics (Addison-Wesley Publ. Co., Reading, 1967).

CHB J. O. Hirschfelder, C. F. Curtiss, and R. B. Bird, Molecular Theory of Gases and Liquids (John Wiley & Sons, New York, 1954).

CM A. Carrington and A. D. McLachlan, Intro. to Magnetic Resonance (Harper & Row, New York, 1967).

CM2 F. Constantineau and E. Magyari, Problems in Quantum Mechanics (Pergamon Press, Oxford, 1971).

CO C. A. Coulson, Valence, 2nd Ed. (Oxford University Press, London, 1961).

CO1 F. A. Cotton, Chemical Applications of Group Theory, 2nd Ed. (Interscience Publishers, New York, 1971).

CS E. U. Condon and G. H. Shortley, The Theory of Atomic Spectra (Cambridge University Press, 1963).

DA1 J. C. Davis, Jr., Advanced Physical Chemistry (Ronald Press, New York, 1965).

DA2 A. S. Davydov, Quantum Mechanics (NEO Press, Ann Arbor, 1966).

DA3 N. Davidson, Statistical Mechanics, 1st Ed. (McGraw-Hill Book Company, New York, 1962).

DE M. Dewar, The Molecular Orbital Theory of Organic Chemistry (McGraw-Hill Book Co., New York, 1969).

DE2 P. Debye, Polar Molecules (Dover Publications, New York, 1929).

DH T. P. Das and E. L. Hahn, Nuclear Quadrupole Resonance (Academic Press, New York, 1958).

DI P. A. M. Dirac, The Principles of Quantum Mechanics, 4th Ed. (Oxford, University Press, 1958).

DU S. Dushman, The Elements of Quantum Mechanics, 1st Ed. (John Wiley and Sons, Inc., 1938).

DW R. H. Dicke and J. P. Wittke, Introduction to Quantum Mechanics (Addison-Wesley Publishing Co., Reading, 1960).

EI R. M. Eisberg, Fundamentals of Modern Physics (John Wiley and Sons, Inc., New York, 1962).

EWK H. Eyring, J. Walter, G. E. Kimball, Quantum Chemistry (John Wiley and Sons, Inc., 1944).

FA J. E. G. Farina, Quantum Theory of Scattering Processes (Pergamon Press, Oxford, 1973).

FE1 R. P. Feynman, R. B. Leighton, and M. Sands, The Feynman Lectures on Physics, Vol. 1 (Mechanics, Radiation, and Heat) (Addison-Wesley Publishing Company, Reading, 1964).

FE2 R. P. Feynman, R. B. Leighton, and M. Sands, The Feynman Lectures on Physics, Vol. 2 (Electromagnetism and Matter) (Addison-Wesley Publishing Company, Reading, 1964).

FE3 R. P. Feynman, R. B. Leighton, and M. Sands, The Feynman Lectures on Physics, Vol. 3 (Quantum Mechanics) (Addison-Wesley, Reading, 1965).

FL1 S. Flügge, Practical Quantum Mechanics I (Springer-Velag, Berlin, 1971).

FO G. R. Fowles, Introduction to Modern Optics (Holt, Rinehart, and Winston, New York, 1968).

FR H. Fraunfelder, The Mössbauer Effect (W. A. Benjamin, New York, 1962).

GO H. Goldstein, Classical Mechanics (Addison-Wesley Publ. Co., Reading, 1959).

HA M. W. Hanna, Quantum Mechanics in Chemistry, 2nd Ed. (W. A. Benjamin, New York, 1969).

HA1 H. F. Hameka, Advanced Quantum Chemistry (Addison-Wesley Publishing Company, Reading, 1965).

HBE J. O. Hirschfelder, W. Byers Brown, and S. J. Epstein, in Advances in Quantum Chemistry, Vol. 1 (Academic Press, New York, 1964) edited by P. Löwdin, p. 255.

HE W. Heisenberg, The Physical Principles of the Quantum Theory (Dover Publications, 1930).

HE2 V. Heine, Group Theory in Quantum Mechanics (Pergamon Press, London, 1960).

HI W. R. Hindmarsh, Atomic Spectra (Pergamon Press, Oxford, 1967).

HL L. Harris and A. L. Loeb, Introduction to Wave Mechanics (McGraw-Hill Book Co., New York, 1963).

HO C. A. Hollingsworth, Vectors, Matrices, and Group Theory for Scientists and Engineers (McGraw-Hill Book Company, New York, 1967).

HS H. P. Hsu, Fourier Analysis (Simon and Schuster, New York, 1970).

HZ1 G. Herzberg, Atomic Spectra and Atomic Structure (Dover Publications, New York, 1944).

HZ2 G. Herzberg, Molecular Spectra and Molecular Structure I (Van Nostrand Company, Princeton, 1950).

HZ3 G. Herzberg, Molecular Spectra and Molecular Structure II (Van Nostrand Company, Princeton, 1945).

JA M. Jammer, Conceptual Development of Quantum Mechanics (McGraw-Hill Book Company, New York, 1966).

JO H. H. Jaffe' and M. Orchin, Theory and Application of Ultraviolet Spectroscopy (John Wiley and Sons, Inc., New York, 1962).

KA W. Kauzmann, Quantum Chemistry (Academic Press, New York, 1957).

KG V. I. Kogan and V. M. Galitsky, Problems in Quantum Mechanics (Prentice Hall, Inc., Englewood Cliffs, N. J., 1963).

KI G. W. King, Spectroscopy and Molecular Structure (Holt, Rinehart, and Winston, New York, 1964).

KP M. Karplus and R. N. Porter, Atoms & Molecules (W. A. Benjamin, Inc., New York, 1970).

KR H. A. Kramers, Quantum Mechanics (Dover Publications, New York, 1957).

LA L. D. Landau and E. M. Lifshitz, Quantum Mechanics (Addison-Wesley Publ. Co., Reading, 1958).

LE1 I. N. Levine, Quantum Chemistry, Vol. 1 (Allyn and Bacon, Boston, 1970).

LE2 I. N. Levine, Quantum Chemistry, Vol. 2 (Allyn and Bacon, Boston, 1970).

LH R. Lynden-Bell and R. K. Harris, Nuclear Magnetic Resonance Spectroscopy (Appleton-Century-Crofts, New York, 1969).

LI S. H. Lin, in Physical Chemistry, Vol. II, Edited by H. Eyring, D. Henderson, and W. Jost (Academic Press, New York, 1967) p. 109.

LO P. O. Löwdin in Advances in Quantum Chemistry, Vol. 3 (Academic Press, New York, 1967) edited by P. O. Löwdin.

MA F. A. Matsen, Vector Spaces and Algebras for Chemistry and Physics (Holt, Rinehart and Winston, 1970).

ME E. Merzbacher, Quantum Mechanics, 2nd Ed. (John Wiley and Sons, Inc., New York, 1970).

MI K. S. Miller, Partial Differential Equations in Engineering Problems (Prentice-Hall, New York, 1935).

MKT J. N. Murrell, S. F. A. Kettle, and J. M. Tedder, Valence Theory (John Wiley and Sons, LTD, 1965).

MM1 H. Margeneau and G. M. Murphy, The Mathematics of Physics and Chemistry, 2nd Ed. (D. van Nostrand Company, Princeton, 1956).

MM2 N. F. Mott and H. S. W. Massey, The Theory of Atomic Collisions, 3rd Ed. (Oxford, 1965).

MO W. J. Moore, Physical Chemistry, 4th Ed. (Prentice-Hall, Inc., Englewood Cliffs, 1972).

NE H. E. Newell, Vector Analysis (McGraw-Hill Book Company, New York, 1955).

PA H. J. Pain, The Physics of Vibrations and Waves (John Wiley and Sons, LTD, 1968).

PA1 L. Page, Introduction to Theoretical Mechanics (Van Nostrand Company, New York, 1952).

PA2 R. G. Parr, The Quantum Theory of Molecular Electronic Structure (W. A. Benjamin, Inc., New York, 1963).

PA3 L. Pauling, The Nature of the Chemical Bond (Cornell University Press, New York, 1960).

PB J. A. Pople and D. L. Beveridge, Approximate Molecular Orbital Theory (McGraw-Hill Book Company, New York, 1970).

PC J. L. Powell and B. Crasemann, Quantum Mechanics (Addison-Wesley Co., Reading, 1961).

PI1 F. L. Pilar, Elementary Quantum Chemistry (McGraw-Hill Book Co., New York, 1968).

PI2 K. S. Pitzer, Quantum Chemistry (Prentice-Hall, Inc., New York, 1953).

PP W. K. H. Panofsky and M. Phillips, Classical Electricity and Magnetism (Addison-Wesley Publishing Company, Reading, 1955).

PR R. D. Present, Kinetic Theory of Gases (McGraw-Hill Book Company, Inc., 1958).

PW L. Pauling and E. B. Wilson, Introduction to Quantum Mechanics (McGraw-Hill Book Co., New York, 1935).

RA D. Rapp, Quantum Mechanics (Holt, Rinehart, and Winston, New York, 1971).

RO1 P. Roman, Advanced Quantum Theory, 1st Ed. (Addison-Wesley Publishing Co., Inc., 1965).

RO2 V. Rojansky, Introductory Quantum Mechanics (Prentice-Hall, Englewood Cliffs, N. J., 1938).

RO3 M. E. Rose, Elementary Theory of Angular Momentum (John Wiley and Sons, New York, 1957).

RU G. S. Rushbrook, Introduction to Statistical Mechanics (Oxford University Press, London, 1957).

S1 J. C. Slater, Quantum Theory of Atomic Structure, Vol. I (McGraw-Hill Book Co., New York, 1960).

S2 J. C. Slater, Quantum Theory of Atomic Structure, Vol. II (McGraw-Hill Book Co., New York, 1960).

S3 J. C. Slater, Quantum Theory of Matter, 2nd Ed. (McGraw-Hill Book Co., New York, 1968).

S4 J. C. Slater, Introduction to Chemical Physics (McGraw-Hill Book Company, New York, 1939).

S5 J. C. Slater, Quantum Theory of Molecules and Solids, Vol. 1 (McGraw-Hill Book Company, New York, 1963).

SA D. S. Saxon, Elementary Quantum Mechanics (Holden-Day, Inc., New York, 1968).

SC1 L. I. Schiff, Quantum Mechanics, 3rd Ed. (McGraw-Hill Book Co., New York, 1968).

SC2 D. S. Schonland, Molecular Symmetry (D. van Nostrand Co., LTD, London, 1965).

SC3 C. J. H. Schutte, The Wave Mechanics of Atoms, Molecules, and Ions (St. Martins Press, New York, 1968).

SC4 C. W. Sherwin, Introduction to Quantum Mechanics (Henry Holt and Co., New York, 1959).

SE H. Semat, Introduction to Atomic and Nuclear Physics, 3rd Ed. (Rinehart and Co., Inc., New York, 1954).

SF J. C. Slater and N. H. Frank, Electromagnetism (McGraw-Hill Book Company, New York, 1947).

SI N.B.S., J. Chem. Ed. 48, 569 (1971).

SL C. P. Slichter, Principles of Magnetic Resonance (Harper & Row, New York, 1963).

SP1 M. R. Spiegel, Mathematical Handbook (McGraw-Hill Book Company, New York, 1968).

SP2 M. R. Spiegel, Advanced Mathematics for Engineers and Scientists (McGraw-Hill Book Company, New York, 1971).

SP3 M. R. Spiegel, Vector Analysis (McGraw-Hill Book Company, New York, 1959).

SY K. R. Symon, Mechanics, 3rd Ed. (Addison-Wesley Publ. Co., Reading, 1971).

TH D. ter Haar, Selected Problems in Quantum Mechanics (Academic Press, New York, 1964).

TI M. Tinkham, Group Theory and Quantum Mechanics (McGraw-Hill Book Company, New York, 1964).

TS C. H. Townes and A. L. Schawlow, Microwave Spectroscopy (McGraw-Hill Book Company, New York, 1955).

VW B. L. Van der Waerden, Sources of Quantum Mechanics (Dover Publications, New York, 1968).

W1 K. B. Wiberg, Physical Organic Chemistry (John Wiley and Sons, Inc., New York, 1964).

WA G. N. Watson, A Treatise on the Theory of Bessel Functions, 2nd Ed. (Cambridge University Press, London, 1966).

WDC E. B. Wilson, Jr., J. C. Decius, and P. C. Cross, Molecular Vibrations (McGraw-Hill Book Company, New York, 1955).

WO M. M. Woolfson, An Introduction to X-Ray Crystallography (Cambridge University Press, London, 1970).

WS R. T. Weidner and R. L. Sells, Elementary Modern Physics (Allyn and Bacon, Boston, 1960).

WY C. R. Wylie, Jr., Advanced Engineering Mathematics, 2nd Ed. (McGraw-Hill Book Company, Inc., New York, 1960).

INDEX